Essentials of Atmospheric and Oceanic Dynamics

This is a modern, introductory textbook on the dynamics of the atmosphere and ocean, with a healthy dose of geophysical fluid dynamics. It will be invaluable for intermediate to advanced undergraduate and graduate students in meteorology, oceanography, mathematics and physics. It is unique in taking the reader from very basic concepts to the forefront of research. It also forms an excellent refresher for researchers in atmospheric science and oceanography. It differs from other books at this level in both style and content: as well as very basic material, it includes some elementary introductions to more advanced topics. The advanced sections can easily be omitted for a more introductory course, as they are clearly marked in the text. Readers who wish to explore these topics in more detail can refer to this book's parent, *Atmospheric and Oceanic Fluid Dynamics: Fundamentals and Large-Scale Circulation*, now in its second edition.

Geoffrey K. Vallis is a professor of applied mathematics at the University of Exeter, UK. Prior to taking up his position there, he taught for many years at Princeton University in the USA. He has carried out research in the atmospheric sciences, oceanography and the planetary sciences, and has published over 100 peer-reviewed journal articles. He is the recipient of various prizes and awards, including the Adrian Gill Prize (Royal Meteorological Society) and the Stanislaw M. Ulam Distinguished Scholar Award (Los Alamos National Laboratory). He is the author of *Atmospheric and Oceanic Fluid Dynamics: Fundamentals and Large-Scale Circulation, Second Edition* (2017, Cambridge University Press).

'Vallis' insights into the fundamentals and applications go a long way towards making otherwise complex topics readily grasped by those willing to study. He does not shy away from mathematics where needed, nor does he smother the reader with mathematics where pedagogically unnecessary. Those making it through this book will be ready to tackle a huge suite of research questions related to atmosphere and ocean fluid mechanics. Hence, this book serves an incredibly important role to the academic community. In a nutshell, we need more smart researchers who are adept at atmosphere and ocean dynamics to help understand how those dynamics are increasingly being affected by humanity's choices.

Essentials of Atmospheric and Oceanic Dynamics (*EAOD*) fills an important niche by offering an articulate and authoritative textbook to be worked through by advanced undergraduates and/or entering graduate students taking courses. The inclusion of exercises in *EAOD* is incredibly valuable for both students and teachers clamouring for more problem sets to test understanding. Whereas Vallis' previous book *Atmospheric and Oceanic Fluid Dynamics* (*AOFD*) is the mother reference, *EAOD* offers a pedagogical entrée for those wishing to test the waters, including some deep waters. I will happily keep both books on my shelf and make use of them for personal study and to support the teaching of geophysical fluid dynamics.

Vallis has a clear writing style that brings the reader into the subject in an authoritative and friendly manner. He is a wise guru and gentle tutor. The subject of ocean and atmosphere fluid mechanics has matured greatly through his efforts at writing *AOFD*. *EAOD* furthers that maturation by allowing for a broader readership to tap into his brain. Well done Geoff!'

- Stephen M. Griffies,
Geophysical Fluid Dynamics Laboratory,
National Oceanic and Atmospheric Administration

'The "big book" [*AOFD*] by Vallis is a treasure, but I suspect that this new *Essentials* is destined to be used much more widely in classrooms. Vallis does a superb job of communicating the peculiar tensions between deductive reasoning and physical intuition that underlie this science. The new book is more approachable but no less rigorous. I especially appreciate how the various equation sets are derived in succinct but meaningful ways in the first few chapters, and then used as tools to explore the dynamics in the chapters that follow. It's almost the perfect introductory textbook on this subject, and I plan to use it in my own courses.'

- Brian E. J. Rose,
University at Albany

'He's done it again. In *Essentials*, Geoff Vallis has produced a text that is useful to the student and the experienced scientist alike. While the content is simplified and shortened compared to its parent text, Vallis now provides even more descriptive explanations to support readers in their quest to navigate the physics of fluid flows. These explanations pair well with the theory, serving as an accessible introduction to students while also supporting the more experienced scientist as they put all of the pieces together. This will certainly be a future favourite for reading groups. Even readers with dog-eared versions of the parent book will want a copy of *Essentials*, for in it Vallis has added an entirely new chapter on planetary atmospheres, allowing the interested reader to venture into outer space to apply their newly honed GFD expertise.'

- Elizabeth A. Barnes,
Colorado State University

'For the past decade, Geoff Vallis' book *Atmospheric and Oceanic Fluid Dynamics* has been the "go to" encyclopaedic resource, but it is too lengthy and comprehensive to use as a course textbook. With this superb new shorter volume, Geoff Vallis provides us with the definitive graduate-level textbook, with just the right balance of essential topics alongside glimpses of more advanced topics at the cutting edge of research. The extensive use of margin notes, diamonds to indicate advanced topics, and a comprehensive set of problems will ensure that *Essentials of Atmospheric and Oceanic Dynamics* has much to offer students and researchers at all levels. The book opens with the quote: "Seek simplicity, accept complexity. Exploit simplification, avoid complication." On all counts, this book succeeds magnificently!'

- David Marshall,
University of Oxford

'As its parent book became the bible of the field, but also grew in size and the number of topics it covered in its latest edition, this new book provides a perfect balance and introduction to the essential topics, giving a quick reference without going into all the details. In the Vallis tradition, it is presented clearly, perfectly packaged, and is well organized for both atmospheric and oceanic fluid dynamics. Its simplicity will make it majestically appealing both for people outside the discipline looking for an accessible, yet complete, introduction, and for students within the field at all levels. The inclusion of planetary atmospheres broadens the scope and makes it appealing to a wider and growing audience. Anyone with a background in physics can get the essentials using this book.'

- Yohai Kaspi,
Weizmann Institute of Science

Essentials of Atmospheric and Oceanic Dynamics

GEOFFREY K. VALLIS
University of Exeter

CAMBRIDGE
UNIVERSITY PRESS

CAMBRIDGE
UNIVERSITY PRESS

Shaftesbury Road, Cambridge CB2 8EA, United Kingdom

One Liberty Plaza, 20th Floor, New York, NY 10006, USA

477 Williamstown Road, Port Melbourne, VIC 3207, Australia

314–321, 3rd Floor, Plot 3, Splendor Forum, Jasola District Centre, New Delhi – 110025, India

103 Penang Road, #05–06/07, Visioncrest Commercial, Singapore 238467

Cambridge University Press is part of Cambridge University Press & Assessment, a department of the University of Cambridge.

We share the University's mission to contribute to society through the pursuit of education, learning and research at the highest international levels of excellence.

www.cambridge.org
Information on this title: www.cambridge.org/9781107692794

DOI: 10.1017/9781107588431

First published 2019 (version 2, August 2022)

A catalogue record for this publication is available from the British Library

ISBN 978-1-107-69279-4 Paperback

Additional resources for this publication at www.cambridge.org/vallisessentials.

Contents

Note: In the text itself more advanced sections are marked with a diamond, ♦, and may be omitted on a first reading. If a section is so marked then the marking applies to all the subsections within it.

Preface

Seek simplicity, accept complexity.
Exploit simplification, avoid complication.

This is an introductory book on the dynamics of atmospheres and oceans, with a healthy dose of geophysical fluid dynamics. It is written roughly at the level of advanced or upper-division undergraduates and beginning graduate students, but parts of it will be accessible to first- or second-year undergraduates and I hope that practising scientists will also find it useful. The book is designed for students and scientists who want an introduction to the subject but who may not want all the detail, at least not yet, and its prerequisites are just familiarity with some vector calculus and basic classical physics. Thus, it is meant to be accessible to non-specialists and students who will not necessarily go on to become professional dynamicists. However, as well as very basic material the book does include some elementary introductions to a few 'advanced' topics, such as the residual circulation and turbulence theory, as well as material on the general circulation of the atmosphere and ocean. The more advanced parts could easily be omitted for a first course and, like difficult ski slopes, are marked with a diamond, ♦. Readers may explore these topics more in the references provided, or in this book's parent, *Atmospheric and Oceanic Fluid Dynamics.* Nearly all the topics in this book, except those in the chapter on planetary atmospheres, are dealt with in greater detail there.

What is in the book

The book is divided into three Parts. The first, and longest, provides the foundation for the study of the dynamics of the atmosphere and ocean. It does not assume any prior knowledge of fluid dynamics or thermodynamics, although readers who have such knowledge may be able to skim Chapter 1. The rest of Part I provides an introduction to 'geophysical fluid dynamics', the subject that remains at the heart of atmospheric and oceanic dynamics and without which the subject would be largely qualitative and/or computational. Here we discuss the effects of rotation and stratification, leading into shallow water theory and the quasi-geostrophic and planetary-geostrophic equations. Rossby waves, gravity waves, baroclinic instability and elementary treatments of wave–mean-flow interaction and turbulence round out Part I.

Parts II and III focus on the large scale dynamics and circulation of the atmosphere and ocean, respectively. Our main focus in both Parts is what is sometimes called 'the general circulation', meaning the large-scale quasi-steady and/or time-averaged circulation, but this circulation depends on the effects of time-dependent eddies — the atmosphere's Ferrel Cell may be considered to be 'driven' by the effects of baroclinic instability and Rossby waves. And the El Niño phenomenon, described in the final chapter, is explicitly time dependent. One feature of this book that is not in the parent book is a chapter discussing some of the general principles of planetary atmospheres, a topic of increasing interest because of the new, sometimes quite spectacular, observations of the planets in our Solar System and beyond.

How to use the book

The contents of the book are about enough for a two-term course in atmosphere–ocean dynamics. A term-long, first course in geophysical fluid dynamics could, for example, be based on Part I, omitting some of the earlier or later chapters depending on the students' backgrounds and interests. A term-long course in atmospheric and/or oceanic circulation could be based on Part II and/or Part III, supplementing the material with review articles or research papers as needed, perhaps using data sets to look at the real world (and other planets, if Chapter 13 is to be studied). Alternatively, one could combine aspects of Parts I and II, or Parts I and III, to construct an 'Atmospheric Dynamics' or 'Oceanic Dynamics' course.

If the book is to be used for self-study it could simply be read from beginning to end, although many other pathways are possible and may be preferable. Parts II and III depend on the material in Part I, but the material is reasonably self-contained, and readers who already have some knowledge of geophysical fluid dynamics should feel free to start at a later chapter, or with Part II or Part III. A few problems are collected at the end of some chapters; these are designed to test understanding as well as to fill in gaps and extend the material in the book itself. Many other problems at varying levels of difficulty can be found on the web site of this book, which can easily be found with a search engine. The reader will also see a number of margin notes throughout the book, rather like the ones to the left. The book itself was typeset using LaTeX with Crimson fonts for text, Cronos Pro for sans serif and Minion Math for equations.

Margin notes that are set in a roman (i.e., upright) font emphasize or expand on something that is in the main text.

Margin notes set in italics are asides or historical anecdotes.

I would like to thank Matt Lloyd, Zoë Pruce and Richard Smith at Cambridge University Press for their expert guidance through the writing and production process, as well as many colleagues and students — too many to list, but they know who they are — for their many comments, corrections and criticisms. If you, the reader, have other comments, major or minor, do please contact me.

Part I

Geophysical Fluids

CHAPTER

1 Fluid Fundamentals

FLUIDS, LIKE SOLIDS, move if they are pushed and they warm if they are heated. But, unlike solids, they flow and deform. In this chapter we establish the governing equations of motion for a fluid, with particular attention to air and seawater — the fluids of the atmosphere and ocean, respectively. Readers who already have knowledge of fluid dynamics may skim this chapter and begin reading more seriously at Chapter 2, where we begin to look at the effects of rotation and stratification.

The fluid dynamical equations of motion determine the evolution of a fluid. The equations are based on Newton's laws of motion and the laws of thermodynamics, and embody the principles of conservation of momentum, energy and mass. Initial conditions and boundary conditions are needed to solve the equations.

1.1 Time Derivatives for Fluids

1.1.1 Field and Material Viewpoints

In solid-body mechanics one is normally concerned with the position and momentum of an identifiable object, such as a football or a planet, as it moves through space. In principle we could treat fluids the same way and try to follow the properties of individual fluid parcels as they flow along, perhaps getting hotter or colder as they move. This perspective is known as the *material* or *Lagrangian* viewpoint. However, in fluid dynamical problems we generally would like to know what the values of velocity, density and so on are at *fixed points* in space as time passes. A weather forecast we care about tells us how warm it will be where we live and, if we are given that, we may not care where a particular fluid parcel comes from or where it subsequently goes. Since the fluid is a continuum, this knowledge is equivalent to knowing how the fields of the dynamical variables evolve in space and time. This viewpoint is known as the *field* or *Eulerian* viewpoint.

Although the field viewpoint will often turn out to be the most practically useful, the material description is invaluable both in deriving the equations and in the subsequent insight it frequently provides. This is because the important quantities from a fundamental point of view are often those which are associated with a given fluid element: it is these which directly enter Newton's laws of motion and the thermodynamic equations. It is thus important to have a relationship between the rate of change of quantities associated with a given fluid element and the local rate of change of a field. The material derivative (also called the advective derivative or Lagrangian derivative) provides this relationship.

The Lagrangian viewpoint is named for the Franco-Italian J. L. Lagrange (1736–1813), one of the most renowned mathematicians of his time. The Eulerian point of view is named for Leonhard Euler (1707–1783), the great Swiss mathematician. In fact, Euler is also largely responsible for the Lagrangian view, but the attribution became tangled over time.

1.1.2 The Material Derivative of a Fluid Property

A *fluid element* is an infinitesimal, indivisible, piece of fluid — effectively a very small fluid parcel of fixed mass. The *material derivative,* or the *Lagrangian derivative,* is the rate of change of a property (such as temperature or momentum) of a particular fluid element or finite mass of fluid; that is, it is the total time derivative of a property of a piece of fluid.

Let us suppose that a fluid is characterized by a given velocity field $\boldsymbol{v}(\boldsymbol{x}, t)$, which determines its velocity throughout. Let us also suppose that the fluid has another property φ, and let us seek an expression for the rate of change of φ of a fluid element. Since φ is changing in time and in space we use the chain rule,

$$\delta\varphi = \frac{\partial\varphi}{\partial t}\delta t + \frac{\partial\varphi}{\partial x}\delta x + \frac{\partial\varphi}{\partial y}\delta y + \frac{\partial\varphi}{\partial z}\delta z = \frac{\partial\varphi}{\partial t}\delta t + \delta\boldsymbol{x}\cdot\nabla\varphi. \tag{1.1}$$

This is true in general for any δt, δx, etc. The total time derivative is then

$$\frac{d\varphi}{dt} = \frac{\partial\varphi}{\partial t} + \frac{d\boldsymbol{x}}{dt}\cdot\nabla\varphi. \tag{1.2}$$

If this equation is to provide a material derivative we must identify the time derivative in the second term on the right-hand side with the rate of change of position of a fluid element, namely its velocity. Hence, the material derivative of the property φ is

$$\frac{d\varphi}{dt} = \frac{\partial\varphi}{\partial t} + \boldsymbol{v}\cdot\nabla\varphi. \tag{1.3}$$

The right-hand side expresses the material derivative in terms of the local rate of change of φ plus a contribution arising from the spatial variation of φ, experienced only as the fluid parcel moves. Because the material derivative is so common, and to distinguish it from other derivatives, we denote it by the operator $\mathrm{D}/\mathrm{D}t$. Thus, the material derivative of the field φ is

$$\frac{\mathrm{D}\varphi}{\mathrm{D}t} = \frac{\partial\varphi}{\partial t} + (\boldsymbol{v}\cdot\nabla)\varphi. \tag{1.4}$$

The brackets in the last term of this equation are helpful in reminding us that $(\boldsymbol{v}\cdot\nabla)$ is an operator acting on φ. The operator $\partial/\partial t + (\boldsymbol{v}\cdot\nabla)$ is the *Eulerian representation of the Lagrangian derivative as applied to a field.*

Material derivative of vector field

The material derivative may act on a vector field $\boldsymbol{b}$, in which case

$$\frac{\mathrm{D}\boldsymbol{b}}{\mathrm{D}t} = \frac{\partial \boldsymbol{b}}{\partial t} + (\boldsymbol{v} \cdot \nabla)\boldsymbol{b}. \tag{1.5}$$

In Cartesian coordinates this is

$$\frac{\mathrm{D}\boldsymbol{b}}{\mathrm{D}t} = \frac{\partial \boldsymbol{b}}{\partial t} + u\frac{\partial \boldsymbol{b}}{\partial x} + v\frac{\partial \boldsymbol{b}}{\partial y} + w\frac{\partial \boldsymbol{b}}{\partial z}, \tag{1.6}$$

and for a particular component of $\boldsymbol{b}$, b^x say,

$$\frac{\mathrm{D}b^x}{\mathrm{D}t} = \frac{\partial b^x}{\partial t} + u\frac{\partial b^x}{\partial x} + v\frac{\partial b^x}{\partial y} + w\frac{\partial b^x}{\partial z}, \tag{1.7}$$

and similarly for b^y and b^z. In coordinate systems other than Cartesian the advective derivative of a vector is not simply the sum of the advective derivatives of its components, because the coordinate vectors themselves change direction with position; this will be important when we deal with spherical coordinates.

The Eulerian derivative is the rate of change of a property at a fixed location in space. The material derivative is the rate of change of a property of a given piece of fluid, which may be moving and so changing its position.

1.1.3 Material Derivative of a Volume

The volume that a given, unchanging, mass of fluid occupies is deformed and advected by the fluid motion, and there is no reason why it should remain constant. Rather, the volume will change as a result of the movement of each element of its bounding material surface, and in particular it will change if there is a non-zero normal component of the velocity at the fluid surface. That is, if the volume of some fluid is $\int \mathrm{d}V$, then

$$\frac{\mathrm{D}}{\mathrm{D}t}\int_V \mathrm{d}V = \int_S \boldsymbol{v} \cdot \mathrm{d}\boldsymbol{S}, \tag{1.8}$$

where the subscript V indicates that the integral is a definite integral over some finite volume V, and the limits of the integral are functions of time since the volume is changing. The integral on the right-hand side is over the closed surface, S, bounding the volume. Although intuitively apparent (to some), this expression may be derived more formally using Leibniz's formula for the rate of change of an integral whose limits are changing. Using the divergence theorem on the right-hand side, (1.8) becomes

$$\frac{\mathrm{D}}{\mathrm{D}t}\int_V \mathrm{d}V = \int_V \nabla \cdot \boldsymbol{v}\,\mathrm{d}V. \tag{1.9}$$

The rate of change of the volume of an infinitesimal fluid element of volume ΔV is obtained by taking the limit of this expression as the volume tends to zero, giving

$$\lim_{\Delta V \to 0} \frac{1}{\Delta V}\frac{\mathrm{D}\Delta V}{\mathrm{D}t} = \nabla \cdot \boldsymbol{v}. \tag{1.10}$$

We will often write such expressions informally as

$$\frac{\mathrm{D}\Delta V}{\mathrm{D}t} = \Delta V \nabla \cdot \boldsymbol{v}, \tag{1.11}$$

with the limit implied.

Consider now the material derivative of some fluid property, ξ say, multiplied by the volume of a fluid element, ΔV. Such a derivative arises when ξ is the amount per unit volume of ξ-substance — the mass density or the amount of a dye per unit volume, for example. Then we have

$$\frac{\mathrm{D}}{\mathrm{D}t}(\xi \Delta V) = \xi \frac{\mathrm{D}\Delta V}{\mathrm{D}t} + \Delta V \frac{\mathrm{D}\xi}{\mathrm{D}t}. \tag{1.12}$$

Using (1.11) this becomes

$$\frac{\mathrm{D}}{\mathrm{D}t}(\xi \Delta V) = \Delta V \left(\xi \nabla \cdot \boldsymbol{v} + \frac{\mathrm{D}\xi}{\mathrm{D}t} \right), \tag{1.13}$$

and the analogous result for a finite fluid volume is just

$$\frac{\mathrm{D}}{\mathrm{D}t} \int_V \xi \,\mathrm{d}V = \int_V \left(\xi \nabla \cdot \boldsymbol{v} + \frac{\mathrm{D}\xi}{\mathrm{D}t} \right) \mathrm{d}V. \tag{1.14}$$

This expression is to be contrasted with the Eulerian derivative for which the volume, and so the limits of integration, are fixed and we have

$$\frac{\mathrm{d}}{\mathrm{d}t} \int_V \xi \,\mathrm{d}V = \int_V \frac{\partial \xi}{\partial t} \,\mathrm{d}V. \tag{1.15}$$

Now consider the material derivative of a fluid property φ multiplied by the mass of a fluid element, $\rho \Delta V$, where ρ is the fluid density. Such a derivative arises when φ is the amount of φ-substance per unit mass (note, for example, that the momentum of a fluid element is $\rho \boldsymbol{v} \Delta V$). The material derivative of $\varphi \rho \Delta V$ is given by

$$\frac{\mathrm{D}}{\mathrm{D}t}(\varphi \rho \Delta V) = \rho \Delta V \frac{\mathrm{D}\varphi}{\mathrm{D}t} + \varphi \frac{\mathrm{D}}{\mathrm{D}t}(\rho \Delta V). \tag{1.16}$$

But $\rho \Delta V$ is just the mass of the fluid element, and that is constant — that is how a fluid element is defined. Thus the second term on the right-hand side vanishes and

$$\frac{\mathrm{D}}{\mathrm{D}t}(\varphi \rho \Delta V) = \rho \Delta V \frac{\mathrm{D}\varphi}{\mathrm{D}t} \qquad \text{and} \qquad \frac{\mathrm{D}}{\mathrm{D}t} \int_V \varphi \rho \,\mathrm{d}V = \int_V \rho \frac{\mathrm{D}\varphi}{\mathrm{D}t} \,\mathrm{d}V, \tag{1.17a,b}$$

where (1.17b) applies to a finite volume. That expression may also be derived more formally using Leibniz's formula for the material derivative of an integral, and the result also holds when φ is a vector. The result is quite different from the corresponding Eulerian derivative, in which the volume is kept fixed; in that case we have:

$$\frac{\mathrm{d}}{\mathrm{d}t} \int_V \varphi \rho \,\mathrm{d}V = \int_V \frac{\partial}{\partial t}(\varphi \rho) \,\mathrm{d}V. \tag{1.18}$$

Various material and Eulerian derivatives are summarized in the shaded box on the facing page.

Material and Eulerian Derivatives

The material derivatives of a scalar (φ) and a vector ($\boldsymbol{b}$) field are given by:

$$\frac{\mathrm{D}\varphi}{\mathrm{D}t} = \frac{\partial \varphi}{\partial t} + \boldsymbol{v} \cdot \nabla\varphi, \qquad \frac{\mathrm{D}\boldsymbol{b}}{\mathrm{D}t} = \frac{\partial \boldsymbol{b}}{\partial t} + (\boldsymbol{v} \cdot \nabla)\boldsymbol{b}. \tag{D.1}$$

Various material derivatives of integrals are:

$$\frac{\mathrm{D}}{\mathrm{D}t}\int_V \varphi \,\mathrm{d}V = \int_V \left(\frac{\mathrm{D}\varphi}{\mathrm{D}t} + \varphi\nabla\cdot\boldsymbol{v}\right)\mathrm{d}V = \int_V \left(\frac{\partial\varphi}{\partial t} + \nabla\cdot(\varphi\boldsymbol{v})\right)\mathrm{d}V, \tag{D.2}$$

$$\frac{\mathrm{D}}{\mathrm{D}t}\int_V \mathrm{d}V = \int_V \nabla\cdot\boldsymbol{v}\,\mathrm{d}V, \tag{D.3}$$

$$\frac{\mathrm{D}}{\mathrm{D}t}\int_V \rho\varphi\,\mathrm{d}V = \int_V \rho\frac{\mathrm{D}\varphi}{\mathrm{D}t}\,\mathrm{d}V. \tag{D.4}$$

These formulae also hold if φ is a vector. The Eulerian derivative of an integral is:

$$\frac{\mathrm{d}}{\mathrm{d}t}\int_V \varphi\,\mathrm{d}V = \int_V \frac{\partial\varphi}{\partial t}\,\mathrm{d}V, \tag{D.5}$$

so that

$$\frac{\mathrm{d}}{\mathrm{d}t}\int_V \mathrm{d}V = 0 \quad\text{and}\quad \frac{\mathrm{d}}{\mathrm{d}t}\int_V \rho\varphi\,\mathrm{d}V = \int_V \frac{\partial\rho\varphi}{\partial t}\,\mathrm{d}V. \tag{D.6}$$

1.2 THE MASS CONTINUITY EQUATION

In classical mechanics mass is absolutely conserved and in solid-body mechanics we normally do not need an explicit equation of mass conservation. However, in fluid mechanics a fluid may flow into and away from a particular location, and fluid density may change, and we need an equation to describe that change.

1.2.1 An Eulerian Derivation

We first derive the mass conservation equation from an Eulerian point of view; that is, our reference frame is fixed in space and the fluid flows through it. Consider an infinitesimal, rectangular cuboid, control volume, $\Delta V = \Delta x\Delta y\Delta z$ that is fixed in space, as in Fig. 1.1. Fluid moves into or out of the volume through its surface, including through its faces in the y–z plane of area $\Delta A = \Delta y\Delta z$ at coordinates x and $x + \Delta x$. The accumulation of fluid within the control volume due to motion in the x-direction is

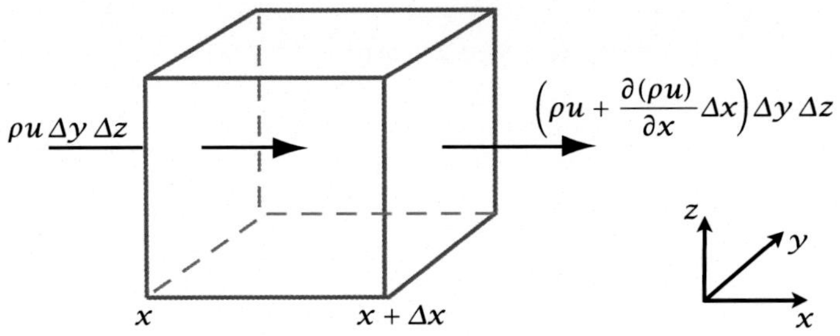

Fig. 1.1: Mass conservation in an Eulerian cuboid control volume. The mass convergence, $-\partial(\rho u)/\partial x$ (plus contributions from the y and z directions), must be balanced by a density increase equal to $\partial\rho/\partial t$.

evidently

$$\Delta y \Delta z[(\rho u)(x, y, z) - (\rho u)(x + \Delta x, y, z)] = -\left.\frac{\partial(\rho u)}{\partial x}\right|_{x,y,z} \Delta x \, \Delta y \, \Delta z. \quad (1.19)$$

To this must be added the effects of motion in the y- and z-directions, namely

$$-\left[\frac{\partial(\rho v)}{\partial y} + \frac{\partial(\rho w)}{\partial z}\right]\Delta x \, \Delta y \, \Delta z. \quad (1.20)$$

This net accumulation of fluid must be accompanied by a corresponding increase of fluid mass within the control volume. This is

$$\frac{\partial}{\partial t}(\text{density} \times \text{volume}) = \Delta x \, \Delta y \, \Delta z \frac{\partial \rho}{\partial t}, \quad (1.21)$$

because the volume is constant. Thus, because mass is conserved, (1.19), (1.20) and (1.21) give

$$\Delta x \, \Delta y \, \Delta z \left[\frac{\partial \rho}{\partial t} + \frac{\partial(\rho u)}{\partial x} + \frac{\partial(\rho v)}{\partial y} + \frac{\partial(\rho w)}{\partial z}\right] = 0. \quad (1.22)$$

The quantity in square brackets must be zero and we therefore have

$$\frac{\partial \rho}{\partial t} + \nabla \cdot (\rho \boldsymbol{v}) = 0. \quad (1.23)$$

This is called the *mass continuity equation* for it recognizes the continuous nature of the mass field in a fluid. There is no diffusion term in (1.23), no term like $\kappa \nabla^2 \rho$. This is because mass is transported by the macroscopic movement of molecules; even if this motion appears diffusion-like, any net macroscopic molecular motion constitutes, by definition, a velocity field.

Neither (1.23) nor the derivation that leads to it depends in any way on Cartesian geometry; a more general vector derivation using an arbitrary control volume is left as an easy exercise for the reader.

1.2.2 Mass Continuity via the Material Derivative

We now derive the mass continuity equation (1.23) from a material perspective. This is the most fundamental approach of all since the principle of mass conservation states simply that the mass of a given element of fluid is, by definition of the element, constant. Thus, consider a small mass of fluid of density ρ and volume ΔV. Then conservation of mass may be represented by

$$\frac{\mathrm{D}}{\mathrm{D}t}(\rho \Delta V) = 0. \tag{1.24}$$

Both the density and the volume of the parcel may change, so

$$\Delta V \frac{\mathrm{D}\rho}{\mathrm{D}t} + \rho \frac{\mathrm{D}\Delta V}{\mathrm{D}t} = \Delta V \left(\frac{\mathrm{D}\rho}{\mathrm{D}t} + \rho \nabla \cdot \boldsymbol{v} \right) = 0, \tag{1.25}$$

where the second expression follows using (1.11). Since the volume element is non-zero the term in brackets must vanish and

$$\frac{\mathrm{D}\rho}{\mathrm{D}t} + \rho \nabla \cdot \boldsymbol{v} = 0. \tag{1.26}$$

After expansion of the first term this becomes identical to (1.23). (A slightly more formal way to derive this result uses (1.14) with ξ replaced by ρ.) Summarizing, equivalent partial differential equations representing conservation of mass are

$$\frac{\mathrm{D}\rho}{\mathrm{D}t} + \rho \nabla \cdot \boldsymbol{v} = 0, \qquad \frac{\partial \rho}{\partial t} + \nabla \cdot (\rho \boldsymbol{v}) = 0. \tag{1.27a,b}$$

1.2.3 Incompressible Fluids

An incompressible fluid is sometimes defined as one whose density is not affected by pressure. This definition may usefully be generalized to mean a fluid whose density is very nearly constant (and so also not affected by temperature or composition) such that the mass continuity equation takes the form (1.30). That equation is normally a very good approximation for seawater, less so for air.

A near-universal property of liquids is that their density is nearly constant; that is, they are essentially *incompressible.* If we write the density as

$$\rho(x, y, z, t) = \rho_0 + \delta\rho(x, y, z, t), \tag{1.28}$$

where ρ_0 is a constant, then a truly incompressible fluid has $\delta\rho = 0$. No fluid is incompressible in this strict sense so we relax the meaning slightly and simply require $|\delta\rho| \ll \rho_0$. When this is satisfied the mass continuity equation, (1.27a) takes on a different form. Equation (1.27a) may be written, without approximation, as

$$\frac{\mathrm{D}\delta\rho}{\mathrm{D}t} + (\rho_0 + \delta\rho)\nabla \cdot \boldsymbol{v} = 0. \tag{1.29}$$

If the fluid is incompressible then the terms involving $\delta\rho$ are much smaller than those involving ρ_0 and hence may be neglected, giving

$$\nabla \cdot \boldsymbol{v} = 0. \tag{1.30}$$

This is the mass continuity equation for an incompressible fluid, and its satisfaction may be taken as the defining quality of an incompressible fluid. The *prognostic* equation, (1.27) has become a *diagnostic* equation.

1.3 THE MOMENTUM EQUATION

The viscous form of the fluid dynamical equations of motion was established by Claude-Louis-Marie-Henri Navier (1785–1836) a French civil engineer, and George Stokes (1819–1903), an Anglo-Irish applied mathematician, who further elucidated viscous effects. Prior to their work the great Swiss mathematician Leonard Euler (1707–1783) had established the general form of the fluid equations for an inviscid incompressible flow, namely the Euler equations.

The momentum equation is a partial differential equation that describes how the velocity or momentum of a fluid responds to internal and imposed forces. We derive it here using material methods, with a very heuristic treatment of the terms representing pressure and viscous forces.

1.3.1 Advection

Let $\boldsymbol{m}(x, y, z, t)$ be the momentum-density field (momentum per unit volume) of the fluid. Thus, $\boldsymbol{m} = \rho\boldsymbol{v}$ and the total momentum of a volume of fluid is given by the volume integral $\int_V \boldsymbol{m}\,\mathrm{d}V$. Now, for a fluid the rate of change of momentum of an identifiable fluid mass is given by the material derivative, and by Newton's second law this is equal to the force acting on it. Thus,

$$\frac{\mathrm{D}}{\mathrm{D}t}\int_V \rho\boldsymbol{v}\,\mathrm{d}V = \int_V \boldsymbol{F}\,\mathrm{d}V, \tag{1.31}$$

where $\boldsymbol{F}$ is the force per unit volume. Now, using (1.17b) (with φ replaced by $\boldsymbol{v}$) to transform the left-hand side of (1.31), we obtain

$$\int_V \left(\rho\frac{\mathrm{D}\boldsymbol{v}}{\mathrm{D}t} - \boldsymbol{F}\right)\mathrm{d}V = 0. \tag{1.32}$$

Because the volume is arbitrary the integrand itself must vanish and we obtain

$$\rho\frac{\mathrm{D}\boldsymbol{v}}{\mathrm{D}t} = \boldsymbol{F} \quad \text{or} \quad \frac{\partial\boldsymbol{v}}{\partial t} + (\boldsymbol{v}\cdot\nabla)\boldsymbol{v} = \frac{\boldsymbol{F}}{\rho}, \tag{1.33a,b}$$

having used (1.5) to expand the material derivative. We have thus obtained an expression for how a fluid accelerates if subject to known forces. As well as external forces (like gravity), a stress arises from the direct contact between one fluid parcel and another, giving rise to pressure and viscous forces, sometimes referred to as *contact* forces.

1.3.2 Pressure and Viscous Forces

The forces due to pressure and viscosity are 'contact forces' arising because of the inter-molecular forces and/or collisions in a fluid. The net pressure force on a fluid element is proportional to the gradient of pressure.

Pressure

Within or at the boundary of a fluid the pressure is the normal force per unit area due to the collective action of molecular motion. Thus

$$\mathrm{d}\widehat{\boldsymbol{F}}_p = -p\,\mathrm{d}\boldsymbol{S}, \tag{1.34}$$

where p is the pressure, $\widehat{\boldsymbol{F}}_p$ is the pressure force and $\mathrm{d}\boldsymbol{S}$ an infinitesimal surface element. If we grant ourselves this intuitive notion, it is a simple matter to assess the influence of pressure on a fluid, for the pressure force on a volume of fluid is the integral of the pressure over the its boundary and so

$$\widehat{\boldsymbol{F}}_p = -\int_S p\,\mathrm{d}\boldsymbol{S}. \tag{1.35}$$

The minus sign arises because the pressure force is directed inwards, whereas $\boldsymbol{S}$ is a vector normal to the surface and directed outwards. Applying a form of the divergence theorem to the right-hand side gives

$$\widehat{\boldsymbol{F}}_p = -\int_V \nabla p \, \mathrm{d}V, \tag{1.36}$$

where the volume V is bounded by the surface S. The pressure gradient force per unit volume, $\boldsymbol{F}_p$, is therefore just $-\nabla p$.

Viscosity

The effects of viscosity are apparent in many situations — the flow of treacle or volcanic lava are obvious examples. The viscous force per unit volume is approximately equal to $\mu \nabla^2 \boldsymbol{v}$, where μ is the coefficient of viscosity. With the pressure and viscous terms the momentum equation becomes,

Equation (1.37) is sometimes called the Navier–Stokes equation. If viscosity is absent the equation is the Euler equation. Sometimes these names are taken as applying to the complete set of equations of motion.

$$\frac{\partial \boldsymbol{v}}{\partial t} + (\boldsymbol{v} \cdot \nabla)\boldsymbol{v} = -\frac{1}{\rho}\nabla p + \nu \nabla^2 \boldsymbol{v} + \boldsymbol{F}_b, \tag{1.37}$$

where $\nu = \mu/\rho$ is the *kinematic viscosity* and $\boldsymbol{F}_b$ represents body forces (per unit mass) such as gravity, $\boldsymbol{g}$. For most large-scale flows in the atmosphere and ocean the viscous term is, in fact, neglibly small.

1.3.3 The Hydrostatic Approximation

Neglecting viscocity, the vertical component (the component parallel to the gravitational force, $\boldsymbol{g}$) of the momentum equation is

$$\frac{\mathrm{D}w}{\mathrm{D}t} = -\frac{1}{\rho}\frac{\partial p}{\partial z} - g, \tag{1.38}$$

where w is the vertical component of the velocity and $\boldsymbol{g} = -g\hat{\mathbf{k}}$. If the fluid is static the gravitational term is balanced by the pressure term and we have

$$\frac{\partial p}{\partial z} = -\rho g, \tag{1.39}$$

and this relation is known as *hydrostatic balance*, or hydrostasy. It is clear in this case that the pressure at a point is given by the weight of the fluid above it, provided that $p = 0$ at the top of the fluid. The flow need not be static for hydrostasy to hold — equation (1.39) is a good *approximation* to (1.38) provided that the vertical acceleration, $\mathrm{D}w/\mathrm{D}t$, is sufficiently small compared to gravity, which is nearly always the case in both atmosphere and ocean except in intense storms. However, because the pressure also appears in the horizontal momentum equations, hydrostatic balance must be *very* well satisfied to ensure that (1.39) provides an accurate enough pressure to determine the horizontal pressure gradients, a point we return to in Section 3.2.

Hydrostatic balance is an approximation to the vertical momentum equation, valid for large-scale motion in both atmosphere and ocean and normally very well satisfied for flows of horizontal scales greater than a few tens of kilometres. It is one of the most fundamental and useful approximations in atmospheric and oceanic dynamics.

Table 1.1: Various thermodynamic parameters used in ideal gas theory, with the specific values being those for dry air.

Symbol	Description	Value
k_B	Boltzmann constant	$1.38 \times 10^{-23}\,\mathrm{J\,K^{-1}}$
N_A	Avogadro constant	$6.02214076 \times 10^{23}\,\mathrm{mol^{-1}}$
R^*	universal gas constant ($= k_B N_A$)	$8.31\,\mathrm{J\,K^{-1}\,mol^{-1}}$
μ	molar mass of dry air	$29 \times 10^{-3}\,\mathrm{kg\,mol^{-1}}$
R	specific gas constant ($= R^*/\mu$)	$287\,\mathrm{J\,kg^{-1}\,K^{-1}}$
c_v	specific heat capacity at const. volume	$717\,\mathrm{J\,kg^{-1}\,K^{-1}}$
c_p	specific heat capacity at const. pressure	$1004\,\mathrm{J\,kg^{-1}\,K^{-1}}$
c_s	sound speed at $T = 273\,\mathrm{K}$	$331\,\mathrm{m\,s^{-1}}$

1.4 The Equation of State

In three dimensions the momentum and continuity equations provide four equations, but contain five unknowns — three components of velocity, density and pressure. Obviously other equations are needed, and an *equation of state* is an expression that diagnostically relates the various thermodynamic variables to each other. Most commonly the equation of state is written in a form that relates temperature, density, pressure and composition, and such an equation is known as the *thermal equation of state,* and it differs from fluid to fluid. In this book we will mainly be dealing with an ideal gas (for the atmosphere) or with seawater (for the ocean). The composition of air varies slightly with water vapour content and the composition of seawater varies slightly with salinity.

1.4.1 Ideal Gas

For an ideal gas of constant composition the equation of state is commonly written as

$$pV = Nk_BT = nR^*T, \tag{1.40}$$

where k_B is Boltzmann's constant, N is the total number of molecules in the volume V, R^* is the universal gas constant and n is the number of moles in that volume, where a mole is the amount of substance that contains Avogadro's number of elementary units. The two expressions on the right-hand side of (1.40) are equivalent because $N = nN_A$ and $R^* = k_BN_A$, where N_A is the Avogadro constant (see Table 1.1 and margin note). For fluid dynamical purposes we divide (1.40) by the total mass, $M = n\mu$, where μ is the molar mass (the mass per mole, often referred to as the molecular weight) of the gas, and obtain

A *mole* is the amount of substance that contains exactly Avogadro's number, N_{AN}, of elementary entities (usually atoms or molecules), and $N_{AN} \equiv 6.02214076 \times 10^{23}$, by definition. The (quasi-dimensional) Avogadro constant, N_A, is the number of elementary units per mole, that number being Avogadro's number. A mole is almost the same as the atomic or molecular weight, or molar mass, in grams. Thus, a mole of molecular oxygen (O_2) has a mass of very nearly 32 grams.

$$p = \rho RT, \tag{1.41}$$

where $R = R^*/\mu$ is the *specific gas constant,* which varies from substance to substance. For dry air $R = 287\,\mathrm{J\,kg^{-1}\,K^{-1}}$. Air has virtually constant composition except for variations in water vapour; these variations make R a weak function of the water vapour content but we regard R as a constant. Finally, it is common in fluid dynamics to work with the inverse of

Symbol	Description	Value
ρ_0	reference density	1.027×10^3 kg m^{-3}
α_0	reference specific volume	9.738×10^{-4} m^3 kg^{-1}
T_0	reference temperature	283 K
S_0	reference salinity	35 ppt = 35 g kg^{-1}
c_{s0}	reference sound speed	1490 m s^{-1}
β_T	thermal expansion coefficient	1.67×10^{-4} K^{-1}
β_S	haline contraction coefficient	0.78×10^{-3} ppt^{-1}
β_p	compressibility coefficient (= α_0/c_{s0}^2)	4.39×10^{-10} m s^2 kg^{-1}
c_{p0}	specific heat capacity at const. pressure	3986 J kg^{-1} K^{-1}

Table 1.2: Various thermodynamic and equation-of-state parameters appropriate for seawater, as used in the equation of state (1.42) and elsewhere. The unit ppt is parts per thousand by weight, or g/kg.

density, or specific volume (i.e., volume per unit mass), $\alpha = 1/\rho$, whence the equation of state becomes $p\alpha = RT$.

1.4.2 Seawater

Water is nearly incompressible: its density changes very little with temperature, salinity, or pressure. However, these variations, small as they are, *are* important in oceanography for they are allow the ocean currents to transport large quantities of heat in the great ocean gyres and in deep abyssal currents. There is no accurate, simple equation of state but for many purposes we can approximate it as

$$\rho = \rho_0 \left[1 - \beta_T(T - T_0) + \beta_S(S - S_0) + \beta_p(p - p_0)\right], \qquad (1.42)$$

where β_T, β_S and β_p are empirical parameters and S_0, T_0 and p_0 are constants, and usually we take $p_0 = 0$. (A still more accurate equation is required for quantitative oceanography.) Typical values of these parameters are given in Table 1.2. The parameter β_p is related to the speed of sound, c_s, given by $c_s^2 = (\partial p/\partial \rho)$. Using this result and (1.42) gives, to a good approximation, $\beta_p = 1/\rho_0 c_s^2$, and $c_s \approx 1500\,\mathrm{m\,s^{-1}}$. *None* of the terms in (1.42) give rise to large variations in density in the ocean.

Density variations *are* important for generating ocean currents but they are *not* important in the mass continuity equation for seawater, and to a very good approximation that equation may be written as $\nabla \cdot \boldsymbol{v} = 0$.

Unfortunately (perhaps) the equation of state introduces another unknown, temperature, into our equation set. We thus have to introduce another physical principle — one coming from thermodynamics — to obtain a complete set of equations, as we now explore.

1.5 THERMODYNAMICS

1.5.1 A Few Fundamentals

The *first law of thermodynamics* states that the internal energy, I, of a body may change because of work done by or on it, or because of a heat input, or because of a change in its chemical composition. We will neglect the last effect so that

$$dI = đQ + đW, \qquad (1.43)$$

where $đW$ is the work done on the body, $đQ$ is the heat input to the body and dI is the change in internal energy, and we take all these quantities to be per unit mass. (Heating arises from such things as radiation and conduction, and work done occurs when a body is compressed.) The quantities on the right-hand side (with a đ) are 'imperfect' differentials or infinitesimals: Q and W are not functions of the state of a body, and the internal energy cannot be regarded as the sum of a 'heat' and a 'work'. That is, we should think of heat and work as having meaning only as fluxes of energy, or rates of energy input, and not as amounts of energy; their sum changes the internal energy of a body, which *is* a function of its state. However, both the heat input and the work done are related to state variables, as follows:

A state variable is, by definition, a function of the state of a fluid, or more generally the state of any body. The internal energy and the entropy are both state variables. Heat and work done are *not* state variables; they have meaning only as inputs or fluxes of energy.

Heat input: Although heat is not itself a state function, there is a state function that responds directly to heating and this is the *entropy*. Specifically, in an infinitesimal quasi-static or reversible process, if an amount of heat $đQ$ (per unit mass) is externally supplied then the specific entropy η will change according to

$$T\,d\eta = đQ. \tag{1.44}$$

That is to say, the entropy changes by an amount equal to the heat input divided by the temperature.

Work done: The work done on a body during a reversible process is equal to the pressure times its change in volume, and if the work is positive then the volume change is negative. Thus if an infinitesimal amount of work $đW$ (per unit mass) is applied to a body then its thermodynamic state will change according to

$$-\,p\,d\alpha = đW, \tag{1.45}$$

where, we recall, $\alpha = 1/\rho$ is the specific volume of the fluid and p is the pressure.

Putting equations (1.43)–(1.45) together we have

$$dI = T\,d\eta - p\,d\alpha. \tag{1.46}$$

This expression is called *the fundamental thermodynamic relation*. Let's see how it applies to a fluid.

1.5.2 Thermodynamic Equation for a Fluid

Let us suppose that the heating of a fluid arises from external agents, such as radiation from the sun, and that the changes in density and internal energy are consequences of this heating. It then makes sense to write (1.46) as

$$dI + p d\alpha = đQ. \tag{1.47}$$

Now, (1.47) applies to a particular fluid element, not to a particular location. The rate of change of the internal energy and volume are thus obtained by taking the material derivative giving

$$\frac{\mathrm{D}I}{\mathrm{D}t} + p\frac{\mathrm{D}\alpha}{\mathrm{D}t} = \dot{Q} \qquad \text{or} \qquad \frac{\mathrm{D}I}{\mathrm{D}t} - \frac{p}{\rho^2}\frac{\mathrm{D}\rho}{\mathrm{D}t} = \dot{Q}, \tag{1.48}$$

where $\dot{Q}$ is the rate of heat input. We can now use the mass conservation equation, (1.27a), to rewrite the time derivative of density to give

$$\frac{\mathrm{D}I}{\mathrm{D}t} + \frac{p}{\rho}\nabla\cdot\boldsymbol{v} = \dot{Q}. \tag{1.49}$$

This is the 'thermodynamic equation' of a fluid, and it tells us how the internal energy responds to heating.

An ideal gas

In an ideal gas the internal energy is a function of temperature alone, and is given by

$$I = c_v T, \tag{1.50}$$

where c_v is the heat capacity at constant volume. For the gases in Earth's atmosphere, c_v itself is, to a very good approximation, a constant and (1.49) becomes

$$c_v\frac{\mathrm{D}T}{\mathrm{D}t} + \frac{p}{\rho}\nabla\cdot\boldsymbol{v} = \dot{Q}. \tag{1.51}$$

Equation (1.51) is perhaps the most commonly used form of the thermodynamic equation in the atmospheric sciences.

We can rewrite (1.47) as

$$\mathrm{d}I + \mathrm{d}(\alpha p) - \alpha\,\mathrm{d}p = đQ \qquad \text{or} \qquad \mathrm{d}(c_v T) + \mathrm{d}(RT) - \alpha\,\mathrm{d}p = đQ, \tag{1.52a,b}$$

In an ideal gas, c_p, c_v and R are functions only of temperature and they are related by $c_p - c_v = R$. In Earth's atmosphere they are all in fact very nearly constant.

where the second expression holds for an ideal gas. We let $c_p = c_v + R$ and (1.52b) then becomes, for constant c_p,

$$c_p\mathrm{d}T - \alpha\,\mathrm{d}p = đQ \qquad \text{implying} \qquad c_p\frac{\mathrm{D}T}{\mathrm{D}t} - \alpha\frac{\mathrm{D}p}{\mathrm{D}t} = \dot{Q}, \tag{1.53a,b}$$

From (1.53a), we can see that c_p is just the heat capacity at constant pressure. Equation (1.53b) is equivalent to (1.51), although perhaps not as widely used. The complete set of equations of motion for a fluid are summarized on the next page.

1.6 POTENTIAL TEMPERATURE AND ENTROPY

When a fluid is heated its entropy increases, obeying $T\mathrm{d}\eta = đQ$ as in (1.44). Taking the material derivative gives

$$T\frac{\mathrm{D}\eta}{\mathrm{D}t} = \dot{Q}. \tag{1.54}$$

The Equations of Motion for a Fluid

The momentum equation

The momentum equation is an evolution equation for velocity and embodies Newton's second law, namely that force equals mass times acceleration. It is

$$\frac{\mathrm{D}\boldsymbol{v}}{\mathrm{D}t} = -\frac{1}{\rho}\nabla p + \nu\nabla^2\boldsymbol{v} + \boldsymbol{g}, \tag{F.1}$$

where $\boldsymbol{g}$ is the gravitational force per unit mass.

The mass continuity equation

The mass continuity equation embodies the principle of conservation of mass. Two equivalent forms are

$$\frac{\mathrm{D}\rho}{\mathrm{D}t} + \rho\nabla\cdot\boldsymbol{v} = 0, \qquad \frac{\partial\rho}{\partial t} + \nabla\cdot(\rho\boldsymbol{v}) = 0. \tag{F.2}$$

If the fluid is incompressible the above equations become

$$\nabla\cdot\boldsymbol{v} = 0. \tag{F.3}$$

The thermodynamic equation

The thermodynamic equation is a statement of the first law of thermodynamics and is

$$\frac{\mathrm{D}I}{\mathrm{D}t} + \frac{p}{\rho}\nabla\cdot\boldsymbol{v} = \dot{Q}, \tag{F.4}$$

where $\dot{Q}$ is the total heating. In an ideal gas $I = c_v T$ and $p = \rho RT$ and the equation becomes

$$\frac{\mathrm{D}T}{\mathrm{D}t} + \frac{RT}{c_v}\nabla\cdot\boldsymbol{v} = \frac{\dot{Q}}{c_v} = \kappa\nabla^2 T + J, \tag{F.5}$$

and here we split the heating into a diffusion of temperature and an external source, J.

Equation of state

The equation of state is a diagnostic equation that connects pressure, density, temperature and, if the constituents of the fluid vary, composition. For a single-component ideal gas the equation of state is

$$p = \rho RT. \tag{F.6}$$

For seawater an approximate equation of state that gives density as a function of pressure, temperature and salinity is

$$\rho = \rho_0\left[1 - \beta_T(T - T_0) + \beta_S(S - S_0) + \beta_p(p - p_0)\right]. \tag{F.7}$$

For quantitative oceanography we need a more accurate equation of state and that necessitates the inclusion of nonlinear terms.

This is a form of the thermodynamic equation that could be used *instead* of the internal energy equation (1.49), but to do so involves relating entropy to the other thermodynamic variables with an equation of state of the form

$$\eta = \eta(T, p),$$

or, in general, any two of pressure, density and temperature. (Entropy is also a function of composition but we treat that as constant.) We can obtain such an expression for an ideal gas, as follows.

Using the ideal gas equation of state we may write the first law of thermodynamics in two equivalent ways, namely

$$c_v \mathrm{d}T + p\mathrm{d}\alpha = đQ \qquad \text{or} \qquad c_p \mathrm{d}T - \alpha\, \mathrm{d}p = đQ. \tag{1.56a,b}$$

Then, since the entropy increase obeys $T\mathrm{d}\eta = đQ$, we have

$$c_v \mathrm{d}T + p\mathrm{d}\alpha = T\mathrm{d}\eta \qquad \text{or} \qquad c_p \mathrm{d}T - \alpha\, \mathrm{d}p = T\mathrm{d}\eta. \tag{1.57a,b}$$

These are both forms of the fundamental thermodynamic relation, applied to an ideal gas. Using the equation of state in the form $\alpha = RT/p$ the above equations may be written

$$c_v \frac{\mathrm{d}T}{T} - R\frac{\mathrm{d}\rho}{\rho} = \mathrm{d}\eta \qquad \text{or} \qquad c_p \frac{\mathrm{d}T}{T} - R\frac{\mathrm{d}p}{p} = \mathrm{d}\eta. \tag{1.58a,b}$$

We can integrate (1.58) to give two equivalent explicit expressions for the entropy in an ideal gas, namely

$$\eta = c_v \log T - R\log\rho + \text{constant} = c_p \log T - R\log p + \text{constant}. \tag{1.59}$$

As always in classical mechanics, entropy contains an arbitrary constant.

For convenience we now define a quantity, θ, called *potential temperature* such that

$$\eta = c_p \ln\theta, \tag{1.60}$$

so that $\mathrm{d}\eta = c_p(\mathrm{d}\theta/\theta)$. Using this expression in (1.57b) and integrating gives

$$c_p \log T - R\log p = c_p \log\theta + \text{constant}. \tag{1.61}$$

If we fold the constant of integration into the pressure term this equation can be written as

$$\theta = T\left(\frac{p_R}{p}\right)^{R/c_p}, \tag{1.62}$$

where p_R is a constant. And given (1.60), we can write (1.54) as

$$c_p \frac{\mathrm{D}\theta}{\mathrm{D}t} = \frac{\theta}{T}\dot{Q}. \tag{1.63}$$

This form of the thermodynamic equation is equivalent to (1.51) and in some ways is a simpler expression, although it is valid only for an ideal gas — obtaining an analogous expression for seawater is more difficult because of the nonlinearity of the true seawater equation of state and because c_p for seawater is not a constant.

Although entropy responds directly to the heating, it is *not* a measure of the heat content of a body, nor does any such measure properly exist. Note, for example, that the entropy increase depends on the temperature at which heat is added. Similarly, a body does not contain a certain amount of work. Rather, both work done and heating change the internal energy.

1.6.1 Meaning of Potential Temperature

We introduced potential temperature as a measure of the entropy of a fluid and that is how it is best regarded. However, it has a useful physical interpretation, as follows. Suppose a parcel of constant composition is moved adiabatically from one location to another at a different pressure, and by adiabatic we mean that no heat enters or leaves the parcel; that is, $\dot{Q} = 0$. In a reversible process (such as fluid flow) the entropy and potential temperature are then conserved. However, the temperature of the parcel *will* change, because the fluid may be compressed or expand, as (1.56) tells us. Consider a fluid parcel at some pressure p_1 with temperature T_1 and potential temperature θ_1. Now move that parcel adiabatically to a pressure p_R. The temperature of the parcel changes but its potential temperature does not, so that the final potential temperature is just θ_1. The final temperature of the parcel, T_2 say, is equal to θ_1, because $T = \theta$ when $p = p_R$. Thus, the final temperature of the parcel is equal to its initial potential temperature. We may thus say that *the potential temperature of a parcel is equal to the temperature that a parcel will acquire if taken adiabatically to a standard pressure, p_R.* This statement is commonly taken as the *definition* of potential temperature, and if we begin here the connection with entropy is then made by realizing that in an adiabatic process at constant composition the entropy also does not change, and so is a function of potential temperature.

Potential temperature is the temperature that a fluid parcel will have if taken adiabatically and at constant composition to a reference pressure. Potential density is the analogous quantity for density. Both of these quantities are functions of the entropy of a parcel, and functions of the entropy alone if composition is fixed.

Potential temperature is sometimes a more convenient variable to use than entropy, and its temperature-like quality gives it an intuitive appeal; still, entropy is the more fundamental variable. Finally, although p_R may be chosen arbitrarily, in most atmospheric applications $p_R = 1000\,\text{hPa}$, which is the approximate pressure at sea level. Thus, potential temperature is the temperature that a parcel achieves when adiabatically brought to the surface; the potential temperature is higher than the *in situ* temperature because the parcel is compressed as it descends, and the compression increases the internal energy, and hence increases the temperature, of the parcel.

1.6.2 Potential Density

By analogy with potential temperature, the *potential density*, ρ_θ, is the density that a fluid parcel would have if moved adiabatically and at constant composition to a reference pressure, p_R. It is a useful quantity because it turns out to be a measure of the stability of a fluid parcel with respect to convection, in both air and water, as we will discover later.

Ideal gas

Suppose a fluid with temperature T and at pressure p is moved to a pressure p_R, where its temperature becomes equal to θ. For an ideal gas its potential density at pressure p is equal to the density it has at p_R so that

$$\rho_\theta = \frac{p_R}{R\theta} = \frac{p_R}{RT}\left(\frac{p}{p_R}\right)^{R/c_p} = \rho\left(\frac{p_R}{p}\right)^{c_v/c_p}, \tag{1.64}$$

using $RT = p/\rho$ to obtain the last expression.

Seawater

There is no corresponding exact expression for potential density of seawater. However, we can obtain a good approximation by first noting that density is almost constant, and then Taylor-expanding the density around the density at the reference level, at which $T = \theta$ and $p = p_R$. At first order we have

$$\rho(p) \approx \rho(p_R) + (p - p_R)\frac{\partial \rho}{\partial p}. \tag{1.65}$$

The first term on the right-hand side is, by definition, the potential density and the derivative in the second term is the inverse of the square of speed of sound, which is nearly constant in seawater. Thus

$$\rho_\theta \approx \rho - \frac{1}{c_s^2}(p - p_R). \tag{1.66}$$

Because the speed of sound is a measurable quantity, (1.66) is a very useful practical expression for potential density, although because there are small variations in the speed of sound the expression is not especially accurate for problems involving large depth variations.

1.7 THE ENERGY BUDGET

The total energy of a fluid includes the kinetic, potential and internal energies. Both the fluid flow and pressure forces will move energy from place to place, but we nevertheless expect that total energy will be conserved. Let us therefore see what form energy conservation takes in a fluid.

Energy is transported from place to place within a fluid both by advection and by the pressure force, which does work on a fluid.

1.7.1 An Energy Equation

We begin with the inviscid momentum equation with a time-independent potential Φ,

$$\rho\frac{\mathrm{D}\boldsymbol{v}}{\mathrm{D}t} = -\nabla p - \rho\nabla\Phi. \tag{1.67}$$

In a uniform gravitational field $\Phi = gz$ but we can be a little more general in our derivation. Now take the dot product of (1.67) with $\boldsymbol{v}$ and obtain an equation for the evolution of kinetic energy,

$$\frac{1}{2}\rho\frac{\mathrm{D}\boldsymbol{v}^2}{\mathrm{D}t} = -\boldsymbol{v}\cdot\nabla p - \rho\boldsymbol{v}\cdot\nabla\Phi = -\nabla\cdot(p\boldsymbol{v}) + p\nabla\cdot\boldsymbol{v} - \rho\boldsymbol{v}\cdot\nabla\Phi. \tag{1.68}$$

The internal energy equation for adiabatic flow is

$$\rho\frac{\mathrm{D}I}{\mathrm{D}t} = -p\nabla\cdot\boldsymbol{v}. \tag{1.69}$$

Finally, and somewhat trivially, the potential satisfies

$$\rho\frac{\mathrm{D}\Phi}{\mathrm{D}t} = \rho\boldsymbol{v}\cdot\nabla\Phi. \tag{1.70}$$

Adding (1.68), (1.69) and (1.70) we obtain

$$\rho \frac{\mathrm{D}}{\mathrm{D}t}\left(\frac{1}{2}\boldsymbol{v}^2 + I + \Phi\right) = -\nabla \cdot (p\boldsymbol{v}), \tag{1.71}$$

which, on expanding the material derivative and using the mass conservation equation, becomes

$$\frac{\partial}{\partial t}\left[\rho\left(\frac{1}{2}\boldsymbol{v}^2 + I + \Phi\right)\right] + \nabla \cdot \left[\rho\boldsymbol{v}\left(\frac{1}{2}\boldsymbol{v}^2 + I + \Phi + p/\rho\right)\right] = 0. \tag{1.72}$$

This may be written

$$\frac{\partial E}{\partial t} + \nabla \cdot [\boldsymbol{v}(E + p)] = 0, \tag{1.73}$$

where $E = \rho(\boldsymbol{v}^2/2 + I + \Phi)$ is the total energy per unit volume of the fluid. with contributions from the kinetic energy ($\rho\boldsymbol{v}^2/2$), the internal energy (ρI) and the potential energy ($\rho\Phi$). Equation (1.73) is the energy equation for an unforced, inviscid and adiabatic, compressible fluid. The energy flux term vanishes when integrated over a closed domain with rigid boundaries, implying that the total energy is conserved. However, there can be an exchange of energy between kinetic, potential and internal components. It is the divergent term, $\nabla \cdot \boldsymbol{v}$, that connects the kinetic energy equation, (1.68), and the internal energy equation, (1.69). In an incompressible fluid this term is absent, and the internal energy is divorced from the other components of energy. This consideration will be important when we consider the Boussinesq equations in Section 2.5. Note finally that the flux of energy, $\boldsymbol{F}_E = \boldsymbol{v}(E + p)$ is not equal to the velocity times the energy; rather, energy is also transferred by pressure. We may write the energy flux as

$$\boldsymbol{F}_E = \rho\boldsymbol{v}\left(\frac{\boldsymbol{v}^2}{2} + \Phi + h\right), \tag{1.74}$$

where $h = I + p/\rho$ is the enthalpy. That is, the local rate of change of energy is determined by the fluxes of kinetic energy, potential energy and *enthalpy*, not internal energy, because enthalpy can take into account the work done by the pressure.

The flux of energy is given by the velocity times the Bernoulli function, where the Bernoulli function is a combination of the kinetic, potential and internal energies and the pressure. It is equal to the sum of the kinetic energy, potential energy and enthalpy. The Bernoulli function is constant along streamlines in steady flow.

Bernoulli's theorem

The quantity

$$B = \left(E + \frac{p}{\rho}\right) = \left(\frac{1}{2}\boldsymbol{v}^2 + I + \Phi + p\alpha\right) = \left(\frac{1}{2}\boldsymbol{v}^2 + h + \Phi\right), \tag{1.75}$$

is the general form of the Bernoulli function, equal to the sum of the kinetic energy, the potential energy and the enthalpy. Equation (1.73) may be written as

$$\frac{\partial E}{\partial t} + \nabla \cdot (\rho\boldsymbol{v}B) = 0. \tag{1.76}$$

The Bernoulli function itself is not conserved, even for adiabatic flow. However, for steady flow $\nabla \cdot (\rho \boldsymbol{v}) = 0$, and the $\partial/\partial t$ terms vanish so that (1.76) may be written $\boldsymbol{v} \cdot \nabla B = 0$, or even $\mathrm{D}B/\mathrm{D}t = 0$. The Bernoulli function is then a constant along streamlines, a result commonly known as Bernoulli's theorem. For adiabatic flow at constant composition we also have $\mathrm{D}\theta/\mathrm{D}t = 0$. Thus, steady flow is both along surfaces of constant θ and along surfaces of constant B, and the vector

$$\boldsymbol{l} = \nabla\theta \times \nabla B \tag{1.77}$$

is parallel to streamlines.

Bernoulli's theorem was developed mainly by Daniel Bernoulli (1700–1782). The Bernoulli family produced several (at least eight) talented mathematicians over three generations in the seventeenth and eighteenth centuries, and is often regarded as the most mathematically distinguished family of all time.

1.7.2 Energy Conservation for Constant Density Fluids

If the density of a fluid is constant the derivation goes through just as above, but with an important simplification. The divergence of the velocity is zero, and so the $\nabla \cdot \boldsymbol{v}$ term does not appear on the right-hand side of the kinetic energy equation, (1.68). The right-hand side of the internal energy equation, (1.69), is then zero and the internal energy and the kinetic energy are then de-coupled. The kinetic energy equation straightforwardly becomes

$$\frac{\partial K}{\partial t} + \nabla \cdot (\boldsymbol{v}B) = 0, \tag{1.78}$$

where $K = \boldsymbol{v}^2/2$ and, here, $B = \phi + \Phi + \boldsymbol{v}^2/2$ where $\phi = p/\rho_0$ is the 'kinematic pressure'. The Bernoulli function does not contain the internal energy and the total *kinetic* energy is conserved.

1.7.3 Viscous Effects

We might expect that viscosity will always act to reduce the kinetic energy of a flow, and here we demonstrate this for a constant density fluid satisfying

$$\frac{\mathrm{D}\boldsymbol{v}}{\mathrm{D}t} = -\nabla(\phi + \Phi) + \nu\nabla^2\boldsymbol{v}. \tag{1.79}$$

The energy equation becomes

$$\frac{\mathrm{d}\widehat{E}}{\mathrm{d}t} \equiv \frac{\mathrm{d}}{\mathrm{d}t}\int_V E\,\mathrm{d}V = \mu\int_V \boldsymbol{v}\cdot\nabla^2\boldsymbol{v}\,\mathrm{d}V. \tag{1.80}$$

The right-hand side is negative definite. To see this we use the vector identity

$$\nabla \times (\nabla \times \boldsymbol{v}) = \nabla(\nabla \cdot \boldsymbol{v}) - \nabla^2\boldsymbol{v}, \tag{1.81}$$

and because $\nabla \cdot \boldsymbol{v} = 0$ we have $\nabla^2\boldsymbol{v} = -\nabla \times \boldsymbol{\omega}$, where $\boldsymbol{\omega} \equiv \nabla \times \boldsymbol{v}$. Thus,

$$\frac{\mathrm{d}\widehat{E}}{\mathrm{d}t} = -\mu\int_V \boldsymbol{v}\cdot(\nabla\times\boldsymbol{\omega})\,\mathrm{d}V = -\mu\int_V \boldsymbol{\omega}\cdot(\nabla\times\boldsymbol{v})\,\mathrm{d}V = -\mu\int_V \boldsymbol{\omega}^2\,\mathrm{d}V, \tag{1.82}$$

after integrating by parts, providing $\boldsymbol{v}\times\boldsymbol{\omega}$ vanishes at the boundary. Thus, viscosity acts to extract kinetic energy from the flow. The loss of kinetic energy reappears as an irreversible warming of the fluid and the total energy of the fluid is still conserved. The warming effect is small in Earth's atmosphere but it can be large in other planets.

Notes and References

Many books on fluid dynamics go into more detail about the equations of motion than we have, and Kundu *et al.* (2015) and Acheson (1990) both provide accessible introductions. The official definitions of the Avogadro constant and mole will change in May 2019 (as part of a larger redefinition of various SI units) and from then on a mole will bear no reference to carbon-12 (which it does in the old definition), although the change has no consequences for us. We use the new definitions in the text. The Avogadro constant is then defined as $6.02214076 \times 10^{23}\ \mathrm{mol}^{-1}$ and is to be regarded as a dimensional constant; Avogadro's number is the nondimensional number with the same numerical value, and a mole is the amount of substance containing Avogadro's number of elementary units.

Problems

1.1 Show that the derivative of an integral is given by

$$\frac{\mathrm{d}}{\mathrm{d}t}\int_{x_1(t)}^{x_2(t)} \varphi(x,t)\,\mathrm{d}x = \int_{x_1}^{x_2} \frac{\partial \varphi}{\partial t}\,\mathrm{d}x + \frac{\mathrm{d}x_2}{\mathrm{d}t}\varphi(x_2,t) - \frac{\mathrm{d}x_1}{\mathrm{d}t}\varphi(x_1,t). \tag{P1.1}$$

By generalizing to three dimensions show that the material derivative of an integral of a fluid property is given by

$$\frac{\mathrm{D}}{\mathrm{D}t}\int_V \varphi(\boldsymbol{x},t)\,\mathrm{d}V = \int_V \frac{\partial \varphi}{\partial t}\,\mathrm{d}V + \int_S \varphi\boldsymbol{v}\cdot \mathrm{d}\boldsymbol{S} = \int_V \left[\frac{\partial \varphi}{\partial t} + \nabla\cdot(\boldsymbol{v}\varphi)\right]\mathrm{d}V, \tag{P1.2}$$

where the surface integral ($\int_S$) is over the surface bounding the volume V. Hence deduce that

$$\frac{\mathrm{D}}{\mathrm{D}t}\int_V \rho\varphi\,\mathrm{d}V = \int_V \rho\frac{\mathrm{D}\varphi}{\mathrm{D}t}\,\mathrm{d}V. \tag{P1.3}$$

1.2 (*a*) If molecules move quasi-randomly, why is there no diffusion term in the mass continuity equation?

(*b*) Suppose that a fluid contains a binary mixture of dry air and water vapour. Show that the change in mass of a parcel of air due to the diffusion of water vapour is exactly balanced by the diffusion of dry air in the opposite direction.

1.3 If it is momentum, not velocity, that responds when a force is applied (according to Newton's second law), why is the (inviscid) momentum equation given by $\rho\mathrm{D}\boldsymbol{v}/\mathrm{D}t = -\nabla p$ and not $\mathrm{D}(\rho\boldsymbol{v})/\mathrm{D}t = -\nabla p$?

1.4 Using the observed value of molecular diffusion of heat in water, estimate how long it would take for a temperature anomaly to mix from the top of the ocean to the bottom, assuming that molecular diffusion alone is responsible. Comment on whether you think the real ocean has reached equilibrium after the last ice age (which ended about 12,000 years ago).

1.5 Show that viscosity will dissipate kinetic energy in a compressible fluid.

1.6 (*a*) Suppose that a sealed, insulated container consists of two compartments, and that one of them is filled with an ideal gas and the other is a vacuum. The partition separating the compartments is removed. How does the temperature of the gas change? (Answer: it stays the same. Explain.) Obtain an expression for the final potential temperature, in terms of the initial temperature of the gas and the volumes of the two compartments.

(*b*) A dry parcel that is ascending adiabatically through the atmosphere will generally cool as it moves to lower pressure and expands, and its potential temperature stays the same. How can this be consistent with your answer to part (*a*)?

Reconcile your answers with the first law of thermodynamics for an ideal gas, namely that

$$đQ = T\,\mathrm{d}\eta = c_p \frac{\mathrm{d}\theta}{\theta} = \mathrm{d}I + đW = c_v\,\mathrm{d}T + p\,\mathrm{d}\alpha. \qquad \text{(P1.4)}$$

1.7 Beginning with the expression for potential temperature for a simple ideal gas, $\theta = T(p_R/p)^\kappa$, where $\kappa = R/c_p$, show that

$$\mathrm{d}\theta = (\theta/T)(\mathrm{d}T - (\alpha/c_p)\,\mathrm{d}p), \qquad \text{(P1.5)}$$

and that the first law of thermodynamics may be written as

$$đQ = T\,\mathrm{d}\eta = c_p(T/\theta)\mathrm{d}\theta. \qquad \text{(P1.6)}$$

1.8 Show that adiabatic flow in an ideal gas satisfies $p\rho^{-\gamma}$ = constant, where $\gamma = c_p/c_v$.

1.9 Show that for an ideal gas in hydrostatic balance, changes in dry static energy ($M = c_pT + gz$) and potential temperature (θ) are related by $\delta M = c_p(T/\theta)\delta\theta$. (The quantity c_pT/θ is known as the 'Exner function'.)

1.10 Using the equation of state for seawater given in the text, or a more accurate one obtained from the literature, estimate the fractional change in density of the world's oceans due to changes in pressure, salinity and temperature.

1.11 Using an accurate equation of state obtained from the literature (for example, from Vallis (2017) or IOC *et al.* (2010)) estimate by how much (1.42) is in error in the world's oceans.

CHAPTER

2 Equations for a Rotating Planet

PLANETS ARE ALMOST SPHERES. They also rotate. Here we consider how the equations of motion are affected by these facts, first by looking at how rotation affects the dynamics and then by expressing the equations in spherical coordinates.

2.1 EQUATIONS IN A ROTATING FRAME OF REFERENCE

Newton's second law of motion, that the rate of change of momentum of a body is proportional to the imposed force, applies in so-called inertial frames of reference that are either stationary or moving only with a constant rectilinear velocity relative to the distant galaxies. Now Earth spins around its axis once a day, so the surface of the Earth is *not* an inertial frame. Nevertheless, it is very convenient to describe the motion of the atmosphere or ocean relative to Earth's surface rather than in some inertial frame. How we do that is the subject of this section.

2.1.1 Rate of Change of a Vector

Consider first a vector $\boldsymbol{C}$ of constant length rotating relative to an inertial frame at a constant angular velocity $\boldsymbol{\Omega}$. Then, in a frame rotating with that same angular velocity it appears stationary and constant. If in a small interval of time δt the vector $\boldsymbol{C}$ rotates through a small angle $\delta\lambda$ then the change in $\boldsymbol{C}$, as perceived in the inertial frame, is given by (see Fig. 2.1)

$$\delta \boldsymbol{C} = |\boldsymbol{C}| \cos\vartheta \, \delta\lambda \, \boldsymbol{m}, \tag{2.1}$$

where the vector $\boldsymbol{m}$ is the unit vector in the direction of change of $\boldsymbol{C}$, which is perpendicular to both $\boldsymbol{C}$ and $\boldsymbol{\Omega}$. But the rate of change of the angle λ is just, by definition, the angular velocity so that $\delta\lambda = |\boldsymbol{\Omega}|\delta t$ and

$$\delta \boldsymbol{C} = |\boldsymbol{C}||\boldsymbol{\Omega}| \sin\widehat{\vartheta} \, \boldsymbol{m} \, \delta t = \boldsymbol{\Omega} \times \boldsymbol{C} \, \delta t, \tag{2.2}$$

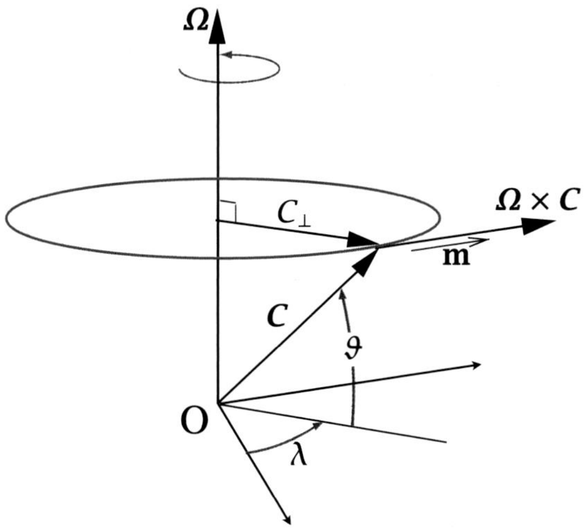

Fig. 2.1: A vector $\boldsymbol{C}$ rotating at an angular velocity $\boldsymbol{\Omega}$. It appears to be a constant vector in the rotating frame, whereas in the inertial frame it evolves according to $(\mathrm{d}\boldsymbol{C}/\mathrm{d}t)_I = \boldsymbol{\Omega} \times \boldsymbol{C}$.

using the definition of the vector cross-product, where $\widehat{\vartheta} = (\pi/2 - \vartheta)$ is the angle between $\boldsymbol{\Omega}$ and $\boldsymbol{C}$. Thus

$$\left(\frac{\mathrm{d}\boldsymbol{C}}{\mathrm{d}t}\right)_I = \boldsymbol{\Omega} \times \boldsymbol{C}, \tag{2.3}$$

where the left-hand side is the rate of change of $\boldsymbol{C}$ as perceived in the inertial frame.

Now consider a vector $\boldsymbol{B}$ that changes in the inertial frame. In a small time δt the change in $\boldsymbol{B}$ as seen in the rotating frame is related to the change seen in the inertial frame by

$$(\delta \boldsymbol{B})_I = (\delta \boldsymbol{B})_R + (\delta \boldsymbol{B})_{\mathrm{rot}}, \tag{2.4}$$

where the terms are, respectively, the change seen in the inertial frame, the change due to the vector itself changing as measured in the rotating frame, and the change due to the rotation. Using (2.2) $(\delta \boldsymbol{B})_{\mathrm{rot}} = \boldsymbol{\Omega} \times \boldsymbol{B}\,\delta t$, and so the rates of change of the vector $\boldsymbol{B}$ in the inertial and rotating frames are related by

$$\left(\frac{\mathrm{d}\boldsymbol{B}}{\mathrm{d}t}\right)_I = \left(\frac{\mathrm{d}\boldsymbol{B}}{\mathrm{d}t}\right)_R + \boldsymbol{\Omega} \times \boldsymbol{B}. \tag{2.5}$$

This relation applies to a vector $\boldsymbol{B}$ that, as measured at any one time, is the same in both inertial and rotating frames.

2.1.2 Velocity and Acceleration in a Rotating Frame

The velocity of a body is not measured to be the same in the inertial and rotating frames, so care must be taken when applying (2.5) to velocity. First apply (2.5) to $\boldsymbol{r}$, the position of a particle, to obtain

$$\left(\frac{\mathrm{d}\boldsymbol{r}}{\mathrm{d}t}\right)_I = \left(\frac{\mathrm{d}\boldsymbol{r}}{\mathrm{d}t}\right)_R + \boldsymbol{\Omega} \times \boldsymbol{r} \tag{2.6}$$

or

$$\boldsymbol{v}_I = \boldsymbol{v}_R + \boldsymbol{\Omega} \times \boldsymbol{r}. \tag{2.7}$$

We refer to $\boldsymbol{v}_R$ and $\boldsymbol{v}_I$ as the relative and inertial velocity, respectively, and (2.7) relates the two. Apply (2.5) again, this time to the velocity $\boldsymbol{v}_R$ to give

$$\left(\frac{\mathrm{d}\boldsymbol{v}_R}{\mathrm{d}t}\right)_I = \left(\frac{\mathrm{d}\boldsymbol{v}_R}{\mathrm{d}t}\right)_R + \boldsymbol{\Omega} \times \boldsymbol{v}_R, \tag{2.8}$$

or, using (2.7)

$$\left(\frac{\mathrm{d}}{\mathrm{d}t}(\boldsymbol{v}_I - \boldsymbol{\Omega} \times \boldsymbol{r})\right)_I = \left(\frac{\mathrm{d}\boldsymbol{v}_R}{\mathrm{d}t}\right)_R + \boldsymbol{\Omega} \times \boldsymbol{v}_R, \tag{2.9}$$

or

$$\left(\frac{\mathrm{d}\boldsymbol{v}_I}{\mathrm{d}t}\right)_I = \left(\frac{\mathrm{d}\boldsymbol{v}_R}{\mathrm{d}t}\right)_R + \boldsymbol{\Omega} \times \boldsymbol{v}_R + \frac{\mathrm{d}\boldsymbol{\Omega}}{\mathrm{d}t} \times \boldsymbol{r} + \boldsymbol{\Omega} \times \left(\frac{\mathrm{d}\boldsymbol{r}}{\mathrm{d}t}\right)_I. \tag{2.10}$$

Then, noting that

$$\left(\frac{\mathrm{d}\boldsymbol{r}}{\mathrm{d}t}\right)_I = \left(\frac{\mathrm{d}\boldsymbol{r}}{\mathrm{d}t}\right)_R + \boldsymbol{\Omega} \times \boldsymbol{r} = (\boldsymbol{v}_R + \boldsymbol{\Omega} \times \boldsymbol{r}), \tag{2.11}$$

and assuming that the rate of rotation is constant, (2.10) becomes

$$\left(\frac{\mathrm{d}\boldsymbol{v}_R}{\mathrm{d}t}\right)_R = \left(\frac{\mathrm{d}\boldsymbol{v}_I}{\mathrm{d}t}\right)_I - 2\boldsymbol{\Omega} \times \boldsymbol{v}_R - \boldsymbol{\Omega} \times (\boldsymbol{\Omega} \times \boldsymbol{r}). \tag{2.12}$$

This equation may be interpreted as follows. The term on the left-hand side is the rate of change of the relative velocity as measured in the rotating frame. The first term on the right-hand side is the rate of change of the inertial velocity as measured in the inertial frame (the inertial acceleration, which is, by Newton's second law, equal to the force on a fluid parcel divided by its mass). The second and third terms on the right-hand side (including the minus signs) are the *Coriolis force* and the *centrifugal force* per unit mass. Neither of these are usually regarded as true forces — they may be thought of as quasi-forces (i.e., 'as if' forces); that is, when a body is observed from a rotating frame it behaves as if unseen forces are present that affect its motion.

Centrifugal force

If $\boldsymbol{r}_\perp$ is the perpendicular distance from the axis of rotation (see Fig. 2.1 and substitute $\boldsymbol{r}$ for $\mathbf{C}$), then, because $\boldsymbol{\Omega}$ is perpendicular to $\boldsymbol{r}_\perp$, $\boldsymbol{\Omega} \times \boldsymbol{r} = \boldsymbol{\Omega} \times \boldsymbol{r}_\perp$. Then, using the vector identity $\boldsymbol{\Omega} \times (\boldsymbol{\Omega} \times \boldsymbol{r}_\perp) = (\boldsymbol{\Omega} \cdot \boldsymbol{r}_\perp)\boldsymbol{\Omega} - (\boldsymbol{\Omega} \cdot \boldsymbol{\Omega})\boldsymbol{r}_\perp$ and noting that the first term is zero, we see that the centrifugal force per unit mass is just given by

$$\boldsymbol{F}_{\mathrm{ce}} = -\boldsymbol{\Omega} \times (\boldsymbol{\Omega} \times \boldsymbol{r}) = \Omega^2 \boldsymbol{r}_\perp. \tag{2.13}$$

This may usefully be written as the gradient of a scalar potential,

$$\boldsymbol{F}_{\mathrm{ce}} = -\nabla \Phi_{\mathrm{ce}}. \tag{2.14}$$

where $\Phi_{\mathrm{ce}} = -(\Omega^2 r_\perp^2)/2 = -(\boldsymbol{\Omega} \times \boldsymbol{r}_\perp)^2/2$.

Coriolis force

The Coriolis force per unit mass is given by

$$\boldsymbol{F}_{\text{Co}} = -2\boldsymbol{\Omega} \times \boldsymbol{v}_R. \tag{2.15}$$

We consider the effects of the Coriolis force extensively, but for now we just note three basic properties:

(i) There is no Coriolis force on bodies that are stationary in the rotating frame.

(ii) The Coriolis force acts to deflect moving bodies at right angles to their direction of travel.

(iii) The Coriolis force does no work on a body because it is perpendicular to the velocity, and so $\boldsymbol{v}_R \cdot (\boldsymbol{\Omega} \times \boldsymbol{v}_R) = 0$.

The Coriolis force is named for Gaspard-Gustave de Coriolis (1792–1843) who discussed the effect in an engineering context in 1835, although the basic effect may have been first recognized (as were so many things) by Leonhard Euler (1707–1783). The (now-called) Coriolis term is also contained in Laplace's tidal equations, formulated in 1776, published in English in Laplace (1832). William Ferrel (1817–1891) was perhaps the first to appreciate the effect of the force on Earth's circulation, identifying and discussing the relevant term ($2\Omega v \sin \vartheta$) in Laplace's equations.

2.1.3 Equations of Motion in a Rotating Frame

Momentum equation

Since (2.12) simply relates the accelerations of a particle in the inertial and rotating frames, then in the rotating frame of reference the three-dimensional momentum equation may be written

$$\frac{\mathrm{D}\boldsymbol{v}}{\mathrm{D}t} + 2\boldsymbol{\Omega} \times \boldsymbol{v} = -\frac{1}{\rho}\nabla p - \nabla\Phi_{\text{ce}} + \boldsymbol{g}, \tag{2.16}$$

where all velocities and accelerations are measured with respect to the inertial frame. Since the centrifugal term does not vary with the fluid motion we can incorporate it into gravitational force, $\boldsymbol{g}$, so giving an 'effective gravity' that varies slightly with position over Earth's surface.

Mass continuity and the thermodynamic equation

The mass conservation equation and the thermodynamic equation are unchanged in a rotating frame. To see this consider the material derivative of some variable, φ, such as temperature or density. The material derivative is just the rate of change of φ of an identifiable fluid parcel and that clearly does not depend on the reference frame. Thus, without further ado, we can write

$$\left(\frac{\mathrm{D}\varphi}{\mathrm{D}t}\right)_R = \left(\frac{\mathrm{D}\varphi}{\mathrm{D}t}\right)_I, \tag{2.17}$$

where the material derivatives are $(\mathrm{D}\varphi/\mathrm{D}t)_R = (\partial\varphi/\partial t)_R + \boldsymbol{v}_R \cdot \nabla\varphi$ and $(\mathrm{D}\varphi/\mathrm{D}t)_I = (\partial\varphi/\partial t)_I + \boldsymbol{v}_I \cdot \nabla\varphi$. The individual terms differ in the two frames; that is $(\partial\varphi/\partial t)_R \neq (\partial\varphi/\partial t)_I$, but the material derivatives are equal.

Further, the divergence operator is the same in the inertial and rotating frame. Using (2.7), we have that

$$\nabla \cdot \boldsymbol{v}_I = \nabla \cdot (\boldsymbol{v}_R + \boldsymbol{\Omega} \times \boldsymbol{r}) = \nabla \cdot \boldsymbol{v}_R \tag{2.18}$$

since $\nabla \cdot (\boldsymbol{\Omega} \times \boldsymbol{r}) = 0$. Thus, using (2.17) and (2.18), the mass conservation equation (1.27b) may be written

$$\frac{\mathrm{D}\rho}{\mathrm{D}t} + \rho\nabla \cdot \boldsymbol{v}_R = 0, \tag{2.19}$$

and similarly for the thermodynamic equation, where all observables are measured in the *rotating* frame. That is, the mass conservation equation and the thermodynamic equation are unaltered by the presence of rotation. More generally, the evolution equation for a scalar whose value is the same in rotating and inertial frames is unaltered by rotation.

2.2 ✦ EQUATIONS OF MOTION IN SPHERICAL COORDINATES

Since Earth, and most other planets that we know about, are very nearly spherical we cast the equations in spherical coordinates. We will only very rarely use the full spherical form of the equations but our approximations begin with that form. The trusting reader may skim the derivations and go straight to the main results, given in (2.34), (2.35) and (2.40).

2.2.1 Some Identities in Spherical Coordinates

The location of a point is given by the coordinates (λ, ϑ, r) where λ is the angular distance eastwards (i.e., longitude), ϑ is angular distance polewards (i.e., latitude) and r is the radial distance from the centre of the Earth — see Fig. 2.2. (In some other fields of study co-latitude is used as a spherical coordinate.) If a is the radius of the Earth, then we also define $z = r - a$. At a given location we may also define the Cartesian increments $(\delta x, \delta y, \delta z) = (r \cos\vartheta\, \delta\lambda, r\delta\vartheta, \delta r)$.

For a scalar quantity φ the material derivative in spherical coordinates is

$$\frac{\mathrm{D}\varphi}{\mathrm{D}t} = \frac{\partial\varphi}{\partial t} + \frac{u}{r\cos\vartheta}\frac{\partial\varphi}{\partial\lambda} + \frac{v}{r}\frac{\partial\varphi}{\partial\vartheta} + w\frac{\partial\varphi}{\partial r}, \tag{2.20}$$

where the velocity components corresponding to the coordinates (λ, ϑ, r) are

$$(u, v, w) \equiv \left(r\cos\vartheta\frac{\mathrm{D}\lambda}{\mathrm{D}t}, r\frac{\mathrm{D}\vartheta}{\mathrm{D}t}, \frac{\mathrm{D}r}{\mathrm{D}t} \right). \tag{2.21}$$

That is, u is the zonal velocity, v is the meridional velocity and w is the vertical velocity. If we define $(\hat{\mathbf{i}}, \hat{\mathbf{j}}, \hat{\mathbf{k}})$ to be the unit vectors in the direction of increasing (λ, ϑ, r) then

$$\boldsymbol{v} = \hat{\mathbf{i}}u + \hat{\mathbf{j}}v + \hat{\mathbf{k}}w. \tag{2.22}$$

Note also that $\mathrm{D}r/\mathrm{D}t = \mathrm{D}z/\mathrm{D}t$.

The divergence of a vector $\boldsymbol{B} = \hat{\mathbf{i}}B^\lambda + \hat{\mathbf{j}}B^\vartheta + \hat{\mathbf{k}}B^r$ is

$$\nabla\cdot\boldsymbol{B} = \frac{1}{\cos\vartheta}\left[\frac{1}{r}\frac{\partial B^\lambda}{\partial\lambda} + \frac{1}{r}\frac{\partial}{\partial\vartheta}(B^\vartheta\cos\vartheta) + \frac{\cos\vartheta}{r^2}\frac{\partial}{\partial r}(r^2 B^r)\right]. \tag{2.23}$$

The vector gradient of a scalar is

$$\nabla\varphi = \hat{\mathbf{i}}\frac{1}{r\cos\vartheta}\frac{\partial\varphi}{\partial\lambda} + \hat{\mathbf{j}}\frac{1}{r}\frac{\partial\varphi}{\partial\vartheta} + \hat{\mathbf{k}}\frac{\partial\varphi}{\partial r}. \tag{2.24}$$

The Laplacian of a scalar is

$$\nabla^2\varphi \equiv \nabla\cdot\nabla\varphi = \frac{1}{r^2\cos\vartheta}\left[\frac{1}{\cos\vartheta}\frac{\partial^2\varphi}{\partial\lambda^2} + \frac{\partial}{\partial\vartheta}\left(\cos\vartheta\frac{\partial\varphi}{\partial\vartheta}\right) + \cos\vartheta\frac{\partial}{\partial r}\left(r^2\frac{\partial\varphi}{\partial r}\right)\right]. \tag{2.25}$$

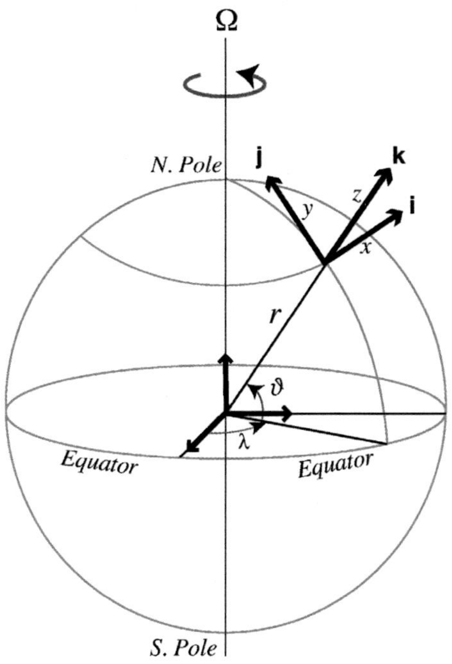

Fig. 2.2: The spherical coordinate system. The orthogonal unit vectors $\hat{\mathbf{i}}$, $\hat{\mathbf{j}}$ and $\hat{\mathbf{k}}$ point in the direction of increasing longitude λ, latitude ϑ, and altitude z. Locally, one may apply a Cartesian system with variables x, y and z measuring distances along $\hat{\mathbf{i}}$, $\hat{\mathbf{j}}$ and $\hat{\mathbf{k}}$.

The curl of a vector is

$$\operatorname{curl} \boldsymbol{B} = \nabla \times \boldsymbol{B} = \frac{1}{r^2 \cos\vartheta} \begin{vmatrix} \hat{\mathbf{i}}\, r\cos\vartheta & \hat{\mathbf{j}}\, r & \hat{\mathbf{k}} \\ \partial/\partial\lambda & \partial/\partial\vartheta & \partial/\partial r \\ B^\lambda r\cos\vartheta & B^\vartheta r & B^r \end{vmatrix}. \tag{2.26}$$

The vector Laplacian $\nabla^2 \boldsymbol{B}$ (used for example when calculating viscous terms in the momentum equation) may be obtained from the vector identity

$$\nabla^2 \boldsymbol{B} = \nabla(\nabla \cdot \boldsymbol{B}) - \nabla \times (\nabla \times \boldsymbol{B}). \tag{2.27}$$

Only in Cartesian coordinates does this take the simple form

$$\nabla^2 \boldsymbol{B} = \frac{\partial^2 \boldsymbol{B}}{\partial x^2} + \frac{\partial^2 \boldsymbol{B}}{\partial y^2} + \frac{\partial^2 \boldsymbol{B}}{\partial z^2}. \tag{2.28}$$

The expansion in spherical coordinates is of itself, to most eyes, rather uninformative.

♦ *Rate of change of unit vectors*

In spherical coordinates the defining unit vectors are $\hat{\mathbf{i}}$, the unit vector pointing eastwards, parallel to a line of latitude; $\hat{\mathbf{j}}$ is the unit vector pointing polewards, parallel to a meridian; and $\hat{\mathbf{k}}$, the unit vector pointing radially outward. The directions of these vectors change with location, and in fact this is the case in nearly all coordinate systems, with the notable exception of the Cartesian one, and thus their material derivative is not zero.

One way to evaluate these changes is to first obtain the effective rotation rate $\boldsymbol{\Omega}_{\text{flow}}$, relative to the Earth, of a unit vector as it moves with the flow, and then apply (2.3). Specifically, let the fluid velocity be $\boldsymbol{v} = (u, v, w)$. The meridional component, v, produces a displacement $r\delta\vartheta = v\delta t$, and this gives rise to a local effective vector rotation rate around the local zonal axis of $-(v/r)\hat{\mathbf{i}}$, the minus sign arising because a displacement in the direction of the north pole is produced by negative rotational displacement around the $\hat{\mathbf{i}}$ axis. Similarly, the zonal component, u, produces a displacement $\delta\lambda r\cos\vartheta = u\delta t$ and so an effective rotation rate, about the Earth's rotation axis, of $u/(r\cos\vartheta)$. Now, a rotation around the Earth's rotation axis may be written as (see Fig. 2.3)

$$\boldsymbol{\Omega} = \Omega(\hat{\mathbf{j}}\cos\vartheta + \hat{\mathbf{k}}\sin\vartheta). \tag{2.29}$$

If the scalar rotation rate is not Ω but is $u/(r\cos\vartheta)$, then the vector rotation rate is

$$\frac{u}{r\cos\vartheta}(\hat{\mathbf{j}}\cos\vartheta + \hat{\mathbf{k}}\sin\vartheta) = \hat{\mathbf{j}}\frac{u}{r} + \hat{\mathbf{k}}\frac{u\tan\vartheta}{r}. \tag{2.30}$$

Thus, the total rotation rate of a vector that moves with the flow is

$$\boldsymbol{\Omega}_{\text{flow}} = -\hat{\mathbf{i}}\frac{v}{r} + \hat{\mathbf{j}}\frac{u}{r} + \hat{\mathbf{k}}\frac{u\tan\vartheta}{r}. \tag{2.31}$$

Applying (2.3) to (2.31), we find

$$\frac{\mathrm{D}\hat{\mathbf{i}}}{\mathrm{D}t} = \boldsymbol{\Omega}_{\text{flow}} \times \hat{\mathbf{i}} = \frac{u}{r\cos\vartheta}(\hat{\mathbf{j}}\sin\vartheta - \hat{\mathbf{k}}\cos\vartheta), \tag{2.32a}$$

$$\frac{\mathrm{D}\hat{\mathbf{j}}}{\mathrm{D}t} = \boldsymbol{\Omega}_{\text{flow}} \times \hat{\mathbf{j}} = -\hat{\mathbf{i}}\frac{u}{r}\tan\vartheta - \hat{\mathbf{k}}\frac{v}{r}, \tag{2.32b}$$

$$\frac{\mathrm{D}\hat{\mathbf{k}}}{\mathrm{D}t} = \boldsymbol{\Omega}_{\text{flow}} \times \hat{\mathbf{k}} = \hat{\mathbf{i}}\frac{u}{r} + \hat{\mathbf{j}}\frac{v}{r}. \tag{2.32c}$$

2.2.2 Equations of Motion

We are now in a position to write down the equations of motion on a spherical planet.

Mass conservation and the thermodynamic equation

The mass conservation equation, (1.27a), expanded in spherical coordinates, is

$$\begin{aligned}\frac{\partial\rho}{\partial t} &+ \frac{u}{r\cos\vartheta}\frac{\partial\rho}{\partial\lambda} + \frac{v}{r}\frac{\partial\rho}{\partial\vartheta} + w\frac{\partial\rho}{\partial r} \\ &+ \frac{\rho}{r\cos\vartheta}\left[\frac{\partial u}{\partial\lambda} + \frac{\partial}{\partial\vartheta}(v\cos\vartheta) + \frac{1}{r}\frac{\partial}{\partial r}(wr^2\cos\vartheta)\right] = 0.\end{aligned} \tag{2.33}$$

Equivalently, using the form (1.27b), this is

$$\frac{\partial\rho}{\partial t} + \frac{1}{r\cos\vartheta}\frac{\partial(u\rho)}{\partial\lambda} + \frac{1}{r\cos\vartheta}\frac{\partial}{\partial\vartheta}(v\rho\cos\vartheta) + \frac{1}{r^2}\frac{\partial}{\partial r}(r^2 w\rho) = 0. \tag{2.34}$$

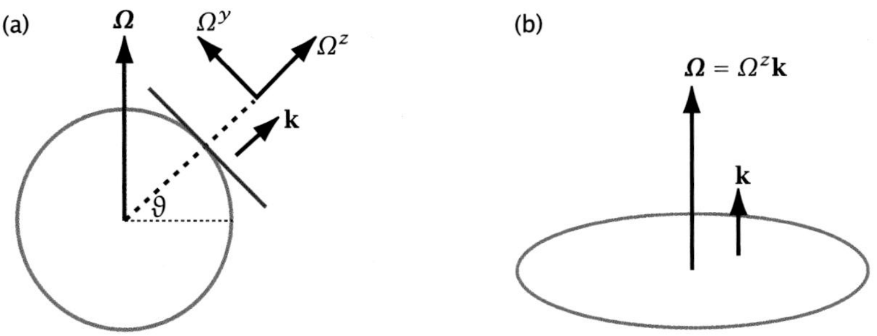

Fig. 2.3: (a) On the sphere the rotation vector $\boldsymbol{\Omega}$ can be decomposed into two components, one in the local vertical and one in the local horizontal, pointing toward the pole. That is, $\boldsymbol{\Omega} = \Omega_y \hat{\mathbf{j}} + \Omega_z \hat{\mathbf{k}}$ where $\Omega_y = \Omega \cos\vartheta$ and $\Omega_z = \Omega \sin\vartheta$. In geophysical fluid dynamics, the rotation vector in the local vertical is often the more important component in the horizontal momentum equations. On a rotating disk, (b), the rotation vector $\boldsymbol{\Omega}$ is parallel to the local vertical $\hat{\mathbf{k}}$.

The thermodynamic equation, in potential temperature form, is just an advection equation so that using (2.20), its (adiabatic) spherical coordinate form is

$$\frac{\mathrm{D}\theta}{\mathrm{D}t} = \frac{\partial\theta}{\partial t} + \frac{u}{r\cos\vartheta}\frac{\partial\theta}{\partial\lambda} + \frac{v}{r}\frac{\partial\theta}{\partial\vartheta} + w\frac{\partial\theta}{\partial r} = 0, \tag{2.35}$$

and similarly for tracers such as water vapour or salt.

Momentum equation

Recall that the inviscid momentum equation is:

$$\frac{\mathrm{D}\boldsymbol{v}}{\mathrm{D}t} + 2\boldsymbol{\Omega}\times\boldsymbol{v} = -\frac{1}{\rho}\nabla p - \nabla\Phi, \tag{2.36}$$

where Φ is the geopotential. In spherical coordinates the directions of the coordinate axes change with position and so the component expansion of (2.36) is

$$\frac{\mathrm{D}\boldsymbol{v}}{\mathrm{D}t} = \frac{\mathrm{D}u}{\mathrm{D}t}\hat{\mathbf{i}} + \frac{\mathrm{D}v}{\mathrm{D}t}\hat{\mathbf{j}} + \frac{\mathrm{D}w}{\mathrm{D}t}\hat{\mathbf{k}} + u\frac{\mathrm{D}\hat{\mathbf{i}}}{\mathrm{D}t} + v\frac{\mathrm{D}\hat{\mathbf{j}}}{\mathrm{D}t} + w\frac{\mathrm{D}\hat{\mathbf{k}}}{\mathrm{D}t} \tag{2.37a}$$

$$= \frac{\mathrm{D}u}{\mathrm{D}t}\hat{\mathbf{i}} + \frac{\mathrm{D}v}{\mathrm{D}t}\hat{\mathbf{j}} + \frac{\mathrm{D}w}{\mathrm{D}t}\hat{\mathbf{k}} + \boldsymbol{\Omega}_{\text{flow}}\times\boldsymbol{v}, \tag{2.37b}$$

using (2.32). Using either (2.37a) and the expressions for the rates of change of the unit vectors given in (2.32), or (2.37b) and the expression for $\boldsymbol{\Omega}_{\text{flow}}$ given in (2.31), (2.37) becomes

$$\frac{\mathrm{D}\boldsymbol{v}}{\mathrm{D}t} = \hat{\mathbf{i}}\left(\frac{\mathrm{D}u}{\mathrm{D}t} - \frac{uv\tan\vartheta}{r} + \frac{uw}{r}\right) + \hat{\mathbf{j}}\left(\frac{\mathrm{D}v}{\mathrm{D}t} + \frac{u^2\tan\vartheta}{r} + \frac{vw}{r}\right) + \hat{\mathbf{k}}\left(\frac{\mathrm{D}w}{\mathrm{D}t} - \frac{u^2+v^2}{r}\right). \tag{2.38}$$

Using the definition of a vector cross-product, the Coriolis term is:

$$2\boldsymbol{\Omega}\times\boldsymbol{v} = \begin{vmatrix} \hat{\mathbf{i}} & \hat{\mathbf{j}} & \hat{\mathbf{k}} \\ 0 & 2\Omega\cos\vartheta & 2\Omega\sin\vartheta \\ u & v & w \end{vmatrix}$$

$$= \hat{\mathbf{i}}\,(2\Omega w\cos\vartheta - 2\Omega v\sin\vartheta) + \hat{\mathbf{j}}\,2\Omega u\sin\vartheta - \hat{\mathbf{k}}\,2\Omega u\cos\vartheta. \tag{2.39}$$

Using (2.38) and (2.39), and the gradient operator given by (2.24), the momentum equation (2.36) becomes:

$$\frac{\mathrm{D}u}{\mathrm{D}t} - \left(2\Omega + \frac{u}{r\cos\vartheta}\right)(v\sin\vartheta - w\cos\vartheta) = -\frac{1}{\rho r\cos\vartheta}\frac{\partial p}{\partial\lambda}, \tag{2.40a}$$

$$\frac{\mathrm{D}v}{\mathrm{D}t} + \frac{wv}{r} + \left(2\Omega + \frac{u}{r\cos\vartheta}\right)u\sin\vartheta = -\frac{1}{\rho r}\frac{\partial p}{\partial\vartheta}, \tag{2.40b}$$

$$\frac{\mathrm{D}w}{\mathrm{D}t} - \frac{u^2+v^2}{r} - 2\Omega u\cos\vartheta = -\frac{1}{\rho}\frac{\partial p}{\partial r} - g. \tag{2.40c}$$

The terms involving Ω are called 'Coriolis terms'.

2.2.3 Primitive Equations

The primitive equations were first written down in the early twentieth century by Vilhelm Bjerknes (see the note on page 338). Most numerical climate models still use those equations although there is a tendency for very high resolution models to use the full Navier–Stokes equations instead, in the endless quest for fewer approximations and more accuracy.

When considering large-scale flow in the atmosphere, (2.40) may be usefully simplified in three ways:

(i) By making the hydrostatic assumption.
(ii) By neglecting the vertical velocity where it appears alongside a horizontal velocity, because $|w| \ll |u|, |v|$.
(iii) By taking the distance to the centre of Earth to be constant, a, and replacing r by a and $\partial/\partial r$ by $\partial/\partial z$.

These simplifications ultimately arise because of the thinness of the atmosphere compared to the radius of the planet, and approximations *(ii)* and *(iii)*, which should be taken together or not at all, are called the *thin-shell* or *shallow-atmosphere* approximation. The above equations become

$$\frac{\mathrm{D}u}{\mathrm{D}t} - 2\Omega v\sin\vartheta + \frac{uv\tan\vartheta}{a} = -\frac{1}{\rho a\cos\vartheta}\frac{\partial p}{\partial\lambda}, \tag{2.41a}$$

$$\frac{\mathrm{D}v}{\mathrm{D}t} + 2\Omega u\sin\vartheta + \frac{u^2\tan\vartheta}{a} = -\frac{1}{\rho a}\frac{\partial p}{\partial\vartheta}, \tag{2.41b}$$

$$\frac{\partial p}{\partial z} = -\rho g, \tag{2.41c}$$

where $\mathrm{D}/\mathrm{D}t = (\partial/\partial t + (u\cos\vartheta/a)\partial/\partial\lambda + (v/a)\partial/\partial\vartheta + w\partial/\partial z)$. These equations, along with the mass continuity and thermodynamic equations, are known as the *primitive equations* and are widely used in the numerical modelling of large-scale flows in atmospheres and oceans. Noting the ubiquity of $2\Omega\sin\vartheta$, we define the *Coriolis parameter*, $f \equiv 2\Omega\sin\vartheta$.

2.3 CARTESIAN APPROXIMATIONS: THE TANGENT PLANE

2.3.1 The f-plane

Although the rotation of the Earth is central for many dynamical phenomena, the sphericity of the Earth is not always so, especially for scales smaller than global; it is then convenient to use a locally Cartesian representation of the equations. Referring to the red line in Fig. 2.3 we define a plane tangent to the surface of the Earth at a latitude ϑ_0, and then use a Cartesian coordinate system (x, y, z) to describe motion on that plane.

For small excursions on the plane, $(x, y, z) \approx (a\lambda \cos\vartheta_0, a(\vartheta-\vartheta_0), z)$. Consistently, the velocity is $\boldsymbol{v} = (u, v, w)$, so that u,v and w are the components of the velocity *in the tangent plane,* in approximately the east–west, north–south and vertical directions, respectively.

The momentum equations for flow in this plane are then, without making the shallow-atmosphere approximation,

$$\frac{\partial u}{\partial t} + (\boldsymbol{v}\cdot\nabla)u + 2(\Omega^y w - \Omega^z v) = -\frac{1}{\rho}\frac{\partial p}{\partial x}, \tag{2.42a}$$

$$\frac{\partial v}{\partial t} + (\boldsymbol{v}\cdot\nabla)v + 2(\Omega^z u - \Omega^x w) = -\frac{1}{\rho}\frac{\partial p}{\partial y}, \tag{2.42b}$$

$$\frac{\partial w}{\partial t} + (\boldsymbol{v}\cdot\nabla)w + 2(\Omega^x v - \Omega^y u) = -\frac{1}{\rho}\frac{\partial p}{\partial z} - g, \tag{2.42c}$$

where the rotation vector $\boldsymbol{\Omega} = \Omega^x\hat{\mathbf{i}} + \Omega^y\hat{\mathbf{j}} + \Omega^z\hat{\mathbf{k}}$, with $(\hat{\mathbf{i}}, \hat{\mathbf{j}}, \hat{\mathbf{k}})$ being the unit vectors in the x, y, and z directions, $\Omega^x = 0$, $\Omega^y = \Omega\cos\vartheta_0$ and $\Omega^z = \Omega\sin\vartheta_0$. We now ignore the components of $\boldsymbol{\Omega}$ not in the direction of the local vertical; this is called the 'traditional approximation' and here it may be regarded as a Cartesian analogue of the shallow-atmosphere approximation. (Note that all the omitted terms are multiplied by or produce a vertical velocity.) The above equations become

$$\frac{\mathrm{D}u}{\mathrm{D}t} - f_0 v = -\frac{1}{\rho}\frac{\partial p}{\partial x}, \qquad \frac{\mathrm{D}v}{\mathrm{D}t} + f_0 u = -\frac{1}{\rho}\frac{\partial p}{\partial y}, \qquad \frac{\mathrm{D}w}{\mathrm{D}t} = -\frac{1}{\rho}\frac{\partial p}{\partial z} - g, \tag{2.43a,b,c}$$

where $f_0 = 2\Omega^z = 2\Omega\sin\vartheta_0$. Defining the horizontal velocity vector $\boldsymbol{u} = (u, v, 0)$, the first two of these equations may be written as

$$\frac{\mathrm{D}\boldsymbol{u}}{\mathrm{D}t} + \boldsymbol{f}_0 \times \boldsymbol{u} = -\frac{1}{\rho}\nabla_z p, \tag{2.44}$$

where $\mathrm{D}\boldsymbol{u}/\mathrm{D}t = \partial\boldsymbol{u}/\partial t + \boldsymbol{v}\cdot\nabla\boldsymbol{u}$, $\boldsymbol{f}_0 = 2\Omega\sin\vartheta_0\hat{\mathbf{k}} = f_0\hat{\mathbf{k}}$. These equations are the same as the momentum equations in a system in which the rotation vector is aligned with the local vertical, as in Fig. 2.3b; we have made the *f-plane* approximation. In the vertical direction there is often a good balance between the pressure gradient force and gravity and if so we make the hydrostatic approximation and (2.43c) becomes $\partial p/\partial z = -\rho g$. This equation plus (2.43a,b) give us the primitive equations on the f-plane.

The f-plane approximation is a rather severe approximation that is appropriate for motion on scales much smaller than global. It allows for the effect of rotation, but not for the change of Coriolis parameter with latitude. The f-plane approximation and the hydrostatic approximation are independent of each other.

2.3.2 The Beta-Plane

The magnitude of the vertical component of rotation varies with latitude, and this has important dynamical consequences. We can approximate this effect by allowing the effective rotation vector to vary on the tangent plane. Thus, noting that, for small variations in latitude,

$$f = 2\Omega\sin\vartheta \approx 2\Omega\sin\vartheta_0 + 2\Omega(\vartheta - \vartheta_0)\cos\vartheta_0, \tag{2.45}$$

then on the tangent plane we may mimic this by allowing the Coriolis parameter to vary as

$$f = f_0 + \beta y, \tag{2.46}$$

The beta-plane approximation allows for inclusion of the important *dynamical* effects of a spherical planet, in particular the variation of the Coriolis parameter, without the complication of spherical coordinates. The beta-plane greatly simplifies many calculations, although it does not properly account for the *geometric* effects of sphericity.

where $f_0 = 2\Omega \sin \vartheta_0$ and $\beta = \partial f/\partial y = (2\Omega \cos \vartheta_0)/a$. This important approximation is known as the *beta-plane*, or *β-plane*, approximation; it captures the the most important *dynamical* effects of sphericity, without the complicating *geometric* effects, which are not essential to describe many phenomena. The momentum equations (2.43) are unaltered except that f_0 is replaced by $f_0 + \beta y$ to represent a varying Coriolis parameter. Thus, sphericity combined with rotation is dynamically equivalent to a *differentially rotating* system. For future reference, we write down the β-plane horizontal momentum equations:

$$\frac{\mathrm{D}\boldsymbol{u}}{\mathrm{D}t} + \boldsymbol{f} \times \boldsymbol{u} = -\frac{1}{\rho}\nabla_z p, \tag{2.47}$$

where $\boldsymbol{f} = (f_0 + \beta y)\hat{\mathbf{k}}$. In component form this equation becomes

$$\frac{\mathrm{D}u}{\mathrm{D}t} - fv = -\frac{1}{\rho}\frac{\partial p}{\partial x}, \qquad \frac{\mathrm{D}v}{\mathrm{D}t} + fu = -\frac{1}{\rho}\frac{\partial p}{\partial y}. \tag{2.48a,b}$$

The mass conservation, thermodynamic and hydrostatic equations in the β-plane approximation are the same as the usual Cartesian, f-plane, forms of those equations, and are summarized in the box on the next page.

2.4 Density Variations in the Atmosphere and Ocean

We now change gear and look at how density varies in the vertical. We will find that we can further simplify the equations of motion by making the Boussinesq approximation or, a little counter-intuitively, use pressure as our vertical coordinate.

2.4.1 Variation of Density in the Atmosphere

To estimate how density varies in the atmosphere let us suppose temperature is a constant, T_0, which is obviously a very rough approximation, and invoke hydrostatic balance . We then have $\partial p/\partial z = -\rho g$ where, since the atmosphere is an ideal gas, $p = \rho R T_0$. The density variation with height is then given by

$$\frac{\partial \rho}{\partial z} = \left(\frac{\partial \rho}{\partial p}\right)_T \frac{\partial p}{\partial z} = \left(\frac{1}{RT_0}\right)\frac{\partial p}{\partial z} = \frac{-\rho g}{RT_0}, \tag{2.49}$$

using hydrostasy for the last equality. Solving (2.49) gives

$$\rho = \rho_0 \exp\left(\frac{-z}{H_\rho}\right), \tag{2.50}$$

where $H_\rho = RT_0/g$ is the *scale height* of the atmosphere. The calculation is inaccurate if the temperature changes significantly with height but the qualitative result remains, namely that density changes are negligible only if we concern ourselves with motion much less than a scale height. One branch of meteorology where the incompressibility condition is applied with some quantitative justification is boundary-layer meteorology.

Equations on a Tangent Plane

On a tangent plane, and using Cartesian coordinates, the equations of motion with no forcing or friction are:

$$x\text{-momentum:} \qquad \frac{Du}{Dt} - fv = -\frac{1}{\rho}\frac{\partial p}{\partial x}, \tag{T.1}$$

$$y\text{-momentum:} \qquad \frac{Dv}{Dt} + fu = -\frac{1}{\rho}\frac{\partial p}{\partial y}, \tag{T.2}$$

$$z\text{-momentum:} \qquad \frac{Dw}{Dt} = -\frac{1}{\rho}\frac{\partial p}{\partial z} - g, \tag{T.3a}$$

$$\text{or hydrostatic balance:} \qquad \frac{\partial p}{\partial z} = -g\rho, \tag{T.3b}$$

$$\text{mass continuity:} \qquad \frac{\partial \rho}{\partial t} + \nabla \cdot (\rho \boldsymbol{v}) = 0, \tag{T.3}$$

$$\text{thermodynamic:} \qquad \frac{D\theta}{Dt} = 0. \tag{T.4}$$

In these equations $D/Dt = \partial/\partial t + u\partial/\partial x + v\partial/\partial y + w\partial/\partial z$ and $\nabla\cdot(\rho \boldsymbol{v}) = \partial(\rho u)/\partial x + \partial(\rho v)/\partial y + \partial(\rho w)/\partial z$. On the f-plane $f = f_0$, whereas on the β-plane $f = f_0 + \beta y$.

However, we can often suppose that density does not vary if we are mainly seeking a qualitative understanding of some phenomenon, as we explore in later chapters.

2.4.2 Variation of Density in the Ocean

The variations of density in the ocean are due to three effects: compression of water by pressure (which we denote as $\Delta_p \rho$); thermal expansion of water if its temperature changes ($\Delta_T \rho$); and haline contraction if its salinity changes ($\Delta_S \rho$). How big are these? An appropriate equation of state to approximately evaluate these effects is the linear one,

$$\rho = \rho_0 \left[1 - \beta_T(T - T_0) + \beta_S(S - S_0) + \beta_p(p - p_0)\right], \tag{2.51}$$

where $\beta_T \approx 2 \times 10^{-4}\,\mathrm{K}^{-1}$, $\beta_S \approx 10^{-3}\,\mathrm{g/kg}^{-1}$ and $\beta_p = 4.4 \times 10^{-10}\,\mathrm{m\,s^{-2}\,kg^{-1}}$ (see Table 1.2). The three effects may then be evaluated as follows.:

Pressure compressibility. We have $\Delta_p \rho \approx \beta_p \rho_0 \Delta p = \Delta p / c_s^2 \approx \rho_0 g H / c_s^2$, where H is the depth and we have used the hydrostatic approximation to evaluate the pressure change. Thus,

$$\frac{|\Delta_p \rho|}{\rho_0} \approx \frac{gH}{c_s^2} \sim 4 \times 10^{-2}, \tag{2.52}$$

for $H = 8\,\mathrm{km}$. The quantity $c_s^2/g \approx 200\,\mathrm{km}$ is the density scale height of the ocean. Thus, the pressure at the bottom of the ocean (in the deep trenches), enormous as it is, is insufficient to compress the water enough to make a significant change in its density.

Thermal expansion. We have $\Delta_T\rho \approx -\beta_T\rho_0\Delta T$ and therefore

$$\frac{|\Delta_T\rho|}{\rho_0} \ll 1 \qquad \text{if} \qquad \beta_T\Delta T \ll 1. \tag{2.53}$$

For $\Delta T = 20\,\mathrm{K}$, $\beta_T\Delta T \approx 4\times10^{-3}$, and evidently we would require temperature differences of order β_T^{-1}, or 5000 K to obtain order one variations in density. Boiling water is just a couple of percent lighter than water just above freezing point.

Saline contraction. We have $\Delta_S\rho \approx \beta_S\rho_0\Delta S$ and therefore

$$\frac{|\Delta_S\rho|}{\rho_0} \approx \beta_S\Delta S \sim 5\times10^{-3}, \tag{2.54}$$

for salinity changes of 5 g/kg. (The salinity of the Dead Sea is about 350 g/kg, about 10 times higher than the open ocean, and the water is about 25% heavier than normal seawater, but the variations in salinity there are still too small to lead to significant density changes.)

Evidently, fractional density changes in the ocean from pressure, temperature and salinity changes are indeed very small. (There is one other process associated with variations in density, in both ocean and atmosphere, and that is sound. However, sound waves are nearly always oceanographically and meteorologically unimportant and we shall consider them no further.)

The Boussinesq equations are named for the Frenchman Joseph Boussinesq, who used his eponymous approximation in 1903, although similar approximations were used in 1879 by the German scientist Anton Oberbeck.

2.5 THE BOUSSINESQ EQUATIONS

In the ocean the density variations are quite small but these variations are crucial to the formation of ocean currents. We would like to take advantage of the near incompressibility of water, but we would like to allow density to vary where needed, and the *Boussinesq approximation,* which leads to the Boussinesq equations, is an approximation that allows us to do just that. In fact, although density variations are large in the atmosphere, even there the Boussinesq equations capture many of the important large-scale atmospheric phenomena in an appealingly economical way.

Since density variations are presumptively small we write

$$\rho = \rho_0 + \delta\rho(x, y, z, t), \tag{2.55}$$

where ρ_0 is a constant and we assume that $|\delta\rho| \ll \rho_0$; this is the key assumption in the approximation.

Associated with the constant density is a reference pressure, $p_0(z)$, that is in hydrostatic balance with that density, so that

$$p = p_0(z) + \delta p(x, y, z, t), \qquad \text{where} \qquad \frac{\mathrm{d}p_0}{\mathrm{d}z} = -g\rho_0. \tag{2.56a,b}$$

2.5.1 Momentum Equations

Without approximation, the momentum equation can be written as

$$(\rho_0 + \delta\rho)\frac{\mathrm{D}\boldsymbol{v}}{\mathrm{D}t} = -\nabla\delta p - \frac{\partial p_0}{\partial z}\hat{\mathbf{k}} - g(\rho_0 + \delta\rho)\hat{\mathbf{k}}, \tag{2.57}$$

and using (2.56a) this becomes, again without approximation,

$$(\rho_0 + \delta\rho)\frac{\mathrm{D}\boldsymbol{v}}{\mathrm{D}t} = -\nabla\delta p - g\delta\rho\hat{\mathbf{k}}. \tag{2.58}$$

If density variations are small this equation becomes

$$\frac{\mathrm{D}\boldsymbol{v}}{\mathrm{D}t} = -\nabla\phi + b\hat{\mathbf{k}}, \tag{2.59}$$

where $\phi = \delta p/\rho_0$ is the kinematic pressure (but without the contribution from the hydrostatic basic-state pressure p_0, and often we shall just refer to ϕ as the pressure) and $b = -g\,\delta\rho/\rho_0$ is the *buoyancy*. We should not and do not neglect the term $g\,\delta\rho$, for there is no reason to believe it to be small: $\delta\rho$ may be small, but g is big! Equation (2.59) is the momentum equation in the Boussinesq approximation.

The Boussinesq equations are appropriate for nearly incompressible fluids in a gravitational field. Variations in density are small, but gravitational effects are big. The variations of density in the momentum equation are then ignored *except* when associated with the gravity term.

2.5.2 Mass continuity

The unapproximated mass continuity equation is

$$\frac{\mathrm{D}\delta\rho}{\mathrm{D}t} + (\rho_0 + \delta\rho)\nabla\cdot\boldsymbol{v} = 0. \tag{2.60}$$

Because density variations are preumptively small we may approximate this equation by

$$\nabla\cdot\boldsymbol{v} = 0, \tag{2.61}$$

the same as for a constant density fluid. This equation *absolutely does not* allow one to go back and use (2.60) to say that $D\delta\rho/Dt = 0$; the evolution of density is given by the thermodynamic equation along with an equation of state, and this should not be confused with the mass conservation equation.

The mass continuity equation in the Boussinesq equations is that for an incompressible fluid, namely $\nabla\cdot\boldsymbol{v} = 0$.

2.5.3 Boussinesq Thermodynamics

The Boussinesq equations are closed by the addition of an equation of state, a thermodynamic equation and, if needs be, an evolution equation for salinity. There are various levels of approximation for the thermodynamic equation, and the simplest one is to begin with (1.49) and, because the flow is nearly incompressible, neglect the divergence term. We also approximate the internal energy by $c_v T$ (and in a liquid $c_v \approx c_p$) and obtain

$$\frac{\mathrm{D}T}{\mathrm{D}t} = \dot{T}, \tag{2.62}$$

Boussinesq Equations

The simple Boussinesq equations are, for an inviscid fluid:

momentum equations: $$\frac{D\boldsymbol{v}}{Dt} = -\nabla\phi + b\hat{\mathbf{k}}, \qquad \text{(B.1)}$$

mass conservation equation: $$\nabla\cdot\boldsymbol{v} = 0, \qquad \text{(B.2)}$$

buoyancy equation: $$\frac{Db}{Dt} = \dot{b}. \qquad \text{(B.3)}$$

The buoyancy is related to density by $b = -g\delta\rho/\rho_0$, and in an ideal gas it is approximately related to potential temperature by $b = g\delta\theta/\theta_0$. A more general form of the equations replaces the buoyancy equation by:

thermodynamic equation: $$\frac{DT}{Dt} = \dot{T}, \qquad \text{(B.4)}$$

salinity equation: $$\frac{DS}{Dt} = \dot{S}, \qquad \text{(B.5)}$$

equation of state: $$b = b(T, S, p), \qquad \text{(B.6)}$$

where $p = -\rho_0 gz$ is the hydrostatic pressure.

where $\dot{T} = \dot{Q}/c_v$ and $\dot{Q}$ is the heating. If variations of salinity are also of concern (as they are in the ocean) then we may include an evolution equation for salinity,

$$\frac{DS}{Dt} = \dot{S}, \qquad (2.63)$$

where the right-hand side represents the sources, sinks and diffusive transfer of salt. We then use an equation of state to relate the temperature and salinity to the density and hence buoyancy, and using (1.42) we have

$$b = g\left[\beta_T(T - T_0) + \beta_S(S - S_0) - gz/c_s^2\right], \qquad (2.64)$$

where to compute the pressure term we have taken $p_0 = 0$, $\beta_p = -1/(\rho_0 c_s^2)$ (with c_s being the sound speed) and we use hydrostasy, $p = -\rho_0 gz$.

In laboratory settings where salinity and the pressure effects on density are not important we may use the simple linear equation of state, namely $\rho = \rho_0\left[1 - \beta_T(T - T_0)\right]$ to relate temperature to density and hence buoyancy, giving

$$\frac{Db}{Dt} = \dot{b}, \qquad (2.65)$$

where $\dot{b} = \dot{Q}g\beta_T/c_v$. The 'simple' Boussinesq equations use (2.59), (2.61) and (2.65). In atmospheric applications the buoyancy is taken to be approximately related to the potential temperature by $b = g\delta\theta/\theta_0$, where θ_0 is a constant. All this is summarized in the table above.

2.5.4 Energetics of the Boussinesq System

In a uniform gravitational field but with no other forcing or dissipation, we write the simple Boussinesq equations as

$$\frac{\mathrm{D}\boldsymbol{v}}{\mathrm{D}t} = b\hat{\mathbf{k}} - \nabla\phi, \qquad \nabla \cdot \boldsymbol{v} = 0, \qquad \frac{\mathrm{D}b}{\mathrm{D}t} = 0. \tag{2.66a,b,c}$$

Using (2.66a) and (2.66b) the kinetic energy evolution is given by, after some manipulation,

$$\frac{1}{2}\frac{\mathrm{D}\boldsymbol{v}^2}{\mathrm{D}t} = bw - \nabla \cdot (\phi \boldsymbol{v}). \tag{2.67}$$

Now, the material derivative of z is given by $\mathrm{D}z/\mathrm{D}t = w$ and using this and (2.66c) gives

$$\frac{\mathrm{D}}{\mathrm{D}t}(bz) = wb. \tag{2.68}$$

Subtracting (2.68) from (2.67) and expanding the material derivative gives

$$\frac{\partial}{\partial t}\left(\frac{1}{2}\boldsymbol{v}^2 - bz\right) + \nabla \cdot \left[\boldsymbol{v}\left(\frac{1}{2}\boldsymbol{v}^2 - bz + \phi\right)\right] = 0. \tag{2.69}$$

This constitutes an energy equation for the Boussinesq system, and may be compared to (1.72). The energy density (divided by ρ_0) is just $\boldsymbol{v}^2/2 - bz$. What does the term bz represent? Its integral (multiplied by ρ_0) is the potential energy of the flow minus that of the basic state, or $\int g(\rho - \rho_0)z\,\mathrm{d}z$, and if there were a heating term on the right-hand side of (2.66c) it would directly provide a source of potential energy. The internal energy does not directly enter into the Boussinesq energy equation because the incompressibility condition prevents the conversion between internal and kinetic energy — the term $p\nabla \cdot \boldsymbol{v}$ in (1.68) and (1.69) is zero for an incompressible fluid.

2.5.5 Anelastic Equations

The Boussinesq assumption of incompressibility is too strong for quantitative atmospheric calculations that involve motion having a vertical extent of order a scale height or more. The anelastic equations partially relax that assumption by supposing that the density has a background state, $\tilde{\rho}(z)$, that varies only in the vertical, so that the total density varies as $\rho(x, y, z, t) = \tilde{\rho}(z) + \rho'(x, y, z, t)$. (The background state is often chosen such that the vertical gradient of potential temperature is zero.) The resulting equations of motion are very similar to the Boussinesq set except that the mass continuity equation becomes

$$\frac{\partial u}{\partial x} + \frac{\partial v}{\partial y} + \frac{1}{\tilde{\rho}}\frac{\partial(\tilde{\rho}w)}{\partial z} = 0. \tag{2.70}$$

For an ideal gas, the buoyancy in the anelastic equations is related to the potential temperature (rather than directly to density) by $b = g\delta\theta/\theta_0$, where θ_0 is a constant.

2.6 PRESSURE COORDINATES

Although using z as a vertical coordinate is a natural choice it is not the only option. Any variable that has a one-to-one correspondence with z in the vertical, so any variable that varies monotonically with z, could be used. In the atmosphere pressure almost always falls monotonically with height, and using pressure instead of z as a vertical coordinate provides a useful simplification of the mass conservation and geostrophic relations. as well as a more direct connection with observations, which are often taken at fixed values of pressure.

Pressure coordinates use pressure instead of z as the vertical coordinate. Although this seems strange, pressure coordinates have proven enormously useful in the atmospheric sciences, especially (but not only) in conjunction with the hydrostatic approximation.

2.6.1 General Vertical Coordinates

First consider a general vertical coordinate, ξ. Any variable Ψ that is a function of the coordinates (x, y, z, t) may be expressed instead in terms of (x, y, ξ, t) by considering ξ to be a function of (x, y, z, t). Derivatives with respect to z and ξ are related by

$$\frac{\partial \Psi}{\partial \xi} = \frac{\partial \Psi}{\partial z}\frac{\partial z}{\partial \xi} \quad \text{and} \quad \frac{\partial \Psi}{\partial z} = \frac{\partial \Psi}{\partial \xi}\frac{\partial \xi}{\partial z}. \tag{2.71a,b}$$

Horizontal derivatives in the two coordinate systems are related by the chain rule,

$$\left(\frac{\partial \Psi}{\partial x}\right)_\xi = \left(\frac{\partial \Psi}{\partial x}\right)_z + \left(\frac{\partial z}{\partial x}\right)_\xi \frac{\partial \Psi}{\partial z}. \tag{2.72}$$

The material derivative in ξ coordinates may be derived by noting that the 'vertical velocity' in ξ coordinates is $\mathrm{D}\xi/\mathrm{D}t$, just as the vertical velocity in z coordinates is $w = \mathrm{D}z/\mathrm{D}t$. We can thus write

$$\frac{\mathrm{D}\Psi}{\mathrm{D}t} = \left(\frac{\partial \Psi}{\partial t}\right)_{x,y,\xi} + \boldsymbol{u}\cdot\nabla_\xi \Psi + \dot{\xi}\frac{\partial \Psi}{\partial \xi}, \tag{2.73}$$

where ∇_ξ is the gradient operator at constant ξ. The operator $\mathrm{D}/\mathrm{D}t$ is physically the same in z or ξ coordinates because it is the total derivative of some property of a fluid parcel, and this is independent of the coordinate system. However, the individual terms within it differ between coordinate systems.

2.6.2 Application to Pressure

In pressure coordinates the analogue of the vertical velocity is $\omega \equiv Dp/Dt$, and the advective derivative itself is given by

$$\frac{\mathrm{D}}{\mathrm{D}t} = \frac{\partial}{\partial t} + \boldsymbol{u}\cdot\nabla_p + \omega\frac{\partial}{\partial p}. \tag{2.74}$$

Note, though, that the advective derivative is the same operator as it is in height coordinates, since it is just the total derivative of a given fluid parcel; it is just written with different coordinates.

To obtain an expression for the pressure force, now let $\xi = p$ in (2.72) and apply the relationship to p itself to give

$$0 = \left(\frac{\partial p}{\partial x}\right)_z + \left(\frac{\partial z}{\partial x}\right)_p \frac{\partial p}{\partial z}, \tag{2.75}$$

which, using the hydrostatic relationship, gives

$$\left(\frac{\partial p}{\partial x}\right)_z = \rho\left(\frac{\partial \Phi}{\partial x}\right)_p, \tag{2.76}$$

where $\Phi = gz$ is the *geopotential*. Thus, the horizontal pressure force in the momentum equations is

$$\frac{1}{\rho}\nabla_z p = \nabla_p \Phi, \tag{2.77}$$

where the subscripts on the gradient operator indicate that the horizontal derivatives are taken at constant z or constant p. The horizontal momentum equation thus becomes

$$\frac{\mathrm{D}\boldsymbol{u}}{\mathrm{D}t} + \boldsymbol{f} \times \boldsymbol{u} = -\nabla_p \Phi, \tag{2.78}$$

where $\mathrm{D}/\mathrm{D}t$ is given by (2.74). The hydrostatic equation in height coordinates is $\partial p/\partial z = -\rho g$ and in pressure coordinates this becomes

$$\frac{\partial \Phi}{\partial p} = -\alpha \qquad \text{or} \qquad \frac{\partial \Phi}{\partial p} = -\frac{p}{RT}. \tag{2.79}$$

The mass conservation equation simplifies attractively in pressure coordinates, if the hydrostatic approximation is used. Recall that the mass conservation equation can be derived from the material form

$$\frac{\mathrm{D}}{\mathrm{D}t}(\rho\,\delta V) = 0, \tag{2.80}$$

where $\delta V = \delta x\,\delta y\,\delta z$ is a volume element. But by the hydrostatic relationship $\rho\delta z = -(1/g)\delta p$ and thus

$$\frac{\mathrm{D}}{\mathrm{D}t}(\delta x\,\delta y\,\delta p) = 0. \tag{2.81}$$

This is completely analogous to the expression for the material conservation of volume in an incompressible fluid, (1.11). Thus, without further ado, we can write the mass conservation in pressure coordinates as

$$\left(\frac{\partial u}{\partial x}\right)_p + \left(\frac{\partial v}{\partial y}\right)_p + \frac{\partial \omega}{\partial p} \qquad \text{or} \qquad \nabla_p \cdot \boldsymbol{u} + \frac{\partial \omega}{\partial p} = 0, \tag{2.82}$$

where the horizontal derivatives are taken at constant pressure.

The (adiabatic) thermodynamic equation is, of course, still $\mathrm{D}\theta/\mathrm{D}t = 0$, and θ may be related to pressure and temperature using its definition and

Equations of Motion in Pressure and Log-Pressure Coordinates

The adiabatic, inviscid primitive equations in pressure coordinates are:

$$\frac{\mathrm{D}\boldsymbol{u}}{\mathrm{D}t} + \boldsymbol{f} \times \boldsymbol{u} = -\nabla_p \Phi, \tag{P.1}$$

$$\frac{\partial \Phi}{\partial p} = \frac{-RT}{p}, \tag{P.2}$$

$$\nabla_p \cdot \boldsymbol{u} + \frac{\partial \omega}{\partial p} = 0, \tag{P.3}$$

$$\frac{\partial T}{\partial t} + u\frac{\partial T}{\partial x} + v\frac{\partial T}{\partial y} - \omega S_p = 0 \quad \text{or} \quad \frac{\mathrm{D}\theta}{\mathrm{D}t} = 0, \tag{P.4}$$

where $S_p = \kappa T/p - \partial T/\partial p$, $\kappa = R/c_p$ and $\theta = T(p_s/p)^\kappa$ is the potential temperature. The above equations are, respectively, the horizontal momentum equation, the hydrostatic equation, the mass continuity equation and the thermodynamic equation. Using hydrostasy and the ideal gas relation, the thermodynamic equation may also be written as

$$\frac{\partial T}{\partial t} + u\frac{\partial T}{\partial x} + v\frac{\partial T}{\partial y} + \omega\frac{\partial s}{\partial p} = 0, \tag{P.5}$$

where $s = T + gz/c_p$ is the dry static energy divided by c_p.

The corresponding equations in log-pressure coordinates are

$$\frac{\mathrm{D}\boldsymbol{u}}{\mathrm{D}t} + \boldsymbol{f} \times \boldsymbol{u} = -\nabla_Z \Phi, \tag{P.6}$$

$$\frac{\partial \Phi}{\partial Z} = \frac{RT}{H}, \tag{P.7}$$

$$\nabla_Z \cdot \boldsymbol{u} + \frac{1}{\rho_R}\frac{\partial \rho_R W}{\partial z} = 0, \tag{P.8}$$

$$\frac{\partial T}{\partial t} + u\frac{\partial T}{\partial x} + v\frac{\partial T}{\partial y} + W S_Z = 0 \quad \text{or} \quad \frac{\mathrm{D}\theta}{\mathrm{D}t} = 0, \tag{P.9}$$

where $\rho_R = \rho_0 \exp(-Z/H)$ and $S_Z = \kappa T/H + \partial T/\partial Z$. The thermodynamic equation may also be written as

$$\frac{\partial}{\partial t}\frac{\partial \Phi}{\partial Z} + u\frac{\partial}{\partial x}\frac{\partial \Phi}{\partial z} + v\frac{\partial}{\partial y}\frac{\partial \Phi}{\partial z} + W N_*^2 = 0, \tag{P.10}$$

where $N_*^2 = (R/H)S_Z$.

the ideal gas equation to complete the equation set. However, because the hydrostatic equation is written in terms of temperature and not potential temperature it is convenient to write the thermodynamic equation accordingly. To do this we begin with the thermodynamic equation in the form of (1.53b) namely $c_p \mathrm{D}T/\mathrm{D}t - \alpha\,\mathrm{D}p/\mathrm{D}t = 0$. Since $\omega \equiv \mathrm{D}p/\mathrm{D}t$ this equation is simply

$$c_p\frac{\mathrm{D}T}{\mathrm{D}t} - \frac{RT}{p}\omega = 0, \tag{2.83}$$

which is an appropriate thermodynamic equation in pressure coordinates. It is sometimes useful to write the equation as

$$\frac{\partial T}{\partial t} + u\frac{\partial T}{\partial x} + v\frac{\partial T}{\partial y} - \omega S_p = 0, \quad \text{where} \quad S_p = \frac{\kappa T}{p} - \frac{\partial T}{\partial p} = -\frac{T}{\theta}\frac{\partial \theta}{\partial p}, \tag{2.84a,b}$$

using the ideal gas equation and the definition of potential temperature, with $\kappa = R/c_p$. The quantity S_p is a measure of static stability and it is closely related to the buoyancy frequency N.

The main practical difficulty with the pressure-coordinate equations is the lower boundary condition. Using

$$w \equiv \frac{\mathrm{D}z}{\mathrm{D}t} = \frac{\partial z}{\partial t} + \boldsymbol{u}\cdot\nabla_p z + \omega\frac{\partial z}{\partial p}, \tag{2.85}$$

and (2.79), the boundary condition of $w = 0$ at $z = z_s$ becomes

$$\frac{\partial \Phi}{\partial t} + \boldsymbol{u}\cdot\nabla_p \Phi - \alpha\omega = 0 \tag{2.86}$$

at $p(x, y, z_s, t)$. In theoretical or idealized studies, it is common to assume that the lower boundary is in fact a constant pressure surface and simply assume that $\omega = 0$. The pressure coordinate equations (collected together in the shaded box on the preceding page) are very similar in structure to the hydrostatic general Boussinesq equations (see the shaded box on page 38), and in fact a formal one-to-one correspondence may be made, but we shall not explore that here.

2.6.3 Log-Pressure Coordinates

Log-pressure coordinates are a variation of pressure coordinates in which the vertical coordinate is $Z = -H\ln(p/p_R)$ where p_R is a reference pressure (say 1000 hPa) and H is a constant (for example a scale height RT_0/g where T_0 is also a constant) so that Z has units of length. (Uppercase letters are conventionally used for some variables in log-pressure coordinates and are not to be confused with scaling parameters.) The 'vertical velocity' for the system is

$$W \equiv \frac{\mathrm{D}Z}{\mathrm{D}t}, \tag{2.87}$$

and the advective derivative is

$$\frac{\mathrm{D}}{\mathrm{D}t} \equiv \frac{\partial}{\partial t} + \boldsymbol{u}\cdot\nabla_p + W\frac{\partial}{\partial Z}. \tag{2.88}$$

The horizontal momentum equation is unaltered from (2.78), although we use (2.88) to evaluate the advective derivative. It is straightforward to show that the hydrostatic equation becomes

$$\frac{\partial \Phi}{\partial Z} = \frac{RT}{H}. \tag{2.89}$$

The mass continuity equation (2.82) becomes

$$\frac{\partial u}{\partial x} + \frac{\partial v}{\partial y} + \frac{\partial W}{\partial Z} - \frac{W}{H} = 0, \tag{2.90}$$

which may be written as

$$\nabla_Z \cdot \boldsymbol{u} + \frac{1}{\rho_R}\frac{\partial(\rho_R W)}{\partial z} = 0, \tag{2.91}$$

where $\nabla_Z \cdot$ is the divergence at constant Z and $\rho_R = \rho_0 \exp(-Z/H)$.

As with pressure coordinates, it is convenient to write the thermodynamic equation in terms of temperature and not potential temperature and, since $W = -(H/p)\mathrm{D}p/\mathrm{D}t$, (1.53b) becomes

$$c_p \frac{\mathrm{D}T}{\mathrm{D}t} + W\frac{RT}{H} = 0. \tag{2.92}$$

This equation may be written as

$$\frac{\partial T}{\partial t} + u\frac{\partial T}{\partial x} + v\frac{\partial T}{\partial y} + WS_Z = 0, \tag{2.93}$$

where $S_Z = \kappa T/H + \partial T/\partial Z$.

Writing the equations in log-pressure form can be quite revealing. For example, integrating the hydrostatic equation, (2.89) gives, with $\Phi = gz$,

$$z(p_2) - z(p_1) = -\frac{R}{g}\int_{p_1}^{p_2} T \,\mathrm{d}\ln p. \tag{2.94}$$

Thus, the thickness of a layer between two pressure levels is proportional to its average temperature. Also, we see that at constant temperature the geometric height increases linearly with the logarithm of pressure. At a temperature of 240 K the scale height, RT/g, is about 7 km and at 280 K it is 8.2 km. A useful rule of thumb for Earth's atmosphere (and one that holds at 240 K) is that geometric height increases by about 16 km for each factor of ten decrease in pressure, and pressures of 1000 hPa, 100 hPa and 10 hPa roughly correspond to heights of 0 km, 16 km and 32 km, and so on.

Notes and References

The equation sets used in meteorology and oceanography are reviewed by Phillips (1963) and White (2002). Two books that discuss rotating fluid dynamics, with contrasting styles and choices of topics, are Salmon (1998) and Holton & Hakim (2012). See the notes section at the end of Chapter 3 for more references.

Problems

2.1 We wish to derive the momentum equations in a rotating frame of reference, but one in which $\boldsymbol{\Omega}$ is not constant; that is $\mathrm{d}\boldsymbol{\Omega}/\mathrm{d}t \neq 0$. Derive a momentum equation in this frame of reference, identifying clearly the Coriolis term, the centrifugal term, and any additional terms that may arise that are different from the case with $\boldsymbol{\Omega}$ constant. Give a brief interpretation of any new terms, as well as the Coriolis and centrifugal terms.

2.2 *(a)* Show that on Earth we might normally expect the centrifugal term to be much larger than the Coriolis term. Show that if the centrifugal term is incorporated into gravity, and if Earth is a perfect sphere, then gravity is no longer in the local vertical. Estimate the angle by which the apparent gravity differs from the vertical.

(b) If Earth were a perfect sphere, but with otherwise the same distribution of continents and ocean basins, would the distribution of sea level be more or less the same as it is today or would it be radically different?

2.3 At what latitude is the angle between the direction of Newtonian gravity (due to the mass of the Earth) and that of effective gravity (Newtonian gravity plus centrifugal terms) the largest? At what latitudes is this angle zero?

2.4 *(a)* Consider a fluid that obeys the hydrostatic relation $\partial p/\partial z = -\rho g$. If the fluid is an isothermal ideal gas show that the density and pressure both diminish exponentially with height. What is the e-folding height? (This is also called the 'scale height' of the atmosphere.) Obtain an expression for the height, z, as a function of pressure.

(b) Now suppose that the atmosphere has a uniform lapse rate (i.e., $\mathrm{d}T/\mathrm{d}z = -\Gamma =$ constant). Show that the height at a pressure p is given by

$$z = \frac{T_0}{\Gamma}\left[1 - \left(\frac{p_0}{p}\right)^{-R\Gamma/g}\right],$$

where T_0 is the temperature at $z = 0$.

(c) Are the answers you obtained in parts (a) and (b) the same as each other in the isothermal (constant temperature) limit? Explain.

2.5 A fluid at rest evidently satisfies the hydrostatic relation, which says that the pressure at the surface is given by the weight of the fluid above it. Now consider a very deep atmosphere on a small spherical planet. A unit cross-sectional area at the planet's surface lies beneath a column of fluid whose cross-section increases with height, because the total area of the atmosphere increases with distance away from the centre of the planet. Is the pressure at the surface still given by the hydrostatic relation, or is it greater than this because of the increased mass of fluid in the column? If it is still given by the hydrostatic relation, then the pressure at the surface, integrated over the entire area of the planet, is less than the total weight of the fluid. Is this true? Alternatively, the pressure at the surface might be greater than that implied by hydrostatic balance, and if so explain how the hydrostatic relation fails.

2.6 In a self-gravitating spherical fluid, like a star, hydrostatic balance may be written

$$\frac{\partial p}{\partial r} = -\frac{GM(r)}{r^2}\rho, \qquad \text{(P2.1)}$$

where $M(r)$ is the mass interior to a sphere of radius r, and G is a constant. Obtain an expression for the pressure as a function of radius when the fluid (a) has constant density, and (b) is an isothermal ideal gas (if possible). The star is of radius a.

2.7 Show that the inviscid, adiabatic, hydrostatic primitive equations for a compressible fluid conserve a form of energy (kinetic plus potential plus internal), and that the kinetic energy has no contribution from the vertical velocity. Obtain an explicit form for the conserved energy, and provide a physical interpretation for this result. (You may assume Cartesian geometry and a uniform gravitational field in the vertical direction.)

Alternatively, use the hydrostatic Boussinesq equations and again show that the vertical velocity does not contribute. Obtain an explicit form for the conserved energy and interpret your result.

2.8 (*a*) Consider a scalar field, like temperature, T. Explain in words why the material derivative in a rotating frame is equal to the material derivative in the inertial frame; that is, explain why $(\mathrm{D}T/\mathrm{D}t)_I = (\mathrm{D}T/\mathrm{D}t)_R$.

(*b*) The material derivative of a scalar is given by $\partial T/\partial t + (\boldsymbol{v}\cdot\nabla)T$. Show (mathematically, with equations) that the individual terms are different in the rotating and inertial frames (and obtain an expression for how much) but that their sum is the same.

2.9 Begin with the mass conservation in height coordinates, namely $\mathrm{D}\rho/\mathrm{D}t + \rho\nabla\cdot\boldsymbol{v} = 0$. Transform this into pressure coordinates using the chain rule (or otherwise) and derive the pressure coordinate version mass conservation equation in the form $\nabla_p \cdot \boldsymbol{u} + \partial\omega/\partial p = 0$.

2.10 Consider a dry, hydrostatic, ideal-gas atmosphere whose lapse rate is one of constant potential temperature. What is its vertical extent? That is, at what height, if at all, does the density vanish? Discuss the physical realism of such a state.

2.11 Consider the simple Boussinesq equations with heating and viscosity, namely $\mathrm{D}\boldsymbol{v}/\mathrm{D}t = -\nabla\phi + \hat{\mathbf{k}}b + \nu\nabla^2\boldsymbol{v}$, $\nabla\cdot\boldsymbol{v} = 0$, $\mathrm{D}b/\mathrm{D}t = J + \kappa\nabla^2 b.$, where J is an external heat source.

(*a*) Obtain an energy equation similar to that in Section 2.5.4, but now with the terms on the right-hand side that represent viscous and diabatic effects.

(*b*) Over a closed volume and in a statistically steady state, show that the dissipation of kinetic energy must be balanced by a buoyancy source.

(*c*) Show that the heating must occur at a lower level than the cooling (that is, $\int z\dot{Q}\,\mathrm{d}z < 0$ where the integral is over the depth of the domain and $\dot{Q} = J + \kappa\nabla^2 b$) if a kinetic-energy dissipating, statistically steady, circulation is to be maintained.

CHAPTER

3

Dynamics on a Rotating Planet

WE NOW PUT THE EQUATIONS OF MOTION TO USE, and in so doing start our journey into the *dynamics* of fluid motion on a rotating planet. We begin rather gently by way of an introduction to scaling, which is the basis of the art of making sensible approximations.

3.1 A Gentle Introduction to Scaling

The units we use to measure length, velocity and so on are irrelevant to the dynamics, and SI units may not be the most appropriate ones for a given problem. Rather, it is useful to express the equations of motion in terms of 'nondimensional' variables, by which we mean expressing every variable as the ratio of its value to some reference value. We try to choose the reference value as a natural one for a given flow, in order that, where possible, the nondimensional variables are order-unity quantities, and doing this is called *scaling the equations*. Much of the art of fluid dynamics lies in choosing sensible scaling factors for the problem at hand for then the sizes of the various terms become clear, and we here we give a simple, non-rotating, example.

3.1.1 The Reynolds Number

Osborne Reynolds (1842–1912) was an Irish born (Belfast) physicist who was professor of engineering at Manchester University from 1868–1905. He was also one of the first scientists to think about the concept of group velocity.

Consider the constant-density momentum equation in Cartesian coordinates. If a typical velocity is U, a typical length is L, a typical time scale is T, and a typical value of the pressure deviation is Φ, then the approximate sizes of the various terms in the momentum equation are given by

$$\frac{\partial \boldsymbol{v}}{\partial t} + (\boldsymbol{v}\cdot\nabla)\boldsymbol{v} = -\nabla\phi + \nu\nabla^2\boldsymbol{v}, \tag{3.1a}$$

$$\frac{U}{T} \qquad \frac{U^2}{L} \sim \frac{\Phi}{L} \qquad \nu\frac{U}{L^2}. \tag{3.1b}$$

The ratio of the inertial (i.e., the advective) terms to the viscous terms is $(U^2/L)/(\nu U/L^2) = UL/\nu$, and this is the *Reynolds number*. More formally, we can nondimensionalize the momentum equation by writing

$$\widehat{\boldsymbol{v}} = \frac{\boldsymbol{v}}{U}, \qquad \widehat{\boldsymbol{x}} = \frac{\boldsymbol{x}}{L}, \qquad \widehat{t} = \frac{t}{T}, \qquad \widehat{\phi} = \frac{\phi}{\Phi}, \tag{3.2}$$

where the terms with hats on are *nondimensional* values of the variables and the capitalized quantities are known as *scaling values*, and these are the approximate magnitudes of the variables. We now choose the scaling values so that the nondimensional variables are of order unity, or $\widehat{u} = \mathcal{O}(1)$. Thus, for example, we choose U so that $u = \mathcal{O}(U)$, where the notation should be taken to mean that the magnitude of the variable u is approximately U, or that $u \sim U$, and we say that 'u scales like U'.

In this problem, we have no way to scale pressure and time except with the velocity and length scales we have chosen, and the only dimensionally correct choices are then

$$T = \frac{L}{U}, \qquad \Phi = U^2. \tag{3.3}$$

Substituting (3.2) and (3.3) into the momentum equation gives

$$\frac{U^2}{L}\left[\frac{\partial \widehat{\boldsymbol{v}}}{\partial \widehat{t}} + (\widehat{\boldsymbol{v}} \cdot \nabla)\widehat{\boldsymbol{v}}\right] = -\frac{U^2}{L}\nabla\widehat{\phi} + \frac{\nu U}{L^2}\nabla^2\widehat{\boldsymbol{v}}, \tag{3.4}$$

where we use the convention that when ∇ operates on a nondimensional variable it is a nondimensional operator. Equation (3.4) simplifies to

$$\frac{\partial \widehat{\boldsymbol{v}}}{\partial \widehat{t}} + (\widehat{\boldsymbol{v}} \cdot \nabla)\widehat{\boldsymbol{v}} = -\nabla\widehat{\phi} + \frac{1}{Re}\nabla^2\widehat{\boldsymbol{v}}, \tag{3.5}$$

where

$$Re \equiv \frac{UL}{\nu} \tag{3.6}$$

is, again, the Reynolds number. If we have chosen our length and velocity scales sensibly — that is, if we have scaled them properly — each variable in (3.5) is order unity, with the viscous term being multiplied by the parameter $1/Re$. There are two important conclusions:

(i) The ratio of the importance of the inertial terms to the viscous terms is given by the *Reynolds number,* defined by (3.6). In the absence of other forces, such as those due to gravity and rotation, the Reynolds number is the only nondimensional parameter explicitly appearing in the momentum equation. Hence its value, along with the boundary conditions and geometry, controls the behaviour of the system.

(ii) More generally, by scaling the equations of motion appropriately the parameters determining the behaviour of the system become explicit. *Scaling the equations is intelligent nondimensionalization.*

Nondimensionalizing the equations does not, however, absolve the investigator from the responsibility of producing dimensionally correct equations. One should regard nondimensional equations as dimensional equations in units appropriate for the problem at hand.

3.2 HYDROSTATIC BALANCE

Life is too short to solve every complex problem in detail, and the atmospheric and oceanic sciences abound with complex problems. In their usual form the fluid dynamical equations alone are a set of six nonlinear partial differential equations (three momentum equations, a thermodynamic equation, a mass continuity equation and an equation of state) describing velocity, pressure, temperature and density. To solve real-world problems we need to add water vapour or salinity, as well as the equations of radiative transfer. All this makes for a complex system, and to make progress we need to simplify where possible and eliminate unimportant effects. We have already seen how we might do that for fluids of nearly constant density in making the Boussinesq approximation, and we now look at the effects of gravity and rotation and see how these give rise hydrostatic balance and geostrophic balance, the dominant balances in the vertical and horizontal directions, respectively. The corresponding states, hydrostasy and geostrophy, are not exactly realized, but their approximate satisfaction has profound consequences on the behaviour of atmospheres and oceans.

We begin with hydrostatic balance. We first encountered it in Section 1.3.3 but now we take a closer look. We start by scaling the equations, just as we did in the previous section.

3.2.1 Scaling Estimates

Consider the relative sizes of terms in the vertical momentum equation, (2.42c):

$$\frac{W}{T} + \frac{UW}{L} + \frac{W^2}{H} + \Omega U \sim \left|\frac{1}{\rho}\frac{\partial p}{\partial z}\right| + g. \tag{3.7}$$

For most large-scale motion in the atmosphere and ocean the terms on the right-hand side are orders of magnitude larger than those on the left, and therefore must be approximately equal. Explicitly, suppose $W \sim 1\ \mathrm{cm\,s^{-1}}$, $L \sim 10^5$ m, $H \sim 10^3$ m, $U \sim 10\ \mathrm{m\,s^{-1}}$, $T = L/U$. Then by substituting into (3.7) it seems that the pressure term is the only one which could balance the gravitational term, and we are led to approximate (2.42c) by,

$$\frac{\partial p}{\partial z} = -\rho g. \tag{3.8}$$

This equation, which is a vertical momentum equation, is known as *hydrostatic balance*.

However, (3.8) is not always a useful equation! Let us suppose that the density is a constant, ρ_0. We can then write the pressure as

$$p(x, y, z, t) = p_0(z) + p'(x, y, z, t), \qquad \text{where} \qquad \frac{\partial p_0}{\partial z} \equiv -\rho_0 g. \tag{3.9}$$

That is, p_0 and ρ_0 are in hydrostatic balance. On the f-plane, the inviscid vertical momentum equation becomes, without approximation,

$$\frac{\mathrm{D}w}{\mathrm{D}t} = -\frac{1}{\rho_0}\frac{\partial p'}{\partial z}. \tag{3.10}$$

Thus, *for constant density fluids the gravitational term has no dynamical effect:* there is no buoyancy force, and the pressure term in the horizontal momentum equations can be replaced by p'. Hydrostatic balance, and in particular (3.9), is not a useful vertical momentum equation in this case. If the fluid is stratified, we should therefore subtract off the hydrostatic pressure associated with the mean density before we can determine whether hydrostasy is a useful *dynamical* approximation, accurate enough to determine the horizontal pressure gradients. This is automatic in the Boussinesq equations, where the vertical momentum equation is

$$\frac{\mathrm{D}w}{\mathrm{D}t} = -\frac{\partial \phi}{\partial z} + b, \tag{3.11}$$

and the hydrostatic balance of the basic state is already subtracted out. In the more general equation,

$$\frac{\mathrm{D}w}{\mathrm{D}t} = -\frac{1}{\rho}\frac{\partial p}{\partial z} - g, \tag{3.12}$$

we need to compare the advective term on the left-hand side with the pressure variations arising from horizontal flow in order to determine whether hydrostasy is an appropriate vertical momentum equation.

3.2.2 Hydrostatic Balance and the Aspect Ratio

In a Boussinesq fluid we write the horizontal and vertical momentum equations as

$$\frac{\mathrm{D}\boldsymbol{u}}{\mathrm{D}t} + \boldsymbol{f} \times \boldsymbol{u} = -\nabla_z \phi, \qquad \frac{\mathrm{D}w}{\mathrm{D}t} = -\frac{\partial \phi}{\partial z} + b. \tag{3.13a,b}$$

With $\boldsymbol{f} = 0$, (3.13a) implies the scaling

$$\phi \sim U^2. \tag{3.14}$$

If we then use mass conservation, $\nabla_z \cdot \boldsymbol{u} + \partial w/\partial z = 0$, to scale vertical velocity we find

$$w \sim W = \frac{H}{L}U = \alpha U, \tag{3.15}$$

where $\alpha = H/L$ is the aspect ratio. The advective terms in the vertical momentum equation all scale as

$$\frac{\mathrm{D}w}{\mathrm{D}t} \sim \frac{UW}{L} = \frac{U^2 H}{L^2}. \tag{3.16}$$

Using (3.14) and (3.16) the ratio of the advective term to the pressure gradient term in the vertical momentum equations then scales as

$$\frac{|\mathrm{D}w/\mathrm{D}t|}{|\partial\phi/\partial z|} \sim \frac{U^2 H/L^2}{U^2/H} \sim \left(\frac{H}{L}\right)^2. \tag{3.17}$$

Thus, the condition for hydrostasy, that $|Dw/Dt|/|\partial\phi/\partial z| \ll 1$, is

$$\alpha^2 \equiv \left(\frac{H}{L}\right)^2 \ll 1. \tag{3.18}$$

The advective term in the vertical momentum may then be neglected. Thus, *hydrostatic balance arises from a small aspect ratio approximation.*

We can obtain the same result more formally by nondimensionalizing the momentum equations. Using uppercase symbols to denote scaling values we write

$$(x, y) = L(\hat{x}, \hat{y}), \qquad z = H\hat{z}, \qquad \boldsymbol{u} = U\hat{\boldsymbol{u}}, \qquad w = W\hat{w} = \frac{HU}{L}\hat{w},$$
$$t = T\hat{t} = \frac{L}{U}\hat{t}, \qquad \phi = \Phi\hat{\phi} = U^2\hat{\phi}, \qquad b = B\hat{b} = \frac{U^2}{H}\hat{b}, \tag{3.19}$$

where the hatted variables are nondimensional and the scaling for w is suggested by the mass conservation equation, $\nabla_z \cdot \boldsymbol{u} + \partial w/\partial z = 0$. Substituting (3.19) into (3.13) (with $\boldsymbol{f} = 0$) gives us the nondimensional equations

$$\frac{D\hat{\boldsymbol{u}}}{D\hat{t}} = -\nabla\hat{\phi}, \qquad \alpha^2\frac{D\hat{w}}{D\hat{t}} = -\frac{\partial\hat{\phi}}{\partial\hat{z}} + \hat{b}, \tag{3.20a,b}$$

where $D/D\hat{t} = \partial/\partial\hat{t} + \hat{u}\partial/\partial\hat{x} + \hat{v}\partial/\partial\hat{y} + \hat{w}\partial/\partial\hat{z}$ and we use the convention that when ∇ operates on nondimensional quantities the operator itself is nondimensional. From (3.20b) it is clear that hydrostatic balance obtains when $\alpha^2 \ll 1$, that is when the aspect ratio is small.

3.3 GEOSTROPHIC AND THERMAL WIND BALANCE

We now consider the dominant dynamical balance in the horizontal components of the momentum equation. In the horizontal plane (meaning along geopotential surfaces) we find that the Coriolis term is much larger than the advective terms and the dominant balance is between it and the horizontal pressure force. This balance is called *geostrophic balance,* and it occurs when the Rossby number is small, as we now investigate.

3.3.1 The Rossby Number

The *Rossby number* characterizes the importance of rotation in a fluid. It is, essentially, the ratio of the magnitude of the relative acceleration to the Coriolis acceleration, and it is of fundamental importance in geophysical fluid dynamics. It arises from a simple scaling of two of the terms horizontal momentum equation, namely

$$\frac{\partial \boldsymbol{u}}{\partial t} + (\boldsymbol{v}\cdot\nabla)\boldsymbol{u} + \boldsymbol{f}\times\boldsymbol{u} = -\frac{1}{\rho}\nabla_z p, \tag{3.21a}$$

$$\frac{U^2}{L} \qquad\qquad fU \tag{3.21b}$$

The Rossby number, U/fL, is named for C.-G. Rossby (1898–1957), a Swedish scientist who worked for many years in the United States and who was one of the great pioneers of dynamical meteorology in the mid-twentieth century. The Russian meteorologist I. Kibel introduced a similar number in 1940 and the number is sometimes called the Kibel or Rossby–Kibel number.

Table 3.1: Typical scales of large-scale flow in atmosphere and ocean. The choices given are representative of large-scale midlatitude eddying motion in both systems.

Variable	Scaling symbol	Meaning	Atm. value	Oce. value
(x, y)	L	horiz. scale	10^6 m	10^5 m
t	T	time scale	10^5 s (1 day)	10^6 s
(u, v)	U	horiz. velocity	$10\,\mathrm{m\,s^{-1}}$	$0.1\,\mathrm{m\,s^{-1}}$
	Ro	Rossby no., U/fL	0.1	0.01

where U is the approximate magnitude of the horizontal velocity and L is a typical length scale over which that velocity varies. (We assume that $W/H \lesssim U/L$, so that vertical advection does not dominate the advection.) The ratio of the sizes of the advective and Coriolis terms is defined to be the Rossby number,

$$Ro \equiv \frac{U}{fL}. \tag{3.22}$$

If the Rossby number is small then rotation effects are important and possibly dominant and, as the values in Table 3.1 indicate, this is the case for large-scale flow in both ocean and atmosphere.

Another intuitive way to think about the Rossby number is in terms of time scales. The Rossby number based on a time scale is given by

$$Ro_T \equiv \frac{1}{fT}, \tag{3.23}$$

where T is a time scale associated with the dynamics at hand. If the time scale is an advective one, meaning that $T \sim L/U$, then this definition is equivalent to (3.22). Now, $f = 2\Omega \sin\vartheta$, where Ω is the angular velocity of the rotating frame and equal to $2\pi/T_p$ where T_p is the period of rotation (about 24 hours for Earth). Thus,

$$Ro_T = \frac{T_p}{4\pi T \sin\vartheta} = \frac{T_i}{T}, \tag{3.24}$$

where $T_i = 1/f$ is the 'inertial time scale', about three hours in midlatitudes. Thus, for phenomena with time scales much longer than this, such as the motion of the Gulf Stream or a midlatitude atmospheric weather system, the effects of the Earth's rotation can be expected to be important, whereas a short-lived phenomenon, such as a cumulus cloud or tornado, may be oblivious to such rotation.

The notion of geostrophic balance is rooted in work by the Dutch scientist C. D. H. Buys-Ballot in the mid-nineteenth century. His eponymous law states that if a person in the Northern Hemisphere stands with her back to the wind then the atmospheric pressure is low to the left, high to the right. Around the same time the American scientist W. Ferrel used similar ideas in the context of the atmospheric general circulation.

3.3.2 Geostrophic Balance

If the Rossby number is sufficiently small in (3.21a) then the rotation term dominates the nonlinear advection term, and if the time period of the motion scales advectively then the rotation term also dominates the local time derivative. The only term that can then balance the rotation term is

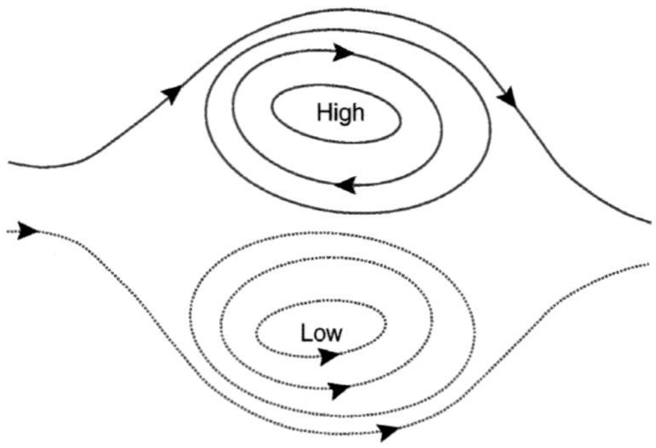

Fig. 3.1: Sketch of geostrophic flow with a positive value of the Coriolis parameter f. Flow is parallel to the lines of constant pressure (isobars). Cyclonic flow is anticlockwise around a low pressure region and anticyclonic flow is clockwise around a high. If f were negative, as in the Southern Hemisphere, (anti)cyclonic flow would be (anti)clockwise.

the pressure term, and therefore we must have

$$\boldsymbol{f} \times \boldsymbol{u} \approx -\frac{1}{\rho}\nabla_z p, \tag{3.25}$$

or, in Cartesian component form

$$fu \approx -\frac{1}{\rho}\frac{\partial p}{\partial y}, \qquad fv \approx \frac{1}{\rho}\frac{\partial p}{\partial x}. \tag{3.26}$$

This balance is known as *geostrophic balance*, and its consequences are profound, giving geophysical fluid dynamics a special place in the broader field of fluid dynamics. We *define* the geostrophic velocity, (u_g, v_g) by the expressions

$$fu_g = -\frac{1}{\rho}\frac{\partial p}{\partial y}, \qquad fv_g = \frac{1}{\rho}\frac{\partial p}{\partial x}, \tag{3.27}$$

and for low Rossby number flow $u \approx u_g$ and $v \approx v_g$. In spherical coordinates the geostrophic velocity is

$$fu_g = -\frac{1}{\rho a}\frac{\partial p}{\partial \vartheta}, \qquad fv_g = \frac{1}{a\rho\cos\vartheta}\frac{\partial p}{\partial \lambda}, \tag{3.28}$$

where $f = 2\Omega\sin\vartheta$, a is the radius of the planet, and ϑ and λ are latitude and longitude respectively. The same expressions hold in both hemispheres, although in the Southern Hemisphere f is negative.

Geostrophic balance has a number of immediate ramifications:

- Geostrophic flow is parallel to lines of constant pressure (isobars). If $f > 0$ the flow is anticlockwise round a region of low pressure and clockwise around a region of high pressure (see Fig. 3.1).
- If the Coriolis force is constant and if the density does not vary in the horizontal the geostrophic flow is horizontally non-divergent and

$$\nabla_z \cdot \boldsymbol{u}_g = \frac{\partial u_g}{\partial x} + \frac{\partial v_g}{\partial y} = 0. \tag{3.29}$$

We may then define the *geostrophic streamfunction*, ψ, by

$$\psi \equiv \frac{p}{f_0 \rho_0}, \qquad \text{whence} \qquad u_g = -\frac{\partial \psi}{\partial y}, \quad v_g = \frac{\partial \psi}{\partial x}. \tag{3.30}$$

The vertical component of vorticity, ζ, is then given by

$$\zeta = \hat{\mathbf{k}} \cdot \nabla \times \boldsymbol{v} = \frac{\partial v}{\partial x} - \frac{\partial u}{\partial y} = \nabla_z^2 \psi. \tag{3.31}$$

- If the Coriolis parameter is not constant, then cross-differentiating (3.27) gives, for constant density geostrophic flow,

$$v_g \frac{\partial f}{\partial y} + f \nabla_z \cdot \boldsymbol{u}_g = 0, \tag{3.32}$$

which, using the mass continuity equation $\nabla_z \cdot \boldsymbol{u} = -\partial w/\partial z$, gives

$$\beta v_g = f \frac{\partial w}{\partial z}. \tag{3.33}$$

where $\beta \equiv \partial f/\partial y = 2\Omega \cos \vartheta / a$. This expression is particularly important in the theory of ocean circulation, as we will discover in Chapter 14.

- The scaling for the kinematic pressure is no longer given by $\Phi = U^2$ (or, for pressure itself, $P = \rho U^2$) as in (3.19). Rather, since $\boldsymbol{f} \times \boldsymbol{u} \approx -\rho^{-1} \nabla p$, the scaling for the pressure and buoyancy, B, are given by

$$\Phi = fUL, \quad P = \rho f UL \quad \text{and} \quad B = \frac{fUL}{H}. \tag{3.34}$$

3.3.3 Taylor–Proudman Effect

If $\beta = 0$, then (3.33) implies that the vertical velocity is not a function of height. In fact, in that case none of the components of velocity vary with height if density is also constant, as we now show. If the flow is exactly in geostrophic and hydrostatic balance then the horizontal and vertical momentum equations become

$$v = \frac{1}{f_0} \frac{\partial \phi}{\partial x}, \quad u = -\frac{1}{f_0} \frac{\partial \phi}{\partial y}, \quad \frac{\partial \phi}{\partial z} = -g. \tag{3.35a,b,c}$$

Differentiating (3.35a,b) with respect to z, and using (3.35c) yields

$$\frac{\partial v}{\partial z} = \frac{-1}{f_0} \frac{\partial g}{\partial x} = 0, \qquad \frac{\partial u}{\partial z} = \frac{1}{f_0} \frac{\partial g}{\partial y} = 0. \tag{3.36}$$

Noting that the geostrophic velocities are horizontally non-divergent (that is, $\partial u/\partial x + \partial v/\partial y = 0$), and using mass continuity then gives $\partial w/\partial z = 0$. Thus, none of the velocity components vary with height.

If there is a solid horizontal boundary anywhere in the fluid, for example at the surface, then $w = 0$ at that surface and thus $w = 0$ everywhere. Hence the motion occurs in planes that lie perpendicular to the axis of rotation, and the flow is effectively two dimensional. This result is known as the *Taylor–Proudman effect*, namely that for constant density flow in geostrophic and hydrostatic balance the vertical derivatives of the horizontal and the vertical velocities are zero. At zero Rossby number, if the vertical velocity is zero somewhere in the flow, it is zero everywhere in that vertical column; furthermore, the horizontal flow has no vertical shear, and the fluid moves like a slab. The effects of rotation have provided a *stiffening* of the fluid in the vertical.

The Taylor–Proudman effect is named for G. I. Taylor and I. Proudman who wrote papers developing the result in 1921 and 1916, respectively. The effect is sometimes called the Taylor–Proudman 'theorem', but it is more usefully thought of as a physical effect, with manifestations even when the conditions for its satisfaction are not precisely met — which they never are.

In neither the atmosphere nor the ocean do we observe precisely such vertically coherent flow, mainly because of the effects of stratification. However, it is typical of geophysical fluid dynamics that the assumptions underlying a derivation are not fully satisfied, yet there are manifestations of it in real flow. For example, one might have naïvely expected, because $\partial w/\partial z = -\nabla_z \cdot \boldsymbol{u}$, that the scales of the various variables would be related by $W/H \sim U/L$. However, if the flow is rapidly rotating we expect that the horizontal flow will be in near geostrophic balance and therefore nearly divergence free; thus $\nabla_z \cdot \boldsymbol{u} \ll U/L$, and $W \ll HU/L$.

3.3.4 Thermal Wind Balance

Thermal wind balance arises by combining the geostrophic and hydrostatic approximations, and this is most easily done in the context of the Boussinesq equations, or in pressure coordinates. Beginning with the Boussinesq equations, geostrophic balance may be written

$$-fv_g = -\frac{\partial \phi}{\partial x}, \qquad fu_g = -\frac{\partial \phi}{\partial y}. \tag{3.37a,b}$$

Combining these relations with hydrostatic balance, $\partial\phi/\partial z = b$, gives

$$f\frac{\partial v_g}{\partial z} = \frac{\partial b}{\partial x}, \qquad f\frac{\partial u_g}{\partial z} = -\frac{\partial b}{\partial y}. \tag{3.38a,b}$$

These equations represent *thermal wind balance,* and the vertical derivative of the geostrophic wind is the 'thermal wind'.

If the density or buoyancy is constant then the right-hand sides of (3.38) are zero and there is no shear, recovering the Taylor–Proudman result. But suppose that the temperature falls in the poleward direction. Then thermal wind balance implies that the (eastward) wind will increase with height — just as is observed in the atmosphere! In general, a vertical shear of the horizontal wind is associated with a horizontal temperature gradient, and this is one of the most simple and far-reaching effects in geophysical fluid dynamics. The underlying physical mechanism is illustrated in Fig. 3.2.

Geostrophic and thermal wind balance in pressure coordinates

In pressure coordinates geostrophic balance is just

$$\boldsymbol{f} \times \boldsymbol{u}_g = -\nabla_p \Phi, \tag{3.39}$$

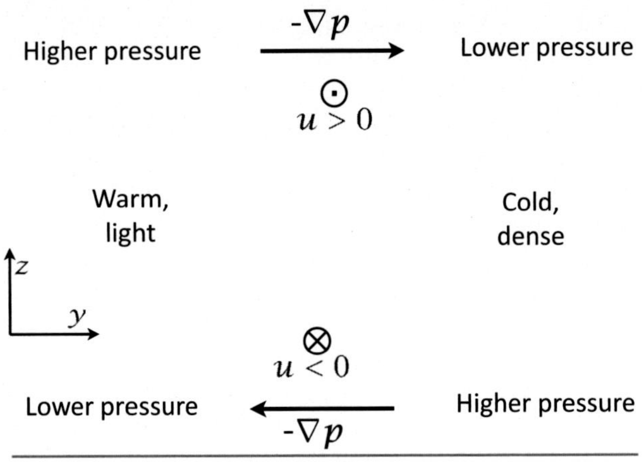

Fig. 3.2: The mechanism of thermal wind. A cold fluid is denser than a warm fluid, so by hydrostasy the vertical pressure gradient is greater where the fluid is cold. Thus, pressure gradients form as shown, where 'higher' and 'lower' mean relative to the average at that height. The horizontal pressure gradients are balanced by the Coriolis force, producing (for $f > 0$) the horizontal winds shown. Only the wind *shear* is given by the thermal wind.

where Φ is the geopotential and ∇_p is the gradient operator taken at constant pressure. If f is constant, it follows from (3.39) that the geostrophic wind is non-divergent on pressure surfaces. Taking the vertical derivative of (3.39) (that is, its derivative with respect to p) and using the hydrostatic equation, $\partial\Phi/\partial p = -\alpha$, gives the thermal wind equation

$$\boldsymbol{f} \times \frac{\partial \boldsymbol{u}_g}{\partial p} = \nabla_p \alpha = \frac{R}{p}\nabla_p T, \tag{3.40}$$

where the last equality follows using the ideal gas equation and because the horizontal derivative is at constant pressure. In component form this is

$$-f\frac{\partial v_g}{\partial p} = \frac{R}{p}\frac{\partial T}{\partial x}, \qquad f\frac{\partial u_g}{\partial p} = \frac{R}{p}\frac{\partial T}{\partial y}. \tag{3.41}$$

In log-pressure coordinates, with $Z = -H\ln(p/p_R)$, thermal wind is

$$\boldsymbol{f} \times \frac{\partial \boldsymbol{u}_g}{\partial Z} = -\frac{R}{H}\nabla_Z T. \tag{3.42}$$

The effect in all these cases is the same: a horizontal temperature gradient, or a temperature gradient along an isobaric surface, is accompanied by a vertical shear of the horizontal wind.

3.3.5 Scaling for Vertical Velocity

If the Coriolis parameter is constant then the horizontal components of flows that are in geostrophic balance have zero horizontal divergence ($\nabla_x \cdot \boldsymbol{u} = 0$) and zero vertical velocity. We can therefore expect that any flow with small Rossby number will have a 'small' vertical velocity. Let us make this statement more precise using the rotating Boussinesq equations, (3.13) with constant Coriolis parameter. Let $\boldsymbol{u} = \boldsymbol{u}_g + \boldsymbol{u}_a$ where the geostrophic flow satisfies $\boldsymbol{f}_0 \times \boldsymbol{u}_g = -\nabla\phi$. The horizontal momentum equation, with corresponding scales for each term, then becomes

$$\frac{\partial \boldsymbol{u}}{\partial t} + \boldsymbol{u}\cdot\nabla\boldsymbol{u} + w\frac{\partial \boldsymbol{u}}{\partial z} + \boldsymbol{f}_0 \times \boldsymbol{u}_a = 0, \tag{3.43}$$

$$\frac{U^2}{L} \quad \frac{U^2}{L} \quad \frac{WU}{H} \quad f_0 U_a. \tag{3.44}$$

This equation suggests a scaling for the ageostrophic flow of

$$U_a = \frac{U}{f_0 L} U = Ro U. \tag{3.45}$$

That is, the ageostrophic flow is a Rossby number smaller (at least) than the geostrophic flow. To obtain a scaling for the vertical velocity we look to the mass continuity equation written in the form

$$\frac{\partial w}{\partial z} = -\nabla \cdot \boldsymbol{u}_a, \tag{3.46}$$

since only the ageostrophic flow has a divergence. Equations (3.45) and (3.46) suggest the scaling

$$W = Ro \frac{HU}{L}. \tag{3.47}$$

That is, the vertical velocity is of order Rossby number smaller than an estimate based purely on the mass continuity equation would suggest.

If the Coriolis parameter is not constant then the geostrophic flow itself is divergent and this induces a vertical velocity, as in (3.33). The scaling for vertical velocity is now

$$W = \frac{\beta}{f} HU = \widetilde{\beta} \frac{HU}{L}, \tag{3.48}$$

where $\widetilde{\beta} = \beta L / f$ is less than one for all flows except those with a truly global scale.

♦ *Scaling for hydrostatic balance in rotating flow*

We have seen that if the flow is close to geostrophic balance then the vertical velocity is smaller than it would be otherwise, and we might thus expect hydrostatic balance to be even more likely to obtain. To show this, we use the scalings of (3.19) except that, using (3.47) and (3.34),

$$w = \frac{RoHU}{L}\widehat{w}, \qquad \phi = \Phi\widehat{\phi} = fUL\widehat{\phi}, \qquad b = B\widehat{b} = \frac{fUL}{H}\widehat{b}, \tag{3.49}$$

Carrying through a similar procedure to that in Section 3.2.2 (for details see Problem 3.4) we now obtain the nondimensional vertical momentum equation:

$$Ro^2 \alpha^2 \frac{\mathrm{D}\widehat{w}}{\mathrm{D}\widehat{t}} = -\frac{\partial \widehat{\phi}}{\partial \widehat{z}} - \widehat{b}. \tag{3.50}$$

We note the additional factor Ro^2 in the material derivative, implying that a rapidly rotating fluid is *more* likely to be in hydrostatic balance than a non-rotating fluid, other conditions being equal. Thus, weather scale systems in the atmosphere as well as the great ocean gyres are likely to be in very good hydrostatic balance, and for such flows the conceptual and practical simplifications afforded by the hydrostatic approximation can hardly be overstated.

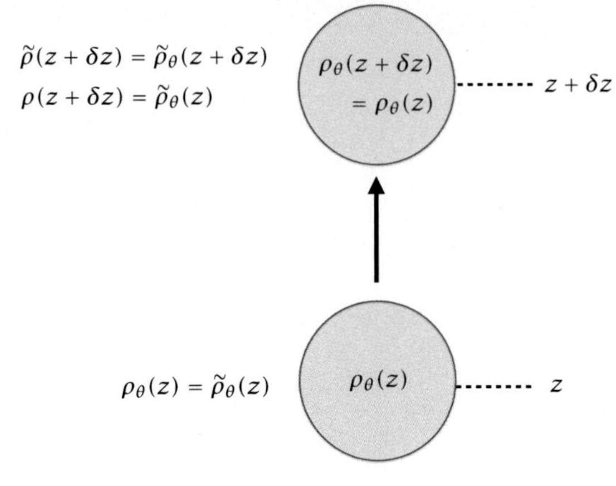

Fig. 3.3: A parcel is adiabatically displaced upward from level z to $z + \delta z$. A tilde denotes the value in the environment, and variables without tildes are those in the parcel.

The parcel preserves its potential density, ρ_θ, which it takes from the environment at level z. If $z + \delta z$ is the reference level, the potential density there is equal to the actual density. The parcel's stability is determined by the difference between its density and the environmental density, as in (3.51). If the difference is positive the displacement is stable, and conversely.

3.4 Static Instability and the Parcel Method

Is hydrostatic balance stable? Consider a thought experiment in which we take a fluid parcel and we displace it vertically. Does it come back to its original location as if it were attached to a spring? And if so, does it oscillate around its original position? Or, once displaced, does the fluid parcel keep on going like a ball released from the top of a hill? To answer this question, consider a fluid initially at rest in a constant gravitational field, and therefore in hydrostatic balance. Suppose that a small parcel of the fluid is adiabatically displaced upwards by the small distance δz, and that the fluid parcel then assumes the pressure of its environment. If after the displacement the parcel is lighter than its environment, it will accelerate upwards, because the upward pressure gradient force is now greater than the downward gravity force on the parcel; that is, the parcel is *buoyant* (a manifestation of Archimedes' principle) and the fluid is *statically unstable.* If on the other hand the fluid parcel finds itself heavier than its surroundings, the downward gravitational force will be greater than the upward pressure force and the fluid will sink back towards its original position and an oscillatory motion will develop. Such an equilibrium is *statically stable.* Using such simple parcel arguments we now develop criteria for the stability of the environmental profile.

3.4.1 Stability and the Profile of Potential Density

Consider the case of a stationary fluid whose density varies with altitude. We denote this background state with a tilde, as in $\tilde{\rho}(z)$. We then displace a fluid parcel adiabatically a small distance from z to δz, as in Fig. 3.3. In such a displacement it is the *potential density* ρ_θ (not the actual density) that is materially conserved, because potential density takes into account the effects of pressure compressibility. Let us use the pressure at level $z + \delta z$ as the reference level, where potential density equals *in situ* density.

The parcel starting at z takes on the potential density of its environment so that $\rho_\theta(z) = \tilde{\rho}_\theta(z)$ and it preserves this as it rises, so that

$\rho_\theta(z + \delta z) = \rho_\theta(z)$. But since $z + \delta z$ is the reference level, the *in situ* density of the displaced parcel, $\rho(z + \delta z)$, is equal to its potential density $\rho_\theta(z + \delta z)$, which is equal to $\widetilde{\rho}_\theta(z)$. Thus, at $z + \delta z$, the environment has *in situ* density equal to $\widetilde{\rho}_\theta(z + \delta z)$ and the parcel has *in situ* density equal to $\widetilde{\rho}_\theta(z)$. Putting this together in a single equation, the difference in density between the parcel and its environment at $z + \delta z$ is given by

$$\begin{aligned}\delta\rho &= \rho(z + \delta z) - \widetilde{\rho}(z + \delta z) = \rho_\theta(z + \delta z) - \widetilde{\rho}_\theta(z + \delta z) \\ &= \rho_\theta(z) - \widetilde{\rho}_\theta(z + \delta z) = \widetilde{\rho}_\theta(z) - \widetilde{\rho}_\theta(z + \delta z),\end{aligned} \tag{3.51}$$

and therefore

$$\delta\rho = -\frac{\partial \widetilde{\rho}_\theta}{\partial z}\delta z, \tag{3.52}$$

where the derivative on the right-hand side is the environmental gradient of potential density, in the vertical direction. If the right-hand side is positive, the parcel is heavier than its surroundings and the displacement is stable. That is, *the stability of a parcel of fluid is determined by the gradient of the locally-referenced potential density.*

The conditions for stability are thus:

$$\begin{aligned}\text{Stability}: \quad & \frac{\partial \widetilde{\rho}_\theta}{\partial z} < 0, \\ \text{Instability}: \quad & \frac{\partial \widetilde{\rho}_\theta}{\partial z} > 0.\end{aligned} \tag{3.53a,b}$$

The equation of motion of the fluid parcel is then given by a direct application of Newton's second law, that the mass times the acceleration is given by the force acting on the parcel. The force is equal to g times the above-derived buoyancy difference so that

$$\frac{\partial^2 \delta z}{\partial t^2} = \frac{g}{\rho}\left(\frac{\partial \widetilde{\rho}_\theta}{\partial z}\right)\delta z = -N^2 \delta z, \tag{3.54}$$

where, noting that $\rho(z) = \widetilde{\rho}_\theta(z)$ to within $O(\delta z)$,

$$N^2 = -\frac{g}{\widetilde{\rho}_\theta}\left(\frac{\partial \widetilde{\rho}_\theta}{\partial z}\right). \tag{3.55}$$

A parcel that is displaced in a stably stratified fluid will thus oscillate at the frequency N, known as the buoyancy frequency or Brunt–Väisälä frequency, after its discovers. The above expression for the buoyancy frequency is a general one, true in both liquids and gases in a constant gravitational field. The quantity $\widetilde{\rho}_\theta$ is the *locally-referenced* potential density of the environment. The reference level is not important for the atmosphere (as we see below), but it is for the ocean particularly in the presence of salinity: parcels at the same level with the same *in situ* density may have different potential densities if their salinity differs. For fresh water in a laboratory setting potential density is virtually equal to *in situ* density.

3.4.2 Gaseous Atmospheres

Buoyancy frequency

In the atmosphere potential density is related to potential temperature by $\rho_\theta = p_R/(\theta R)$, where p_R is the reference level for potential temperature. Using this expression in (3.55) gives

$$N^2 = \frac{g}{\widetilde{\theta}}\left(\frac{\partial \widetilde{\theta}}{\partial z}\right), \tag{3.56}$$

where $\widetilde{\theta}$ refers to the environmental profile of potential temperature. The reference value p_R does not appear, and we are free to choose this value arbitrarily — the surface pressure is a common choice. The conditions for stability, (3.53), then correspond to $N^2 > 0$ for stability and $N^2 < 0$ for instability. On average the atmosphere is stable and in the troposphere (the lowest several kilometres of the atmosphere) the average N is about $0.01\,\mathrm{s}^{-1}$, with a corresponding period, $(2\pi/N)$, of about 10 minutes. In the stratosphere (which lies above the troposphere) N^2 is a few times higher than this.

Dry adiabatic lapse rate

The dry adiabatic lapse rate, Γ_d is the rate at which temperature falls in the vertical (that is, $-\partial T/\partial z$) that corresponds to a neutral profile of potential temperature or entropy, $\partial\theta/\partial z = 0$ in a dry atmosphere. In an ideal gas $\Gamma_d = g/c_p$.

The negative of the rate of change of the (real) temperature in the vertical is known as the *temperature lapse rate*, or often just the lapse rate, and denoted Γ. The lapse rate corresponding to $\partial\theta/\partial z = 0$ is called the *dry adiabatic lapse rate* and denoted Γ_d. Using $\theta = T(p_0/p)^{R/c_p}$ and $\partial p/\partial z = -\rho g$ we find that the lapse rate and the potential temperature lapse rate are related by

$$\frac{T}{\theta}\frac{\partial\theta}{\partial z} = \frac{\partial T}{\partial z} + \frac{g}{c_p}, \tag{3.57}$$

so that the dry adiabatic lapse rate is given by

$$\Gamma_d = \frac{g}{c_p}. \tag{3.58}$$

The conditions for static stability corresponding to (3.53) are thus:

$$\begin{aligned} &\text{Stability}: \quad \frac{\partial\widetilde{\theta}}{\partial z} > 0, \quad \text{or} \quad -\frac{\partial\widetilde{T}}{\partial z} < \Gamma_d, \\ &\text{Instability}: \quad \frac{\partial\widetilde{\theta}}{\partial z} < 0, \quad \text{or} \quad -\frac{\partial\widetilde{T}}{\partial z} > \Gamma_d. \end{aligned} \tag{3.59a,b}$$

The observed lapse rate in Earth's atmosphere, Γ, is often about than $6\,\mathrm{K\,km^{-1}}$ whereas the dry adiabatic lapse rate is about $10\,\mathrm{K\,km^{-1}}$. Why the discrepancy? Why is the atmosphere so apparently stable? One reason is that the atmosphere contains water vapour. If a moist parcel rises, then, as it enters a cooler environment, water vapour may condense releasing more heat and leading to more ascent; because of this, a moist atmosphere may be unstable when a dry atmosphere is stable. The atmosphere will then take on a lapse rate that is just critical to moist convection, and this moist adiabatic lapse rate much less that the dry adiabatic one. We defer a more detailed treatment to Chapter 11.

3.4.3 Liquid Oceans

No simple, accurate, analytic expression is available for computing static stability in the ocean because of the complicated equation of state. However, we can obtain an approximate expression that captures the main effects. In an adiabatic displacement

$$\delta\rho_\theta = \delta\rho - \frac{1}{c_s^2}\delta p = 0. \tag{3.60}$$

If the fluid is hydrostatic $\delta p = -\rho g \delta z$ so that if a parcel is displaced adiabatically its density changes according to

$$\left(\frac{\partial \rho}{\partial z}\right)_{\rho_\theta} = -\frac{\rho g}{c_s^2}. \tag{3.61}$$

If a parcel is displaced a distance δz upwards then the density difference between it and its new surroundings is

$$\delta\rho = -\left[\left(\frac{\partial \rho}{\partial z}\right)_{\rho_\theta} - \left(\frac{\partial \widetilde{\rho}}{\partial z}\right)\right]\delta z = \left[\frac{\rho g}{c_s^2} + \left(\frac{\partial \widetilde{\rho}}{\partial z}\right)\right]\delta z, \tag{3.62}$$

where the tilde denotes the background environmental field. It follows that the stratification is given by

$$N^2 = -g\left[\frac{g}{c_s^2} + \frac{1}{\widetilde{\rho}}\left(\frac{\partial \widetilde{\rho}}{\partial z}\right)\right]. \tag{3.63}$$

This expression holds for both liquids and gases, and for ideal gases it is the same as (3.56) (as a little algebra will show, using $c_s^2 = \gamma p/\rho$). In the ocean the factor of g/c_s^2 is small but not negligible. It is a slightly destabilising factor in the sense that a density profile with an *in situ* density that increases with depth is not necessarily stable. Rather, a profile is stable only if the density increases with depth more rapidly than potential density gradient, equal to $-(\partial\widetilde{\rho}/\partial z + g/c_s^2)$. In liquids, a good approximation is to use a reference value ρ_0 for the undifferentiated density in the denominator. On average the ocean is statically stable, with typical values of N in the upper ocean being about $0.01\,\mathrm{s}^{-1}$, falling to $0.001\,\mathrm{s}^{-1}$ in the more homogeneous abyssal ocean. These frequencies correspond to periods of about 10 and 100 minutes, respectively.

Notes and References

Two standard texts on geophysical fluid dynamics are those by Gill (1982) and Pedlosky (1987a). The book by Cushman-Roisin & Beckers (2011) also includes some material on numerical approaches.

Problems

3.1 Consider two-dimensional, incompressible, fluid flow in a rotating frame of reference on the f-plane. Linearize the equations about a state of rest to obtain the momentum equations:

$$\frac{\partial u}{\partial t} - fv = -\frac{\partial \phi}{\partial x}, \qquad \frac{\partial v}{\partial t} + fu = -\frac{\partial \phi}{\partial y}. \tag{P3.1}$$

(*a*) Ignore the pressure term and determine the solution to the resulting equations. Show that the speed of fluid parcels is constant. Show that the trajectory of the fluid parcels is a circle with radius $|U|/f$, where $|U|$ is the fluid speed. (These solutions are inertial oscillations.)
(*b*) What is the period of oscillation of a fluid parcel?
(*c*) If parcels travel in straight lines in inertial frames, why is the answer to (b) not equal to the rotation period of the frame of reference? (See also Problem 4.3.)

3.2 Consider a dry, hydrostatic, ideal-gas atmosphere whose lapse rate is one of constant potential temperature. What is its vertical extent? That is, at what height does the density vanish? Is this result a problem for the any of the assumptions we have made in the first three chapters?

3.3 Consider a rapidly rotating (i.e., in near geostrophic balance) Boussinesq fluid on the f-plane.
(*a*) Show that the pressure divided by the density scales as $\phi \sim fUL$.
(*b*) Show that the horizontal divergence of the geostrophic wind vanishes. Thus, argue that the scaling $W \sim UH/L$ is an *overestimate* for the magnitude of the vertical velocity. Obtain a scaling estimate for the magnitude of vertical velocity in rapidly rotating flow.
(*c*) Using these results, or otherwise, discuss whether hydrostatic balance is more or less likely to hold in a rotating flow than in non-rotating flow.

3.4 Using the Boussinesq equations (or the fully compressible equations if you really want to):
(*a*) Obtain a criterion for the satisfaction of hydrostasy in rotating flow. Specifically, derive (3.50), showing all the steps in your derivation.
(*b*) Now suppose the flow is stratified and the thermodynamic equation is $\mathrm{D}b/\mathrm{D}t + N^2 w = 0$. Obtain an expression similar to (3.50) for the satisfaction of hydrostasy in stratified, rotating flow.

3.5 Using approximate but realistic values for the observed stratification, calculate the buoyancy period for (a) the midlatitude troposphere, (b) the stratosphere, (c) the oceanic thermocline, (d) the oceanic abyss.

3.6 What is the dry adiabatic lapse rate and buoyancy period for Venus, Mars and Jupiter? Obtain the appropriate physical parameters from other sources and make any assumptions you wish but state them clearly and justify them.

3.7 (*a*) Estimate the magnitude of the zonal thermal wind 5 km above the surface in the midlatitude atmosphere in summer and winter using approximate values for the meridional temperature gradient in Earth's atmosphere.
(*b*) Repeat the exercise for Venus and Mars.
(*c*) Repeat the exercise for Earth's ocean, 1 km below the surface.

CHAPTER

4 Shallow Water Equations

THE *SHALLOW WATER EQUATIONS* are a set of equations that describe, not surprisingly, a shallow layer of fluid, and in particular one that is in hydrostatic balance and has constant density. The equations are useful for two reasons:

(i) They are a simpler set of equations than the full three-dimensional ones, and so allow for a much more straightforward analysis of sometimes complex problems.

(ii) In spite of their simplicity, the equations provide a reasonably realistic representation of a variety of phenomena in atmospheric and oceanic dynamics.

Put simply, the shallow water equations are a very useful *model* for geophysical fluid dynamics. Let's dive head first into the equations and see what they can do for us.

4.1 SHALLOW WATER EQUATIONS OF MOTION

The shallow water equations apply, by definition, to a fluid layer of constant density in which the horizontal scale of the flow is much greater than the layer depth, and which have a free surface at the top (or sometimes at the bottom). Because the fluid is of constant density the fluid motion is fully determined by the momentum and mass continuity equations, and because of the assumed small aspect ratio the hydrostatic approximation is well satisfied, as we discussed in Section 3.2.2. Thus, consider a fluid above which is another fluid of negligible density, as illustrated in Fig. 4.1. Our notation is that $\boldsymbol{v} = u\hat{\mathbf{i}} + v\hat{\mathbf{j}} + w\hat{\mathbf{k}}$ is the three-dimensional velocity and $\boldsymbol{u} = u\hat{\mathbf{i}} + v\hat{\mathbf{j}}$ is the horizontal velocity, $h(x, y)$ is the thickness of the liquid column, H is its mean height, and η is the height of the free surface. In a flat-bottomed container $\eta = h$, whereas in general $h = \eta - \eta_B$, where η_B is the height of the floor of the container.

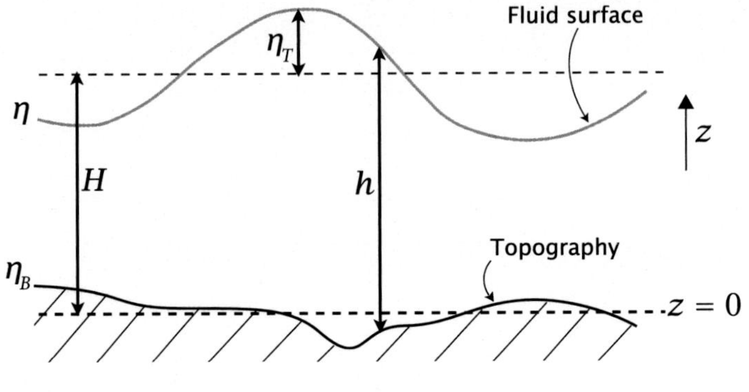

Fig. 4.1: A shallow water system where h is the thickness of a water column, H its mean thickness, η the height of the free surface and η_B is the height of the lower, rigid, surface above some arbitrary origin, typically chosen such that the average of η_B is zero. The quantity η_B is the deviation free surface height so we have $\eta = \eta_B + h = H + \eta_T$.

4.1.1 Momentum Equations

The vertical momentum equation is just the hydrostatic equation,

$$\frac{\partial p}{\partial z} = -\rho_0 g, \tag{4.1}$$

and, because density is assumed constant, we may integrate this to

$$p(x, y, z, t) = -\rho_0 g z + p_o. \tag{4.2}$$

The key assumption underlying the shallow water equations is that of a small aspect ratio, so that $H/L \ll 1$, where H is the fluid depth and L the horizontal scale of motion. This gives rise to the hydrostatic approximation, and this in turn leads to the z-independence of the velocity field and the 'sloshing' nature of the flow.

At the top of the fluid, $z = \eta$, the pressure is determined by the weight of the overlying fluid and this is negligible. Thus, $p = 0$ at $z = \eta$, giving

$$p(x, y, z, t) = \rho_0 g(\eta(x, y, t) - z). \tag{4.3}$$

The consequence of this is that *the horizontal gradient of pressure is independent of height*. That is

$$\nabla_z p = \rho_0 g \nabla_z \eta, \qquad \text{where} \qquad \nabla_z = \hat{\mathbf{i}}\frac{\partial}{\partial x} + \hat{\mathbf{j}}\frac{\partial}{\partial y}. \tag{4.4}$$

(In the rest of this chapter we drop the subscript z unless that causes ambiguity; the three-dimensional gradient operator is denoted by ∇_3. We also mostly use Cartesian coordinates, but the shallow water equations may certainly be applied over a spherical planet.) The horizontal momentum equations therefore become

$$\frac{\mathrm{D}\boldsymbol{u}}{\mathrm{D}t} = -\frac{1}{\rho_0}\nabla p = -g\nabla\eta. \tag{4.5}$$

The right-hand side of this equation is independent of the vertical coordinate z. Thus, if the flow is initially independent of z, it must stay so. (This z-independence is unrelated to that arising from the rapid rotation necessary for the Taylor–Proudman effect.) The velocities u and v are functions of x, y and t only, and the horizontal momentum equation is therefore

$$\frac{\mathrm{D}\boldsymbol{u}}{\mathrm{D}t} = \frac{\partial \boldsymbol{u}}{\partial t} + u\frac{\partial \boldsymbol{u}}{\partial x} + v\frac{\partial \boldsymbol{u}}{\partial y} = -g\nabla\eta. \tag{4.6}$$

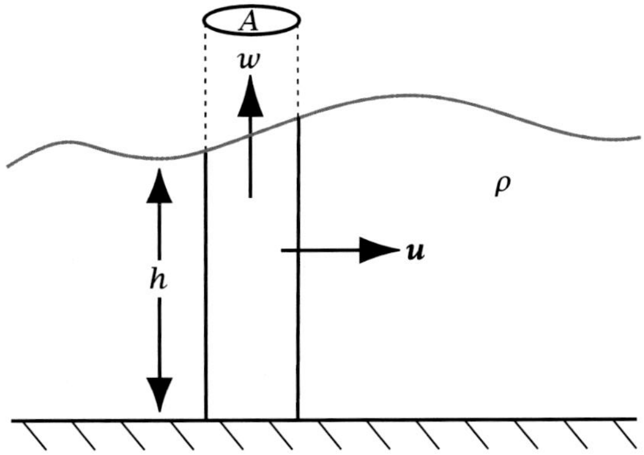

Fig. 4.2: The mass budget for a column of area A in a flat-bottomed shallow water system. The fluid leaving the column is $\oint \rho h \boldsymbol{u} \cdot \boldsymbol{n}\, dl$ where $\boldsymbol{n}$ is the unit vector normal to the boundary of the fluid column. There is a non-zero vertical velocity at the top of the column if the mass convergence into the column is non-zero.

In the presence of rotation, (4.6) easily generalizes to

$$\frac{D\boldsymbol{u}}{Dt} + \boldsymbol{f} \times \boldsymbol{u} = -g\nabla\eta, \tag{4.7}$$

where $\boldsymbol{f} = f\hat{\mathbf{k}}$. Just as with the fully three-dimensional equations, f may be constant or may vary with latitude, so that on a spherical planet $f = 2\Omega \sin\vartheta$ and on the β-plane $f = f_0 + \beta y$.

4.1.2 Mass Continuity Equation

The mass contained in a fluid column of height h and cross-sectional area A is given by $\int_A \rho_0 h \, \mathrm{d}\boldsymbol{A}$ (see Fig. 4.2). If there is a net flux of fluid across the column boundary (by advection) then this must be balanced by a net increase in the mass in A, and therefore a net increase in the height of the water column. The mass convergence into the column is given by

$$F_m = \text{mass flux in} = -\int_S \rho_0 \boldsymbol{u} \cdot \mathrm{d}\boldsymbol{S}, \tag{4.8}$$

where S is the area of the vertical boundary of the column. The surface area of the column is composed of elements of area $h\boldsymbol{n}\,\delta l$, where δl is a line element circumscribing the column and $\boldsymbol{n}$ is a unit vector perpendicular to the boundary, pointing outwards. Thus (4.8) becomes

$$F_m = -\oint \rho_0 h \boldsymbol{u} \cdot \boldsymbol{n} \, \mathrm{d}l. \tag{4.9}$$

Using the divergence theorem in two dimensions, (4.9) simplifies to

$$F_m = -\int_A \nabla \cdot (\rho_0 \boldsymbol{u} h) \, \mathrm{d}A, \tag{4.10}$$

where the integral is over the cross-sectional area of the fluid column (looking down from above). This is balanced by the local increase in height of the water column, given by

$$F_m = \frac{\mathrm{d}}{\mathrm{d}t} \int \rho_0 \, \mathrm{d}V = \frac{\mathrm{d}}{\mathrm{d}t} \int_A \rho_0 h \, \mathrm{d}A = \int_A \rho_0 \frac{\partial h}{\partial t} \, \mathrm{d}A. \tag{4.11}$$

The Shallow Water Equations

For a single-layer fluid, and including the Coriolis term, the inviscid shallow water equations are:

$$\text{momentum:} \qquad \frac{D\boldsymbol{u}}{Dt} + \boldsymbol{f} \times \boldsymbol{u} = -g\nabla\eta, \qquad \text{(SW.1)}$$

$$\text{mass continuity:} \qquad \frac{Dh}{Dt} + h\nabla \cdot \boldsymbol{u} = 0, \qquad \text{(SW.2)}$$

$$\text{or} \qquad \frac{\partial h}{\partial t} + \nabla \cdot (h\boldsymbol{u}) = 0, \qquad \text{(SW.3)}$$

where $\boldsymbol{u}$ is the horizontal velocity, h is the total fluid thickness, η is the height of the upper free surface, and h and η are related by

$$h(x, y, t) = \eta(x, y, t) - \eta_B(x, y), \qquad \text{(SW.4)}$$

where η_B is the height of the lower surface (the bottom topography). The material derivative is

$$\frac{D}{Dt} = \frac{\partial}{\partial t} + \boldsymbol{u} \cdot \nabla = \frac{\partial}{\partial t} + u\frac{\partial}{\partial x} + v\frac{\partial}{\partial y}, \qquad \text{(SW.5)}$$

with the rightmost expression holding in Cartesian coordinates.

Because ρ_0 is constant, the balance between (4.10) and (4.11) leads to

$$\int_A \left[\frac{\partial h}{\partial t} + \nabla \cdot (\boldsymbol{u}h)\right] dA = 0, \qquad (4.12)$$

and because the area is arbitrary the integrand itself must vanish, whence,

$$\frac{\partial h}{\partial t} + \nabla \cdot (\boldsymbol{u}h) = 0 \qquad \text{or} \qquad \frac{Dh}{Dt} + h\nabla \cdot \boldsymbol{u} = 0. \qquad \text{(4.13a,b)}$$

This derivation holds whether or not the lower surface is flat. If it is, then $h = \eta$, and if not $h = \eta - \eta_B$. Equations (4.7) and (4.13) form a complete set, summarized in the shaded box above.

4.1.3 Reduced Gravity Equations

Consider now a single shallow moving layer of fluid *on top* of a deep, quiescent fluid layer (Fig. 4.3), and beneath a fluid of negligible inertia. This configuration is often used as a model of the upper ocean: the upper layer represents flow in perhaps the upper few hundred metres of the ocean, the lower layer being the near-stagnant abyss. If we turn the model upside-down we have a perhaps slightly less realistic model of the atmosphere: the lower layer represents motion in the troposphere above which lies an

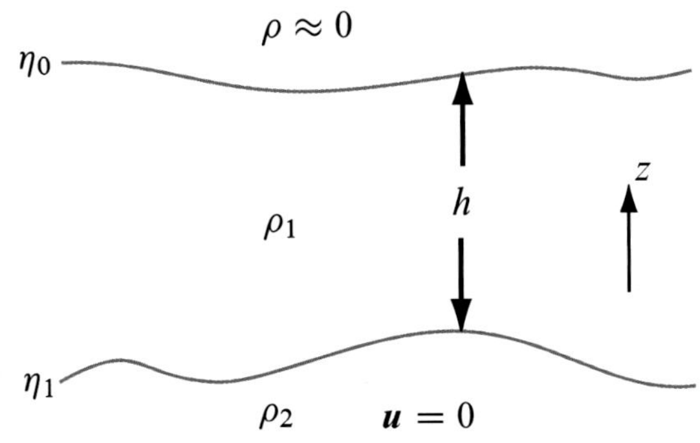

Fig. 4.3: The reduced gravity shallow water system. An active layer lies over a deep, denser, quiescent layer. In a common variation the upper surface is held flat by a rigid lid, and $\eta_0 = 0$.

inactive stratosphere. The equations of motion are virtually the same in both cases, but for definiteness we'll think about the oceanic case.

The pressure in the upper layer is given by integrating the hydrostatic equation down from the upper surface. Thus, at a height z in the upper layer

$$p_1(z) = g\rho_1(\eta_0 - z), \tag{4.14}$$

where η_0 is the height of the upper surface. Hence, everywhere in the upper layer,

$$\frac{1}{\rho_1}\nabla p_1 = g\nabla\eta_0, \tag{4.15}$$

and the momentum equation is

$$\frac{\mathrm{D}\boldsymbol{u}}{\mathrm{D}t} + \boldsymbol{f} \times \boldsymbol{u} = -g\nabla\eta_0. \tag{4.16}$$

In the lower layer the pressure is also given by the weight of the fluid above it. Thus, at some level z in the lower layer,

$$p_2(z) = \rho_1 g(\eta_0 - \eta_1) + \rho_2 g(\eta_1 - z). \tag{4.17}$$

But if this layer is motionless the horizontal pressure gradient in it is zero and therefore

$$\rho_1 g\eta_0 = -\rho_1 g'\eta_1 + \text{constant}, \tag{4.18}$$

where $g' = g(\rho_2 - \rho_1)/\rho_1$ is the *reduced gravity*, and in the ocean $\rho_2 - \rho_1)/\rho \ll 1$ and $g' \ll g$. The momentum equation becomes

$$\frac{\mathrm{D}\boldsymbol{u}}{\mathrm{D}t} + \boldsymbol{f} \times \boldsymbol{u} = g'\nabla\eta_1. \tag{4.19}$$

The equations are completed by the usual mass conservation equation,

$$\frac{\mathrm{D}h}{\mathrm{D}t} + h\nabla\cdot\boldsymbol{u} = 0, \tag{4.20}$$

where $h = \eta_0 - \eta_1$. Because $g \gg g'$, (4.18) shows that surface displacements are *much smaller* than the displacements at the interior interface. We see this in the real ocean where the mean interior isopycnal displacements may be several tens of metres but variations in the mean height of the ocean surface are of the order of centimetres.

4.2 CONSERVATION PROPERTIES

There are two common types of conservation property in fluids: (i) material invariants; and (ii) integral invariants. Material invariance occurs when a property (φ say) is conserved on each fluid element, and so obeys the equation $\mathrm{D}\varphi/\mathrm{D}t = 0$. An integral invariant is one that is conserved after an integration over some, usually closed, volume; energy is an example. The simplicity of the shallow water equations allows us to transparently see how these arise.

4.2.1 Energy Conservation: an Integral Invariant

Since we have made various simplifications in deriving the shallow water system, it is not self-evident that energy should be conserved, or indeed what form the energy takes. The kinetic energy density (*KE*), meaning the kinetic energy per unit area, is $\rho_0 h\boldsymbol{u}^2/2$. The potential energy density (*PE*) of the fluid is

$$PE = \int_0^h \rho_0 g z \, \mathrm{d}z = \frac{1}{2}\rho_0 g h^2. \tag{4.21}$$

The factor ρ_0 appears in both kinetic and potential energies and, because it is a constant, we will omit it. For algebraic simplicity we also assume the bottom is flat, at $z = 0$.

Energy is moved from place to place by pressure forces as well as advection, and is not a material invariant. However, in total energy is conserved and it is an integral invariant.

Using the mass conservation equation (4.13b) we obtain an equation for the evolution of potential energy density, namely

$$\frac{\mathrm{D}}{\mathrm{D}t}\frac{gh^2}{2} + gh^2\nabla\cdot\boldsymbol{u} = 0 \tag{4.22a}$$

or

$$\frac{\partial}{\partial t}\frac{gh^2}{2} + \nabla\cdot\left(\boldsymbol{u}\frac{gh^2}{2}\right) + \frac{gh^2}{2}\nabla\cdot\boldsymbol{u} = 0. \tag{4.22b}$$

From the momentum and mass continuity equations we obtain an equation for the evolution of kinetic energy density, namely

$$\frac{\mathrm{D}}{\mathrm{D}t}\frac{h\boldsymbol{u}^2}{2} + \frac{\boldsymbol{u}^2 h}{2}\nabla\cdot\boldsymbol{u} = -g\boldsymbol{u}\cdot\nabla\frac{h^2}{2} \tag{4.23a}$$

or

$$\frac{\partial}{\partial t}\frac{h\boldsymbol{u}^2}{2} + \nabla\cdot\left(\boldsymbol{u}\frac{h\boldsymbol{u}^2}{2}\right) + g\boldsymbol{u}\cdot\nabla\frac{h^2}{2} = 0. \tag{4.23b}$$

Adding (4.22b) and (4.23b) we obtain

$$\frac{\partial}{\partial t}\frac{1}{2}\left(h\boldsymbol{u}^2 + gh^2\right) + \nabla\cdot\left[\frac{1}{2}\boldsymbol{u}\left(gh^2 + h\boldsymbol{u}^2 + gh^2\right)\right] = 0 \tag{4.24}$$

or

$$\frac{\partial E}{\partial t} + \nabla\cdot\boldsymbol{F} = 0, \tag{4.25}$$

where $E = KE + PE = (h\boldsymbol{u}^2 + gh^2)/2$ is the density of the total energy and $\boldsymbol{F} = \boldsymbol{u}(h\boldsymbol{u}^2/2 + gh^2)$ is the energy flux. If the fluid is confined to a domain bounded by rigid walls, on which the normal component of velocity vanishes, then on integrating (4.24) over that area and using Gauss's theorem, the total energy is seen to be conserved; that is

$$\frac{\mathrm{d}\widehat{E}}{\mathrm{d}t} = \frac{1}{2}\frac{\mathrm{d}}{\mathrm{d}t}\int_A (h\boldsymbol{u}^2 + gh^2)\,\mathrm{d}A = 0. \tag{4.26}$$

Such an energy principle also holds in the case with bottom topography. Note that, as we found in the case for a compressible fluid in Chapter 2, the energy flux in (4.25) is not just the energy density multiplied by the velocity; it contains an additional term $g\boldsymbol{u}h^2/2$, and this represents the energy transfer occurring when the fluid does work against the pressure force.

4.2.2 Potential Vorticity: a Material Invariant

The vorticity of a fluid, denoted $\boldsymbol{\omega}$, is defined to be the curl of the velocity field. Let us also define the shallow water vorticity, $\boldsymbol{\omega}^*$, as the curl of the horizontal velocity. We therefore have:

$$\boldsymbol{\omega} \equiv \nabla \times \boldsymbol{v}, \qquad \boldsymbol{\omega}^* \equiv \nabla \times \boldsymbol{u}. \tag{4.27}$$

Because $\partial u/\partial z = \partial v/\partial z = 0$, only the vertical component of $\boldsymbol{\omega}^*$ is non-zero and

$$\boldsymbol{\omega}^* = \hat{\mathbf{k}}\left(\frac{\partial v}{\partial x} - \frac{\partial u}{\partial y}\right) = \hat{\mathbf{k}}\zeta. \tag{4.28}$$

Considering first the non-rotating case, we use the vector identity

$$(\boldsymbol{u}\cdot\nabla)\boldsymbol{u} = \frac{1}{2}\nabla(\boldsymbol{u}\cdot\boldsymbol{u}) - \boldsymbol{u}\times(\nabla\times\boldsymbol{u}), \tag{4.29}$$

to write the momentum equation, (4.7) as

$$\frac{\partial \boldsymbol{u}}{\partial t} + \boldsymbol{\omega}^* \times \boldsymbol{u} + \boldsymbol{f}\times\boldsymbol{u} = -\nabla\left(g\eta + \frac{1}{2}\boldsymbol{u}^2\right). \tag{4.30}$$

To obtain an evolution equation for the vorticity we take the curl of (4.30), and make use of the vector identity

$$\begin{aligned}\nabla\times(\boldsymbol{\omega}^*\times\boldsymbol{u}) &= (\boldsymbol{u}\cdot\nabla)\boldsymbol{\omega}^* - (\boldsymbol{\omega}^*\cdot\nabla)\boldsymbol{u} + \boldsymbol{\omega}^*\nabla\cdot\boldsymbol{u} - \boldsymbol{u}\nabla\cdot\boldsymbol{\omega}^*\\ &= (\boldsymbol{u}\cdot\nabla)\boldsymbol{\omega}^* + \boldsymbol{\omega}^*\nabla\cdot\boldsymbol{u},\end{aligned} \tag{4.31}$$

using the fact that $\nabla\cdot\boldsymbol{\omega}^*$ is the divergence of a curl and therefore zero, and $(\boldsymbol{\omega}^*\cdot\nabla)\boldsymbol{u} = 0$ because $\boldsymbol{\omega}^*$ is perpendicular to the surface in which $\boldsymbol{u}$ varies. A similar expression holds for $\boldsymbol{f}$ so that taking the curl of (4.30) gives

$$\frac{\partial\zeta}{\partial t} + (\boldsymbol{u}\cdot\nabla)(\zeta+f) = -(\zeta+f)\nabla\cdot\boldsymbol{u}, \tag{4.32}$$

where $\zeta = \hat{\mathbf{k}}\cdot\boldsymbol{\omega}^*$ and f varies with latitude (and on the beta-plane $f = f_0 + \beta y$). Since f does not vary with time we can write (4.32) as

$$\frac{\mathrm{D}}{\mathrm{D}t}(\zeta+f) = -(\zeta+f)\nabla\cdot\boldsymbol{u}. \tag{4.33}$$

The mass conservation equation, (4.13b) may be written as

$$-(\zeta + f)\nabla \cdot \boldsymbol{u} = \frac{\zeta + f}{h}\frac{\mathrm{D}h}{\mathrm{D}t}, \tag{4.34}$$

and using this equation and (4.32) we obtain

$$\frac{\mathrm{D}}{\mathrm{D}t}(\zeta + f) = \frac{\zeta + f}{h}\frac{\mathrm{D}h}{\mathrm{D}t}, \tag{4.35}$$

which is equivalent to

$$\frac{\mathrm{D}Q}{\mathrm{D}t} = 0 \qquad \text{where} \qquad Q = \left(\frac{\zeta + f}{h}\right). \tag{4.36}$$

The important quantity Q is known as the *potential vorticity*, and (4.36) is the potential vorticity equation.

4.3 SHALLOW WATER WAVES

Let us now look at the gravity waves that occur in shallow water. To isolate the essence we consider waves in a single fluid layer, with a flat bottom and a free upper surface, in which gravity provides the sole restoring force.

4.3.1 Non-Rotating Shallow Water Waves

Given a flat bottom the fluid thickness is equal to the free surface displacement (Fig. 4.1), and taking the basic state of the fluid to be at rest we let

$$h(x, y, t) = H + h'(x, y, t) = H + \eta'(x, y, t), \tag{4.37a}$$

$$\boldsymbol{u}(x, y, t) = \boldsymbol{u}'(x, y, t). \tag{4.37b}$$

The mass conservation equation, (4.13b), then becomes

$$\frac{\partial \eta'}{\partial t} + (H + \eta')\nabla \cdot \boldsymbol{u}' + \boldsymbol{u}' \cdot \nabla \eta' = 0, \tag{4.38}$$

and neglecting squares of small quantities this yields the linear equation

$$\frac{\partial \eta'}{\partial t} + H\nabla \cdot \boldsymbol{u}' = 0. \tag{4.39}$$

Similarly, linearizing the momentum equation, (4.7) with $\boldsymbol{f} = 0$, yields

$$\frac{\partial \boldsymbol{u}'}{\partial t} = -g\nabla \eta'. \tag{4.40}$$

Eliminating velocity by differentiating (4.39) with respect to time and taking the divergence of (4.40) leads to

$$\frac{\partial^2 \eta'}{\partial t^2} - gH\nabla^2 \eta' = 0, \tag{4.41}$$

which may be recognized as a wave equation. We can find the dispersion relationship for this by substituting the trial solution

$$\eta' = \mathrm{Re}\,\tilde{\eta}\mathrm{e}^{\mathrm{i}(\boldsymbol{k}\cdot\boldsymbol{x}-\omega t)}, \tag{4.42}$$

where $\tilde{\eta}$ is a complex constant, $\boldsymbol{k} = k\hat{\mathbf{i}} + l\hat{\mathbf{j}}$ is the horizontal wavenumber and Re indicates that the real part of the solution should be taken. If, for simplicity, we restrict attention to the one-dimensional problem, with no variation in the y-direction, then substituting into (4.41) leads to the dispersion relationship

$$\omega = \pm ck, \tag{4.43}$$

where $c = \sqrt{gH}$; that is, the wave speed is proportional to the square root of the mean fluid depth and is independent of the wavenumber — the waves are dispersionless. The general solution is a superposition of all such waves, with the amplitudes of each wave (or Fourier component) being determined by the Fourier decomposition of the initial conditions.

Because the waves are dispersionless, the general solution can be written as

$$\eta'(x,t) = \frac{1}{2}\left[F(x-ct) + F(x+ct)\right], \tag{4.44}$$

where $F(x)$ is the height field at $t = 0$. From this, it is easy to see that the shape of an initial disturbance is preserved as it propagates both to the right and to the left at speed c.

4.3.2 Rotating Shallow Water (Poincaré) Waves

We now consider the effects of rotation on shallow water waves. Linearizing the rotating, flat-bottomed f-plane shallow water equations, (SW.1) and (SW.2) on page 66, about a state of rest we obtain

$$\frac{\partial u'}{\partial t} - f_0 v' = -g\frac{\partial \eta'}{\partial x}, \qquad \frac{\partial v'}{\partial t} + f_0 u' = -g\frac{\partial \eta'}{\partial y}, \tag{4.45a,b}$$

$$\frac{\partial \eta'}{\partial t} + H\left(\frac{\partial u'}{\partial x} + \frac{\partial v'}{\partial y}\right) = 0. \tag{4.45c}$$

To obtain a dispersion relationship we let

$$(u, v, \eta) = (\tilde{u}, \tilde{v}, \tilde{\eta})\mathrm{e}^{\mathrm{i}(\boldsymbol{k}\cdot\boldsymbol{x}-\omega t)}, \tag{4.46}$$

and substitute into (4.45), giving

$$\begin{pmatrix} -\mathrm{i}\omega & -f_0 & \mathrm{i}gk \\ f_0 & -\mathrm{i}\omega & \mathrm{i}gl \\ \mathrm{i}Hk & \mathrm{i}Hl & -\mathrm{i}\omega \end{pmatrix}\begin{pmatrix} \tilde{u} \\ \tilde{v} \\ \tilde{\eta} \end{pmatrix} = 0. \tag{4.47}$$

This homogeneous equation has non-trivial solutions only if the determinant of the matrix vanishes, and that condition gives

$$\omega(\omega^2 - f_0^2 - c^2K^2) = 0, \tag{4.48}$$

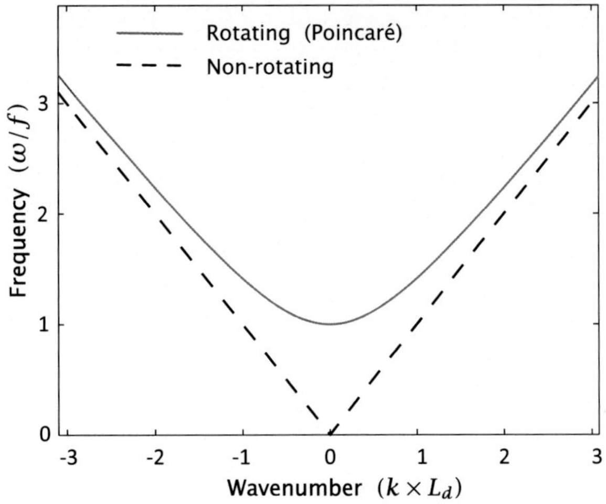

Fig. 4.4: Dispersion relation for Poincaré waves and non-rotating shallow water waves. Frequency is scaled by the Coriolis frequency f, and wavenumber by the inverse deformation radius $\sqrt{gH}/f$. For small wavenumbers the frequency of the Poincaré waves is approximately f, and for high wavenumbers it asymptotes to that of non-rotating waves.

where $K^2 = k^2 + l^2$ and $c^2 = gH$. There are two classes of solution to (4.48). The first is simply $\omega = 0$, i.e., time-independent flow corresponding to geostrophic balance in (4.45). Because geostrophic balance gives a divergence-free velocity field for a constant Coriolis parameter the equations are satisfied by a time-independent solution. The second set of solutions gives the dispersion relation

$$\omega^2 = f_0^2 + c^2(k^2 + l^2), \tag{4.49}$$

or

$$\omega^2 = f_0^2 + gH(k^2 + l^2). \tag{4.50}$$

The corresponding waves are known as *Poincaré* waves, and the dispersion relationship is illustrated in Fig. 4.4. Note that the frequency is always greater than the Coriolis frequency f_0. There are two interesting limits:

(Jules) Henri Poincaré (1854–1912) was a remarkable French mathematician, physicist and philosopher. As well as the Poincaré waves of this chapter he is also known as one of the founders of what is now known as chaos theory — he knew that the weather was inherently, and not just in practice, unpredictable.

(i) The short wave limit. If

$$K^2 \gg \frac{f_0^2}{gH}, \tag{4.51}$$

where $K^2 = k^2 + l^2$, then the dispersion relationship reduces to that of the non-rotating case (4.43). This condition is equivalent to requiring that the wavelength be much shorter than the *deformation radius*, defined by

$$L_d = \frac{\sqrt{gH}}{f}. \tag{4.52}$$

Specifically, if $l = 0$ and $\lambda = 2\pi/k$ is the wavelength, the condition is $\lambda^2 \ll L_d^2(2\pi)^2$. The numerical factor of $(2\pi)^2$ is more than an order of magnitude, so care must be taken when deciding if the condition is satisfied in particular cases. Furthermore, the wavelength must still be longer than the depth of the fluid, otherwise the shallow water condition is not met.

(ii) The long wave limit. If

$$K^2 \ll \frac{f_0^2}{gH}, \tag{4.53}$$

that is if the wavelength is much longer than the deformation radius L_d, then the dispersion relationship is

$$\omega = f_0. \tag{4.54}$$

These waves are known as *inertial oscillations*. The equations of motion giving rise to them are

$$\frac{\partial u'}{\partial t} - f_0 v' = 0, \qquad \frac{\partial v'}{\partial t} + f_0 u' = 0, \tag{4.55}$$

which are equivalent to material equations for free particles in a rotating frame, unconstrained by pressure forces.

4.3.3 Kelvin Waves

The Kelvin wave is a particular type of gravity wave that exists in the presence of both rotation and a lateral boundary. Suppose there is a solid boundary at $y = 0$; general harmonic solutions in the y-direction are not allowable, as these would not satisfy the condition of no normal flow at the boundary. Do any wave-like solutions exist? The answer is yes, and to show that we begin with the linearized shallow water equations, namely

The affirmative answer to the question in the text was provided by William Thomson in the nineteenth century. Thomson later became Lord Kelvin, taking the name of the river flowing near his university in Glasgow, and the waves are now eponymously known as Kelvin waves.

$$\frac{\partial u'}{\partial t} - f_0 v' = -g\frac{\partial \eta'}{\partial x}, \qquad \frac{\partial v'}{\partial t} + f_0 u' = -g\frac{\partial \eta'}{\partial y}, \tag{4.56a,b}$$

$$\frac{\partial \eta'}{\partial t} + H\left(\frac{\partial u'}{\partial x} + \frac{\partial v'}{\partial y}\right) = 0. \tag{4.56c}$$

The fact that $v' = 0$ at $y = 0$ suggests that we look for a solution with $v' = 0$ everywhere, whence these equations become

$$\frac{\partial u'}{\partial t} = -g\frac{\partial \eta'}{\partial x}, \qquad f_0 u' = -g\frac{\partial \eta'}{\partial y}, \qquad \frac{\partial \eta'}{\partial t} + H\frac{\partial u'}{\partial x} = 0. \tag{4.57a,b,c}$$

Equations (4.57a) and (4.57c) lead to the standard wave equation

$$\frac{\partial^2 u'}{\partial t^2} = c^2\frac{\partial^2 u'}{\partial x^2}, \tag{4.58}$$

where $c = \sqrt{gH}$, the usual wave speed of shallow water waves. The solution of (4.58) is

$$u' = F_1(x + ct, y) + F_2(x - ct, y), \tag{4.59}$$

with corresponding surface displacement

$$\eta' = \sqrt{H/g}\left[-F_1(x + ct, y) + F_2(x - ct, y)\right]. \tag{4.60}$$

The solution represents the superposition of two waves, one (F_1) travelling in the negative x-direction, and the other in the positive x-direction. To obtain the y dependence of these functions we use (4.57b) which gives

$$\frac{\partial F_1}{\partial y} = \frac{f_0}{\sqrt{gH}} F_1, \qquad \frac{\partial F_2}{\partial y} = -\frac{f_0}{\sqrt{gH}} F_2, \tag{4.61}$$

with solutions

$$F_1 = F(x + ct)\mathrm{e}^{y/L_d}, \qquad F_2 = G(x - ct)\mathrm{e}^{-y/L_d}, \tag{4.62}$$

where $L_d = \sqrt{gH}/f_0$ is the radius of deformation. If we consider flow in the half-plane in which $y > 0$, then for positive f_0 the solution F_1 grows exponentially away from the wall, and so fails to satisfy the condition of boundedness at infinity. It thus must be eliminated, leaving the general solution

$$u' = \mathrm{e}^{-y/L_d} G(x - ct), \qquad v' = 0, \qquad \eta' = \sqrt{H/g}\,\mathrm{e}^{-y/L_d} G(x - ct). \tag{4.63a,b,c}$$

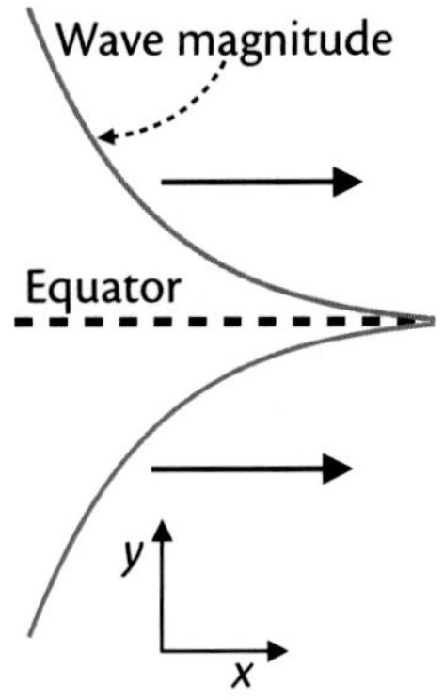

Fig. 4.5: Kelvin waves propagating eastward at the equator and decaying rapidly away to either side.

These are Kelvin waves and they decay exponentially away from the boundary. In general, for f_0 positive (negative) the boundary is to the right (left) of the wave direction. Given a constant Coriolis parameter, we could also have obtained a solution on a meridional wall, again with the wave moving with the wall on its the right if f_0 is positive. The equator also acts like a wall, with Kelvin waves propagating eastward along it, decaying to either side, as in Fig. 4.5. (The details of the solution for equatorial Kelvin waves differ from the case with $f = f_0$, because now $f = 0$ at the equator itself, but the structure is similar to the solution above.)

4.4 GEOSTROPHIC ADJUSTMENT

Geostrophic balance occurs in the shallow water equations when the Rossby number U/fL is small and the Coriolis term dominates the advective terms in the momentum equation. In the single-layer shallow water equations the geostrophic flow is:

$$\boldsymbol{f} \times \boldsymbol{u}_g = -g\nabla\eta. \tag{4.64}$$

Thus, the geostrophic velocity is proportional to the slope of the surface, as sketched in Fig. 4.6. The result arises because by hydrostatic balance the slope of an interfacial surface is directly related to the difference in pressure gradient on either side, so that in geostrophic balance the velocity is related to the slope.

But consider the more general question, *why* are the atmosphere and ocean close to geostrophic balance? Suppose that the initial state were not balanced, that it had small scale motion and sharp pressure gradients. How would the system evolve? It turns out that all of these sharp gradients would radiate away, leaving behind a geostrophically balanced state. The process is called *geostrophic adjustment*, and it occurs quite generally in rotating fluids, whether stratified or not. The shallow water equations are an ideal set of equations to explore this process, as we now see.

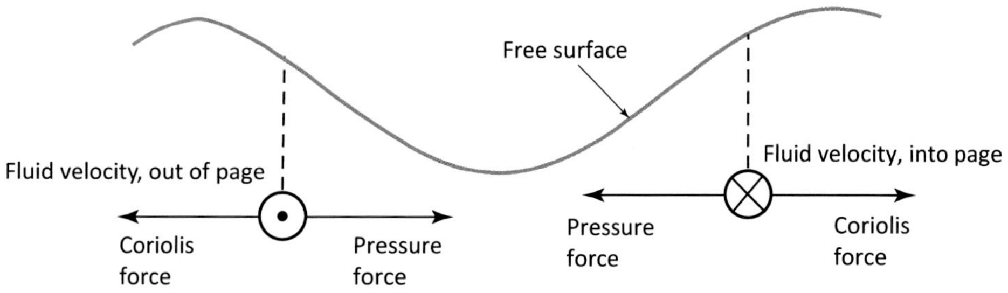

Fig. 4.6: Geostrophic flow in a shallow water system, with a positive value of the Coriolis parameter f, as in the Northern Hemisphere. The pressure force is directed down the gradient of the height field, and this can be balanced by the Coriolis force if the fluid velocity is at right angles to it. If f were negative, the geostrophic flow would be reversed.

4.4.1 Posing the Problem

Let us consider the free evolution of a single shallow layer of fluid whose initial state is manifestly unbalanced, and we suppose that surface displacements are small so that the evolution of the system is described by the linearized shallow equations of motion. These are

$$\frac{\partial \boldsymbol{u}}{\partial t} + \boldsymbol{f} \times \boldsymbol{u} = -g\nabla\eta, \qquad \frac{\partial \eta}{\partial t} + H\nabla \cdot \boldsymbol{u} = 0, \tag{4.65a,b}$$

where η is the free surface displacement and H is the mean fluid depth, and we omit the primes on the linearized variables.

4.4.2 Non-Rotating Flow

We consider first the non-rotating problem set, with little loss of generality, in one dimension. We suppose that initially the fluid is at rest but with a simple discontinuity in the height field so that

$$\eta(x, t=0) = \begin{cases} +\eta_0 & x < 0 \\ -\eta_0 & x > 0, \end{cases} \tag{4.66}$$

and $u(x, t = 0) = 0$ everywhere. We can realize these initial conditions physically by separating two fluid masses of different depths by a thin dividing wall, and then quickly removing the wall. What is the subsequent evolution of the fluid? The general solution to the linear problem is given by (4.44) where the functional form is determined by the initial conditions so that here

$$F(x) = \eta(x, t=0) = -\eta_0 \operatorname{sgn}(x). \tag{4.67}$$

Equation (4.44) states that this initial pattern is propagated to the right and to the left. That is, two discontinuities in fluid height move to the right and left at a speed $c = \sqrt{gH}$. Specifically, the solution is

$$\eta(x, t) = -\frac{1}{2}\eta_0[\operatorname{sgn}(x + ct) + \operatorname{sgn}(x - ct)]. \tag{4.68}$$

The initial conditions may be much more complex than a simple front, but, because the waves are dispersionless, the solution is still simply a sum

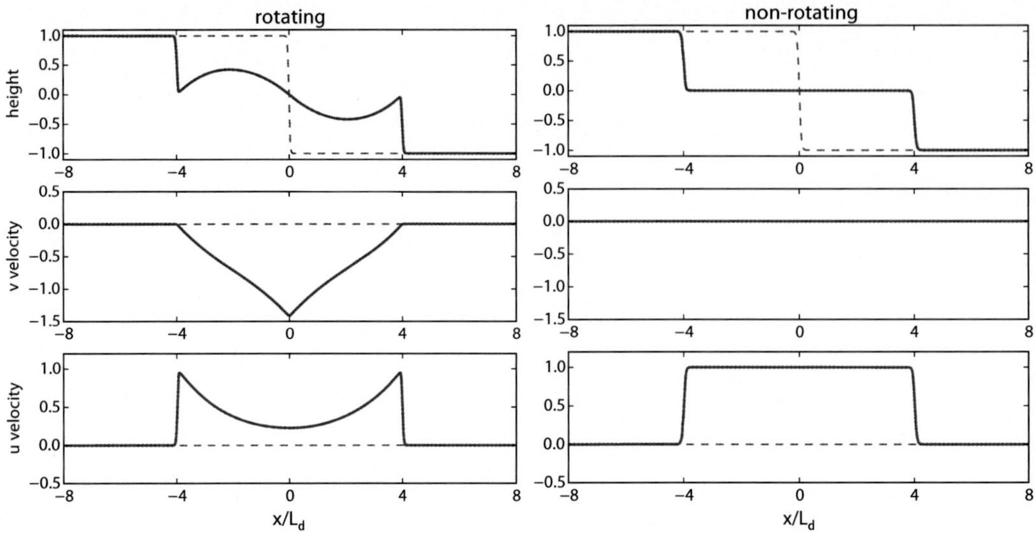

Fig. 4.7: The solutions of the shallow water equations obtained by numerically integrating the equations of motion with and without rotation. The panels show snapshots of the state of the fluid (solid lines) soon after being released from a stationary initial state (red dashed lines) with a height discontinuity. The rotating flow is evolving toward an end state similar to Fig. 4.8 whereas the non-rotating flow will eventually become stationary. In the non-rotating case L_d is defined using the rotating parameters.

of the translation of those initial conditions to the right and to the left at speed c. The velocity field in this class of problem is obtained from

$$\frac{\partial u}{\partial t} = -g\frac{\partial \eta}{\partial x}, \tag{4.69}$$

which gives, using (4.44),

$$u = -\frac{g}{2c}[F(x+ct) - F(x-ct)]. \tag{4.70}$$

Consider the case with initial conditions given by (4.66). At a given location, away from the initial disturbance, the fluid remains at rest and undisturbed until the front arrives. After the front has passed, the fluid surface is again undisturbed and the velocity is uniform and non-zero. Specifically:

$$\eta = \begin{cases} -\eta_0 \mathrm{sgn}(x) \\ 0 \end{cases} \qquad u = \begin{cases} 0 & |x| > ct \\ (\eta_0 g/c) & |x| < ct. \end{cases} \tag{4.71}$$

The solution with a discontinuity in the height field, and zero initial velocity, is illustrated in the right-hand panels in Fig. 4.7. The front propagates in either direction from the discontinuity and, in this case, the final velocity, as well as the fluid displacement, is zero. That is, the disturbance is radiated completely away.

4.4.3 Rotating Flow

Rotation makes a profound difference to the adjustment problem of the shallow water system, because a steady, adjusted, solution can exist with non-zero gradients in the height field — the associated pressure gradients being balanced by the Coriolis force — and potential vorticity conservation provides a powerful constraint on the fluid evolution. In a rotating shallow fluid that conservation is represented by

$$\frac{\partial Q}{\partial t} + \boldsymbol{u} \cdot \nabla Q = 0, \tag{4.72}$$

where $Q = (\zeta + f)/h$. In the linear case with constant Coriolis parameter, (4.72) becomes

$$\frac{\partial q}{\partial t} = 0, \qquad q = \left(\zeta - f_0 \frac{\eta}{H}\right). \tag{4.73}$$

This equation may be obtained either from the linearized velocity and mass conservation equations, (4.65), or from (4.72) directly. In the latter case we can write

$$Q = \frac{\zeta + f_0}{H + \eta} \approx \frac{1}{H}(\zeta + f_0)\left(1 - \frac{\eta}{H}\right) \approx \frac{1}{H}\left(f_0 + \zeta - f_0 \frac{\eta}{H}\right) = \frac{f_0}{H} + \frac{q}{H}, \tag{4.74}$$

having used $f_0 \gg |\zeta|$ and $H \gg |\eta|$. The term f_0/H is a constant and so dynamically unimportant, as is the H^{-1} factor multiplying q. Further, the advective term $\boldsymbol{u} \cdot \nabla Q$ becomes $\boldsymbol{u} \cdot \nabla q$, and this is second order in perturbed quantities and so is neglected. Thus, making these approximations, (4.72) reduces to (4.73). The potential vorticity field is therefore fixed in space! Of course, this was also true in the non-rotating case where the fluid is initially at rest. Then $q = \zeta = 0$ and the fluid remains irrotational throughout the subsequent evolution of the flow. However, this is rather a weak constraint on the subsequent evolution of the fluid; it does nothing, for example, to prevent the conversion of all the potential energy to kinetic energy. In the rotating case the potential vorticity is non-zero, and potential vorticity conservation and geostrophic balance are all we need to infer the final steady state, assuming it exists, without solving for the details of the flow evolution, as we now see.

With an initial condition for the height field given by (4.66), the initial potential vorticity is given by

$$q(x, y) = \begin{cases} -f_0\eta_0/H & x < 0 \\ f_0\eta_0/H & x > 0, \end{cases} \tag{4.75}$$

and this remains unchanged throughout the adjustment process. The final steady state is then the solution of the equations

$$\zeta - f_0 \frac{\eta}{H} = q(x, y), \qquad f_0 u = -g \frac{\partial \eta}{\partial y}, \qquad f_0 v = g \frac{\partial \eta}{\partial x}, \tag{4.76a,b,c}$$

where $\zeta = \partial v/\partial x - \partial u/\partial y$. Because the Coriolis parameter is constant, the velocity field is horizontally non-divergent and we may define a streamfunction $\psi = g\eta/f_0$. Equations (4.76) then reduce to

$$\left(\nabla^2 - \frac{1}{L_d^2}\right)\psi = q(x, y), \tag{4.77}$$

where $L_d = \sqrt{gH}/f_0$ is the radius of deformation, as in (4.52), sometimes called the 'Rossby radius'. It is a naturally occurring length scale in problems involving both rotation and gravity, and arises in a slightly different form in stratified fluids.

If an unbalanced flow is left to freely evolve, gravity waves will propagate quickly away from the disturbance. However, the balanced component of the flow is constrained by potential vorticity conservation and it can only evolve advectively, and so much more slowly. Thus, after some time, only the balanced part of the flow remains, with the unbalanced flow having been radiated away and dissipated. This process is *geostrophic adjustment*.

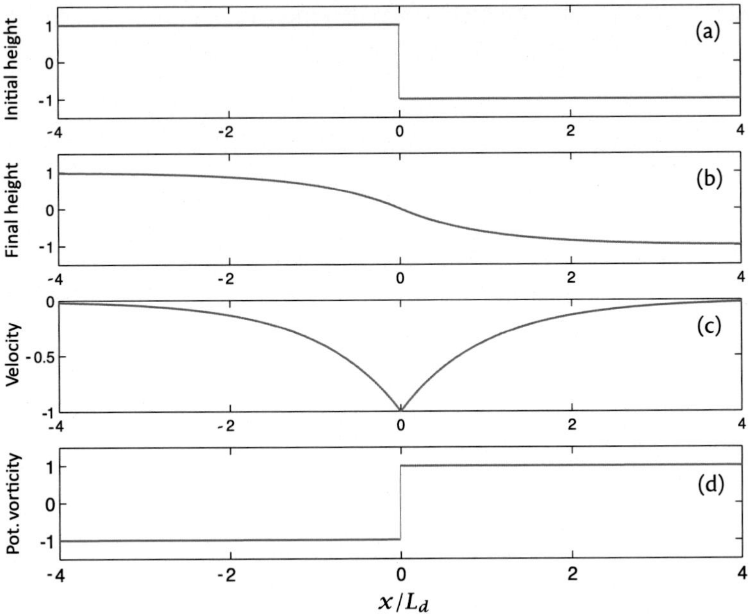

Fig. 4.8: Solutions of a linear geostrophic adjustment problem. (a) Initial height field, given by (4.66) with $\eta_0 = 1$. (b) Equilibrium (final) height field, η given by (4.79) and $\eta = f_0\psi/g$. (c) Equilibrium geostrophic velocity, normal to the gradient of the height field, given by (4.80). (d) Potential vorticity, given by (4.75), and this does not evolve. The distance, x is nondimensionalized by the deformation radius L_d and the velocity by $\eta_0(g/f_0L_d)$. Changes to the initial state occur within $\mathcal{O}(L_d)$ of the initial discontinuity.

The initial conditions (4.75) admit of a nice analytic solution, for the flow will remain uniform in y, and (4.77) reduces to

$$\frac{\partial^2\psi}{\partial x^2} - \frac{1}{L_d^2}\psi = \frac{f_0\eta_0}{H}\mathrm{sgn}(x). \tag{4.78}$$

We solve this separately for $x > 0$ and $x < 0$ and then match the solutions and their first derivatives at $x = 0$, also imposing the condition that the velocity decays to zero as $x \to \pm\infty$. The solution is

$$\psi = \begin{cases} -(g\eta_0/f_0)(1 - \mathrm{e}^{-x/L_d}) & x > 0 \\ +(g\eta_0/f_0)(1 - \mathrm{e}^{x/L_d}) & x < 0. \end{cases} \tag{4.79}$$

The velocity field associated with this is obtained from (4.76b,c), and is

$$u = 0, \qquad v = -\frac{g\eta_0}{f_0 L_d}\mathrm{e}^{-|x|/L_d}. \tag{4.80}$$

The velocity is perpendicular to the slope of the free surface, and a jet forms along the initial discontinuity, as illustrated in Fig. 4.8.

The important point of this problem is that the variations in the height and field are not radiated away to infinity, as in the non-rotating problem. Rather, potential vorticity conservation constrains the influence of the adjustment to within a deformation radius (we see now why this name is appropriate) of the initial disturbance. This property is a general one in geostrophic adjustment — it also arises if the initial condition consists of a velocity jump. The time evolution of the rotating flow, obtained by a numerical integration, is illustrated in the left-hand panels of Fig. 4.7.

Fronts propagate away at a speed $\sqrt{gH} = 1$, just as in the non-rotating case, but in the rotating flow they leave behind a geostrophically balanced state with a non-zero meridional velocity.

4.5 ♦ A VARIATIONAL PERSPECTIVE ON ADJUSTMENT

In the non-rotating problem, all of the initial potential energy is eventually radiated away to infinity. In the rotating problem, the final state contains both potential and kinetic energy, mostly trapped within a deformation radius of the initial disturbance, because potential vorticity conservation on parcels prevents all of the energy being dispersed. This suggests that it may be informative to think of the geostrophic adjustment problem as a *variational problem:* we seek to minimize the energy consistent with the conservation of potential vorticity. We stay in the linear approximation in which, because the advection of potential vorticity is neglected, potential vorticity remains constant at each point.

The energy of the flow, $\widehat{E}$, is given by the sum of potential and kinetic energies, namely

$$\widehat{E} = \int (H\boldsymbol{u}^2 + g\eta^2)\,\mathrm{d}A, \tag{4.81}$$

(where $\mathrm{d}A = \mathrm{d}x\,\mathrm{d}y$) and the potential vorticity field is

$$q = \zeta - f_0\frac{\eta}{H} = (v_x - u_y) - f_0\frac{\eta}{H}, \tag{4.82}$$

where the subscripts x, y denote derivatives. The problem is then to extremize the energy subject to potential vorticity conservation. This is a constrained problem in the calculus of variations, sometimes called an *isoperimetric* problem because of its origins in maximizing the area of a surface for a given perimeter.

The mathematical problem is to extremize the integral

$$I = \int \left\{H(u^2 + v^2) + g\eta^2 + \lambda(x, y)\left[(v_x - u_y) - f_0\eta/H\right]\right\}\,\mathrm{d}A, \tag{4.83}$$

where $\lambda(x, y)$ is a Lagrange multiplier, undetermined at this stage. It is a function of space: if it were a constant, the integral would merely extremize energy subject to a given integral of potential vorticity, and rearrangements of potential vorticity (which here we wish to disallow) would leave the integral unaltered.

As there are three independent variables there are three Euler–Lagrange equations that must be solved in order to minimize I. These are

$$\begin{gathered}
\frac{\partial L}{\partial \eta} - \frac{\partial}{\partial x}\frac{\partial L}{\partial \eta_x} - \frac{\partial}{\partial y}\frac{\partial L}{\partial \eta_y} = 0, \\
\frac{\partial L}{\partial u} - \frac{\partial}{\partial x}\frac{\partial L}{\partial u_x} - \frac{\partial}{\partial y}\frac{\partial L}{\partial u_y} = 0, \qquad \frac{\partial L}{\partial v} - \frac{\partial}{\partial x}\frac{\partial L}{\partial v_x} - \frac{\partial}{\partial y}\frac{\partial L}{\partial v_y} = 0,
\end{gathered} \tag{4.84}$$

where L is the integrand on the right-hand side of (4.83). Substituting the expression for L into (4.84) gives, after a little algebra,

$$2g\eta - \frac{\lambda f_0}{H} = 0, \qquad 2Hu + \frac{\partial \lambda}{\partial y} = 0, \qquad 2Hv - \frac{\partial \lambda}{\partial x} = 0, \tag{4.85}$$

and then eliminating λ gives the simple relationships

$$u = -\frac{g}{f_0}\frac{\partial \eta}{\partial y}, \qquad v = \frac{g}{f_0}\frac{\partial \eta}{\partial x}. \tag{4.86}$$

These are the equations of geostrophic balance! Thus, in the linear approximation, *geostrophic balance is the minimum energy state for a given field of potential vorticity.*

Notes and References

The shallow water equations are sometimes called the Saint-Venant equations after Adhémar Jean Claude Barré de Saint-Venant (1797–1886) who wrote down a form of the equations in 1871. Pierre-Simon Laplace (1749–1825) previously wrote down a linear version of the equations on the sphere, now known as Laplace's tidal equations, in 1776, and one would not be surprised if Euler knew about the equations before that.

Problems

4.1 Using the shallow water equations:

(a) A cylindrical column of air at 30° latitude with radius 100 km expands horizontally (shrinking in depth accordingly) to twice its original radius. If the air is initially at rest, what is the mean tangential velocity at the perimeter after the expansion?

(b) An air column at 60°N with zero relative vorticity ($\zeta = 0$) stretches from the surface to the tropopause, which we assume is a rigid lid, at 10 km. The air column moves zonally on to a plateau 2.5 km high. What is its relative vorticity? Suppose it then moves southwards to 30°N, staying on the plateau. What is its vorticity?

4.2 Show that the vertical velocity within a shallow water system is given by

$$w = \frac{z - \eta_b}{h}\frac{\mathrm{D}h}{\mathrm{D}t} + \frac{\mathrm{D}\eta_b}{\mathrm{D}t}. \tag{P4.1}$$

Interpret this result, showing that it gives sensible answers at the top and bottom of the fluid layer.

4.3 In the long-wave limit of Poincaré waves, fluid parcels behave as free-agents; that is, like free solid particles moving in a rotating frame unencumbered by pressure forces. Why then, is their frequency given by $\omega = f = 2\Omega$ where Ω is the rotation rate of the coordinate system, and not by Ω itself? Do particles that are stationary or move in a straight line in the inertial frame of reference satisfy the dispersion relationship for Poincaré waves in this limit? Explain. [See also Durran (1993) and Egger (1999).]

4.4 Linearize the f-plane shallow water system about a state of rest. Suppose that there is an initial disturbance given in the general form

$$\eta = \iint \widetilde{\eta}_{k,l} e^{i(kx+ly)} \, dk \, dl, \qquad \text{(P4.2)}$$

where η is the deviation surface height and the Fourier coefficients $\widetilde{\eta}_{k,l}$ are given, and that the initial velocity is zero.

(a) Obtain the geopotential field at the completion of geostrophic adjustment, and show that the deformation scale is a natural length scale in the problem.

(b) Show that the change in total energy during the adjustment is always less than or equal to zero. Neglect any initial divergence.
N.B. Because the problem is linear, the Fourier modes do not interact.

4.5 If energy conservation is one of the most basic physical laws, how can energy be lost in geostrophic adjustment?

4.6 *Geostrophic adjustment of a velocity jump.* In which we consider the evolution of the linearized f-plane shallow water equations in an infinite domain.

(a) Show that the *linearized* potential vorticity, q', for the shallow water system is given by

$$q' = \zeta' - f_0 \frac{h'}{H}, \qquad \text{(P4.3)}$$

using standard notation.

(b) If the flow is in geostrophic balance show that the relative vorticity is given by

$$\zeta' = \nabla^2 \psi,$$

where $\psi = gh'/f_0$. Hence show that the potential vorticity is then given by

$$q' = \nabla^2 \psi - \frac{1}{L_d^2}\psi,$$

and write down an expression for L_d.

(c) Suppose that initially the fluid surface is flat, the zonal velocity is zero and the meridional velocity is given by

$$v(x) = v_0 \, \mathrm{sgn}(x). \qquad \text{(P4.4)}$$

Find the equilibrium height and velocity fields at $t = \infty$.

(d) What are the initial and final kinetic and potential energies?

Partial solution:
The potential vorticity is $q = \zeta - f_0\eta/H$, so that the initial state and final state are both given by

$$q = 2v_0 \, \delta(x). \qquad \text{(P4.5)}$$

(Why?) The final state streamfunction is thus given by $(\partial^2/\partial x^2 - L_d^{-2})\,\psi = q$, with solution $\psi = \psi_0 \exp(x/L_d)$ and $\psi = \psi_0 \exp(-x/L_d)$ for $x < 0$ and $x > 0$, where $\psi_0 = -L_d v_0$ (why?), and $\eta = f_0\psi/g$. The energy is $E = \int (Hv^2 + g\eta^2)/2 \, dx$. The initial KE is infinite, the initial PE is zero, and the final state has $PE = KE = gL_d\eta_0^2/4$; that is, the energy is equipartitioned between kinetic and potential energies.

4.7 In the shallow water equations show that, if the flow is approximately geostrophically balanced, the energy at large scales is predominantly potential energy and the energy at small scales is predominantly kinetic energy. Define precisely what 'large scale' and 'small scale' mean in this context.

4.8 In the shallow water geostrophic adjustment problem, show that at large scales the velocity essentially adjusts to the height field, and that at small scales the height field essentially adjusts to the velocity field. Your derivation may be detailed and mathematical, but explain the result at the end in words and in physical terms.

CHAPTER

5

Geostrophic Theory

GEOSTROPHIC AND HYDROSTATIC BALANCE are the two dominant balances in meteorology and oceanography and in this chapter we exploit these balances to derive various simplified sets of equation. The 'problem' with the full equations is that they are *too* complete, and they contain motions that we don't always care about — sound waves and gravity waves for example. If we can eliminate these modes from the outset then our path toward understanding is not littered with obstacles.

Our specific goal is to derive various sets of 'geostrophic equations', in particular the planetary-geostrophic and quasi-geostrophic equations, by making use of the fact that geostrophic and hydrostatic balance are closely satisfied. We do this first for the shallow water equations and then for the stratified, three-dimensional equations. We will use the Boussinesq equations, but a treatment in pressure coordinates would be very similar. The bottom topography, η_B, can be an unneeded complication in the derivations below and readers may wish to simplify by setting $\eta_B = 0$.

5.1 Scaling the Shallow Water Equations

In order to simplify the equations of motion we first *scale* them — we choose the scales we wish to describe, and then determine the approximate sizes of the terms in the equations. We then eliminate the small terms and derive a set of equations that is simpler than the original set but that consistently describes motion of the chosen scale. With the odd exception, we will denote the scales of variables by capital letters; thus, if L is a typical length scale of the motion we wish to describe, and U is a typical velocity scale, then

$$\begin{aligned} (x, y) &\sim L \qquad \text{or} \qquad (x, y) = \mathcal{O}(L), \\ (u, v) &\sim U \qquad \text{or} \qquad (u, v) = \mathcal{O}(U), \end{aligned} \tag{5.1}$$

and similarly for the other variables in the equations.

We then write the equations of motion in a nondimensional form by writing the variables as

$$(x, y) = L(\widehat{x}, \widehat{y}), \qquad (u, v) = U(\widehat{u}, \widehat{v}), \tag{5.2}$$

where the hatted variables are nondimensional and, by supposition, are $\mathcal{O}(1)$. The various terms in the momentum equation then scale as:

$$\frac{\partial \boldsymbol{u}}{\partial t} + \boldsymbol{u} \cdot \nabla \boldsymbol{u} + \boldsymbol{f} \times \boldsymbol{u} = -g \nabla \eta, \tag{5.3a}$$

$$\frac{U}{T} \qquad \frac{U^2}{L} \qquad fU \sim g\frac{\mathcal{H}}{L}, \tag{5.3b}$$

where the ∇ operator acts in the x–y plane and $\mathcal{H}$ is the amplitude of the variations in the surface displacement. We choose an 'advective scale' for time, meaning that $T = L/U$ and $t = \hat{t}L/U$, and the time derivative then scales the same way as the advection. The ratio of the advective term to the rotational term in the momentum equation (5.3) is $(U^2/L)/(fU) = U/fL$; this is the Rossby number that we previously encountered.

The geostrophic theory of this chapter applies when the Rossby number is small. On Earth the theory is generally appropriate for large-scale flow in the mid- and high latitude atmosphere and ocean. On other planets the applicability of geostrophic theory depends on how rapidly the planet rotates and how big it is. Venus has a rotation rate some 200 times slower than Earth and the Rossby number of the large-scale circulation is quite large. Jupiter rotates much faster than Earth (a Jupiter day is about 10 hours), and the planet is also much bigger, and the Rossby number remains small even close to the equator.

We are interested in flows for which the Rossby number is small, in which case the Coriolis term is largely balanced by the pressure gradient. From (5.3b), variations in η scale according to

$$\mathcal{H} = \frac{fUL}{g} = Ro\frac{f^2L^2}{g} = RoH\frac{L^2}{L_d^2}, \tag{5.4}$$

where $L_d = \sqrt{gH}/f$ is the deformation radius and H is the mean depth of the fluid. The ratio of variations in fluid height to the total fluid height thus scales as

$$\frac{\mathcal{H}}{H} \sim Ro\frac{L^2}{L_d^2}. \tag{5.5}$$

Now, the thickness of the fluid, h, may be written as the sum of its mean and a deviation, h_D

$$h = H + h_D = H + (\eta_T - \eta_B), \tag{5.6}$$

where, referring to Fig. 4.1, η_B is the height of the bottom topography and η_T is the height of the fluid above its mean value. Given the scalings above, the deviation height of the fluid may be written as

$$\eta_T = Ro\frac{L^2}{L_d^2}H\widehat{\eta}_T \quad \text{and} \quad \eta = H + \eta_T = H\left(1 + Ro\frac{L^2}{L_d^2}\widehat{\eta}_T\right), \tag{5.7}$$

where $\widehat{\eta}_T$ is the $\mathcal{O}(1)$ nondimensional value of the surface height deviation. We apply the same scalings to h itself and, if $h_D = h - H = \eta_T - \eta_B$ is the deviation of the thickness from its mean value, then

$$h = H + h_D = H\left(1 + Ro\frac{L^2}{L_d^2}\widehat{h}_D\right), \tag{5.8}$$

where $\widehat{h}_D$ is the nondimensional deviation thickness of the fluid layer.

Nondimensional momentum equation

If we use (5.7) and (5.8) to scale height variations, (5.2) to scale lengths and velocities, and an advective scaling for time, then, and since $\nabla\hat{\eta} = \nabla\hat{\eta}_T$, the momentum equation (5.3) becomes

$$Ro\left[\frac{\partial \hat{\boldsymbol{u}}}{\partial \hat{t}} + (\hat{\boldsymbol{u}}\cdot\nabla)\hat{\boldsymbol{u}}\right] + \hat{\boldsymbol{f}}\times\hat{\boldsymbol{u}} = -\nabla\hat{\eta}, \tag{5.9}$$

where $\hat{\boldsymbol{f}} = \hat{\mathbf{k}}\hat{f} = \hat{\mathbf{k}}f/f_0$, where f_0 is a representative value of the Coriolis parameter. (If f is a constant, then $\hat{f} = 1$, but it is informative to explicitly write $\hat{f}$ in the equations. Also, where the operator ∇ operates on a nondimensional variable, the differentials are taken with respect to the nondimensional variables $\hat{x}, \hat{y}$.) All the variables in (5.9) will now be assumed to be of order unity, and the Rossby number multiplying the local time derivative and the advective terms indicates the smallness of those terms. By construction, the dominant balance in (5.9) is the geostrophic balance between the last two terms.

Nondimensional mass continuity (height) equation

The (dimensional) mass continuity equation can be written as

$$\frac{\mathrm{D}h}{\mathrm{D}t} + h\nabla\cdot\boldsymbol{u} = 0 \quad \text{or} \quad \frac{1}{H}\frac{\mathrm{D}h_D}{\mathrm{D}t} + \left(1 + \frac{h_D}{H}\right)\nabla\cdot\boldsymbol{u} = 0, \tag{5.10}$$

since $\mathrm{D}h/\mathrm{D}t = \mathrm{D}h_D/\mathrm{D}t$. Using (5.2) and (5.8) the above equation may be written

$$Ro\left(\frac{L}{L_d}\right)^2\frac{\mathrm{D}\hat{h}_D}{\mathrm{D}\hat{t}} + \left[1 + Ro\left(\frac{L}{L_d}\right)^2\hat{h}_D\right]\nabla\cdot\hat{\boldsymbol{u}} = 0. \tag{5.11}$$

Equations (5.9) and (5.11) are the nondimensional versions of the full shallow water equations of motion. Since the Rossby number is small we might expect that some terms in this equation can be eliminated with little loss of accuracy, depending on the size of the second nondimensional parameter, $(L/L_d)^2$, as we now explore.

5.2 GEOSTROPHIC SHALLOW WATER EQUATIONS

5.2.1 Planetary-Geostrophic Equations

We now derive simplified equation sets that are appropriate in particular parameter regimes, beginning with an equation set appropriate for the very largest scales. Specifically, we take

$$Ro \ll 1, \qquad \frac{L}{L_d} \gg 1 \quad \text{such that} \quad Ro\left(\frac{L}{L_d}\right)^2 = \mathcal{O}(1). \tag{5.12}$$

The first inequality implies we are considering flows in geostrophic balance. The second inequality means we are considering flows much larger than the deformation radius. The ratio of the deformation radius to scale of motion of the fluid is called the Burger number; that is, $Bu \equiv L_d/L$, so here we are considering small Burger-number flows.

The smallness of the Rossby number means that we can neglect the material derivative in the momentum equation, (5.9), leaving geostrophic balance. Thus, in dimensional form, the momentum equation may be written, in vectorial or component forms, as

$$\boldsymbol{f} \times \boldsymbol{u} = -\nabla \eta,$$
or
$$fv = g\frac{\partial \eta}{\partial x}, \qquad fu = -g\frac{\partial \eta}{\partial y}. \tag{5.13}$$

The *planetary-geostrophic* equations are appropriate for geostrophically balanced flow at very large scales. In the shallow water version, they consist of the full mass conservation equation along with geostrophic balance.

Looking now at the mass continuity equation, (5.11), we see that there are no small terms that can be eliminated. Thus, we have simply the full mass conservation equation,

$$\frac{\partial h}{\partial t} + \nabla \cdot (h\boldsymbol{u}) = 0, \tag{5.14}$$

where h and η are related by $\eta = \eta_B + h$, where η_B is the height of the bottom topography. Equations (5.13) and (5.14) form the *planetary geostrophic shallow water equations*. There is *only one time derivative* in the equations, so there can be no gravity waves. The system is evolved purely through the mass continuity equation, and the flow field is *diagnosed* from the height field.

Planetary-geostrophic potential vorticity

In the (full) shallow water equations potential vorticity is conserved, meaning that

$$\frac{\mathrm{D}}{\mathrm{D}t}\left(\frac{\zeta + f}{h}\right) = 0. \tag{5.15}$$

In the planetary-geostrophic equations we can use (5.13) and (5.14) to show that this conservation law becomes

$$\frac{\mathrm{D}}{\mathrm{D}t}\left(\frac{f}{h}\right) = 0, \tag{5.16}$$

as might be expected since ζ is smaller than f by a factor of the Rossby number. An alternate derivation of the planetary-geostrophic equations is to go directly from (5.15) to (5.16), by virtue of the smallness of the Rossby number, and then simply use (5.16) instead of (5.14) as the evolution equation.

5.2.2 Quasi-Geostrophic Equations

The *quasi-geostrophic equations* are appropriate for scales of the same order as the deformation radius, and so for

$$Ro \ll 1, \quad \frac{L}{L_d} = \mathcal{O}(1) \quad \text{so that} \quad Ro\left(\frac{L}{L_d}\right)^2 \ll 1. \tag{5.17}$$

Since the Rossby number is small the momentum equations again reduce to geostrophic balance, namely (5.13). In the mass continuity equation, we now eliminate all terms involving Rossby number to give

$$\nabla \cdot \boldsymbol{u} = 0. \tag{5.18}$$

Neither geostrophic balance nor (5.18) are prognostic equations, and it appears we have derived an uninteresting, static, set of equations. In fact we haven't gone far enough, since nothing in our derivation says that these quantities do not evolve. To see this, let us suppose that the Coriolis parameter is nearly constant, which is physically consistent with the idea that scales of motion are comparable to the deformation scale. Geostrophic balance with a constant Coriolis parameter gives

$$f_0 u = -g\frac{\partial \eta}{\partial y}, \quad -f_0 v = -g\frac{\partial \eta}{\partial z}, \quad \text{giving} \quad \nabla \cdot \boldsymbol{u} = 0. \tag{5.19}$$

That it to say, the geostrophic flow is divergence-free and we therefore should not suppose that $\nabla \cdot \boldsymbol{u} = 0$ is the dominant term in the height equation.

However, with a little more care we can in fact derive a set of equations that evolves under these conditions, and that furthermore is extraordinarily useful, for it describes the flow on the scales of motion corresponding to weather. We make three explicit assumptions:

(i) The Rossby number is small and the flow is in near geostrophic balance.

(ii) The scales of motion are similar to the deformation scales, so that $L \sim L_d$ and $Ro(L/L_d)^2 \ll 1$.

(iii) Variations of the Coriolis parameter are small, so that $f = f_0 + \beta y$ where $\beta y \ll f_0$.

The velocity is then equal to a geostrophic component, $\boldsymbol{u}_g$ plus an ageostrophic component, $\boldsymbol{u}_a$ where $|\boldsymbol{u}_g| \gg |\boldsymbol{u}_a|$ and the geostrophic velocity satisfies

$$f_0 \times \boldsymbol{u}_g = -g\nabla\eta, \tag{5.20}$$

which, because of the use of a constant Coriolis parameter (assumption *(iii)*), implies $\nabla \cdot \boldsymbol{u}_g = 0$.

We proceed from the shallow water vorticity equation which, as in (4.32), is

$$\frac{\partial \zeta}{\partial t} + (\boldsymbol{u} \cdot \nabla)(\zeta + f) = -(\zeta + f)\nabla \cdot \boldsymbol{u}. \tag{5.21}$$

The quasi-geostrophic equations are appropriate for geostrophically balanced flow at so-called synoptic scales, or weather scales. This scale is mainly determined by the Rossby radius of deformation which is about 1000 km in the atmosphere and 100 km (and less in high latitudes) in the ocean.

A consequence of (5.20) is that the right-hand side contains only the ageostrophic velocity, which is small, and since ζ is smaller than f by a factor of the Rossby number we can ignore $\zeta \nabla \cdot \boldsymbol{u}$ and take f to be equal to f_0. On the left-hand side the velocities are well-approximated by using the geostrophic flow, so that we have

$$\frac{\partial \zeta_g}{\partial t} + (\boldsymbol{u}_g \cdot \nabla)(\zeta_g + f) = -f_0 \nabla \cdot \boldsymbol{u}_a, \tag{5.22}$$

where on the left-hand side f can be replaced by βy.

We now use the mass continuity equation to obtain an expression for the divergence. From (5.10) the mass continuity equation is

$$\frac{\mathrm{D}h_D}{\mathrm{D}t} + (H + h_D)\nabla \cdot \boldsymbol{u} = 0, \tag{5.23}$$

and since $H \gg h_D$ (using (5.11), H is bigger by a factor $(L_d/L)^2 Ro^{-1}$), the equation becomes

$$\frac{\mathrm{D}}{\mathrm{D}t}(\eta - \eta_B) + H\nabla \cdot \boldsymbol{u}_a = 0. \tag{5.24}$$

Combining (5.22) and (5.23) gives

$$\frac{\mathrm{D}}{\mathrm{D}t}\left(\zeta_g + f - \frac{f_0(\eta - \eta_B)}{H}\right) = 0. \tag{5.25}$$

It appears that we have two dynamical variables here, ζ_g and η. However, they are related through geostrophic balance, and the fact that the geostrophic flow is non-divergent. Thus, we may define a streamfunction ψ such that $u_g = -\partial\psi/\partial y, v_g = \partial\psi/\partial x$, whence $\partial u/\partial x + \partial v/\partial y = 0$. The vorticity and height fields are related to the streamfunction by

$$\zeta_g = \frac{\partial v}{\partial x} - \frac{\partial u}{\partial y} = \nabla^2\psi, \qquad \text{and} \qquad \eta = \frac{f_0\psi}{g}, \tag{5.26a,b}$$

where the second relation comes from geostrophic balance. Equation (5.25) may then be written as

$$\frac{\mathrm{D}q}{\mathrm{D}t} = 0, \qquad q = \nabla^2\psi + \beta y - \frac{\psi}{L_d^2} + \frac{f_0\eta_B}{H}, \tag{5.27}$$

where $L_d = \sqrt{gH}/f_0$ and the variable q is the *quasi-geostrophic potential vorticity*.

5.2.3 Quasi-Geostrophic Potential Vorticity

Connection to shallow water potential vorticity

The quantity q given by (5.27) is an approximation (except for dynamically unimportant constant additive and multiplicative factors) to the shallow

Both the planetary-geostrophic and the quasi-geostrophic equations can be written in the form of an evolution equation for potential vorticity, along with an 'inversion' to determine the velocity fields and the height field. The difference in the two equation sets lies in the approximations made in the inversion. In the planetary-geostrophic system relative vorticity is ignored and $Q = f/h$. In quasi-geostrophy the variations in the height field are small and $q = \beta y + \zeta - f_0\eta_T/H$. Both equation sets assume a low Rossby number.

water potential vorticity. To see the truth of this statement let us begin with the expression for the shallow water potential vorticity,

$$Q = \frac{f + \zeta}{h}. \tag{5.28}$$

For simplicity we set bottom topography to zero and then $h = H(1+\eta_T/H)$ and assume that η_T/H is small to obtain

$$Q = \frac{f + \zeta}{H(1 + \eta_T/H)} \approx \frac{1}{H}(f + \zeta)\left(1 - \frac{\eta_T}{H}\right) \approx \frac{1}{H}\left(f_0 + \beta y + \zeta - f_0\frac{\eta_T}{H}\right). \tag{5.29}$$

Because f_0/H is a constant it has no effect in the evolution equation, and the quantity given by

$$q = \beta y + \zeta - f_0\frac{\eta_T}{H} \tag{5.30}$$

is materially conserved. Using geostrophic balance we have $\zeta = \nabla^2\psi$ and $\eta_T = f_0\psi/g$ so that (5.30) is identical (except for η_B) to (5.27).

The approximations needed to go from (5.28) to (5.30) are the same as those used in our earlier, more long-winded, derivation of the quasi-geostrophic equations. That is, we assumed that f itself is nearly constant, and that f_0 is much larger than ζ, equivalent to a low Rossby number assumption. It was also necessary to assume that $H \gg \eta_T$ to enable the expansion of the height field and this approximation is equivalent to requiring that the scale of motion not be significantly larger than the deformation scale. The derivation is completed by noting that the advection of the potential vorticity should be by the geostrophic velocity alone, and we recover (5.27).

5.3 Scaling in the Continuously-Stratified System

We now apply the same scaling ideas, *mutatis mutandis,* to the stratified primitive equations. We use the hydrostatic Boussinesq equations, which we write as

$$\frac{\mathrm{D}\boldsymbol{u}}{\mathrm{D}t} + \boldsymbol{f} \times \boldsymbol{u} = -\nabla_z\phi, \tag{5.31a}$$

$$\frac{\partial \phi}{\partial z} = b, \tag{5.31b}$$

$$\frac{\mathrm{D}b}{\mathrm{D}t} = 0, \tag{5.31c}$$

$$\nabla \cdot \boldsymbol{v} = 0. \tag{5.31d}$$

We will consider only the case of a flat bottom, but topography is a relatively straightforward extension. Anticipating that the average stratification may not scale in the same way as the deviation from it, let us separate out the contribution of the advection of a reference stratification in (5.31c) by writing

$$b = \tilde{b}(z) + b'(x, y, z, t). \tag{5.32}$$

The thermodynamic equation then becomes

$$\frac{Db'}{Dt} + N^2 w = 0, \tag{5.33}$$

where $N^2 = \partial\tilde{b}/\partial z$ and the advective derivative is still three-dimensional. We then let $\phi = \tilde{\phi}(z) + \phi'$, where $\tilde{\phi}$ is hydrostatically balanced by $\tilde{b}$, and the hydrostatic equation becomes

$$\frac{\partial \phi'}{\partial z} = b'. \tag{5.34}$$

Equations (5.33) and (5.34) replace (5.31c) and (5.31b), and ϕ' is used in (5.31a).

5.3.1 Scaling the Equations

We scale the basic variables by supposing that

$$(x, y) \sim L, \quad (u, v) \sim U, \quad t \sim \frac{L}{U}, \quad z \sim H, \quad f \sim f_0, \quad N \sim N_0, \tag{5.35}$$

where the scaling variables (capitalized, except for f_0) are chosen to be such that the nondimensional variables have magnitudes of the order of unity, and the constant N_0 is a representative value of the stratification. We presume that the scales chosen are such that the Rossby number is small; that is $Ro = U/(f_0 L) \ll 1$. In the momentum equation the pressure term then balances the Coriolis force,

$$|\boldsymbol{f} \times \boldsymbol{u}| \sim |\nabla \phi'|, \tag{5.36}$$

and so the pressure scales as

$$\phi' \sim \Phi = f_o U L. \tag{5.37}$$

Using the hydrostatic relation, (5.37) implies that the buoyancy scales as

$$b' \sim B = \frac{f_0 U L}{H}, \tag{5.38}$$

and from this we obtain

$$\frac{(\partial b'/\partial z)}{N^2} \sim Ro \frac{L^2}{L_d^2}, \tag{5.39}$$

where

$$L_d = \frac{N_0 H}{f_0} \tag{5.40}$$

is the deformation radius in the continuously-stratified fluid, analogous to the quantity $\sqrt{gH}/f_0$ in the shallow water system, and we use the same symbol for both. In the continuously-stratified system, *if the scale of motion is the same as or smaller than the deformation radius, and the Rossby number is small, then the variations in stratification are small.* The choice of scale is the key difference between the planetary-geostrophic and quasi-geostrophic equations.

Finally, at least for now, we nondimensionalize the vertical velocity by using the mass conservation equation,

$$\frac{\partial w}{\partial z} = -\left(\frac{\partial u}{\partial x} + \frac{\partial v}{\partial y}\right), \tag{5.41}$$

with the scaling

$$w \sim W = \frac{UH}{L}. \tag{5.42}$$

This scaling will not necessarily be correct if the flow is geostrophically balanced. In this case we can then estimate w by cross-differentiating geostrophic balance (with $\widetilde{\rho}$ constant for simplicity) to obtain the linear geostrophic vorticity equation and corresponding scaling:

$$\beta v \approx f\frac{\partial w}{\partial z}, \qquad w \sim W = \frac{\beta UH}{f_0}. \tag{5.43a,b}$$

If variations in the Coriolis parameter are large and $\beta \sim f_0/L$, then (5.43b) is the same as (5.42), but if f is nearly constant then $W \ll UH/L$.

Given the scalings above (using (5.42) for w) we nondimensionalize by setting

$$\begin{gathered}(\widehat{x}, \widehat{y}) = L^{-1}(x, y), \qquad \widehat{z} = H^{-1}z, \quad (\widehat{u}, \widehat{v}) = U^{-1}(u, v), \quad \widehat{t} = \frac{U}{L}t, \\ \widehat{w} = \frac{L}{UH}w, \quad \widehat{f} = f_0^{-1}f, \quad \widehat{\phi} = \frac{\phi'}{f_0UL}, \quad \widehat{b} = \frac{H}{f_0UL}b',\end{gathered} \tag{5.44}$$

where the hatted variables are nondimensional. The horizontal momentum and hydrostatic equations then become

$$Ro\frac{\mathrm{D}\widehat{\boldsymbol{u}}}{\mathrm{D}\widehat{t}} + \widehat{\boldsymbol{f}} \times \widehat{\boldsymbol{u}} = -\nabla\widehat{\phi}, \tag{5.45}$$

and

$$\frac{\partial \widehat{\phi}}{\partial \widehat{z}} = \widehat{b}. \tag{5.46}$$

The nondimensional mass conservation equation is simply

$$\nabla \cdot \widehat{\boldsymbol{v}} = \left(\frac{\partial \widehat{u}}{\partial \widehat{x}} + \frac{\partial \widehat{v}}{\partial \widehat{y}} + \frac{\partial \widehat{w}}{\partial \widehat{z}}\right) = 0, \tag{5.47}$$

and the nondimensional thermodynamic equation is

$$\frac{f_0UL}{H}\frac{U}{L}\frac{\mathrm{D}\widehat{b}}{\mathrm{D}\widehat{t}} + \widehat{N}^2N_0^2\frac{HU}{L}\widehat{w} = 0, \tag{5.48}$$

or, re-arranging,

$$Ro\frac{\mathrm{D}\widehat{b}}{\mathrm{D}\widehat{t}} + \left(\frac{L_d}{L}\right)^2\widehat{N}^2\widehat{w} = 0. \tag{5.49}$$

The nondimensional equations are summarized in the box on the following page.

Nondimensional Boussinesq Primitive Equations

The nondimensional, hydrostatic, Boussinesq equations in a rotating frame of reference are:

Horizontal momentum: $$Ro\frac{\mathrm{D}\widehat{\boldsymbol{u}}}{\mathrm{D}\hat{t}} + \widehat{\boldsymbol{f}} \times \widehat{\boldsymbol{u}} = -\nabla\widehat{\phi}, \tag{PE.1}$$

Hydrostatic: $$\frac{\partial\widehat{\phi}}{\partial\widehat{z}} = \hat{b}, \tag{PE.2}$$

Mass continuity: $$\frac{\partial\widehat{u}}{\partial\widehat{x}} + \frac{\partial\widehat{v}}{\partial\widehat{y}} + \frac{\partial\widehat{w}}{\partial\widehat{z}} = 0, \tag{PE.3}$$

Thermodynamic: $$Ro\frac{\mathrm{D}\hat{b}}{\mathrm{D}\hat{t}} + \left(\frac{L_d}{L}\right)^2 \widehat{N}^2\widehat{w} = 0. \tag{PE.4}$$

5.4 PLANETARY-GEOSTROPHIC EQUATIONS FOR STRATIFIED FLUIDS

The *planetary-geostrophic equations* are appropriate for geostrophic flow at large horizontal scales. The specific assumptions we apply to the primitive equations are that:

(*i*) $Ro \ll 1$,

(*ii*) $(L_d/L)^2 \ll 1$. More specifically, $Ro(L/L_d)^2 = \mathcal{O}(1)$.

We also allow f to have its full variation with latitude, although we still use Cartesian coordinates. If we look at the equations in the shaded box above we see that, if we are to retain only the dominant terms, the momentum equation should simply be replaced by geostrophic balance, whereas hydrostatic and mass continuity equations are unaltered.

In the thermodynamic equation both terms are of order Rossby number, and therefore we retain both. This circumstance arises because, by assumption, we are dealing with large scales and on these scales the perturbation buoyancy varies as much as the mean buoyancy itself. The buoyancy equation reverts to the evolution of total buoyancy, namely

$$\frac{\mathrm{D}b}{\mathrm{D}t} = 0, \tag{5.50}$$

where the material derivative is fully three-dimensional. This is the only evolution equation in the system. Thus, to summarize, the *planetary-geostrophic* equations of motion are, in dimensional form,

$$\frac{\mathrm{D}b}{\mathrm{D}t} = 0, \qquad \boldsymbol{f} \times \boldsymbol{u} = -\nabla\phi, \qquad \frac{\partial\phi}{\partial z} = b', \qquad \nabla\cdot\boldsymbol{v} = 0. \tag{5.51}$$

Potential vorticity

Manipulation of (5.51) reveals that we can equivalently write the equations as an evolution equation for potential vorticity. Thus, the evolution equations may be written as

$$\frac{DQ}{Dt} = 0, \qquad Q = f\frac{\partial b}{\partial z}. \tag{5.52}$$

The inversion — i.e., the diagnosis of velocity, pressure and buoyancy — is carried out using the hydrostatic, geostrophic and mass conservation equations, and in a real situation the right-hand sides of the buoyancy and potential vorticity equations would have terms representing heating.

5.4.1 Applicability to the Ocean and Atmosphere

In the atmosphere a typical deformation radius NH/f is about 1000 km. The constraint that the scale of motion be much larger than the deformation radius is thus quite hard to satisfy, since one quickly runs out of room on a planet whose equator-to-pole distance is 10 000 km. Thus, only the largest scales of motion can satisfy the planetary-geostrophic scaling in the atmosphere and we should then also properly write the equations in spherical coordinates. In the ocean the deformation radius is about 100 km, so there is lots of room for the planetary-geostrophic equations to hold, and indeed much of the theory of the large-scale structure of the ocean involves the planetary-geostrophic equations.

5.5 Quasi-Geostrophic Equations for Stratified Fluids

Let us now consider the appropriate equations for geostrophic flow at scales of order the deformation radius.

5.5.1 Scaling and Assumptions

The nondimensionalization and scaling are as before and so the nondimensional equations are those in the shaded box on the facing page. The Coriolis parameter is given by

$$\boldsymbol{f} = (f_0 + \beta y)\,\hat{\mathbf{k}}. \tag{5.53}$$

The *variation* of the Coriolis parameter is now assumed to be small (this is a key difference between the quasi-geostrophic system and the planetary-geostrophic system), and in particular we assume that βy is approximately the size of the relative vorticity, and so is much smaller than f_0. The main assumptions needed to derive the QG system are then:

(i) The Rossby number is small, $Ro \ll 1$.

(ii) Length scales are of the same order as the deformation radius, $L \sim L_d$ or $L/L_d = \mathcal{O}(1)$.

(iii) Variations in Coriolis parameter are small, and specifically $|\beta y| \sim Ro\, f_0$.

Given these assumptions, we can write the horizontal velocity as the sum of a geostrophic component and an ageostrophic one:

$$\boldsymbol{u} = \boldsymbol{u}_g + \boldsymbol{u}_a, \quad \text{where} \quad f_0 \hat{\mathbf{k}} \times \boldsymbol{u}_g = -\nabla\phi \quad \text{and} \quad |\boldsymbol{u}_g| \gg |\boldsymbol{u}_a|. \tag{5.54}$$

Since the Coriolis parameter is constant in our definition of the geostrophic velocity the geostrophic divergence is zero; that is

$$\frac{\partial u_g}{\partial x} + \frac{\partial v_g}{\partial y} = 0. \tag{5.55}$$

The vertical velocity is thus given by the divergence of the *ageostrophic* velocity,

$$\frac{\partial w}{\partial z} = -\frac{\partial u_a}{\partial x} + \frac{\partial v_a}{\partial y}. \tag{5.56}$$

Since the ageostrophic velocity is small, the actual vertical velocity is smaller than the scaling suggested by the mass conservation equation in its original form. That is,

$$W \ll \frac{UH}{L}. \tag{5.57}$$

The quasi-geostrophic equations are probably the most used set of equations in theoretical meteorology, certainly for midlatitude dynamics.

5.5.2 Derivation of Stratified QG Equations

We begin by cross-differentiating the horizontal momentum equation to give, after a few lines of algebra, the vorticity equation:

$$\frac{\mathrm{D}}{\mathrm{D}t}(\zeta + f) = -(\zeta + f)\left(\frac{\partial u}{\partial x} + \frac{\partial v}{\partial y}\right) + \left(\frac{\partial u}{\partial z}\frac{\partial w}{\partial y} - \frac{\partial v}{\partial z}\frac{\partial w}{\partial x}\right). \tag{5.58}$$

We now apply the quasi-geostrophic assumption, namely:

(i) The geostrophic velocity and vorticity are much larger than their ageostrophic counterparts, and therefore we use geostrophic values for the terms on the left-hand side.

(ii) On the right-hand side we keep the horizontal divergence (which is small) only where it is multiplied by the big term f. Furthermore, because f is nearly constant we replace it with f_0.

(iii) The second term in large parentheses on the right-hand side is smaller than the advection terms on the left-hand side by the ratio $[UW/(HL)]/[U^2/L^2] = [W/H]/[U/L] \ll 1$, because w is small, as noted above. We thus neglect it.

Given the above, (5.58) becomes

$$\frac{\mathrm{D}_g}{\mathrm{D}t}(\zeta_g + f) = -f_0\left(\frac{\partial u}{\partial x} + \frac{\partial v}{\partial y}\right) = f_0\frac{\partial w}{\partial z}, \tag{5.59}$$

where the second equality uses mass continuity and $\mathrm{D}_g/\mathrm{D}t = \partial/\partial t + \boldsymbol{u}_g \cdot \nabla$ — note that only the (horizontal) geostrophic velocity does any advecting, because w is so small.

Now consider the three-dimensional thermodynamic equation. Since w is small it only advects the basic state, and the perturbation buoyancy is advected only by the geostrophic velocity. Thus, (5.33) becomes,

$$\frac{\mathrm{D}_g b'}{\mathrm{D}t} + wN^2 = 0. \tag{5.60}$$

We now eliminate w between (5.59) and (5.60), which, after a little algebra, yields

$$\frac{\mathrm{D}_g q}{\mathrm{D}t} = 0, \qquad q = \zeta_g + f + \frac{\partial}{\partial z}\left(\frac{f_0 b'}{N^2}\right). \tag{5.61}$$

Hydrostatic and geostrophic wind balance enable us to write the geostrophic velocity, vorticity, and buoyancy in terms of streamfunction ψ [$= p/(f_0\rho_0)$]:

$$\boldsymbol{u}_g = \hat{\mathbf{k}} \times \nabla\psi, \qquad \zeta_g = \nabla^2\psi, \qquad b' = f_0 \partial\psi/\partial z. \tag{5.62}$$

Thus, we have, now omitting the subscript g,

$$\frac{\mathrm{D}q}{\mathrm{D}t} = 0, \qquad q = \nabla^2\psi + f + f_0^2 \frac{\partial}{\partial z}\left(\frac{1}{N^2}\frac{\partial \psi}{\partial z}\right). \tag{5.63a,b}$$

Only the variable part of f (i.e., βy) is relevant in the second term on the right-hand side of the expression for q. The material derivative may be expressed as

$$\frac{\mathrm{D}q}{\mathrm{D}t} = \frac{\partial q}{\partial t} + \boldsymbol{u} \cdot \nabla q = \frac{\partial q}{\partial t} + J(\psi, q). \tag{5.64}$$

where $J(\psi, q) = \partial\psi/\partial x \, \partial q/\partial y - \partial\psi/\partial y \, \partial q/\partial x$.

The quantity q is known as the *quasi-geostrophic potential vorticity,* and it is conserved when advected by the *horizontal* geostrophic flow. All the other dynamical variables may be obtained from potential vorticity as follows:

(i) Streamfunction, using (5.63b).

(ii) Velocity: $\boldsymbol{u} = \hat{\mathbf{k}} \times \nabla\psi$ [$\equiv \nabla^\perp\psi = -\nabla \times (\hat{\mathbf{k}}\psi)$].

(iii) Relative vorticity: $\zeta = \nabla^2\psi$.

(iv) Perturbation pressure: $\phi = f_0\psi$.

(v) Perturbation buoyancy: $b' = f_0 \partial\psi/\partial z$.

By inspection of (5.63b) we see that a length scale $L_d = NH/f_0$, emerges naturally from the quasi-geostrophic dynamics. It is the scale at which buoyancy and relative vorticity effects contribute equally to the potential vorticity, and is called the *deformation radius;* it is analogous to the quantity $\sqrt{gH}/f_0$ arising in shallow water theory. In the upper ocean, with $N \approx 10^{-2}\,\mathrm{s}^{-1}$, $H \approx 10^3\,\mathrm{m}$ and $f_0 \approx 10^{-4}\,\mathrm{s}^{-1}$, then $L_d \approx 100\,\mathrm{km}$. At high latitudes the ocean is much less stratified and f is somewhat larger, and the deformation radius may be as little as 20 km. In the atmosphere, with

$N \approx 10^{-2}\,\mathrm{s}^{-1}$, $H \approx 10^4\,\mathrm{m}$, then $L_d \approx 1000\,\mathrm{km}$. It is this order of magnitude difference in the deformation scales that accounts for a great deal of the quantitative difference in the dynamics of the ocean and the atmosphere. If we take the limit $L_d \to \infty$ then the stratified quasi-geostrophic equations reduce to

$$\frac{\mathrm{D}q}{\mathrm{D}t} = 0, \qquad q = \nabla^2\psi + f. \tag{5.65}$$

This is the two-dimensional vorticity equation; the high stratification of this limit has suppressed all vertical motion, and variations in the flow become confined to the horizontal plane.

Finally, if we allow density to vary in the vertical and carry through the quasi-geostrophic derivation we find that the quasi-geostrophic potential vorticity is given by

$$q = \nabla^2\psi + f + \frac{f_0^2}{\tilde{\rho}}\frac{\partial}{\partial z}\left(\frac{\tilde{\rho}}{N^2}\frac{\partial\psi}{\partial z}\right), \tag{5.66}$$

where $\tilde{\rho}$ is a reference profile of density and a function of z only, typically decreasing approximately exponentially with height.

5.5.3 Upper and Lower Boundary Conditions and Buoyancy Advection

The solution of the elliptic equation in (5.63b) requires vertical boundary conditions on ψ at the ground and at some upper boundary (e.g., the tropopause), and these are given by use of the thermodynamic equation. For a flat, slippery, rigid surface the vertical velocity is zero so that the thermodynamic equation may be written as

$$\frac{\mathrm{D}b'}{\mathrm{D}t} = 0, \qquad b' = f_0\frac{\partial\psi}{\partial z}, \qquad z = 0, H. \tag{5.67}$$

This equation provides the values of $\partial\psi/\partial z$ at the boundary that are then used when solving (5.63b). A heating term could be added as needed, and if there is no upper boundary (as in the real atmosphere) we would apply a condition that the motion decays to zero with height, but we will not deal with that condition further. If the bottom boundary is not flat then the vertical velocity is non-zero and the thermodynamic equation becomes, at the bottom boundary,

$$\frac{\mathrm{D}}{\mathrm{D}t}\left(\frac{\partial\psi}{\partial z}\right) + wN^2 = 0, \qquad w = \boldsymbol{u}\cdot\nabla\eta_B, \tag{5.68}$$

where η_B is the height of the topography. In practice, the bottom boundary condition is often incorporated into the definition of potential vorticity, as we shall see in the two-level model.

5.6 THE TWO-LEVEL QUASI-GEOSTROPHIC EQUATIONS

The continuously-stratified quasi-geostrophic equations have, as their name suggests, a continuous variation in the vertical. It turns out to be

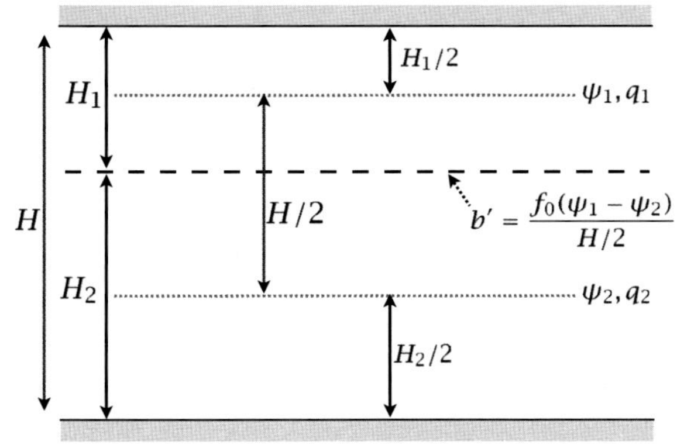

Fig. 5.1: A two-level quasi-geostrophic fluid system with a flat bottom and a rigid lid, with $w = 0$ at both. Here the levels are allowed to be of unequal thicknesses, but in the text we take $H_1 = H_2$.

extraordinarily useful to constrain this variation to that of two levels, in which the velocity is defined at just two vertical levels, and the temperature at just one level in the middle, but keeping the full horizontal variation. The resulting equations are not only algebraically much simpler than the continuous equations but they also capture many of their important features. There are two ways to derive the equations. One is by considering, *ab initio,* the motion of two immiscible shallow layers of fluid, one on top of the other, giving the 'two-layer' quasi-geostrophic equations. The second is by finite-differencing the continuous equations, and this is the procedure we follow here. Both methods give identical answers and we use 'two-layer' and 'two-level' interchangeably throught the book.

5.6.1 Deriving the Equations

The quasi-geostrophic potential vorticity may be written as

$$q = \nabla^2\psi + \beta y + \frac{\partial}{\partial z}\left(\frac{f_0^2}{N^2}\frac{\partial\psi}{\partial z}\right). \tag{5.69}$$

We now apply this equation at the two levels, 1 and 2, in Fig. 5.1, taking N to be constant and $H_1 = H_2 = H/2$ for simplicity (these restrictions may be easily relaxed). Using simple expressions for vertical finite differences gives

$$q_1 = \nabla^2\psi_1 + \beta y + \frac{f_0^2}{N^2}\frac{1}{H/2}\left(\left(\frac{\partial\psi}{\partial z}\right)_{\text{top}} - \frac{\psi_1 - \psi_2}{H/2}\right). \tag{5.70}$$

In this equation $(\partial\psi/\partial z)_{\text{top}}$ is the value of $\partial\psi/\partial z$ at the top of the domain and $2(\psi_1 - \psi_2)/H$ is the value of $\partial\psi/\partial z$ in the middle of the domain. The value at the top is given by the thermodynamic equation, (5.67), and if there is no heating we set $\mathrm{D}/\mathrm{D}t\,(\partial\psi/\partial z)_{\text{top}} = 0$. The potential vorticity equation for the upper level is then given by

$$\frac{\mathrm{D}q_1}{\mathrm{D}t} = 0, \qquad q_1 = \nabla^2\psi_1 + \beta y + \frac{k_d^2}{2}(\psi_2 - \psi_1), \tag{5.71a,b}$$

where $k_d^2 = 8f_0^2/N^2H^2$. In proceeding this way we have built the boundary conditions into the definition of potential vorticity. A thermody-

Quasi-Geostrophic Equations

Continuously-stratified

The adiabatic quasi-geostrophic potential vorticity equation for a Boussinesq fluid on the β-plane, is

$$\frac{Dq}{Dt} = 0, \qquad q = \nabla^2\psi + f\beta y + \frac{\partial}{\partial z}\left(\frac{f_0^2}{N^2}\frac{\partial\psi}{\partial z}\right), \tag{QG.1}$$

where ψ is the streamfunction. The horizontal velocities are given by $(u, v) = (-\partial\psi/\partial y, \partial\psi/\partial x)$. The boundary conditions at the top and bottom are given by the buoyancy equation,

$$\frac{Db}{Dt} = 0, \qquad b = f_0\frac{\partial\psi}{\partial z}. \tag{QG.2}$$

Two-level

The two-level or two-layer quasi-geostrophic equations are

$$\frac{Dq_i}{Dt} = 0, \qquad q_i = \nabla^2\psi_i + \beta y + \frac{k_d^2}{2}(\psi_j - \psi_i), \tag{QG.3}$$

where $i = 1, 2$, denoting the top and bottom levels respectively, and $j = 3 - i$.

namic source and/or frictional terms may readily be added to the right-hand of (5.71a) as needed.

Defining k_d so that $k_d^2/2$ rather than just k_d^2 is used in the two-level quasi-geostrophic potential vorticity is for later algebraic convenience rather than being of fundamental importance.

We apply an exactly analogous procedure to the lower level, and so obtain an expressions for the evolution of the lower layer potential vorticity, namely,

$$\frac{Dq_2}{Dt} = 0, \qquad q_2 = \nabla^2\psi_2 + \beta y + \frac{k_d^2}{2}(\psi_1 - \psi_2), \tag{5.72}$$

Once again the boundary conditions on buoyancy — the finite difference analogue of the buoyancy equations at the top and bottom — are *built in* to the definition of potential vorticity. If topography is present then it has an effect through the buoyancy equation at the bottom and it is incorporated into the definition of the lower level potential vorticity, which becomes

$$q_2 = \nabla^2\psi_2 + \beta y + \frac{k_d^2}{2}(\psi_1 - \psi_2) + \frac{f_0\eta_b}{H_2}. \tag{5.73}$$

This has a similar form to the shallow water potential vorticity with topography, as in (5.27). The two-level equations have proven to be enormously useful in the development of dynamical oceanography and meteorology and we will come back to them in later chapters. The quasi-geostrophic equations are summarized in the box above.

5.7 Frictional Geostrophic Balance and Ekman Layers

5.7.1 Preliminaries

Within a few hundred meters of the ground in the atmosphere, and in the upper and lower tens of metres in the ocean, frictional effects are important and geostrophic balance by itself is not the leading order balance in the momentum equations. Rather, a frictional term is important and the momentum equation becomes, for a Boussinesq fluid,

$$\boldsymbol{f} \times \boldsymbol{u} = -\nabla\phi + \boldsymbol{F}, \tag{5.74}$$

where $\boldsymbol{F}$ is a frictional term and $\nabla\phi$ is the horizontal gradient of pressure, at constant z. The friction ultimately comes from molecular viscosity and is of the form $\boldsymbol{F} = \nu\nabla^2\boldsymbol{u}$. However, in geophysical settings the vertical derivative dominates and it is common to write (5.74) in the form

$$\boldsymbol{f} \times \boldsymbol{u} = -\nabla_z\phi + \frac{1}{\rho_0}\frac{\partial \boldsymbol{\tau}}{\partial z}, \tag{5.75}$$

where $\boldsymbol{\tau}$ is the stress, and since density is assumed constant we shall absorb it into the definition of $\boldsymbol{\tau}$. (The quantity $\boldsymbol{\tau}/\rho_0$ is the 'kinematic stress', but we shall just refer to it as the stress and also denote it $\boldsymbol{\tau}$. Also, in some other fluid dynamical contexts the stress is a tensor but here $\boldsymbol{\tau}$ is a vector.) Commonly, the regions where the stress term is important are *boundary layers* (see Fig. 5.2) and they exist because the fluid takes on values at the boundary that differ from the values in the fluid interior. Boundary layers are ubiquitous in fluids in many circumstances, and if the Coriolis term is important then the region is called an *Ekman layer.* As noted, the stress ultimately arises from molecular forces and $\boldsymbol{\tau} = \nu\partial\boldsymbol{u}/\partial z$ where ν is the coefficient of viscosity, again just including vertical derivatives. However, on the scale of the Ekman layer molecular effects are greatly amplified by the effects of small scale turbulence (as discussed more in Chapter 10) and we commonly replace ν by a much larger 'eddy viscosity', A.

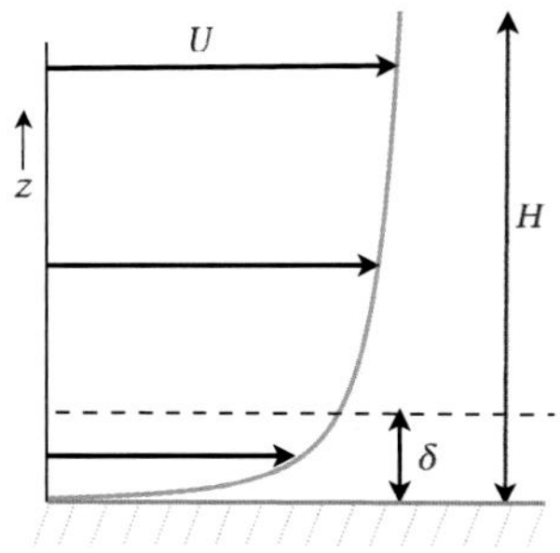

Fig. 5.2: An idealized boundary layer in which fields vary rapidly in order to satisfy the boundary conditions, here that $U = 0$. The scale δ is a measure of the thickness of the boundary layer and H is a typical scale of the variations away from the boundary.

Fridtjof Nansen (1861–1930), the polar explorer and statesman, apparently wanted to understand the motion of pack ice and of his ship, the Fram, *embedded in the ice and which seemed to move with neither the wind nor the surface current. He posed the problem to V. W. Ekman (1874–1954) who then calculated the direction of the flow in the upper ocean and published a paper on the matter in 1905.*

In the atmosphere we imagine that the flow is nearly geostrophic above an Ekman layer, with frictional effects coming from the need to bring the speed of flow down from its high geostrophic value to zero at the ground. In the ocean, on the other hand, the stress largely arises from the wind in the atmosphere blowing over the ocean surface. The stress then decays to zero with depth, and in the deeper ocean the flow again is geostrophic. The stress itself is continuous across the ocean–atmosphere interface, usually with a maximum value at the interface, decaying with height in the atmosphere and with depth in the ocean.

The equations of motion of the Ekman layer are completed by the mass continuity equation, $\nabla\cdot\boldsymbol{v} = 0$, and the hydrostatic equation, $\partial\phi/\partial z = b$. In order to treat the simplest case we take buoyancy to be constant, and in that case, and without any additional loss of generality, we take $b = 0$. The horizontal gradient of pressure is independent of height, which implies that, even though velocity may vary rapidly in the boundary layer the pressure gradient force does not.

5.7.2 Properties of Ekman Layers

We shall now determine three important properties of Ekman layers:

(i) The transport in an Ekman layer.

(ii) The depth of the Ekman layer.

(iii) The vertical velocity at the edge of the Ekman layer.

The importance of these will become clear as we go on.

Transport

We may write the Ekman layer equation (5.74) as

$$\boldsymbol{f} \times (\boldsymbol{u} - \boldsymbol{u}_g) = \frac{\partial \boldsymbol{\tau}}{\partial z} \qquad \text{or equivalently} \qquad f\boldsymbol{u}_a = -\hat{\mathbf{k}} \times \frac{\partial \boldsymbol{\tau}}{\partial z}, \tag{5.76a,b}$$

where $\boldsymbol{u}_g$ is the geostrophic wind such that $\boldsymbol{f} \times \boldsymbol{u}_g = -\nabla\phi$, $\boldsymbol{u}_a = \boldsymbol{u} - \boldsymbol{u}_g$ is the ageostrophic wind and $\hat{\mathbf{k}}$ is the unit vector in the vertical (so that $\boldsymbol{f} = f\hat{\mathbf{k}}$). Consider an ocean with a stress that diminishes with depth from its surface values, $\boldsymbol{\tau}_0$. If we integrate (5.76) over the depth of the Ekman layer (i.e., down from the surface until the stress vanishes) we obtain

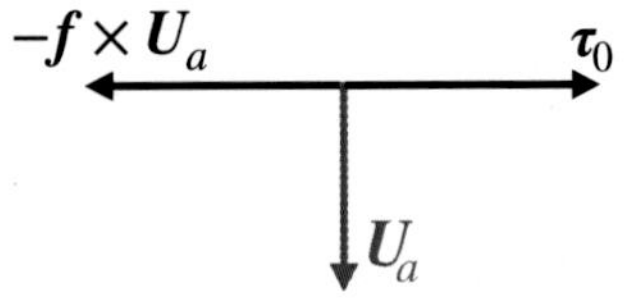

Fig. 5.3: Vertically integrated transport in a Northern Hemisphere ocean Ekman layer. A stress τ_0 at the surface induces an ageostrophic flow, U_a. In order that the Coriolis force on the ageostrophic flow can balance the surface stress the ageostrophic flow must be at right angles to the stress, as illustrated.

$$\int f\boldsymbol{u}_a \,\mathrm{d}z = -\hat{\mathbf{k}} \times \boldsymbol{\tau}_0, \quad \text{or} \quad \int f u_a \,\mathrm{d}z = \tau_0^y, \quad \int f v_a \,\mathrm{d}z = -\tau_0^x, \tag{5.77}$$

where the superscripts denote the x and y components of the stress. That is to say, *the integrated ageostrophic transport in the Ekman layer,* $\boldsymbol{U}_a = \int \boldsymbol{u}_a \,\mathrm{d}z$, *is at right angles to the stress at the surface,* as in Fig. 5.3. This result is independent of the detailed form of the stress, and it gives a partial answer to Nansen's question — if the surface stress is in the direction of the wind, the water beneath it flows at an angle to it, as a consequence of the Coriolis force. More generally, the Ekman transport is the most immediate response of the ocean to the atmosphere and is thus of key importance in the ocean circulation. In the atmosphere we *live* in the Ekman layer, and Ekman layer effects are partly responsible for determining the surface wind.

Depth

Although the transport in the Ekman layer is independent of the detailed form of the stress, the depth — namely the extent of the frictional influence of the surface — is not and to calculate it we use the Ekman-layer equation in the form

$$\boldsymbol{f} \times \boldsymbol{u} = -\nabla\phi + A\frac{\partial^2 \boldsymbol{u}}{\partial z^2}, \tag{5.78}$$

where A is a coefficient of viscosity. Suppose for the moment that the vertical scale of the flow is H; the ratio of size of the frictional term to the Coriolis term is then (A/fH^2), and this ratio defines the Ekman number, Ek:

$$Ek \equiv \left(\frac{A}{fH^2}\right). \tag{5.79}$$

In the fluid interior the Ekman number is usually small. In the Ekman layer itself, however, the viscous terms must be large (otherwise we are not in the Ekman layer), and the vertical scale must therefore be smaller than H. If the viscous and Coriolis terms in (5.78) are required to be the same magnitude we obtain an Ekman layer depth, δ, of

$$\delta \sim \left(\frac{A}{f}\right)^{1/2}. \tag{5.80}$$

The Ekman layer depth decreases with viscosity, becoming a thin boundary layer as $A \to 0$. We also see that the Ekman number can be recast as the ratio of the Ekman layer depth to the vertical scale in the interior; that is, from (5.79) and (5.80), $Ek = (\delta/H)^2$.

Vertical velocity

The vertical velocity at the edge of the Ekman layer may be calculated using the mass continuity equation,

$$\frac{\partial w}{\partial z} = -\left(\frac{\partial u}{\partial x} + \frac{\partial v}{\partial y}\right). \tag{5.81}$$

The right-hand side may be calculated from (5.76) and, noting that divergence of the geostrophic velocity is given by $\nabla \cdot \boldsymbol{v}_g = -\beta v_g/f$, we find that

$$\frac{\partial u}{\partial x} + \frac{\partial v}{\partial y} = -\frac{\beta v_g}{f} + \frac{\partial}{\partial z}\mathrm{curl}_z\left(\frac{\boldsymbol{\tau}}{f}\right), \tag{5.82}$$

where $\mathrm{curl}_z(\boldsymbol{\tau}/f) = \partial_x(\tau^y/f) - \partial_y(\tau^x/f)$. Consider an ocean Ekman layer in which the vertical velocity is zero at the surface and the stress is zero at the bottom of the layer. If we integrate (5.81) over the depth of that layer and use (5.82) we obtain

$$w_E = \mathrm{curl}_z \frac{\boldsymbol{\tau}_0}{f} - \frac{\beta V_g}{f}. \tag{5.83}$$

where $V_g = \int v_g \, dz$ is the integral of the meridional geostrophic velocity over the Ekman layer. Evidently, friction induces a vertical velocity at the edge of the Ekman layer, proportional to the curl of the stress at the surface, and this is perhaps the most used result in Ekman layer theory.

This result is particularly useful for the top Ekman layer in the ocean, where the stress can be regarded as a function of the overlying wind and is largely independent of the flow in the ocean; an equivalent result applies in the atmosphere, but here the surface stress is a function of the geostrophic wind in the free atmosphere. If the curl of the wind stress at the top of the ocean is anticyclonic (i.e., negative if $f > 0$, positive if $f < 0$) then a downward velocity is induced at the base of the Ekman layer, a phenomenon known as *Ekman pumping*, and this is particularly important in setting the structure of the subtropical gyres in the ocean, discussed in Chapter 14.

Notes and References

The first systematic derivation of the quasi-geostrophic equàtions was given by Charney (1948), although aspects of the concept, and the words'quasi-geostrophy', appeared in Durst & Sutcliffe (1938) and Sutcliffe (1947). The various forms of geostrophic equations were brought together in a review article by Phillips (1963) and since then both the quasi-geostrophic and planetary-geostrophic equations have been staples of dynamical meteorology and dynamical oceanography.

Problems

5.1 Do either or both:

(a) Carry through the derivation of the quasi-geostrophic system starting with the anelastic equations and obtain (5.66).

(b) Carry through the derivation of the quasi-geostrophic system in pressure coordinates.

In each case, state the differences between your results and the Boussinesq result.

5.2 (a) The shallow water *planetary geostrophic* equations may be derived by simply omitting ζ in the equation

$$\frac{\mathrm{D}}{\mathrm{D}t}\frac{\zeta + f}{h} = 0 \qquad \text{(P5.1)}$$

by invoking a small Rossby number, so that ζ/f is small. We then relate the velocity field to the height field by hydrostatic balance and obtain:

$$\frac{\mathrm{D}}{\mathrm{D}t}\left(\frac{f}{h}\right) = 0, \qquad fu = -g\frac{\partial h}{\partial y}, \quad fv = g\frac{\partial h}{\partial x}. \qquad \text{(P5.2)}$$

The assumptions of hydrostatic balance and small Rossby number are the same as those used in deriving the quasi-geostrophic equations. Explain nevertheless how some of the assumptions used for quasi-geostrophy are in fact different from those used for planetary-geostrophy, and how the derivations and resulting systems differ from each other. Use any or all of the momentum and mass continuity equations, scaling, nondimensionalization and verbal explanations as needed.

(b) Explain if and how your arguments in part (a) also apply to the stratified equations (using, for example, the Boussinesq equations or pressure coordinates).

5.3 In the derivation of the quasi-geostrophic equations, geostrophic balance leads to the lowest-order horizontal velocity being divergence-free — that is, $\nabla \cdot \boldsymbol{u}_0 = 0$. It seems that this can also be obtained from the mass conservation equation at lowest order. Is this a coincidence? Suppose that the Coriolis parameter varied, and that the momentum equation yielded $\nabla \cdot \boldsymbol{u}_0 \neq 0$. Would there be an inconsistency?

5.4 Consider a wind stress imposed by a mesoscale cyclonic storm (in the atmosphere) given by

$$\boldsymbol{\tau} = -Ae^{-(r/\lambda)^2}(y\hat{\mathbf{i}} - x\hat{\mathbf{j}}), \qquad \text{(P5.3)}$$

where $r^2 = x^2 + y^2$, and A and λ are constants. Also assume constant Coriolis gradient $\beta = \partial f/\partial y$ and constant ocean depth H. In the ocean, find

(a) the Ekman transport, (b) the vertical velocity $w_E(x, y, z)$ below the Ekman layer, (c) the northward velocity $v(x, y, z)$ below the Ekman layer and (d) indicate how you would find the westward velocity $u(x, y, z)$ below the Ekman layer.

5.5 In an atmospheric Ekman layer on the f-plane let us write the momentum equation as

$$\boldsymbol{f} \times \boldsymbol{u} = -\nabla\phi + \frac{1}{\rho_a}\frac{\partial \boldsymbol{\tau}}{\partial z}, \tag{P5.4}$$

where $\boldsymbol{\tau} = A\rho_a \partial \boldsymbol{u}/\partial z$ and A is a constant eddy viscosity coefficient. An *independent* formula for the stress at the ground is $\boldsymbol{\tau} = C\rho_a \boldsymbol{u}$, where C is a constant. Let us take $\rho_a = 1$, and assume that in the free atmosphere the wind is geostrophic and zonal, with $\boldsymbol{u}_g = U\hat{\mathbf{i}}$.

(a) Find an expression for the wind vector at the ground. Discuss the limits $C = 0$ and $C = \infty$. Show that when $C = 0$ the frictionally-induced vertical velocity at the top of the Ekman layer is zero.

(b) Find the vertically integrated horizontal mass flux caused by the boundary layer.

(c) When the stress on the atmosphere is $\boldsymbol{\tau}$, the stress on the ocean beneath is also $\boldsymbol{\tau}$. Why? Show how this is consistent with Newton's third law.

(d) Determine the direction and strength of the surface current, and the mass flux in the oceanic Ekman layer, in terms of the geostrophic wind in the atmosphere, the oceanic Ekman depth and the ratio ρ_a/ρ_o, where ρ_o is the density of the seawater. Include a figure showing the directions of the various winds and currents. How does the boundary-layer mass flux in the ocean compare to that in the atmosphere? (Assume, as needed, that the stress in the ocean may be parameterized with an eddy viscosity.)

Partial solution for (a): A useful trick in Ekman layer problems is to write the velocity as a complex number, $\hat{u} = u + \mathrm{i}v$ and $\hat{u}_g = u_g + \mathrm{i}v_g$. The fundamental Ekman layer equation may then be written as

$$A\frac{\partial^2 \widehat{U}}{\partial z^2} = \mathrm{i} f \widehat{U}, \tag{P5.5}$$

where $\widehat{U} = \hat{u} - \hat{u}_g$. The solution to this is

$$\hat{u} - \hat{u}_g = [\hat{u}(0) - \hat{u}_g] \exp\left[-\frac{(1+\mathrm{i})z}{d}\right], \tag{P5.6}$$

where $d = \sqrt{2A/f}$ and the boundary condition of finiteness at infinity eliminates the exponentially growing solution. The boundary condition at $z = 0$ is $\partial\hat{u}/\partial z = (C/A)\hat{u}$; applying this gives $[\hat{u}(0) - \hat{u}_g]\exp(\mathrm{i}\pi/4) = -Cd\hat{u}(0)/(\sqrt{2}A)$, from which we obtain $\hat{u}(0)$, and the rest of the solution follows.

CHAPTER

6

Rossby Waves

WAVES ARE FAMILIAR TO ALMOST EVERYONE. Gravity waves cover the ocean surface, sound waves allow us to talk and light waves enable us to see. This chapter provides an introduction to their properties, paying particular attention to a wave that is especially important to the large scale flow in both ocean and atmosphere — the Rossby wave. We start with an elementary introduction to wave kinematics, discussing such concepts as phase speed and group velocity. Then, beginning with Section 6.3, we discuss the dynamics of Rossby waves, and this part may be considered to be the natural follow-on from the geostrophic theory of the previous chapter. Rossby waves then reappear frequently in later chapters.

6.1 FUNDAMENTALS AND FORMALITIES

6.1.1 Definitions and Kinematics

A wave is more easily recognized than defined. Loosely speaking, a wave is a propagating disturbance that has a characteristic relationship between its frequency and size, called a *dispersion relation.* To see what all this means, and what a dispersion relation is, suppose that a disturbance, $\psi(\boldsymbol{x}, t)$ (where ψ might be velocity, streamfunction, pressure, etc.), satisfies the equation

$$L(\psi) = 0, \tag{6.1}$$

where L is a linear operator, typically a polynomial in time and space derivatives; one example is $L(\psi) = \partial \nabla^2 \psi / \partial t + \beta \partial \psi / \partial x$. If (6.1) has constant coefficients (if β is constant in this example) then harmonic solutions may often be found that are a superposition of *plane waves,* each of which satisfy

$$\psi = \mathrm{Re}\, \widetilde{\psi} \mathrm{e}^{\mathrm{i}\theta(\boldsymbol{x},t)} = \mathrm{Re}\, \widetilde{\psi} \mathrm{e}^{\mathrm{i}(\boldsymbol{k}\cdot\boldsymbol{x} - \omega t)}, \tag{6.2}$$

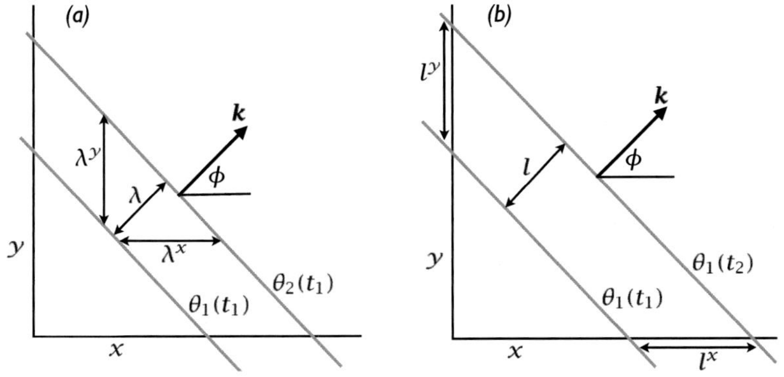

Fig. 6.1: The propagation of a two-dimensional wave. (a) Two lines of constant phase (e.g., two wavecrests) at a time t_1. The wave is propagating in the direction $\boldsymbol{k}$ with wavelength λ. (b) A line of constant phase at two successive times. The phase speed is the speed of advancement of the wavecrest in the direction of travel, and so $c_p = l/(t_2 - t_1)$. The phase speed in the x-direction is the speed of propagation of the wavecrest along the x-axis, and $c_p^x = l^x/(t_2 - t_1) = c_p/\cos\phi$.

where $\widetilde{\psi}$ is a complex constant, θ is the phase, ω is the wave frequency and $\boldsymbol{k}$ is the vector wavenumber (k, l, m) (also written as (k^x, k^y, k^z) or, in subscript notation, k_i). The prefix Re denotes the real part of the expression, but we will drop it if there is no ambiguity.

Waves are characterized by having a particular relationship between the frequency and wavevector known as the *dispersion relation*. This is an equation of the form

$$\omega = \Omega(\boldsymbol{k}), \tag{6.3}$$

where $\Omega(\boldsymbol{k})$, or $\Omega(k_i)$, and meaning $\Omega(k, l, m)$, is some function determined by the form of L in (6.1) and which thus depends on the particular type of wave — the function is different for sound waves, light waves and the Rossby waves and gravity waves we will encounter in this book. Unless it is necessary to explicitly distinguish the function Ω from the frequency ω, we often write $\omega = \omega(\boldsymbol{k})$.

6.1.2 Wave Propagation and Phase Speed

A common property of waves is that they propagate through space with some velocity, which in special cases might be zero. Waves in fluids may carry energy and momentum but do not necessarily transport fluid parcels themselves. Further, it turns out that the speed at which properties like energy are transported (the group speed) may be different from the speed at which the wave crests themselves move (the phase speed). Let's try to understand this statement, beginning with the phase speed. A summary of key results is given on page 107.

Phase speed

Consider the propagation of monochromatic plane waves, for that is all that is needed to introduce the phase speed. Given (6.2) a wave will propagate in the direction of $\boldsymbol{k}$ (Fig. 6.1). At a given instant and location we can align our coordinate axis along this direction, and we write $\boldsymbol{k} \cdot \boldsymbol{x} = Kx^*$, where x^* increases in the direction of $\boldsymbol{k}$ and $K^2 = |\boldsymbol{k}|^2$ is the magnitude of the wavenumber. With this, we can write (6.2) as

$$\psi = \mathrm{Re}\,\widetilde{\psi}\mathrm{e}^{\mathrm{i}(Kx^* - \omega t)} = \mathrm{Re}\,\widetilde{\psi}\mathrm{e}^{\mathrm{i}K(x^* - ct)}, \tag{6.4}$$

where $c = \omega/K$. From this equation it is evident that the phase of the wave propagates at the speed c in the direction of $\boldsymbol{k}$, and we define the *phase speed* by

$$c_p \equiv \frac{\omega}{K}. \tag{6.5}$$

The wavelength of the wave, λ, is the distance between two wavecrests — that is, the distance between two locations along the line of travel whose phase differs by 2π — and evidently this is given by

$$\lambda = \frac{2\pi}{K}. \tag{6.6}$$

In (for simplicity) a two-dimensional wave, and referring to Fig. 6.1, the wavelength and wave vectors in the x- and y-directions are given by,

$$\lambda^x = \frac{\lambda}{\cos\phi}, \qquad \lambda^y = \frac{\lambda}{\sin\phi}, \qquad k^x = K\cos\phi, \quad k^y = K\sin\phi. \tag{6.7}$$

In general, lines of constant phase intersect both the coordinate axes and propagate along them. The speed of propagation along these axes is given by

$$c_p^x = c_p\frac{l^x}{l} = \frac{c_p}{\cos\phi} = c_p\frac{K}{k^x} = \frac{\omega}{k^x}, \qquad c_p^y = c_p\frac{l^y}{l} = \frac{c_p}{\sin\phi} = c_p\frac{K}{k^y} = \frac{\omega}{k^y}, \tag{6.8}$$

using (6.5) and (6.7), and again referring to Fig. 6.1 for notation. The speed of phase propagation along any one of the axes is in general *larger* than the phase speed in the primary direction of the wave. The phase speeds are clearly *not* components of a vector: for example, $c_p^x \neq c_p\cos\phi$. Analogously, the wavevector $\boldsymbol{k}$ is a true vector, whereas the wavelength λ is not.

To summarize, the phase speed and its components are given by

$$c_p = \frac{\omega}{K}, \qquad c_p^x = \frac{\omega}{k^x}, \qquad c_p^y = \frac{\omega}{k^y}. \tag{6.9}$$

6.1.3 The Dispersion Relation

The above description is kinematic, in that it applies to almost any disturbance that has a wavevector and a frequency. The particular *dynamics* of a wave are determined by the relationship between the wavevector and the frequency; that is, by the *dispersion relation.* Once the dispersion relation is known a great many of the properties of the wave follow in a more-or-less straightforward manner. Picking up from (6.3), the dispersion relation is a functional relationship between the frequency and the wavevector of the general form

$$\omega = \Omega(\boldsymbol{k}). \tag{6.10}$$

Perhaps the simplest example of a linear operator that gives rise to waves is the one-dimensional equation

$$\frac{\partial\psi}{\partial t} + c\frac{\partial\psi}{\partial x} = 0. \tag{6.11}$$

Wave Fundamentals

- A wave is a propagating disturbance that has a characteristic relationship between its frequency and size, known as the dispersion relation. Waves typically arise as solutions to a linear problem of the form $L(\psi) = 0$, where L is, commonly, a linear operator in space and time. Two examples are

$$\frac{\partial^2 \psi}{\partial t^2} - c^2 \nabla^2 \psi = 0 \qquad \text{and} \qquad \frac{\partial}{\partial t}\nabla^2 \psi + \beta \frac{\partial \psi}{\partial x} = 0, \tag{WF.1}$$

where the second example gives rise to Rossby waves.

- Solutions to the governing equation are often sought in the form of plane waves that have the form

$$\psi = \operatorname{Re} A e^{\mathrm{i}(\boldsymbol{k}\cdot\boldsymbol{x} - \omega t)}, \tag{WF.2}$$

where A is the wave amplitude, $\boldsymbol{k} = (k, l, m)$ is the wavevector, and ω is the frequency.

- The dispersion relation connects the frequency and wavevector through an equation of the form $\omega = \Omega(\boldsymbol{k})$ where Ω is some function. The relation is normally derived by substituting a trial solution like (WF.2) into the governing equation. For the examples of (WF.1) we obtain $\omega = c^2 K^2$ and $\omega = -\beta k / K^2$ where $K^2 = k^2 + l^2 + m^2$ or, in two dimensions, $K^2 = k^2 + l^2$.

- The phase speed is the speed at which the wave crests move. In the direction of propagation and in the x, y and z directions the phase speeds are given by, respectively,

$$c_p = \frac{\omega}{K}, \qquad c_p^x = \frac{\omega}{k}, \quad c_p^y = \frac{\omega}{l}, \quad c_p^z = \frac{\omega}{m}, \tag{WF.3}$$

where $K = 2\pi/\lambda$ and λ is the wavelength. The wave crests have both a speed (c_p) and a direction of propagation (the direction of $\boldsymbol{k}$), like a vector, but the components defined in (WF.3) are not the components of that vector.

- The group velocity is the velocity at which a wave packet or wave group moves. It is a vector and is given by

$$\boldsymbol{c}_g = \frac{\partial \omega}{\partial \boldsymbol{k}} \quad \text{with components} \quad c_g^x = \frac{\partial \omega}{\partial k}, \quad c_g^y = \frac{\partial \omega}{\partial l}, \quad c_g^z = \frac{\partial \omega}{\partial m}. \tag{WF.4}$$

Most physical quantities of interest are transported at the group velocity.

Substituting a trial solution of the form $\psi = \operatorname{Re} A e^{\mathrm{i}(kx - \omega t)}$ into (6.11) we obtain $(-\mathrm{i}\omega + c\mathrm{i}k)A = 0$, giving the dispersion relation

$$\omega = ck. \tag{6.12}$$

The phase speed of this wave is $c_p = \omega/k = c$. A couple of other examples of governing equations, dispersion relations and phase speeds are:

$$\frac{\partial^2 \psi}{\partial t^2} - c^2 \nabla^2 \psi = 0, \quad \omega^2 = c^2 K^2, \quad c_p = \pm c, \quad c_p^x = \pm \frac{cK}{k}, \quad c_p^y = \pm \frac{cK}{l}, \tag{6.13a}$$

$$\frac{\partial}{\partial t}\nabla^2 \psi + \beta \frac{\partial \psi}{\partial x} = 0, \quad \omega = \frac{-\beta k}{K^2}, \quad c_p = \frac{\omega}{K}, \quad c_p^x = -\frac{\beta}{K^2}, \quad c_p^y = -\frac{\beta k/l}{K^2}, \tag{6.13b}$$

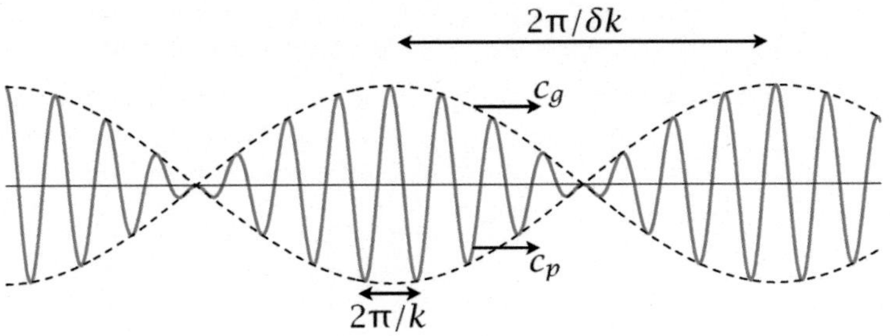

Fig. 6.2: Superposition of two sinusoidal waves with wavenumbers k and $k + \delta k$, producing a wave (solid line) that is modulated by a slowly varying wave envelope or packet (dashed). The envelope moves at the group velocity, $c_g = \partial\omega/\partial k$, and the phase moves at the group speed, $c_p = \omega/k$.

where $K^2 = k^2 + l^2$ and the examples are two-dimensional, with variation in x and y only.

A wave is said to be *nondispersive* if the phase speed is independent of the wavelength. This condition is satisfied for the simple example (6.11) but is manifestly not satisfied for (6.13b), and these waves (Rossby waves, in fact) are *dispersive*. Waves of different wavelengths then travel at different speeds so that a group of waves will spread out — disperse — even if the medium is homogeneous. When a wave is dispersive there is another characteristic speed at which the waves propagate, the group velocity, and we come to this shortly.

Most media are inhomogeneous, but if the medium varies sufficiently slowly in space and time — and in particular if the variations are slow compared to the wavelength and period — we may still have a *local* dispersion relation between frequency and wavevector,

$$\omega = \Omega(\boldsymbol{k}; \boldsymbol{x}, t), \tag{6.14}$$

where x and t are slowly varying parameters. We resume our discussion of this topic in Section 6.5, but before that we introduce the group velocity.

6.2 Group Velocity

Group velocity seems to have been first articulated in about 1841 by the Irish mathematician and physicist William Rowan Hamilton (1806–1865), who is also remembered for his formulation of 'Hamiltonian mechanics'. Hamilton was largely motivated by optics, and it was George Stokes, Osborne Reynolds and John Strutt (also known as Lord Rayleigh) who further developed and generalized the idea in fluid dynamics in the nineteenth and early twentieth centuries.

Information and energy do not, in general, propagate at the phase speed. Rather, most quantities of interest propagate at the *group velocity*, a quantity of enormous importance in wave theory. Roughly speaking, group velocity is the velocity at which a packet or a group of waves will travel, whereas the individual wave crests travel at the phase speed. To introduce the idea we will consider the superposition of plane waves, noting that a truly monochromatic plane wave already fills all space uniformly so that there can be no propagation of energy from place to place.

6.2.1 Superposition of Two Waves

Consider the linear superposition of two waves. Limiting attention to the one-dimensional case, consider a disturbance that is the sum of two waves,

$$\psi = \mathrm{Re}\,\widetilde{\psi}(e^{i(k_1 x - \omega_1 t)} + e^{i(k_2 x - \omega_2 t)}). \tag{6.15}$$

Let us further suppose that the two waves have similar wavenumbers and frequency, and, in particular, that $k_1 = k + \Delta k$ and $k_2 = k - \Delta k$, and $\omega_1 = \omega + \Delta\omega$ and $\omega_2 = \omega - \Delta\omega$. With this, (6.15) becomes

$$\begin{aligned}\psi &= \mathrm{Re}\,\widetilde{\psi}\mathrm{e}^{\mathrm{i}(kx-\omega t)}[\mathrm{e}^{\mathrm{i}(\Delta kx-\Delta\omega t)} + \mathrm{e}^{-\mathrm{i}(\Delta kx-\Delta\omega t)}] \\ &= 2\,\mathrm{Re}\,\widetilde{\psi}\mathrm{e}^{\mathrm{i}(kx-\omega t)}\cos(\Delta k\,x - \Delta\omega\,t).\end{aligned} \tag{6.16}$$

The resulting disturbance, illustrated in Fig. 6.2 has two aspects: a rapidly varying component, with wavenumber k and frequency ω, and a more slowly varying envelope, with wavenumber Δk and frequency $\Delta\omega$. The envelope modulates the fast oscillation, and moves with velocity $\Delta\omega/\Delta k$; in the limit $\Delta k \to 0$ and $\Delta\omega \to 0$ this is the *group velocity*, $c_g = \partial\omega/\partial k$. Group velocity is equal to the phase speed, ω/k, only when the frequency is a linear function of wavenumber. The energy in the disturbance moves at the group velocity — note that the node of the envelope moves at the speed of the envelope and no energy can cross the node. These concepts generalize to more than one dimension, and if the wavenumber is the three-dimensional vector $\boldsymbol{k} = (k, l, m)$ then the three-dimensional envelope propagates at the group velocity given by

$$\boldsymbol{c}_g = \frac{\partial\omega}{\partial\boldsymbol{k}} \equiv \left(\frac{\partial\omega}{\partial k}, \frac{\partial\omega}{\partial l}, \frac{\partial\omega}{\partial m}\right). \tag{6.17}$$

The group velocity, $\boldsymbol{c}_g$ is given by $\boldsymbol{c}_g = \partial\omega/\partial\boldsymbol{k}$ or, in Cartesian components, $\boldsymbol{c}_g = (\partial\omega/\partial k, \partial\omega/\partial l, \partial\omega/\partial m)$. It is a vector, and it gives the velocity at which wave packets, and hence energy, are propagated. The phase speed, $c_p = \omega/k$, is the speed at which the phase of an individual wave moves. For some waves the group velocity and phase speed may be in opposite directions, so that the wave crests appear to move backwards through the wave packets.

The group velocity is also written as $\boldsymbol{c}_g = \nabla_{\boldsymbol{k}}\omega$ or, in subscript notation, $c_{gi} = \partial\Omega/\partial k_i$, with the subscript i denoting the component of a vector.

The above derivation can be generalized to apply to the superposition of many waves. It also applies when the medium through which the waves propagate is not homogeneous, provided that changes occur on a longer space scale than the wavelength of the waves. Energy and most other physically meaningful properties of waves travel at the group velocity, as explicitly shown for Rossby waves in Section 9.2.2.

6.3 ROSSBY WAVE ESSENTIALS

Rossby waves are the most prominent wave in the atmosphere and ocean on large scales, although gravity waves sometimes rival them. They can best be described using the quasi-geostrophic equations, as follows.

6.3.1 The Linear Equation of Motion

The relevant equation of motion is the inviscid, adiabatic potential vorticity equation in the quasi-geostrophic system, as discussed in Chapter 5, namely

$$\frac{\partial q}{\partial t} + \boldsymbol{u}\cdot\nabla q = 0, \tag{6.18}$$

where $q(x, y, z, t)$ is the potential vorticity and $\boldsymbol{u}(x, y, z, t)$ is the horizontal velocity. The velocity is in turn related to a streamfunction by $u = -\partial\psi/\partial y$, $v = \partial\psi/\partial x$, and the potential vorticity is some function

The Rossby wave, the most well-known wave in meteorology and oceanography, is named for Carl-Gustav Rossby who described the essential dynamics in Rossby (1939). The waves are in fact present in the solutions of Hough (1898) although in a rather oblique form and Rossby is normally given credit for having discovered them.

of the streamfunction, which might differ from system to system. Two examples, one applying to a continuously-stratified system and the second to a single layer, are

$$q = f + \zeta + \frac{\partial}{\partial z}\left(\frac{f_0^2}{N^2}\frac{\partial \psi}{\partial z}\right) \qquad \text{and} \qquad q = \zeta + f - k_d^2\psi, \qquad (6.19\text{a,b})$$

where $\zeta = \nabla^2\psi$ is the relative vorticity and $k_d = 1/L_d$ is the inverse radius of deformation for a shallow water system.

Definitions of k_d and L_d differ from source to source (and book to book), often by factors of 2 or π or similar. Caveat lector.

We now *linearize* (6.18); that is, we suppose that the flow consists of a time-independent component (the 'basic state') plus a perturbation, with the perturbation being small compared with the mean flow. The basic state must satisfy the time-independent equation of motion, and it is common and useful to linearize about a zonal flow, $\overline{u}(y,z)$. The basic state is then purely a function of y and so we can write

$$q = \overline{q}(y,z) + q'(x,y,t), \qquad \psi = \overline{\psi}(y,z) + \psi'(x,y,z,t), \qquad (6.20)$$

with a similar notation for the other variables, and $\overline{u} = -\partial\overline{\psi}/\partial y$ and $\overline{v} = 0$. Substituting into (6.18) gives, without approximation,

$$\frac{\partial q'}{\partial t} + \overline{\boldsymbol{u}}\cdot\nabla\overline{q} + \overline{\boldsymbol{u}}\cdot\nabla q' + \boldsymbol{u}'\cdot\nabla\overline{q} + \boldsymbol{u}'\cdot\nabla q' = 0. \qquad (6.21)$$

The primed quantities are presumptively small so we neglect terms involving their products. Further, we are assuming that we are linearizing about a state that is a solution of the equations of motion, so that $\overline{\boldsymbol{u}}\cdot\nabla\overline{q} = 0$. Finally, since $\overline{v} = 0$ (since $\int \partial\psi/\partial x \, \mathrm{d}x = 0$) and $\partial\overline{q}/\partial x = 0$ we obtain

$$\frac{\partial q'}{\partial t} + U\frac{\partial q'}{\partial x} + v'\frac{\partial \overline{q}}{\partial y} = 0, \qquad (6.22)$$

where $U \equiv \overline{u}$. This equation or one very similar appears very commonly in studies of Rossby waves. Let us first consider the simple example of waves in a single layer.

6.3.2 Waves in a Single Layer

Consider a system obeying (6.18) and (6.19b). The dynamics are more easily illustrated on a Cartesian β-plane for which $f = f_0 + \beta y$, and since f_0 is a constant it does not appear in our subsequent derivations.

Infinite deformation radius

If the scale of motion is much less than the deformation scale then we make the approximation that $k_d = 0$ and the vorticity equation may be written as

$$\frac{\partial \zeta}{\partial t} + \boldsymbol{u}\cdot\nabla\zeta + \beta v = 0. \qquad (6.23)$$

We linearize about a constant zonal flow, U, by writing

$$\frac{\partial}{\partial t}\nabla^2\psi' + U\frac{\partial \nabla^2\psi'}{\partial x} + \beta\frac{\partial \psi'}{\partial x} = 0. \tag{6.24}$$

This equation is just a single-layer version of (6.22), with $\partial\overline{q}/\partial y = \beta$, $q' = \nabla^2\psi'$ and $v' = \partial\psi'/\partial x$.

The coefficients in (6.24) are not functions of y or z; this is not a requirement for wave motion to exist but it does enable solutions to be found more easily. Let us seek solutions in the form of a plane wave, namely

$$\psi' = \mathrm{Re}\,\widetilde{\psi}\mathrm{e}^{\mathrm{i}(kx+ly-\omega t)}, \tag{6.25}$$

where $\widetilde{\psi}$ is a complex constant. Solutions of this form are valid in a domain with doubly-periodic boundary conditions; solutions in a channel can be obtained using a meridional variation of $\sin ly$, with no essential changes to the dynamics. The amplitude of the oscillation is given by $\widetilde{\psi}$ and the phase by $kx+ly-\omega t$, where k and l are the x- and y-wavenumbers and ω is the frequency of the oscillation.

Substituting (6.25) into (6.24) yields

$$[(-\omega + Uk)(-K^2) + \beta k]\widetilde{\psi} = 0, \tag{6.26}$$

where $K^2 = k^2 + l^2$. For non-trivial solutions the above equation implies

$$\omega = Uk - \frac{\beta k}{K^2}, \tag{6.27}$$

and this is the *dispersion relation* for barotropic Rossby waves. Evidently the velocity U Doppler shifts the frequency by the amount Uk. The components of the phase speed and group velocity are given by, respectively,

$$c_p^x \equiv \frac{\omega}{k} = U - \frac{\beta}{K^2}, \qquad c_p^y \equiv \frac{\omega}{l} = U\frac{k}{l} - \frac{\beta k}{K^2 l}, \tag{6.28a,b}$$

and

$$c_g^x \equiv \frac{\partial\omega}{\partial k} = U + \frac{\beta(k^2 - l^2)}{(k^2 + l^2)^2}, \qquad c_g^y \equiv \frac{\partial\omega}{\partial l} = \frac{2\beta kl}{(k^2 + l^2)^2}. \tag{6.29a,b}$$

The phase speed in the absence of a mean flow is *westward,* with waves of longer wavelengths travelling more quickly, and the eastward current speed required to hold the waves of a particular wavenumber stationary (i.e., $c_p^x = 0$) is $U = \beta/K^2$. The background flow U evidently just provides a uniform shift to the phase speed, and (in this case) can be transformed away by a change of coordinate. The x-component of the group velocity may also be written as the sum of the phase speed plus a positive quantity, namely

$$c_g^x = c_p^x + \frac{2\beta k^2}{(k^2 + l^2)^2}. \tag{6.30}$$

This means that the zonal group velocity for Rossby wave packets moves eastward relative to its zonal phase speed. A stationary wave ($c_p^x = 0$) has

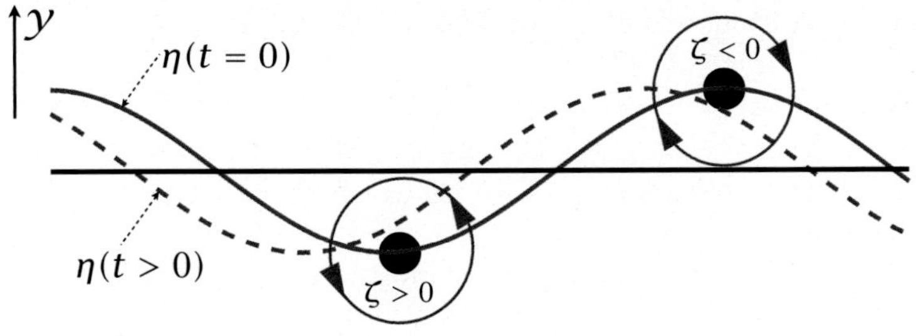

Fig. 6.3: A two-dimensional (x–y) Rossby wave. An initial disturbance displaces a material line at constant latitude (the straight horizontal line) to the solid line marked $\eta(t = 0)$. Conservation of potential vorticity, $\beta y + \zeta$, leads to the production of relative vorticity, ζ, as shown. The associated velocity field (arrows on the circles) then advects the fluid parcels, and the material line evolves into the dashed line with the phase propagating westward.

an eastward group velocity, and this has implications for the 'downstream development' of wave packets, but we will not pursue that topic here.

We see from (6.29) the group velocity is negative (westward) if the x-wavenumber is sufficiently small compared to the y-wavenumber. Essentially, *long waves move information westward and short waves move information eastward,* and this is a common property of Rossby waves. The x-component of the phase speed, on the other hand, is always westward relative to the mean flow.

Finite deformation radius

For a finite deformation radius the basic state $\Psi = -Uy$ is still a solution of the original equations of motion, but the potential vorticity corresponding to this state is $q = Uyk_d^2 + \beta y$ and its gradient is $\nabla q = (\beta + Uk_d^2)\hat{\mathbf{j}}$. The linearized equation of motion is thus

$$\left(\frac{\partial}{\partial t} + U\frac{\partial}{\partial x}\right)(\nabla^2\psi' - \psi' k_d^2) + (\beta + Uk_d^2)\frac{\partial \psi'}{\partial x} = 0. \tag{6.31}$$

Substituting $\psi' = \widetilde{\psi} e^{i(kx+ly-\omega t)}$ we obtain the dispersion relation,

$$\omega = \frac{k(UK^2 - \beta)}{K^2 + k_d^2} = Uk - k\frac{\beta + Uk_d^2}{K^2 + k_d^2}. \tag{6.32}$$

It is clear from the second form of the above equation that the uniform velocity field no longer provides just a simple Doppler shift of the frequency, nor does it provide a uniform addition to the phase speed. This is because the current does not just provide a uniform translation, but, if k_d is non-zero, it also modifies the basic potential vorticity gradient, as explored further in Problem 6.1.

6.3.3 The Mechanism of Rossby Waves

The fundamental mechanism underlying Rossby waves may be understood as follows. Consider a material line of stationary fluid parcels along a line of constant latitude, and suppose that some disturbance causes their displacement to the line marked $\eta(t = 0)$ in Fig. 6.3. In the displacement, the potential vorticity of the fluid parcels is conserved, and in the simplest

Essentials of Rossby Waves

- Rossby waves owe their existence to a gradient of potential vorticity in the fluid. If a fluid parcel is displaced, it conserves its potential vorticity and so its relative vorticity will in general change. The relative vorticity creates a velocity field that displaces neighbouring parcels, whose relative vorticity changes and so on.
- A common source of a potential vorticity gradient is differential rotation, or the β-effect, and in this case the associated Rossby waves are called *planetary waves*. In the presence of non-zero β the ambient potential vorticity increases northward and the phase of the Rossby waves propagates westward. Topography is another source of potential vorticity gradients. In general, Rossby waves propagate to the left of the direction of increasing potential vorticity.
- A common equation of motion for Rossby waves is

$$\frac{\partial q'}{\partial t} + U\frac{\partial q'}{\partial x} + v'\frac{\partial \overline{q}}{\partial y} = 0, \tag{RW.1}$$

with an overbar denoting the basic state and a prime a perturbation. In the case of a single layer of fluid with no mean flow this equation becomes

$$\frac{\partial}{\partial t}(\nabla^2 + k_d^2)\psi' + \beta\frac{\partial \psi'}{\partial x} = 0, \tag{RW.2}$$

with dispersion relation

$$\omega = \frac{-\beta k}{k^2 + l^2 + k_d^2}. \tag{RW.3}$$

- In the absence of a mean flow (i.e., $U = 0$), the phase speed in the zonal direction ($c_p^x = \omega/k$) is always negative, or westward, and is larger for large waves. For (RW.3) the components of the group velocity are given by

$$c_g^x = \frac{\beta(k^2 - l^2 - k_d^2)}{(k^2 + l^2 + k_d^2)^2}, \qquad c_g^y = \frac{2\beta kl}{(k^2 + l^2 + k_d^2)^2}. \tag{RW.4}$$

The group velocity is westward if the zonal wavenumber is sufficiently small, and eastward if the zonal wavenumber is sufficiently large.

case of barotropic flow on the β-plane the potential vorticity is the absolute vorticity, $\beta y + \zeta$. Thus, in either hemisphere, a northward displacement leads to the production of negative relative vorticity and a southward displacement leads to the production of positive relative vorticity. The relative vorticity gives rise to a velocity field which, in turn, advects the parcels in the material line in the manner shown, and the wave propagates westward.

In more complicated situations, such as flow in two layers, considered below, or in a continuously-stratified fluid, the mechanism is essentially the same. A displaced fluid parcel carries with it its potential vorticity and, in the presence of a potential vorticity gradient in the basic state, a potential vorticity anomaly is produced. The potential vorticity anomaly produces a velocity field (an example of potential vorticity inversion) which further displaces the fluid parcels, leading to the formation of a Rossby wave. The vital ingredient is a basic state potential vorticity gradient, such as that provided by the change of the Coriolis parameter with latitude.

6.4 ROSSBY WAVES IN STRATIFIED QUASI-GEOSTROPHIC FLOW

We now consider the dynamics of linear waves in stratified quasi-geostrophic flow on a β-plane, with a resting basic state. The interior flow is governed by the potential vorticity equation, (5.63), and linearizing this about a constant mean zonal flow, U, gives

$$\left(\frac{\partial}{\partial t} + U\frac{\partial}{\partial x}\right)\left[\nabla^2\psi' + \frac{\partial}{\partial z}\left(\frac{f_0^2}{N^2}\frac{\partial \psi'}{\partial z}\right)\right] + \beta\frac{\partial \psi'}{\partial x} = 0, \tag{6.33}$$

where, for simplicity, we suppose that f_0^2/N^2 does not vary with z.

6.4.1 Dispersion Relation and Group Velocity

Deferring the issues of boundary conditions for a moment, let us seek solutions of the form

$$\psi' = \mathrm{Re}\,\widetilde{\psi}\mathrm{e}^{\mathrm{i}(kx+ly+mz-\omega t)}, \tag{6.34}$$

where $\widetilde{\psi}$ is a constant. Substituting (6.34) into (6.33) gives, after a couple of lines of elementary algebra, the dispersion relation

$$\omega = Uk - \frac{\beta k}{k^2 + l^2 + m^2 f_0^2/N^2}. \tag{6.35}$$

As in most wave problems the frequency is, by convention, a positive quantity.

In reality we usually have to satisfy a boundary condition at the top and bottom of the fluid and these are determined by the thermodynamic equation, (5.67), which after linearization becomes

$$\left(\frac{\partial}{\partial t} + U\frac{\partial}{\partial x}\right)\left(\frac{\partial \psi'}{\partial z}\right) + N^2 w = 0. \tag{6.36}$$

If the boundaries are flat, rigid, surfaces then $w = 0$ at those boundaries suggesting that instead of (6.34) we choose a streamfunction of the form

$$\psi' = \mathrm{Re}\,\widetilde{\psi}\mathrm{e}^{\mathrm{i}(kx+ly-\omega t)}\cos\left(\frac{m_j \pi z}{H}\right), \qquad m_j = 1, 2\ldots, \tag{6.37}$$

which satisfies $\partial\psi'/\partial z = 0$ at $z = 0, H$. The dispersion relation is the same as (6.35) with m equal to $m_j\pi/H$. If the domain is finite in the horizontal direction also — as it nearly always is — then the horizontal wavenumbers are also quantized. For example, if the extent of the domain in the

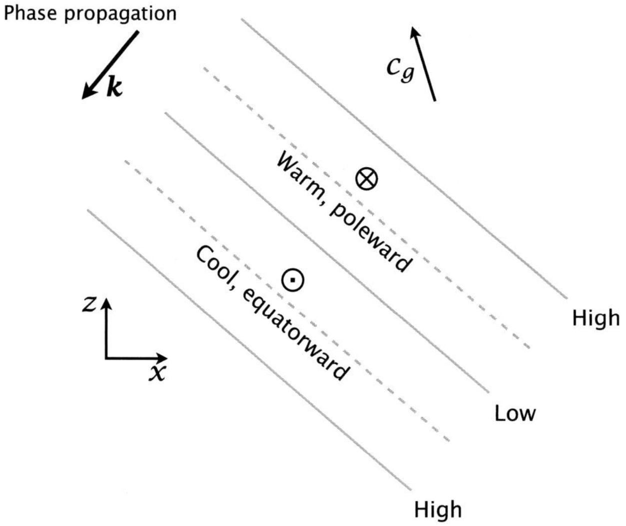

Fig. 6.4: A schematic east-west section of an upwardly propagating Rossby wave. The slanting lines are lines of constant phase and 'high' and 'low' refer to the pressure or streamfunction values. Both k and m are negative so the phase lines are oriented up and to the west. The phase propagates westward and downward, but the group velocity is upward.

x-direction is L_x and the flow is periodic in that direction, then k is restricted to the values $k = 2\pi k_j/L_x$ where k_j is an integer, and similarly for l and l_j. (It is common albeit sometimes confusing to refer to both k and k_j as the wavenumber and also to drop the subscript j on k_j with the distinction then made clear by context, and similarly for l and l_j.)

Using (6.35), the three components of the group velocity for these waves are:

$$c_g^x = U + \frac{\beta[k^2 - (l^2 + m^2 f_0^2/N^2)]}{(k^2 + l^2 + m^2 f_0^2/N^2)^2}, \tag{6.38a}$$

$$c_g^y = \frac{2\beta kl}{(k^2 + l^2 + m^2 f_0^2/N^2)^2}, \qquad c_g^z = \frac{2\beta km f_0^2/N^2}{(k^2 + l^2 + m^2 f_0^2/N^2)^2}. \tag{6.38b,c}$$

The propagation in the horizontal is similar to the propagation in a single-layer model. We see also that higher baroclinic modes (bigger m) will have a more westward group velocity. The vertical group velocity is proportional to m, and for waves that propagate signals upward we choose m to have the same sign as k so that c_g^z is positive. If there is no mean flow then the zonal wavenumber k is negative (in order that frequency is positive) and m must then also be negative. Energy then propagates upward but the phase propagates downward! This case is illustrated in Fig. 6.4 and to understand it better let us look at the vertical propagation in more detail.

Vertical propagation of Rossby waves was not discussed by Rossby, who was mainly interested in their properties in a horizontal plane. Rather, it was Charney & Drazin (1961) who first considered the vertical propagation of the waves in a stratified fluid in any detail.

6.4.2 Vertical Propagation of Rossby Waves

Conditions for wave propagation

With a little re-arrangement the dispersion relation, (6.35), may be written as

$$m^2 = \frac{N^2}{f_0^2}\left(\frac{\beta}{U - c} - (k^2 + l^2)\right), \tag{6.39}$$

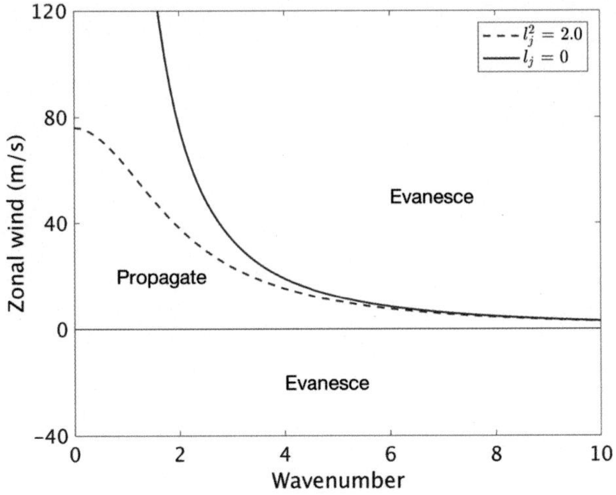

Fig. 6.5: The boundary between propagating waves and evanescent waves as a function of zonal wind and wavenumber, calculated using (6.41). The abscissa is the x-wavenumber, k_j, and the results show the boundary with two values of meridional wavenumber, l_j. Small wavenumbers (large scales) can more easily propagate into the stratosphere, and propagation is inhibited if the zonal wind is too strong or is negative.

where $c = \omega/k$. For waves to propagate upwards we require that $m^2 > 0$: if $m^2 < 0$ the wavenumber is imaginary and the wave amplitude either grows with height (which is unphysical) or is damped. This condition implies

$$0 < U - c < \frac{\beta}{k^2 + l^2}. \tag{6.40}$$

For waves of some given frequency ($\omega = kc$) the above expression provides a condition on U for the vertical propagation of planetary waves. For stationary waves $c = 0$ and the criterion is

Stationary, vertically oscillatory modes can exist only for zonal flows that are eastward and that are less than some critical value $U_c = \beta/(k^2 + l^2)$.

$$0 < U < \frac{\beta}{k^2 + l^2}, \tag{6.41}$$

as illustrated in Fig. 6.5. That is to say, the vertical propagation of stationary Rossby waves occurs only in eastward winds, and winds that are weaker than some critical value, $U_c = \beta/(k^2 + l^2)$ that depends on the scale of the wave. The lower limit, where $U = c$ (and so $U = 0$ if stationary) is known as a critical level and the upper limit is a 'turning level'. Since (6.39) is just a form of the dispersion relation, for any given frequency there is a criterion for propagation given by (6.40), and for any given U there may be a frequency that allows propagation. However, the zero frequency case is often regarded as the most important.

The upper critical velocity, $U_c = \beta/(k^2 + l^2)$, is a function of wavenumber and increases with horizontal wavelength. Thus, for a given eastward flow long waves may penetrate vertically when short waves are trapped, an effect sometimes referred to as 'Charney–Drazin filtering'. An important consequence is that the stratospheric motion is typically of larger scales than that of the troposphere, because Rossby waves tend to be excited first in the troposphere (by baroclinic instability and by flow over topography, among other things), but the shorter waves are trapped and only the longer ones reach the stratosphere. In the summer, the stratospheric winds are often westwards (because polar regions in the stratosphere are warmer than equatorial regions) and *all* waves are trapped in the troposphere; the eastward stratospheric winds that favour vertical penetration occur in the other three seasons, although very strong eastward winds can suppress penetration in mid-winter.

For westward flow, or for sufficiently strong eastward flow, m is imaginary and the waves decay exponentially in the vertical as $\exp(-\alpha z)$ where

$$\alpha = \frac{N}{f_0}\left(k^2 + l^2 - \frac{\beta}{U}\right)^{1/2}, \tag{6.42}$$

and the decay scale is smaller for shorter waves.

An interpretation

One physical way to interpret the propagation criterion is to write (6.41) as

$$0 < |Uk| < \frac{|\beta k|}{k^2 + l^2}, \tag{6.43}$$

allowing for the fact that k may be negative. Now, in a resting medium ($U = 0$) the Rossby wave frequency has a maximum value when $m = 0$ given by

$$\omega = \frac{|\beta k|}{k^2 + l^2}, \tag{6.44}$$

and the minimum frequency is zero. Let us suppose that the stationary Rossby waves are excited by flow of speed U moving over bottom topography. If we move into the frame of reference of the flow then the waves are forced by a moving topography, and the frequency of the forcing is just $|Uk|$. Thus, (6.43) is equivalent to saying that for oscillatory waves to exist *the forcing frequency, $|Uk|$, must lie within the frequency range of vertically propagating Rossby waves.*

6.4.3 Heat Transport and Vertical Propagation

If the group velocity in the z-direction, given by (6.38) is to be positive, then we require the product $km > 0$. This has an important consequence for the heat transport. Remember that the buoyancy b, which is a proxy for temperature, is given by $f_0 \partial\psi/\partial z$, and the northward velocity is $v = \partial\psi/\partial x$. Thus, the northward flux of heat, H say, is given by

$$H = \overline{vb} = f_0 \overline{\frac{\partial\psi}{\partial z}\frac{\partial\psi}{\partial x}}, \tag{6.45}$$

where an overbar denotes a zonal average. To evaluate this expression it is simplest to suppose that $\psi = \psi_0 \cos(kx + ly + mz)$, where ψ_0 is a real constant, in which case we obtain

$$H = f_0 km \psi_0^2 \overline{\sin^2(kx + ly + mz)} = \frac{1}{2} f_0 \psi_0^2 km, \tag{6.46}$$

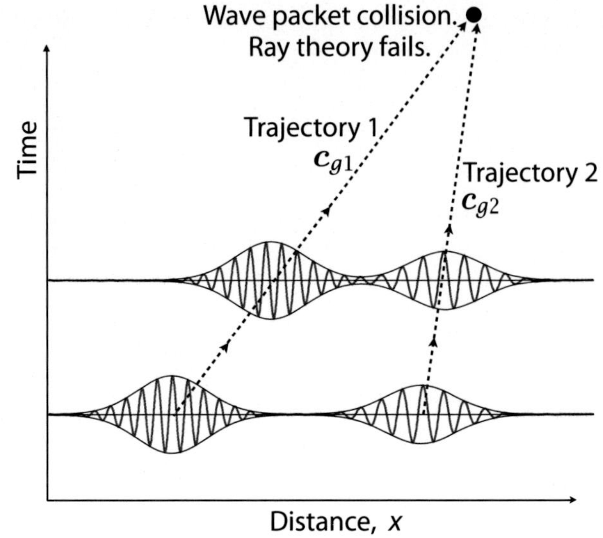

Fig. 6.6: Idealised trajectory of two wavepackets, each with a different wavelength and moving with a different group velocity, as might be calculated using ray theory. If the wave packets collide ray theory must fail. Ray theory gives only the trajectory of the wave packet, not the detailed structure of the waves within a packet.

using the standard result that the average of $\sin^2 x$ over a wavelength is equal to $1/2$. Thus, the heat flux, like the vertical component of the group velocity, is equal to km multiplied by a positive quantity, and we may conclude that an upward propagation of Rossby waves is associated with a polewards heat flux. A moment's thought will reveal that the same conclusion holds in the Southern Hemisphere, even though f_0 is negative there.

6.5 RAY THEORY AND ROSSBY RAYS

In the real world most waves propagate in a medium that is inhomogeneous; for example, in the Earth's atmosphere and ocean the Coriolis parameter varies with latitude and the stratification with height. In these cases it can be hard to obtain the solution of a wave problem by Fourier methods, even approximately. Nonetheless, the idea of signals propagating at the group velocity is a robust one, and we can often obtain some of the information we want — and in particular the trajectory of a wave — using a recipe known as *ray theory*. If the background properties of the medium vary only slowly compared to the wavelength, we assume that the dispersion relation is satisfied *locally*. We then calculate the group velocity, and assume that the trajectory of the wave is along that group velocity.

To implement this recipe, we first obtain a dispersion relation from the governing equation. In the homogeneous case we obtained a dispersion relation $\omega = f(\boldsymbol{k})$; that is the frequency is some function only of the wavenumber. In the inhomogeneous case we proceed the same way, and obtain a dispersion relation *as if* the parameters were fixed, but we then allow the parameters to vary in the dispersion relation; that is, we obtain a dispersion relation of the form $\omega = f(\boldsymbol{k}; \boldsymbol{x}, t)$. For example, for barotropic

Rossby waves, the dispersion relation might now be

$$\omega = U(y) - \frac{\beta(y)}{k^2 + l^2}. \tag{6.47}$$

This is the same as the usual dispersion relation except that U and β are now taken to be slowly varying functions of position. This procedure is valid provided that U and β vary more slowly than the wavelengths of the Rossby waves, which is essentially the same condition used in the WKB approximation, discussed in the appendix below.

The utility of ray theory is that it allows us to calculate the trajectory of a wave packet without fully solving the problem. Thus, suppose that a disturbance somewhere generates Rossby waves — perhaps a sea-surface temperature anomaly in the tropics. The waves propagate away from the disturbance following the group velocity, which we calculate using the dispersion relation. Thus, if the dispersion relation is given by (6.47) then the group velocity is given by

$$c_g^x = U + \frac{\beta(k^2 - l^2)}{(k^2 + l^2)^2}, \qquad c_g^y = \frac{2kl\beta}{(k^2 + l^2)^2}, \tag{6.48a,b}$$

where β and U are functions of space. Thus, the position of the wave packet, $\boldsymbol{x}$, is given by solving (often numerically) the ray equations

$$\frac{\mathrm{d}x}{\mathrm{d}t} = c_g^x, \qquad \frac{\mathrm{d}y}{\mathrm{d}t} = c_g^y, \tag{6.49}$$

where c_g^x and c_g^y are evaluated using (6.48). Figure 6.6 shows schematically how a wave packet might then propagate. In the Rossby-wave case here, the frequency is constant (determined by the source of the waves), as is the x-wavenumber, k, because U and β are only functions of y. As we move along the ray, we recalculate the y-wavenumber, l, at each point using the dispersion relation and then calculate the group velocity at each point, and thence the trajectory. When this is done on Earth, one sometimes can see Rossby waves propagating from the tropics to midlatitude that bring about long-range correlations in weather patterns known as 'teleconnections'.

In 1926, G. Wentzel, H. A. Kramers and L. Brillouin each presented a method to find approximate solutions of the Schrödinger equation in quantum mechanics, and the technique became known as the WKB method. It turns out that the technique had already been discovered by Harold Jeffreys, a mathematical geophysicist, a few years prior to that, and in fact the origins of the method go back to the early nineteenth century. Evidently, methods are named after the last people to discover them.

✦ Appendix A: The WKB Approximation

The WKB method is a way of finding approximate solutions to certain linear differential equations in which the term with the highest derivative is multiplied by a small parameter. In particular, WKB theory can be used to find approximate solutions to wave equations in which the coefficients vary slowly in space or time. Consider an equation of the form

$$\frac{\mathrm{d}^2\xi}{\mathrm{d}z^2} + m^2(z)\xi = 0. \tag{6.50}$$

Such an equation commonly arises in wave problems. If m^2 is positive the equation has wavelike solutions, and if m is constant the solution has the harmonic form

$$\xi = \text{Re}\, A_0 e^{imz}, \tag{6.51}$$

where A_0 is a complex constant. If m varies only slowly with z (meaning that the variations in m only occur on a scale much longer than $1/m$) one might reasonably expect that the harmonic solution above would provide a decent first approximation; that is, we expect the solution to locally look like a plane wave with local wavenumber $m(z)$. However, we might also expect that the solution would not be *exactly* of the form $\exp(im(z)z)$, because the phase of ξ is $\theta(z) = mz$, so that $d\theta/dz = m + z dm/dz \neq m$. Thus, in (6.51) m is not the wavenumber unless m is constant.

The condition that variations in m, or in the wavelength $\lambda \sim m^{-1}$, occur only slowly may be variously expressed as

$$\lambda\left|\frac{\partial \lambda}{\partial z}\right| \ll \lambda \quad \text{or} \quad \left|\frac{\partial m^{-1}}{\partial z}\right| \ll 1 \quad \text{or} \quad \left|\frac{\partial m}{\partial z}\right| \ll m^2. \tag{6.52a,b,c}$$

This condition will generally be satisfied if variations in the background state, or in the medium, occur on a scale much longer than the wavelength.

The Solution

Let us seek solutions of (6.50) in the form

$$\xi = A(z) e^{i\theta(z)}, \tag{6.53}$$

where $A(z)$ and $\theta(z)$ are both real. Using (6.53) in (6.50) yields

$$i\left[2\frac{dA}{dz}\frac{d\theta}{dz} + A\frac{d^2\theta}{dz^2}\right] + \left[A\left(\frac{d\theta}{dz}\right)^2 - \frac{d^2A}{dz^2} - m^2 A\right] = 0. \tag{6.54}$$

The terms in square brackets must each be zero. The WKB approximation is to assume that the amplitude varies sufficiently slowly that $|A^{-1}d^2A/dz^2| \ll m^2$, and hence that the term involving d^2A/dz^2 may be neglected. The real and imaginary parts of (6.54) become

$$\left(\frac{d\theta}{dz}\right)^2 = m^2, \qquad 2\frac{dA}{dz}\frac{d\theta}{dz} + A\frac{d^2\theta}{dz^2} = 0. \tag{6.55a,b}$$

The solution of the first equation above is

$$\theta = \pm \int m \, dz, \tag{6.56}$$

and substituting this into (6.55b) gives

$$2\frac{dA}{dz}m + A\frac{dm}{dz} = 0, \quad \text{with solution} \quad A = A_0 m^{-1/2}, \tag{6.57a,b}$$

where A_0 is a constant. Using (6.56) and (6.57b) in (6.53) gives us

$$\xi(z) = A_0 m^{-1/2} \exp\left(\pm i \int m \, dz\right), \tag{6.58}$$

and this is the WKB solution to (6.50). In terms of real quantities the solution may be written

$$\xi(z) = B_0 m^{-1/2} \cos\left(\int m \, dz\right) + C_0 m^{-1/2} \sin\left(\int m \, dz\right), \qquad (6.59)$$

where B_0 and C_0 are real constants.

Using (6.55a) and the real part of (6.54) we see that the condition for the validity of the approximation is that

$$\left| A^{-1} \frac{d^2 A}{dz^2} \right| \ll m^2, \qquad (6.60a,b)$$

which using (6.57b) is

$$\left| \frac{1}{m^{-1/2}} \frac{d^2 m^{-1/2}}{dz^2} \right| \ll m^2. \qquad (6.61)$$

Equation (6.52) expresses a similar condition to (6.61).

Notes and References

Many books discuss waves in fluids with the ones by LeBlond & Mysak (1980), Pedlosky (2003) and Gill (1982) having a geophysical emphasis. Accessible introductions to WKB theory can be found in Simmonds & Mann (1998) and Holmes (2013). An example of the application of WKB theory to the atmosphere is given in Hoskins & Karoly (1981).

Problems

6.1 *Rossby waves and Galilean invariance in shallow water.* Consider the flat-bottomed shallow-water quasi-geostrophic equations in standard notation,

$$\frac{D}{Dt}\left(\zeta + \beta y - \frac{f_0 \eta}{H}\right) = 0. \qquad (P6.1)$$

(*a*) How is ζ related to η? Express u, v, η and ζ in terms of a streamfunction.

(*b*) Linearize (P6.1) about a state of rest, and show that the resulting system supports two-dimensional Rossby waves. Discuss the limits in which the wavelength is much shorter or much longer than the deformation radius.

(*c*) Now linearize (P6.1) about a *geostrophically balanced state* that is translating uniformly eastwards. This means that: $u = U + u'$ and $\eta = \overline{\eta}(y) + \eta'$, where $\eta(y)$ is in geostrophic balance with U. Obtain an expression for the form of $\overline{\eta}(y)$. Obtain the dispersion relation for Rossby waves in this system. Show that their speed is different from that obtained by adding a constant U to the speed of Rossby waves in part (*b*). The problem is therefore not *Galilean invariant* — why?

6.2 *Horizontal propagation of Rossby waves.* Consider barotropic Rossby waves obeying the dispersion relation $\omega = Uk - \beta k/((k^2 + l^2)$, where U and/or β vary slowly with latitude.

(i) By re-arranging this expression to obtain an expression for the meridional wavenumber, or otherwise, show that the Rossby waves can only propagate in the horizontal if

$$0 < U - c < \frac{\beta}{k^2}, \tag{P6.2}$$

where $c = \omega/k$.

(ii) If the waves approach a latitude where $U = c$ (a 'critical latitude') show that the meridional wavenumber l becomes large but the group velocity in the y-direction becomes small. Show that Rossby waves generated in the atmosphere in midlatitudes are unlikely to propagate into the tropics (where the mean flow is westward).

(iii) Suppose that the Rossby waves approach a latitude where $U - c = \beta/k^2$. Calculate the x- and y-components of the group velocity in this limit and infer that a wave will turn away from such a latitude.

6.3 Consider barotropic Rossby waves with an infinite deformation radius. Obtain an expression for the group velocity in the presence of a uniform eastward mean flow. Show that for stationary waves the group velocity is eastward relative to Earth's surface, and hence deduce that energy propagation is downstream of topographic sources. Is this still true if the deformation radius is finite? Is it true if the mean flow is westward?

6.4 *Rossby waves in the two-level model.*

(a) Consider a two-layer (or two-level) quasi-geostrophic system on a β-plane, and you may consider the layers to be of equal depth. Linearize the equations about a state of rest, and show that they may be transformed into two uncoupled equations. Give a physical interpretation of what these equations represent.

(b) Obtain the dispersion relations for this system (there are two). Show that one of the dispersion relations corresponds to the synchronous, depth-independent motion in the two layers. What does the other one correspond to? Obtain the phases and group velocities of these waves and determine the conditions under which they are eastward or westward.

6.5 In realistic calculations of the vertical propagation of Rossby waves one must take into account the vertical variation of density. Carry through the calculation leading to the Charney–Drazin condition, either using pressure (or log-pressure) coordinates or using the anelastic version of quasi-geostrophy, using (5.66) with $\tilde{\rho} = \rho_0 \exp(-z/H)$. Show that the condition for propagation analogous to (6.41) becomes

$$0 < U < \frac{\beta}{k^2 + l^2 + \gamma^2}, \tag{P6.3}$$

and obtain an expression for γ.

CHAPTER

7 Gravity Waves

WAVES ARISE WHEN A SYSTEM IS PERTURBED and a restoring force tries to bring the system back to equilibrium; the system then overshoots and oscillations ensue. Gravity waves are waves in a fluid in which gravity provides the restoring force. For gravity to have an effect the fluid density must vary, and thus the waves must either exist at a fluid interface or in a stratified fluid — and a fluid interface is just an abrupt form of stratification. It is thus common to think of gravity waves as being either internal waves or surface waves: the former being in the interior of a fluid where the density changes may be continuous and the latter at a fluid interface, and naturally enough the two waves have many similarities. We considered surface waves in the hydrostatic, shallow water case in Chapter 4; now we consider internal waves in the continuously-stratified equations.

Gravity waves are those waves that exist in a fluid for which gravity provides the restoring force. Gravitational waves are a disturbance in the fabric of spacetime caused by accelerating massive bodies, as predicted by the general theory of relativity.

7.1 Internal Waves in a Continuously-Stratified Fluid

Internal gravity waves are waves that are internal to a stratified fluid and that owe their existence to the restoring force of gravity. In this section we will consider the simplest and most fundamental case, that of internal waves in a Boussinesq fluid with constant stratification and no background rotation. To this end, consider a fluid, initially at rest, in which the background buoyancy varies only with height and so the buoyancy frequency, N, is a function only of z. The system satisfies the Boussinesq equations (Section 2.5) and linearizing those equations of motion about this basic state gives the linear momentum equations,

$$\frac{\partial \boldsymbol{u}'}{\partial t} = -\nabla\phi', \qquad \frac{\partial w'}{\partial t} = -\frac{\partial \phi'}{\partial z} + b', \qquad (7.1\text{a,b})$$

and the mass continuity and thermodynamic equations,

$$\frac{\partial u'}{\partial x} + \frac{\partial v'}{\partial y} + \frac{\partial w'}{\partial z} = 0, \qquad \frac{\partial b'}{\partial t} + w'N^2 = 0. \tag{7.1c,d}$$

Our notation is such that $\boldsymbol{u} \equiv u\hat{\mathbf{i}} + v\hat{\mathbf{j}}$, $\boldsymbol{v} \equiv u\hat{\mathbf{i}} + v\hat{\mathbf{j}} + w\hat{\mathbf{k}}$, where $(\hat{\mathbf{i}}, \hat{\mathbf{j}}, \hat{\mathbf{k}})$ are the unit vectors in the x, y and z directions, and the gradient operator is horizontal unless noted. Thus, $\nabla \equiv \hat{\mathbf{i}}\,\partial_x + \hat{\mathbf{j}}\,\partial_y$ and $\nabla_3 \equiv \hat{\mathbf{i}}\,\partial_x + \hat{\mathbf{j}}\,\partial_y + \hat{\mathbf{k}}\,\partial_z$.

A little algebra gives a single equation for w',

$$\left[\frac{\partial^2}{\partial t^2}\left(\nabla^2 + \frac{\partial^2}{\partial z^2}\right) + N^2\nabla^2\right] w' = 0. \tag{7.2}$$

This equation is evidently *not* isotropic. If N^2 is a constant — that is, if the background buoyancy varies linearly with z — then the coefficients of each term are constant, and we may then seek solutions of the form

$$w' = \mathrm{Re}\ \widetilde{w} e^{\mathrm{i}(kx+ly+mz-\omega t)}, \tag{7.3}$$

where Re denotes the real part, a denotation that will frequently be dropped unless ambiguity arises, and other variables oscillate in a similar fashion. Using (7.3) in (7.2) yields the dispersion relation:

$$\omega^2 = \frac{(k^2 + l^2)N^2}{k^2 + l^2 + m^2} = \frac{K^2N^2}{K_3^2}, \tag{7.4}$$

where $K^2 = k^2+l^2$ and $K_3^2 = k^2+l^2+m^2$. The frequency (see Fig. 7.1) is thus always less than N, approaching N for small horizontal scales, $K^2 \gg m^2$. If we neglect pressure perturbations, as in the parcel argument of Section 3.4, then the two equations,

$$\frac{\partial w'}{\partial t} = b', \qquad \frac{\partial b'}{\partial t} + w'N^2 = 0, \tag{7.5}$$

form a closed set, and give $\omega^2 = N^2$.

If the basic state density increases with height then $N^2 < 0$ and the basic state is unstable. The disturbance grows exponentially according to $\exp(\sigma t)$ where

$$\sigma = \mathrm{i}\omega = \pm\frac{K\widetilde{N}}{K_3}, \tag{7.6}$$

where $\widetilde{N}^2 \equiv -N^2$ and $K_3 = \sqrt{K_3^2}$. Most convective activity in the ocean and atmosphere is, ultimately, related to an instability of this form, although of course there are many complicating issues — water vapour in the atmosphere, salt in the ocean, the effects of rotation and so forth.

7.1.1 Hydrostatic Internal Waves

Let us now suppose that the fluid satisfies the hydrostatic Boussinesq equations. The linearized two-dimensional equations of motion become

$$\frac{\partial \boldsymbol{u}'}{\partial t} = -\nabla\phi', \qquad 0 = -\frac{\partial \phi'}{\partial z} + b', \tag{7.7a}$$

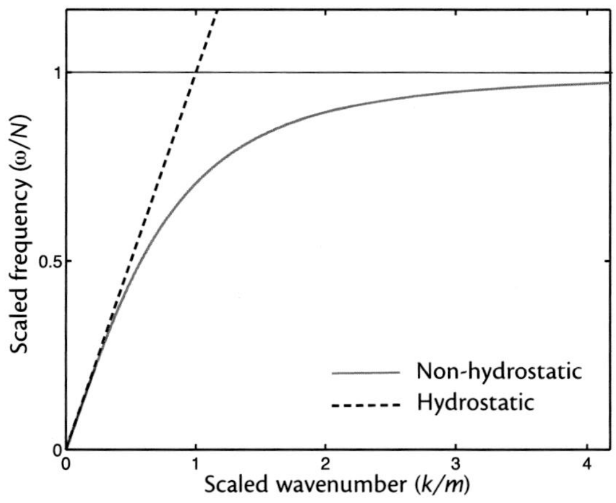

Fig. 7.1: Scaled frequency, ω/N, plotted as a function of scaled horizontal wavenumber, k/m, using the full dispersion relation of (7.4) with $l = 0$ (solid line, asymptoting to unit value for large k/m), and with the hydrostatic dispersion relation (7.8) (dashed line, tending to ∞ for large k/m).

$$\frac{\partial u'}{\partial x} + \frac{\partial v'}{\partial y} + \frac{\partial w'}{\partial z} = 0, \qquad \frac{\partial b'}{\partial t} + w'N^2 = 0, \tag{7.7b}$$

where these are the horizontal and vertical momentum equations, the mass continuity equation and the thermodynamic equation respectively. A little algebra gives the dispersion relation,

$$\omega^2 = \frac{(k^2 + l^2)N^2}{m^2}. \tag{7.8}$$

The frequency and, if N^2 is negative, the growth rate, are unbounded as $K^2/m^2 \to \infty$, and the hydrostatic approximation thus has quite unphysical behaviour for small horizontal scales. Many numerical models of the large-scale circulation in the atmosphere and ocean do make the hydrostatic approximation. In these models convection must be *parameterized;* otherwise, it would simply occur at the smallest scale available, namely the size of the numerical grid, and this type of unphysical behaviour should be avoided. In nonhydrostatic models convection must also be parameterized if the horizontal resolution of the model is too coarse to properly resolve the convective scales.

7.2 Properties of Internal Waves

Internal waves have a number of interesting and counter-intuitive properties — let's discuss them.

7.2.1 The Dispersion Relation

We can write the dispersion relation, (7.4), as

$$\omega = \pm N \cos\vartheta, \tag{7.9}$$

where $\cos^2\vartheta = K^2/(K^2 + m^2)$ so that ϑ is the angle between the three-dimensional wave-vector, $\boldsymbol{k} = k\hat{\mathbf{i}} + l\hat{\mathbf{j}} + m\hat{\mathbf{k}}$, and the horizontal. The frequency is evidently a function only of N and ϑ, and, if this is given, the frequency is not a function of wavelength. This has some interesting consequences for wave reflection, as we see below.

We can also write the dispersion relation, (7.4), as

$$\frac{\omega^2}{N^2 - \omega^2} = \frac{K^2}{m^2}. \tag{7.10}$$

Thus, and consistently with our first point, given the wave frequency the ratio of the vertical to the horizontal wavenumber is fixed.

7.2.2 Polarization Relations

The oscillations of pressure, velocity and buoyancy are, naturally, connected, and we can obtain the relations between them with some simple manipulations. If the pressure field is oscillating like $\phi' = \widetilde{\phi}\exp[\mathrm{i}(\boldsymbol{k}\cdot\boldsymbol{x} - \omega t)] = \widetilde{\phi}\exp[\mathrm{i}(kx+ly+mz-\omega t)]$ then, using (7.1a), the horizontal velocity components satisfy

$$(\widetilde{u}, \widetilde{v}) = (k, l)\,\omega^{-1}\widetilde{\phi}. \tag{7.11}$$

Evidently, since the frequency is real, the velocities are in phase with the pressure. We can obtain similar relations for the other variables and, since all the fields are real, it is convenient to express the relations in terms of sines and cosines. If we choose pressure to vary as a cosine then after some algebra we obtain

$$\phi = \Phi_0 \cos(kx + ly + mz - \omega t), \tag{7.12a}$$

$$(u, v) = (k, l)\frac{\Phi_0}{\omega}\cos(kx + ly + mz - \omega t), \tag{7.12b}$$

$$w = \frac{-K^2}{m\omega}\Phi_0 \cos(kx + ly + mz - \omega t), \tag{7.12c}$$

$$b = \frac{N^2K^2}{m\omega^2}\Phi_0 \sin(kx + ly + mz - \omega t), \tag{7.12d}$$

where Φ_0 is a constant. The vertical velocity is thus in phase with the pressure perturbation, and for regions of positive m (and so with upward phase propagation) regions of high relative pressure are associated with downward fluid motion. The above relations between pressure, buoyancy and velocity are known as *polarization relations.*

7.2.3 Relation between Wave Vector and Velocity

On multiplying (7.12b) and (7.12c) by (k, l) and m, respectively, we see that

$$\boldsymbol{k}\cdot\widetilde{\boldsymbol{v}} = 0, \tag{7.13}$$

where $\boldsymbol{k}$ and $\widetilde{\boldsymbol{v}}$ are three-dimensional vectors. This means that, at any instant, the wave vector is perpendicular to the velocity vector, and the velocity is therefore aligned *along* the direction of the troughs and crests,

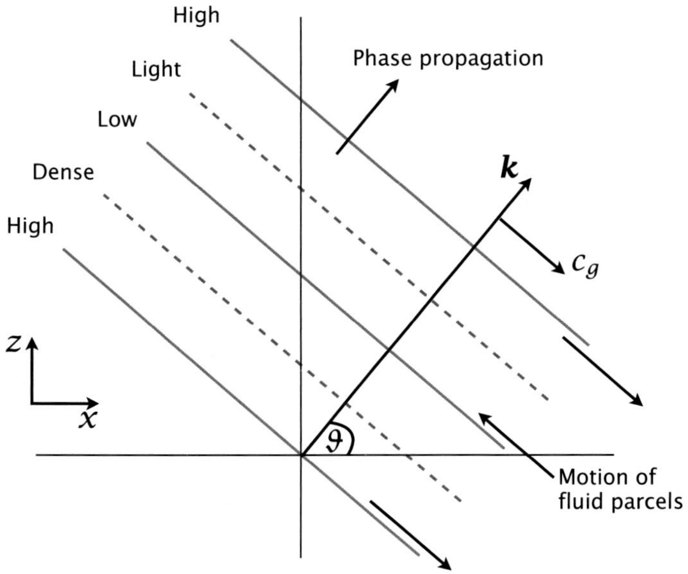

Fig. 7.2: An internal wave propagating in the direction $\boldsymbol{k}$. Both k and m are positive for the wave shown. The solid lines show crests and troughs of constant pressure, and the dashed lines the corresponding crests and troughs of buoyancy (or density). The motion of the fluid parcels is along the lines of constant phase, as shown, and is parallel to the group velocity and perpendicular to the phase speed.

along which there is no pressure gradient. If the wave vector is purely horizontal (i.e., $m = 0$), then the motion is purely vertical and $\omega = N$.

The vertical and horizontal velocities are related to the wavenumbers. If (for simplicity, and with no loss of generality) the motion is in the x–y plane with $v = l = 0$, then it is a corollary of (7.13) that

$$\frac{\widetilde{u}}{\widetilde{w}} = -\frac{m}{k}. \tag{7.14}$$

Furthermore, from (7.3) with $l = 0$, at any given instant all of the perturbation quantities in the wave are constant along the lines $kx+mz$ = constant. Thus, *all fluid parcel motions are parallel to the wave fronts.* Now, since the wave frequency is related to the background buoyancy frequency by $\omega = \pm N \cos\vartheta$, it follows that the fluid parcels oscillate along lines that are at an angle $\vartheta = \cos^{-1}(\omega/N)$ to the vertical. The polarization relations and the group and phase velocities are illustrated in Fig. 7.2. Let us now discuss the wave properties in a little more detail.

7.2.4 A Parcel Argument and Physical Interpretation

Let us consider first the dispersion relation itself and try to derive it more physically, or at least heuristically. Let us suppose there is a wave propagating in the (x, z) plane at some angle ϑ to the horizontal, with fluid parcels moving parallel to the troughs and crests, as in Fig. 7.2. In general the restoring force on a parcel is due to both the pressure gradient and gravity, but along the crests there is no pressure gradient. Referring to Fig. 7.3, for a total displacement Δs the restoring force, F_{res}, in the direc-

Fig. 7.3: Parcel displacements and associated forces in an internal gravity wave in which the parcel displacements are occurring at an angle ϑ to the vertical, as in Fig. 7.2.

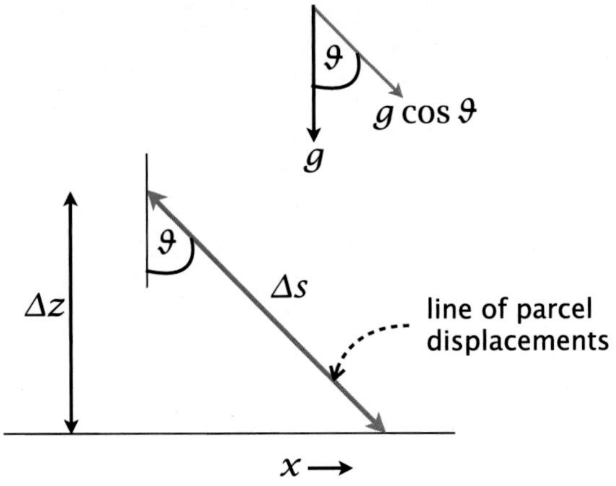

tion of the particle displacement is

$$F_{\text{res}} = g\cos\vartheta\,\Delta\rho = g\cos\vartheta\,\frac{\partial\rho}{\partial z}\Delta z = g\cos\vartheta\,\frac{\partial\rho}{\partial z}\Delta s\cos\vartheta = \rho_0\frac{\partial b}{\partial z}\cos^2\vartheta\,\Delta s, \tag{7.15}$$

noting that $\Delta z = \cos\vartheta\,\Delta s$. The equation of motion of a parcel moving along a trough or crest is therefore

$$\rho_0\frac{\mathrm{d}^2\Delta s}{\mathrm{d}t^2} = -\rho_0 N^2\cos^2\vartheta\,\Delta s, \tag{7.16}$$

which implies a frequency $\omega = N\cos\vartheta$, as in (7.9). One of the $\cos\vartheta$ factors in (7.16) comes from the fact that the parcel displacement is at an angle to the direction of gravity, and the other comes from the fact that the restoring force that a parcel experiences is proportional to $N\cos\vartheta$. (The reader may also wish to refer ahead to Fig. 7.6 and Section 7.3.1 for a similar argument.)

Now consider the wave illustrated in Fig. 7.2. For this wave both k and m are positive, and the frequency is assumed positive by convention to avoid duplicative solutions. The slanting solid and dashed lines are lines of constant phase, and from (7.12) the buoyancy and pressure are 1/4 of a wavelength out of phase. When k and m are both positive the extrema in the buoyancy field lag the extrema in the vertical velocity by $\pi/2$, as illustrated. The perturbation velocities are zero along the lines of extreme buoyancy. This follows because the velocities are in phase with the pressure, which as we noted is out of phase with the buoyancy.

Given the direction of the fluid parcel displacement in Fig. 7.2, the direction of the phase propagation c_p up and to the right may be deduced from the following argument. Buoyancy perturbations arise because of vertical advection of the background stratification, $w'\partial b_0/\partial z = w'N^2$. A local maximum in rising motion, and therefore a tendency to increase the fluid density, is present along the 'Low' line 1/4 wavelength upward and to the right of the 'Dense' phase line. Thus, the density of fluid along the 'Low' phase line increases and the 'Dense' phase line moves upward and to

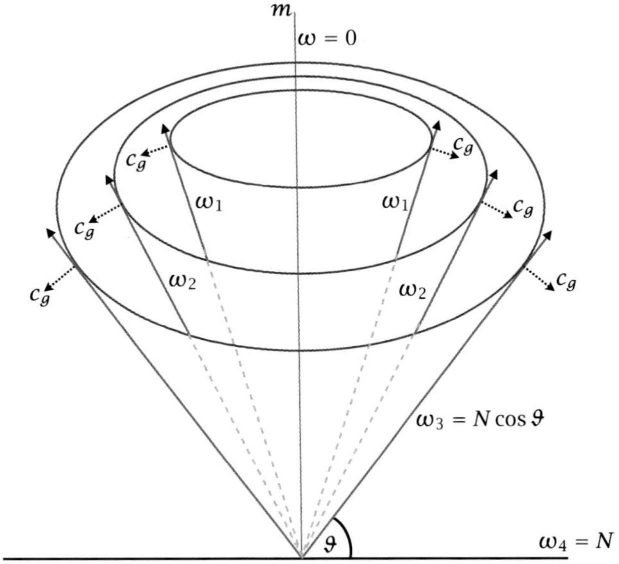

Fig. 7.4: Internal wave cones. The surfaces of constant frequency are cones, defined by the surface that has a constant angle to the horizontal. The wave vector, and so the phase velocity, point along the cone away from the origin, and the frequency of any wave with a wave vector in the cone is $N\cos\vartheta$. The group velocity is at right angles to the cone and pointed in the direction of increasing frequency, as indicated by the arrows on the dashed lines. In the vertical direction the phase speed and group velocity have opposite signs.

the right. If the fluid parcel motion were reversed the pattern of 'High–Dense–Low–Light–High' in Fig. 7.2 would remain the same. However, the downward fluid motion along the 'Low' line would cause the fluid to lose density, and so the phase lines would propagate downward and to the left. Evidently, the wave fronts, or the lines of constant phase, move at right angles to the fluid-parcel trajectories. In the figure we see that the group velocity is denoted as being at right angles to the phase speed, so let's discuss this.

7.2.5 Group Velocity and Phase Speed

As we noted above, the frequency of internal waves is given by $\omega = N\cos\vartheta$, where ϑ is the angle the wave vector makes with the horizontal. This means that the surfaces of constant frequency are *cones*, as illustrated in Fig. 7.4. To evaluate phase and group velocities in a useful way it is convenient to use spherical polar coordinates, as in Fig. 7.5, in which

$$k = K_3\cos\vartheta\cos\lambda, \qquad l = K_3\cos\vartheta\sin\lambda, \qquad m = K_3\sin\vartheta, \tag{7.17}$$

so that $\boldsymbol{k} = K_3(\cos\vartheta\cos\lambda, \cos\vartheta\sin\lambda, \sin\vartheta)$. The angles are ϑ, the angle of the wave vector with the horizontal and λ, which determines the orientation in the horizontal plane. (The notation is similar to the spherical coordinates of Chapter 2 — see Fig. 2.2 — although here ϑ is the angle with the horizontal, not the angle with the equatorial plane. Note also that λ is also used in this chapter for wavelength.) We also note that

$$\sin^2\vartheta = \frac{m^2}{k^2 + l^2 + m^2}, \qquad \cos^2\vartheta = \frac{K^2}{K_3^2} = \frac{k^2 + l^2}{k^2 + l^2 + m^2}, \qquad \tan\lambda = \frac{l}{k}. \tag{7.18}$$

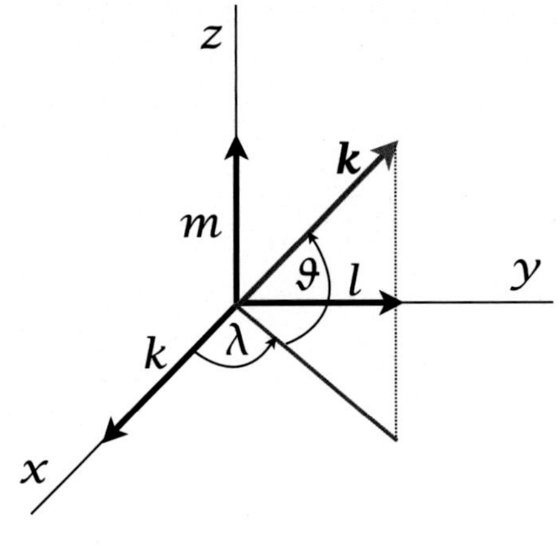

Fig. 7.5: The spherical coordinates used to describe internal waves, as in (7.17). The angle ϑ is the angle of the wave vector with the horizontal, and λ determines the orientation in the horizontal plane. The wave vector $\boldsymbol{k}$ is given by $\boldsymbol{k} = (k, l, m)$, these being the wavenumbers in the direction of increasing (x, y, z), respectively.

In many problems we can align the direction of the wave propagation with the x-axis and take $l = 0$ and $\tan\lambda = 0$.

The phase speed of the internal waves in the direction of the wave vector (sometimes referred to as the phase velocity) is given by

$$c_p = \frac{\omega}{K_3} = \frac{N}{K_3}\cos\vartheta = \frac{NK}{K_3^2}. \tag{7.19}$$

The phase speeds (as conventionally-defined) in the x,y and z directions are

$$c_p^x \equiv \frac{\omega}{k} = \frac{N}{k}\cos\vartheta, \qquad c_p^y \equiv \frac{\omega}{l} = \frac{N}{l}\cos\vartheta, \qquad c_p^z \equiv \frac{\omega}{m} = \frac{N}{m}\cos\vartheta. \tag{7.20a,b,c}$$

As noted in Section 6.1.2, these quantities are the speed of propagation of the wave crests in the respective directions. In general, each speed is *larger* than the phase speed in the direction perpendicular to the wave crests (that is, in the direction of the wave vector), but no information is transmitted at these speeds.

The group velocity is given by

$$\boldsymbol{c}_g = \left(\frac{\partial\omega}{\partial k}, \frac{\partial\omega}{\partial l}, \frac{\partial\omega}{\partial m}\right). \tag{7.21}$$

Using (7.4) we find

$$c_g^x = \frac{\partial\omega}{\partial k} = \frac{Nm}{K_3^2}\frac{km}{KK_3} = \left(\frac{N}{K_3}\sin\vartheta\right)\cos\lambda\sin\vartheta, \tag{7.22a}$$

$$c_g^y = \frac{\partial\omega}{\partial l} = \frac{Nm}{K_3^2}\frac{lm}{KK_3} = \left(\frac{N}{K_3}\sin\vartheta\right)\sin\lambda\sin\vartheta, \tag{7.22b}$$

$$c_g^z = \frac{\partial\omega}{\partial m} = -\frac{Nm}{K_3^2}\frac{K}{K_3} = -\left(\frac{N}{K_3}\sin\vartheta\right)\cos\vartheta. \tag{7.22c}$$

The magnitude of the group velocity is evidently

$$|c_g| = \frac{N}{K_3}\sin\vartheta, \tag{7.23}$$

and the group velocity vector is directed at an angle ϑ to the vertical, as in Fig. 7.4. This angle is perpendicular to the cone itself; that is, the group velocity is perpendicular to the wave vector, as may be verified by taking the dot product of (7.17) and (7.22) which gives

$$\boldsymbol{k} \cdot \boldsymbol{c}_g = 0. \tag{7.24}$$

The group velocity is therefore parallel to the motion of the fluid parcels, as illustrated in Fig. 7.2. Furthermore, because energy propagates with the group velocity, and the latter is *parallel* to lines of constant phase, energy propagates perpendicular to the direction of phase propagation — very different from most other wave types. In the vertical direction we see from (7.20c) and (7.22c) that

$$\frac{\omega}{m}\frac{\partial \omega}{\partial m} = -\frac{N^2}{K_3^2}\cos^2\vartheta < 0. \tag{7.25}$$

That is, the phase speed and the group velocity have opposite signs, meaning that if the wave crests move downward the group moves upward!

7.2.6 Effect of a Mean Flow

Suppose that there is a mean flow, U, in the x-direction, as is common in both atmosphere and ocean. The dispersion relation, (7.4), simply becomes

$$(\omega - Uk)^2 = \frac{K^2 N^2}{K^2 + m^2}. \tag{7.26}$$

The frequency is Doppler shifted, as expected, but the upward propagation of waves is affected in an interesting way. From (7.26) we find that the vertical component of the group velocity may be written as

$$\frac{\partial \omega}{\partial m} = \frac{-m(\omega - Uk)}{K^2 + m^2} = \frac{-mk(c - U)}{K^2 + m^2}, \tag{7.27}$$

where $c = \omega/k$ is the phase speed in the x-direction. If U is not constant but is varying slowly with z then (7.27) still holds, although m itself will also vary slowly with z. We see that the group velocity goes to zero at the location where $U = c$, known as a 'critical line', and the wave stalls.

The physical consequence of group velocity going to zero approaching a critical line is that any dissipation that may be present has more time to act. That is, we can expect a wave to be preferentially dissipated near a critical line, giving up its momentum to the mean flow and its energy to create mixing — the former being important in the atmosphere (for this is the mechanism producing the 'quasi-biennial oscillation') and the latter in the ocean. The mixing caused by internal wave breaking is in part responsible for the global scale meridional overturning circulation in the ocean, which we discuss more in Chapter 15.

The quasi-biennial oscillation, or QBO, is a quasi-periodic reversal of the zonal wind in the equatorial stratosphere, with a period of about 26 months. It is caused by the deposition of momentum in the stratosphere by upward-propagating Rossby and gravity waves.

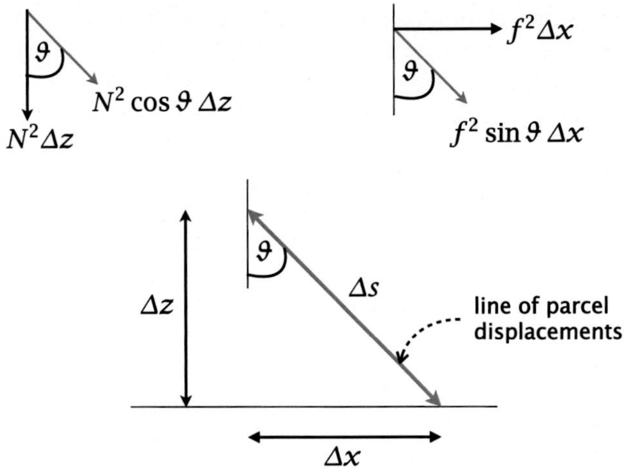

Fig. 7.6: Parcel displacements and associated forces in an inertia-gravity wave in which the parcel displacements are occurring at an angle ϑ to the vertical. Coriolis and buoyancy forces are present, and $\Delta s = \Delta z/\cos\vartheta = \Delta x/\sin\vartheta$.

7.3 ✦ INTERNAL WAVES IN A ROTATING FRAME OF REFERENCE

In the presence of both a Coriolis force and stratification a displaced fluid will feel two restoring forces — one due to gravity and the other to rotation. The first gives rise to gravity waves, as we have discussed, and the second to inertial waves. When the two forces both occur the resulting waves are called *inertia-gravity waves.* The algebra describing them can be complicated so we begin with a simple parcel argument to lay bare the basic dynamics; refer to Section 7.2.4 as needed.

7.3.1 A Parcel Argument

Consider a parcel that is displaced along a slantwise path in the x–z plane, as shown in Fig. 7.6, with a horizontal displacement of Δx and a vertical displacement of Δz. Let us suppose that the fluid is Boussinesq and that there is a stable and uniform stratification given by $N^2 = -g\rho_0^{-1}\partial\rho_0/\partial z = \partial b/\partial z$. Referring to (7.15) as needed, the component of the restoring buoyancy force, F_b say, in the direction of the parcel oscillation is given by (7.15),

$$F_b = -N^2 \cos\vartheta\,\Delta z = -N^2\cos^2\vartheta\,\Delta s. \tag{7.28}$$

The parcel also experiences a restoring Coriolis force, F_C, and the component of this in the direction of the parcel displacement is

$$F_C = -f^2 \sin\vartheta\,\Delta x = -f^2\sin^2\vartheta\,\Delta s. \tag{7.29}$$

Here, and for the rest of the chapter, we denote the Coriolis parameter by f. It should be regarded as a constant in any given problem (so there are no Rossby waves), but its value varies with latitude. Using (7.28) and (7.29) the (Lagrangian) equation of motion for a displaced parcel is

$$\frac{d^2\Delta s}{dt^2} = -(N^2\cos^2\vartheta + f^2\sin^2\vartheta)\Delta s, \tag{7.30}$$

and hence the frequency is given by

$$\omega^2 = N^2 \cos^2\vartheta + f^2 \sin^2\vartheta. \tag{7.31}$$

Now, nearly everywhere in both atmosphere and ocean, $N^2 > f^2$. From (7.31) we then see that the frequency lies in the interval $N^2 > \omega^2 > f^2$. (To see this, put $N = f$ or $f = N$ in (7.31), and use $\sin^2\vartheta + \cos^2\vartheta = 1$. See also (7.43) below.) If the parcel displacements approach the vertical then the Coriolis force diminishes and $\omega \to N$, and similarly $\omega \to f$ as the displacements become horizontal. The ensuing waves are then pure inertial waves.

We can write (7.31) in terms of wavenumbers since, for motion in the x–z plane,

$$\cos^2\vartheta = \frac{k^2}{k^2 + m^2}, \qquad \sin^2\vartheta = \frac{m^2}{k^2 + m^2}, \tag{7.32}$$

where k and m are the horizontal and vertical wavenumbers and $l = 0$. The dispersion relation becomes

$$\omega^2 = \frac{N^2k^2 + f^2m^2}{k^2 + m^2}. \tag{7.33}$$

Let's now move on to a discussion using the linearized equations of motion.

7.3.2 Equations of Motion

In a rotating frame of reference, specifically on an f-plane, the linearized equations of motion are the momentum equations

$$\frac{\partial \boldsymbol{u}'}{\partial t} + \boldsymbol{f}_0 \times \boldsymbol{u}' = -\nabla\phi', \qquad \frac{\partial w'}{\partial t} = -\frac{\partial \phi'}{\partial z} + b', \tag{7.34a,b}$$

and the mass continuity and thermodynamic equations,

$$\frac{\partial u'}{\partial x} + \frac{\partial v'}{\partial y} + \frac{\partial w'}{\partial z} = 0, \qquad \frac{\partial b'}{\partial t} + w'N^2 = 0. \tag{7.34c,d}$$

These are similar to (7.1), with the addition of a Coriolis term in the horizontal momentum equations.

To obtain a single equation for w' we take the horizontal divergence of (7.34a) and use the continuity equation to give

$$\frac{\partial}{\partial t}\left(\frac{\partial w'}{\partial z}\right) + f\zeta' = \nabla^2\phi', \tag{7.35}$$

where $\zeta' \equiv (\partial v'/\partial x - \partial u'/\partial y)$ is the vertical component of the vorticity. We may obtain an evolution equation for that vorticity by taking the curl of (7.34a), giving

$$\frac{\partial \zeta'}{\partial t} = f\frac{\partial w'}{\partial z}. \tag{7.36}$$

Eliminating vorticity between these equations gives

$$\left(\frac{\partial^2}{\partial t^2} + f^2\right)\frac{\partial w'}{\partial z} = \frac{\partial}{\partial t}\nabla^2\phi'. \tag{7.37}$$

We may obtain another equation linking pressure and vertical velocity by eliminating the buoyancy between (7.34b) and (7.34d), so giving

$$\frac{\partial^2 w'}{\partial t^2} + N^2 w' = -\frac{\partial}{\partial t}\frac{\partial \phi'}{\partial z}. \tag{7.38}$$

Eliminating ϕ' between (7.37) and (7.38) gives a single equation for w' analogous to (7.2), namely

$$\left[\frac{\partial^2}{\partial t^2}\left(\nabla^2 + \frac{\partial^2}{\partial z^2}\right) + f^2\frac{\partial^2}{\partial z^2} + N^2\nabla^2\right]w' = 0. \tag{7.39}$$

If we assume a time dependence of the form $w' = \widehat{w}e^{-i\omega t}$, this equation may be written in the sometimes useful form,

$$\frac{\partial^2 \widehat{w}}{\partial z^2} = \left(\frac{N^2 - \omega^2}{\omega^2 - f^2}\right)\nabla^2\widehat{w}. \tag{7.40}$$

7.3.3 Dispersion Relation

If we assume wave solutions to (7.39) of the standard harmonic form, that is $w' = \widetilde{w}\exp[i(kx + ly + mz - \omega t)]$, we obtain the dispersion relation

$$\omega^2 = \frac{f^2 m^2 + (k^2 + l^2)N^2}{k^2 + l^2 + m^2}, \tag{7.41}$$

which is a minor generalization of (7.33). We can also write the dispersion relation as

$$\omega^2 = f^2\sin^2\vartheta + N^2\cos^2\vartheta, \tag{7.42}$$

From the dispersion relation, (7.43), we can see that the frequency lies between N and f, no matter which is larger.

as we have derived already, and which we may write as

$$\omega^2 = f^2 - (f^2 - N^2)\cos^2\vartheta, \quad \text{or} \quad \omega^2 = N^2 - (N^2 - f^2)\sin^2\vartheta, \tag{7.43}$$

where ϑ is the angle of the wavevector with the horizontal, and as noted earlier the frequency therefore lies between N and f. The waves satisfying (7.41) are called inertia-gravity waves and are analogous to surface gravity waves in a rotating frame — that is, Poincaré waves — discussed in Section 4.3.2.

In many atmospheric and oceanic situations $f \ll N$ (in fact typically $N/f \sim 100$, the main exception being weakly stratified near-surface mixed layers and the deep abyss in the ocean) and $f < \omega < N$. From (7.42) the frequency is dependent only on the angle the wavevector makes with the horizontal, and the surfaces of constant frequency again form cones in wavenumber space, although depending on the values of f and ω the frequency does not necessarily decrease monotonically with ϑ as in the non-rotating case.

A few limits

Internal gravity waves with rotation have a few interesting special cases:

(*i*) A purely horizontal wave vector. In this case $m = 0$ and $\omega = N$. The waves are then unaffected by the Earth's rotation. This is because the Coriolis force is (in the f-plane approximation) due to the product of the Coriolis parameter and the horizontal component of the velocity. If the wave vector is horizontal, the fluid velocities are purely vertical and so the Coriolis force vanishes.

(*ii*) A purely vertical wave vector. In this case $\omega = f$, the fluid velocities are horizontal and the fluid parcels do not feel the stratification. The oscillations are then known as *inertial waves,* although they are not inertial in the sense of there being no implied force in an inertial frame of reference. This case and the previous one show that when $\omega = N$ or $\omega = f$ the group velocity is zero.

(*iii*) In the limit $N \to 0$ we have pure inertial waves with a frequency $0 < \omega < f$, and specifically $\omega = f \sin\vartheta$. Similarly, as $f \to 0$ we have pure internal waves, as discussed previously, with $\omega = N\cos\vartheta$.

(*iv*) The hydrostatic limit. Hydrostasy occurs in the limit of large horizontal scales, $k, l \ll m$. If we therefore neglect k^2 and l^2 where they appear with m^2 in (7.41) we obtain

$$\omega^2 = f^2 + N^2\frac{K^2}{m^2} = f^2 + N^2\cos^2\vartheta. \tag{7.44}$$

In this case the frequency of the waves is small compared to the buoyancy frequency, or $\omega^2 N^2$, because $K^2/m^2 \ll 1$. Put another way, low frequencies tend to have a small aspect ratio and be hydrostatic.

7.4 ♦ TOPOGRAPHIC GENERATION OF INTERNAL WAVES

How are internal waves generated? One way that is important in both the ocean and atmosphere is by way of a horizontal flow, such as a mean wind or, in the ocean, a tide or a mesoscale eddy, passing over a topographic feature. This forces the fluid to move up and/or down, so generating an internal wave, commonly known as a mountain wave. In this section we illustrate the mechanism with simple examples of steady, uniform flow with constant stratification over idealized topography.

7.4.1 Sinusoidal Mountain Waves

For simplicity we ignore the effects of the Earth's rotation and pose the problem in two dimensions, x and z, using the Boussinesq approximation. Our goal is to calculate the response to a steady, uniform flow of magnitude U over a sinusoidally varying boundary $h = \tilde{h}\cos kx$ at $z = 0$, as in Fig. 7.7 with $k = 2\pi/\lambda$. The topographic variations are assumed small, so allowing the dynamics to be linearized, which enables an arbi-

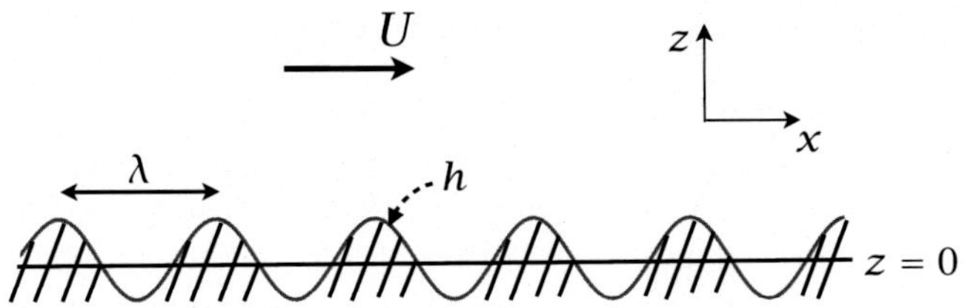

Fig. 7.7: Uniform flow, U, in the x-direction flowing over sinusoidal topography, h. The vertical coordinate is stretched, and in reality $|h| \ll \lambda$.

trarily shaped boundary to be considered by appropriately summing over Fourier modes.

Equation of motion and dispersion relation

The momentum equations, the buoyancy equation, and the mass continuity equation are, respectively,

$$\left(\frac{\partial}{\partial t} + U\frac{\partial}{\partial x}\right)u' = -\frac{\partial \phi}{\partial x}, \qquad \left(\frac{\partial}{\partial t} + U\frac{\partial}{\partial x}\right)w' = -\frac{\partial \phi}{\partial z} + b', \qquad (7.45\text{a,b})$$

$$\left(\frac{\partial}{\partial t} + U\frac{\partial}{\partial x}\right)u' + w'N^2 = 0, \qquad \frac{\partial u'}{\partial x} + \frac{\partial w'}{\partial z} = 0. \qquad (7.45\text{c,d})$$

We henceforth drop the primes on the perturbation quantities.

The dispersion relation is obtained by noting that the equations are the same as with no mean flow, save that $\partial/\partial t$ is replaced by $\partial/\partial t + U\partial/\partial x$. Thus, similarly to (7.26), the dispersion relation is

$$(\omega - Uk)^2 = \frac{k^2 N^2}{k^2 + m^2} \qquad \text{or} \qquad \omega = Uk \pm \frac{kN}{(k^2 + m^2)^{1/2}}. \qquad (7.46\text{a,b})$$

The horizontal and vertical components of the group velocity are then given by

$$c_g^z = \pm \frac{Nmk}{(k^2 + m^2)^{3/2}}, \qquad c_g^x = U \pm \frac{Nm^2}{(k^2 + m^2)^{3/2}}. \qquad (7.47\text{a,b})$$

Steady waves have $\omega = 0$ and if $U > 0$ such waves are associated with the negative root in (7.46b), the positive root in (7.47a) and the negative root in (7.47b). Energy will propagate upwards, and away from the mountain, if c_g^z is positive.

The solution

If we are looking for steady solutions we can follow the recipe of Section 7.1 but with $U\partial/\partial x$ replacing the time derivative. By analogy to (7.2) we find a single equation for w namely

$$\left[U\frac{\partial^2}{\partial x^2}\left(\frac{\partial^2}{\partial x^2} + \frac{\partial^2}{\partial z^2}\right) + N^2\frac{\partial^2}{\partial x^2}\right]w = 0. \qquad (7.48)$$

We take the topography to have the harmonic form $h = \operatorname{Re} h_0 \exp(ikx)$ and its presence is felt via a lower boundary condition, $w = U\partial h/\partial x =$

$\text{Re}\, Uh_0 ik \exp(ikx)$ at $z = 0$. We thus seek solutions to (7.47) of the form

$$w = \text{Re}\, Uh_0 ik e^{i(kx+mz)}. \tag{7.49}$$

The harmonic dependence in z is valid only because we take N to be constant. The value of m is given by the dispersion relation (7.46) with $\omega = 0$, which gives

$$m^2 = \left(\frac{N}{U}\right)^2 - k^2. \tag{7.50}$$

We see that m^2 may be negative, and so m imaginary, if $N^2 < k^2U^2$, so evidently there is a qualitative difference between short waves, with $k^2 > (N/U)^2$, and long waves, with $k^2 < (N/U)^2$. In the atmosphere we might take $U \sim 10\,\text{m}\,\text{s}^{-1}$ and $N \sim 10^{-2}\,\text{s}^{-1}$, in which case $U/N \sim 1\,\text{km}$. In the ocean below the thermocline we might take $U \sim 0.1\,\text{m}\,\text{s}^{-1}$ and $N \sim 10^{-3}\,\text{s}^{-1}$, whence $U/N \sim 100\,\text{m}$.

A useful interpretation of (7.50) arises if, since the mean flow is constant, we consider the problem in the frame of reference of that mean flow. In this case the topography has the form $h = \text{Re}\, h_0 \exp ik(x - Ut)$, and the solution in the moving frame is

Flow over topography is an important source of gravity waves in both atmosphere and ocean, with the flow coming from weather systems in the atmosphere or mesoscale eddies and tides in the atmosphere. Convection is also a significant source of gravity waves in the tropical atmosphere.

$$w = \text{Re}\, Uh_0 ik e^{i(kx+mz+Ukt)}. \tag{7.51}$$

That is, the fluid now oscillates with a frequency $\omega = -Uk$. The condition for propagation, $N^2 < k^2U^2$ is just the same as $\omega^2 < N^2$, which is just the condition for gravity waves to oscillate in a stratified fluid.

Given the solution for w we can use the polarization relations of Section 7.2 to obtain the solutions for perturbation horizontal velocity and pressure. One way to proceed is to pose the problem in the moving frame, with $\omega = -Uk$, and directly use (7.12). Then back in the stationary frame we obtain the solutions

$$w = w_0 e^{i(kx+mz)} = iUkh_0 e^{imz} e^{ikx}, \tag{7.52a}$$

$$u = u_0 e^{i(kx+mz)} = -imUh_0 e^{imz} e^{ikx}, \tag{7.52b}$$

$$\phi = \phi_0 e^{i(kx+mz)} = imU^2 h_0 e^{imz} e^{ikx}, \tag{7.52c}$$

where m is given by (7.50). Let us see what the solutions mean, and if and how waves propagate.

7.4.2 Energy Propagation

The direction of energy propagation is given by the group velocity. For steady waves ($\omega = 0$) we have, using (7.47a),

$$c_g^z = \frac{Nkm}{(k^2 + m^2)^{3/2}} = \frac{Ukm}{k^2 + m^2}. \tag{7.53a,b}$$

This means that for positive U an upward group velocity, and hence upward energy propagation, occur when k and m have the same sign. This property may also be deduced by evaluating the vertical energy flux, $\overline{w\phi}$, where the overbar denotes a zonal average, using (7.52), with appropriate care to take the real part of the fields. The phase speed, $c_p^z = \omega/m$, is zero in the stationary frame and it is $-Uk/m$ in the translating frame.

Short, trapped waves

If the undulations on the boundary are sufficiently short then $k^2 > (N/U)^2$ and m^2 is negative and m is purely imaginary. Writing $m = is$, so that $s^2 = k^2 - (N/U)^2$, the solutions have the form

$$w = \mathrm{Re}\, w_0 \mathrm{e}^{ikx - sz}. \tag{7.54}$$

We must choose the solution with $s > 0$ in order that the solution decays away from the mountain, and internal waves are not propagated into the interior. (If there were a rigid lid or a density discontinuity at the top of the fluid, as at the top of the ocean, then the possibility of reflection would arise and we would seek to satisfy the upper boundary condition with a combination of decaying and amplifying modes.) The above result is entirely consistent with the dispersion relation for internal waves, namely $\omega = N \cos \vartheta$: because $\cos \vartheta < 1$ the frequency ω must be less than N so that if the forcing frequency is higher than N no internal waves will be generated.

Because the waves are trapped waves we do not expect energy to propagate away from the mountains. To verify this, from the polarization relation (7.52) we have

$$w = \frac{k}{mU}\phi = \frac{-ik}{sU}\phi. \tag{7.55}$$

The pressure and the vertical velocity are therefore out of phase by $\pi/2$, and the vertical energy flux, $\overline{w\phi}$ is identically zero. This is consistent with the fact that the energy flux is in the direction of the group velocity; the group velocity is given by (7.53a) and for an imaginary m the real part is zero. A solution to the problem in the short wave limit is shown in the top panels of Fig. 7.8, with $s = 1$ and $m = \mathrm{i}$. (Here and in Fig. 7.9 the lengths and wavenumbers are in units such that the length of the domain is 2π and its height is that shown.)

Long, propagating waves

Suppose now that $k^2 < (N/U)^2$ so that $\omega^2 < N^2$. From (7.50) m is now real and the solution has propagating waves of the form

$$w = w_0 \mathrm{e}^{\mathrm{i}(kx+mz)}, \qquad m^2 = \left(\frac{N}{U}\right)^2 - k^2. \tag{7.56}$$

Vertical propagation is occurring because the forcing frequency is less than the buoyancy frequency. The angle at which fluid parcel oscillations occur is then slanted off the vertical at an angle ϑ such that the forcing frequency is equal to the natural frequency of oscillations at that angle, namely

$$\vartheta = \cos^{-1}\left(\frac{Uk}{N}\right). \tag{7.57}$$

The angle ϑ is also the angle between the wavevector $\boldsymbol{k}$ and the horizontal, as in (7.9), because the wavevector is at right angles to the parcel oscillations. If $Uk = N$ then the fluid parcel oscillations are vertical and, using

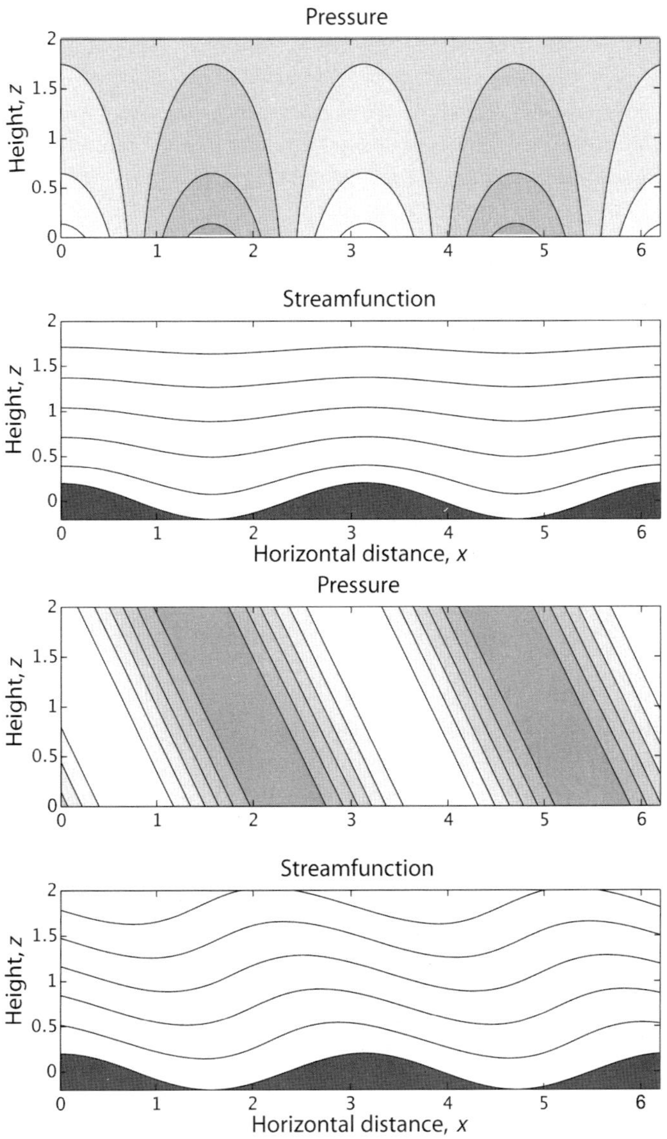

Fig. 7.8: Flow over a sinusoidal ridge in the shortwave limit (top two panels) and the longwave limit (bottom two panels), in a nondimensional domain of length 2π and height 2.

The top two panels show the solutions for the flow over a sinusoidal ridge, using (7.52), in the short wave limit ($Uk > N$) and with $m = \mathrm{i}$. The top panel shows the pressure, with darker grey indicating higher pressure. The second panel shows contours of the total streamfunction, $\psi - Uz$, with flow coming in from the left, and the topography itself (solid). The perturbation amplitude decreases exponentially with height.

The bottom two panels show solutions in the long wave limit ($Uk < N$) with $m = 1$. The pressure is high on the windward side of the topography, and phase lines tilt upstream with height for both pressure and streamfunction.

(7.50), $m = 0$. Thus, although the group velocity is directed vertically, parallel to the fluid parcel oscillations, its magnitude is zero, from (7.53).

Solutions for flow over topography in the long wave limit and $m = 1$ are shown in the lower two panels of Fig. 7.8. The flow is coming in from the left, and the phase lines evidently tilt upstream with height. Lines of constant phase follow $kx + mz$ = constant, and in the solution shown both k and m are positive ($k = m = 1$). Thus, the lines slope back at a slope $x/z = -m/k$, and energy propagates up and to the left. The phase propagation is actually downward in this example, as the reader may confirm. The pressure is high on the upstream side of the mountain, and this provides a drag on the flow — a so-called topographic form drag.

Topographically Generated Gravity Waves (Mountain Waves)

- In both atmosphere and ocean an important mechanism for the generation of gravity waves is flow over bottom topography, and the ensuing waves are sometimes called mountain waves. A canonical case is that of a uniform flow over a sinusoidal topography, with constant stratification. If the flow is in the x-direction and there is no y-variation then the boundary condition is
$$w(x, z = 0) = U\frac{\partial h}{\partial x} = -\mathrm{i}Uk\widetilde{h}. \qquad \text{(MW.1)}$$
Solutions of the problem may be found in the form $w(x, z, t) = w_0 \exp[\mathrm{i}(kx + mz - \omega t)]$, where the boundary condition at $z = 0$ is given by (MW.1), the frequency is given by the internal wave dispersion relation, and the other dynamical fields are obtained using the polarization relations.
- One way to solve the problem is to transform into a frame moving with the background flow, U, provided U is unform. The topography then appears to oscillate with a frequency $-Uk$, and this in turn becomes the frequency of the gravity waves.
- Propagating gravity waves can only be supported if the frequency is less than N, meaning that $Uk < N$. That is, the waves must be sufficiently long and therefore the topography must be of sufficiently large scale.
- When propagating waves exist, energy is propagated upward away from the topography. The topography also exerts a drag on the background flow.
- If the waves are too short they are evanescent, decaying exponentially with height. That is, they are trapped near the topography
- In the presence of rotation the wave frequency must lie between the buoyancy frequency N and the inertial frequency f. If the flow is constant, waves can thus radiate upward if
$$f < Uk < N. \qquad \text{(MW.2)}$$
Thus, both very long waves and very short waves are evanescent.

7.5 ♦ FLOW OVER AN ISOLATED RIDGE

Most mountains are of course not perfect sinusoids, but we can construct a solution for any given topography using a superposition of Fourier modes. In this section we will illustrate the solution for a mountain consisting of a single ridge; the actual solution must usually be obtained numerically, and we will sketch the method and show some results.

7.5.1 Sketch of the Methodology

The methodology to compute a solution is as follows. Consider a topographic profile, $h(x)$, and let us suppose that it is periodic in x over some

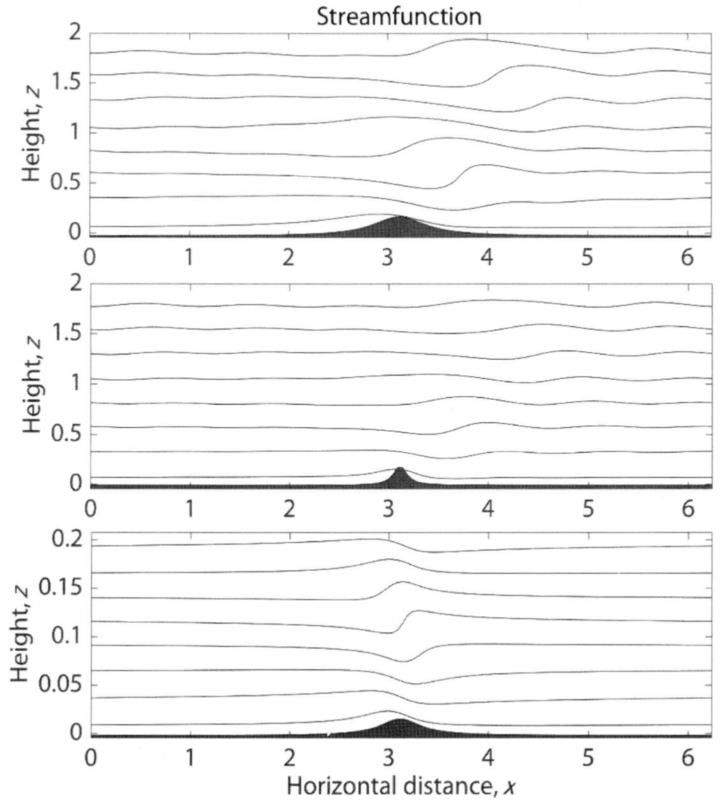

Fig. 7.9: The streamfunction for three solutions for uniform flow over isolated topography. The top panel shows flow over a wide hump, resulting in waves propagating vertically and downstream. The middle panel shows flow over a narrow hump and most of the waves evanesce away from the hump. The bottom panel shows the solution in a hydrostatic regime, very common in the atmosphere and ocean; here the group velocity is purely vertical and the disturbance is directly above the mountain.

distance L. Such a profile can (nearly always) be decomposed into a sum of Fourier coefficients, meaning that we can write

$$h(x) = \sum_k \widetilde{h}_k e^{ikx}, \tag{7.58}$$

where $\widetilde{h}_k$ are the Fourier coefficients. We can obtain the set of $\widetilde{h}_k$ by multiplying (7.58) by e^{-ikx} and integrating over the domain from $x = 0$ to $x = L$, a procedure known as taking the discrete Fourier transform of $h(x)$, and there are standard computer algorithms for doing this efficiently. Once we have obtained the values of $\widetilde{h}_k$ we essentially solve the problem separately for each k in precisely the same manner as we did in the previous section. For *each* k there is a vertical wavenumber given by (7.50), so that for each wavenumber we obtain a solution for pressure of the form $\widetilde{\phi}_k(z)$, and similarly for the other variables. Once we have the solution for each wavenumber, then at each level we sum over all the wavenumbers to obtain the solution in real space; that is, we evaluate

$$\phi(x,z) = \sum_k \widetilde{\phi}_z(z) e^{ikx}; \tag{7.59}$$

that is to say, we take the inverse discrete Fourier transform.

7.5.2 A Witch's Brew

For specificity let us consider the bell-shaped topographic profile

$$h(x) = \frac{h_0 a^2}{a^2 + x^2}, \tag{7.60}$$

The profile is named for Maria Agnesi, 1718–1799, an Italian mathematician and later a theologian who discussed the properties of the curve. The term 'witch' (in 'Witch of Agnesi') seems to be a mistranslation from Italian, or a deliberate and possibly unkind play on words, of versiera, which refers to a curve and not an adversary of God or a she-devil (which would be avversiera). Maria Agnesi, evidently a remarkable person, may have been the first woman appointed to a professorship of mathematics and that was at the University of Bologna.

sometimes called the Witch of Agnesi. Such a profile is composed of *many* (in fact an infinite number of) Fourier coefficients of differing amplitudes. If the profile is narrow (meaning a is small, in a sense made clearer below) then there will be a great many significant coefficients at high wavenumbers. In fact, in the limiting case of an infinitely thin ridge (a delta function) all wavenumbers are present with equal weight, so there are certainly more large wavenumbers than small wavenumbers. However, if a is large, then the contributing wavenumbers are predominantly small.

In the problem of flow over topography the natural horizontal scale of the flow is U/N. If $a \gg U/N$ then the dominant wavenumbers are small and the solution consists of waves propagating upward with little loss of amplitude and phase lines tilting upstream, as illustrated in the top panel of Fig. 7.9. In the case of a narrow ridge, as illustrated in the middle panel of that figure, the perturbation is largely trapped near to the mountain and the perturbation fields largely decay exponentially with height. Nevertheless, because the ridge does contain *some* small wavenumbers, some weak, propagating large-scale disturbances are generated. The fluid acts as a low-pass filter, and the perturbation aloft consists only of large scales.

A hydrostatic solution

If the ridge is sufficiently wide then the solution is essentially hydrostatic, with little dependence of the vertical structure on the horizontal wavenumber; that is, using (7.50) at large scales, $m^2 \approx (N/U)^2$. Furthermore, the x-component of the group velocity is zero, which can be seen from (7.47b) using $m = N/U$ (since m is positive for upward wave propagation and $m \gg k$ by hydrostasy). Explicitly we have

$$c_g^x = U - \frac{Nm^2}{(k^2 + m^2)^{3/2}} \approx U - \frac{N}{m} = U - U = 0. \tag{7.61}$$

Thus, the disturbance appears directly over the mountain, with no downstream propagation, as in the bottom panel of Fig. 7.9. The pattern therefore repeats itself in the vertical at intervals of $2\pi/m$ or $2\pi U/N$, with neither a downstream nor upstream influence. This solution is in fact the most atmospherically relevant one, for it is that produced by atmospheric flow over mountains that have horizontal scales larger than a few kilometres (i.e. $a \gg U/N$). It is also oceanographically relevant for flow over features of greater than a few kilometres, except possibly in regions of very weak stratification and large abyssal currents. Scales that are much larger are filtered by the Coriolis effect, because the frequency must lie between f and N.

Notes and References

The books by Gill (1982) and Sutherland (2010) provide more discussion of gravity waves in general, and Durran (2015) provides more discussion of mountain waves.

Problems

7.1 In the atmosphere gravity waves may be generated by flow over mountains, for example by westerly winds flowing over the Rockies. Suppose that the wind is $10\,\mathrm{m\,s^{-1}}$. Under what circumstances of stratification and width of the mountain range will the disturbance (a) propagate downstream; (b) upstream; (c) appear directly above the mountain; (d) evanesce? Be as quantitative as possible, and comment on the realism and/or likelihood of each case.

7.2 In the ocean, gravity waves may be generated by flow over bottom topography, with the flow generated either by tides, which have a frequency of order hours, or by mesoscale eddies, which have a timescale of weeks and which may have a different strength. Discuss the properties of the gravity waves generated by these two effects. (This is an open-ended question that will require research into the literature.)

7.3 As shown in Section 7.1, the frequency of hydrostatic gravity waves increases without limit as the horizontal scale becomes smaller and smaller.

(a) What is the physical reason behind this effect. (That is, give an argument in words as to why this effect occurs, and why it does not in the nonhydrostatic case.)

(b) In the atmosphere, at what scale does this effect become a practical concern? Suppose we are building a numerical model of the atmosphere; at what grid resolution do we need to be concerned as to whether we have a nonhydrostatic or hydrostatic model?

7.4 *Co-existence of gravity waves and Rossby waves.* Consider the linear shallow-water equations in a rotating frame of reference:

$$\frac{\partial u}{\partial t} - fv = -\frac{\partial \phi}{\partial x}, \qquad \frac{\partial v}{\partial t} + fu = -\frac{\partial \phi}{\partial y}, \qquad \frac{\partial \phi}{\partial t} + c^2\left(\frac{\partial u}{\partial x} + \frac{\partial v}{\partial y}\right) = 0, \tag{P7.1}$$

where $\phi = g'\eta$ and $c^2 = g'H$, η is the free surface height, H is the reference depth of the fluid and g' is the reduced gravity. Show that these equations reduce to a single equation for v, namely

$$\frac{1}{c^2}\frac{\partial^3 v}{\partial t^3} + \frac{f^2}{c^2}\frac{\partial v}{\partial t} - \frac{\partial}{\partial t}\nabla^2 v - \beta\frac{\partial v}{\partial x} = 0, \tag{P7.2}$$

where $\beta = \partial f/\partial y$. Now suppose that f and β are both constant in the above equation and derive the dispersion relation

$$\omega^2 - \frac{\beta k c^2}{\omega} = f_0^2 + c^2(k^2 + l^2). \tag{P7.3}$$

By considering appropriate limits of this equation, or otherwise, show that the system contains both Rossby and gravity waves. Plot the dispersion relation and show that with sensible values of parameters there is a frequency gap between the two kinds of waves.

CHAPTER

8 Instability

WHAT HYDRODYNAMIC STATES OCCUR IN NATURE? Any flow must of course be a solution of the equations of motion, subject to the relevant initial and boundary conditions. There are many steady solutions to the equations of motion — certain purely zonal flows, for example — but the flows we experience are unsteady, time-dependent solutions, not steady solutions. Why should this be? It is because for any steady flow to persist it must be stable to those small perturbations that inevitably arise, but all the steady solutions that are known for the large-scale flow in the Earth's atmosphere and ocean have been found to be unstable.

Our focus in this chapter is on barotropic and baroclinic instability. *Baroclinic instability* (and we will define the term more precisely later on) is an instability that arises in rotating, stratified fluids that are subject to a horizontal temperature gradient. It is the instability that gives rise to the large- and mesoscale motion in the atmosphere and ocean — it produces atmospheric weather systems, for example — and so is, perhaps, the form of hydrodynamic instability that most affects the human condition. *Barotropic instability* is an instability that arises because of the shear in a flow, and may occur in fluids of constant density. It is important to us for two reasons: first, in its own right as an instability mechanism for jets and vortices and as an important process in turbulence; second, many problems in barotropic and baroclinic instability are very similar, so that the solutions and insight we obtain in the often simpler problems in barotropic instability may be useful in the baroclinic problem.

8.1 KELVIN–HELMHOLTZ INSTABILITY

We first consider what is perhaps the simplest physically interesting instance of a fluid-dynamical instability — that of a constant-density flow with a shear perpendicular to the fluid's mean velocity, this being an example of a *Kelvin–Helmholtz instability*. Specifically, we consider two fluid masses of equal density, with an interface at $y = 0$, moving with velocities $-U$ and $+U$ in the x-direction, respectively, as in Fig. 8.1. There is no variation in the basic flow in the z-direction (normal to the page), and we will assume this is also true for the instability. This flow is clearly a solution of the Euler equations.

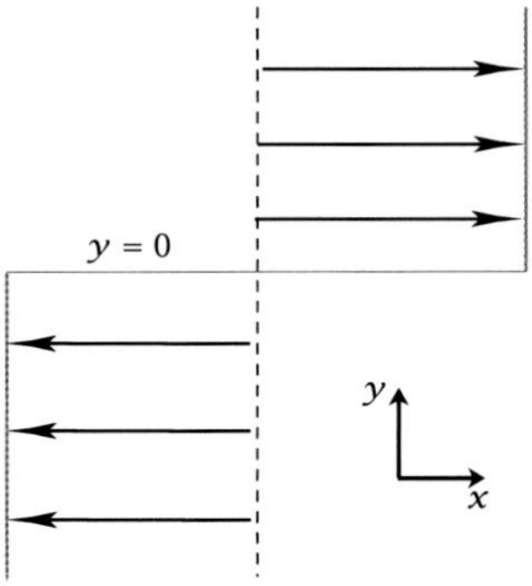

Fig. 8.1: A simple basic state giving rise to shear-flow instability. The velocity profile is discontinuous and the density is uniform. In more general problems in Kelvin–Helmholtz instability the density may also vary.

What happens if the flow is perturbed slightly? If the perturbation is initially small then even if it grows we can, for small times after the onset of instability, neglect the nonlinear interactions in the governing equations because these are the squares of small quantities. The equations determining the evolution of the initial perturbation are then the Euler equations linearized about the steady solution. Thus, denoting perturbation quantities with a prime and basic state variables with capital letters, for $y > 0$ the perturbation satisfies

$$\frac{\partial \boldsymbol{u}'}{\partial t} + U\frac{\partial \boldsymbol{u}'}{\partial x} = -\nabla p', \qquad \nabla\cdot\boldsymbol{u}' = 0, \tag{8.1a,b}$$

and a similar equation holds for $y < 0$, but with U replaced by $-U$. Given periodic boundary conditions in the x-direction, we may seek solutions of the form

$$\phi'(x,y,t) = \mathrm{Re}\sum_k \widetilde{\phi}_k(y)\exp[\mathrm{i}k(x-ct)], \tag{8.2}$$

where ϕ is any field variable (e.g., pressure or velocity), and Re denotes that only the real part should be taken. (Typically we use tildes over variables to denote Fourier-like modes, and we often omit the marker 'Re'.) Because (8.1a) is linear, the Fourier modes do not interact and we may confine attention to just one. Taking the divergence of (8.1a), the left-hand side vanishes and the pressure satisfies Laplace's equation

$$\nabla^2 p' = 0. \tag{8.3}$$

This has solutions in the form

$$p' = \begin{cases} \widetilde{p}_1 \mathrm{e}^{\mathrm{i}kx-ky}\mathrm{e}^{\sigma t} & y > 0, \\ \widetilde{p}_2 \mathrm{e}^{\mathrm{i}kx+ky}\mathrm{e}^{\sigma t} & y < 0, \end{cases} \tag{8.4}$$

where, anticipating the possibility of growing solutions, we have written the time variation in terms of a growth rate, $\sigma = -\mathrm{i}kc$. In general σ is complex: if it has a positive real component, the amplitude of the perturbation will grow and there is an instability; if σ has a non-zero imaginary component, then there will be oscillatory motion, and there may be both oscillatory motion *and* an instability. To obtain the dispersion relationship, we consider the y-component of (8.1a), namely (for $y > 0$)

$$\frac{\partial v_1'}{\partial t} + U\frac{\partial v_1'}{\partial x} = -\frac{\partial p_1'}{\partial y}. \tag{8.5}$$

Substituting a solution of the form $v_1' = \widetilde{v}_1\exp(\mathrm{i}kx + \sigma t)$ yields, with (8.4),

$$(\sigma + \mathrm{i}kU)\widetilde{v}_1 = k\widetilde{p}_1. \tag{8.6}$$

But the velocity normal to the interface is, at the interface, nothing but the rate of change of the position of the interface itself; that is, at $y = +0$

$$v_1 = \frac{\partial \eta'}{\partial t} + U\frac{\partial \eta'}{\partial x}, \tag{8.7}$$

or

$$\widetilde{v}_1 = (\sigma + \mathrm{i}kU)\widetilde{\eta}, \tag{8.8}$$

where η' is the displacement of the interface from its equilibrium position. Using this in (8.6) gives

$$(\sigma + \mathrm{i}kU)^2\widetilde{\eta} = k\widetilde{p}_1. \tag{8.9}$$

The above few equations pertain to motion on the $y > 0$ side of the interface. Similar reasoning on the other side gives (at $y = -0$)

$$(\sigma - \mathrm{i}kU)^2\widetilde{\eta} = -k\widetilde{p}_2. \tag{8.10}$$

But at the interface $p_1 = p_2$, because pressure must be continuous. The dispersion relationship then emerges from (8.9) and (8.10), giving

$$\sigma^2 = k^2U^2. \tag{8.11}$$

This equation has two roots, one of which is positive. Thus, the amplitude of the perturbation grows exponentially, like $\mathrm{e}^{\sigma t}$, and the flow is *unstable*. The instability itself can be seen in the natural world when billow clouds appear wrapped up into spirals: the clouds are acting as tracers of fluid flow, and are a manifestation of the instability at finite amplitude, as seen later in Fig. 8.2.

8.2 INSTABILITY OF PARALLEL SHEAR FLOW

We now consider a little more systematically the instability of parallel shear flows. This is a classic problem in hydrodynamic stability theory, and there are two particular reasons for our own interest:

(i) The instability is an example of *barotropic instability*, which abounds in the ocean and atmosphere. Roughly speaking, barotropic instability arises when a flow is unstable by virtue of its horizontal shear, with gravitational and buoyancy effects being secondary.

(ii) The instability is in many ways analogous to *baroclinic instability*, which is the main instability giving rise to weather systems in the atmosphere and similar phenomena in the ocean.

We restrict attention to two-dimensional, incompressible flow; this illustrates the physical mechanisms in the most transparent way, in part because it allows for the introduction of a streamfunction and the automatic satisfaction of the mass continuity equation. In fact, for parallel two-dimensional shear flows the most unstable disturbances are two-dimensional ones.

The vorticity equation for incompressible two-dimensional flow is just

$$\frac{D\zeta}{Dt} = 0. \tag{8.12}$$

We suppose the basic state to be a parallel flow in the x-direction that may vary in the y-direction. That is

$$\overline{\boldsymbol{u}} = U(y)\hat{\mathbf{i}}. \tag{8.13}$$

The linearized vorticity equation is then

$$\frac{\partial \zeta'}{\partial t} + U\frac{\partial \zeta'}{\partial x} + v'\frac{\partial Z}{\partial y} = 0, \tag{8.14}$$

where $Z = -\partial_y U$. Because the mass continuity equation has the simple form $\partial u'/\partial x + \partial v'/\partial y = 0$, we may introduce a streamfunction ψ such that $u' = -\partial\psi'/\partial y$, $v' = \partial\psi'/\partial x$ and $\zeta' = \nabla^2\psi'$. The linear vorticity equation becomes

$$\frac{\partial \nabla^2\psi'}{\partial t} + U\frac{\partial \nabla^2\psi'}{\partial x} + \frac{\partial Z}{\partial y}\frac{\partial \psi'}{\partial x} = 0. \tag{8.15}$$

The coefficients of the x-derivatives are not themselves functions of x; thus, we may seek solutions that are harmonic functions (sines and cosines) in the x-direction, but the y dependence must remain arbitrary at this stage and we write

$$\psi' = \mathrm{Re}\,\widetilde{\psi}(y)\mathrm{e}^{\mathrm{i}k(x-ct)}. \tag{8.16}$$

The full solution is a superposition of all wavenumbers, but since the problem is linear the waves do not interact and it suffices to consider them separately. If c is purely real then c is the phase speed of the wave; if c has a positive imaginary component then the wave will grow exponentially and is thus *unstable*.

From (8.16) we have

$$u' = \widetilde{u}(y)\mathrm{e}^{\mathrm{i}k(x-ct)} = -\widetilde{\psi}_y\mathrm{e}^{\mathrm{i}k(x-ct)}, \tag{8.17a}$$

$$v' = \widetilde{v}(y)\mathrm{e}^{\mathrm{i}k(x-ct)} = \mathrm{i}k\widetilde{\psi}\mathrm{e}^{\mathrm{i}k(x-ct)}, \tag{8.17b}$$

$$\zeta' = \widetilde{\zeta}(y)\mathrm{e}^{\mathrm{i}k(x-ct)} = (-k^2\widetilde{\psi} + \widetilde{\psi}_{yy})\mathrm{e}^{\mathrm{i}k(x-ct)}, \tag{8.17c}$$

where the y subscript denotes a derivative. Using (8.17) in (8.14) gives

$$(U - c)(\widetilde{\psi}_{yy} - k^2\widetilde{\psi}) - U_{yy}\widetilde{\psi} = 0, \tag{8.18}$$

which is known as *Rayleigh's equation*. It is the linear vorticity equation for disturbances to parallel shear flow, and in the presence of a β-effect it generalizes slightly to

$$(U - c)(\widetilde{\psi}_{yy} - k^2\widetilde{\psi}) + (\beta - U_{yy})\widetilde{\psi} = 0, \tag{8.19}$$

which is known as the Rayleigh–Kuo equation.

John Strutt (1842–1919) became 3rd Baron Rayleigh on the death of his father in 1873 and, in a testament to the enduring British class system, is almost universally known as Lord Rayleigh. He made major contributions in many areas of physics, among them fluid mechanics (including the theory of sound and instability theory), the analysis of the composition of gases (leading to the discovery of argon), and scattering theory.

8.2.1 Piecewise Linear Flows

Although Rayleigh's equation is linear and has a simple form, it is nevertheless quite difficult to analytically solve for an arbitrary smoothly varying profile. It is simpler to consider *piecewise linear* flows, in which U_y is constant over some interval, with U or U_y changing abruptly to another value at a line of discontinuity, as for example in Fig. 8.1. The curvature, U_{yy} is accounted for through the satisfaction of matching conditions, analogous to boundary conditions, at the lines of discontinuity (as in Section 8.1), and solutions in each interval are then exponential functions.

Jump or matching conditions

The idea, then, is to solve the linearized vorticity equation separately in the continuous intervals in which vorticity is constant, matching the solution with that in the adjacent regions. The matching conditions arise from two physical conditions:

(i) That normal stress should be continuous across the interface. For an inviscid fluid this implies that pressure be continuous.

(ii) That the normal velocity of the fluid on either side of the interface should be consistent with the motion of the interface itself.

Let us consider the implications of these two conditions.

(i) Continuity of pressure

The linearized momentum equation in the direction along the interface is:

$$\frac{\partial u'}{\partial t} + U\frac{\partial u'}{\partial x} + v'\frac{\partial U}{\partial y} = -\frac{\partial p'}{\partial x}. \tag{8.20}$$

For normal modes, $u' = -\widetilde{\psi}_y e^{ik(x-ct)}$, $v' = ik\widetilde{\psi}e^{ik(x-ct)}$ and $p' = \widetilde{p}e^{ik(x-ct)}$, and (8.20) becomes

$$ik(U-c)\widetilde{\psi}_y - ik\widetilde{\psi}U_y = -ik\widetilde{p}. \tag{8.21}$$

Because pressure is continuous across the interface we have the first *matching* or *jump condition,*

$$\Delta[(U-c)\widetilde{\psi}_y - \widetilde{\psi}U_y] = 0, \tag{8.22}$$

where the operator Δ denotes the difference in the values of the argument (in square brackets) across the interface. That is, the quantity $(U-c)\widetilde{\psi}_y - \widetilde{\psi}U_y$ is continuous.

We can obtain this condition directly from Rayleigh's equation, (8.19), written in the form

$$[(U-c)\widetilde{\psi}_y - U_y\widetilde{\psi}]_y + [\beta - k^2(U-c)]\widetilde{\psi} = 0. \tag{8.23}$$

Integrating across the interface gives (8.22).

(ii) Material interface condition
At the interface, the normal velocity v is given by the kinematic condition

$$v = \frac{D\eta}{Dt}, \tag{8.24}$$

where η is the interface displacement. The linear version of (8.24) is

$$\frac{\partial \eta'}{\partial t} + U\frac{\partial \eta'}{\partial x} = \frac{\partial \psi'}{\partial x}. \tag{8.25}$$

If the fluid itself is continuous then this equation must hold at either side of the interface, giving two equations and their normal-mode counterparts, namely,

$$\frac{\partial \eta'}{\partial t} + U_1\frac{\partial \eta'}{\partial x} = \frac{\partial \psi_1'}{\partial x} \quad \longrightarrow \quad (U_1 - c)\widetilde{\eta} = \widetilde{\psi}_1, \tag{8.26}$$

$$\frac{\partial \eta'}{\partial t} + U_2\frac{\partial \eta'}{\partial x} = \frac{\partial \psi_2'}{\partial x} \quad \longrightarrow \quad (U_2 - c)\widetilde{\eta} = \widetilde{\psi}_2. \tag{8.27}$$

Material continuity at the interface thus gives the second jump condition

$$\Delta\left[\frac{\widetilde{\psi}}{U - c}\right] = 0. \tag{8.28}$$

That is, $\widetilde{\psi}/(U - c)$ is continuous at the interface. Note that if U is continuous across the interface the condition becomes one of continuity of the normal velocity.

8.2.2 Kelvin–Helmholtz Instability, Revisited

We now use Rayleigh's equation and the jump conditions to consider the situation illustrated in Fig. 8.1; that is, vorticity is everywhere zero except in a thin sheet at $y = 0$. On either side of the interface, Rayleigh's equation is simply

$$(U_i - c)(\partial_{yy}\widetilde{\psi}_i - k^2\widetilde{\psi}_i) = 0, \qquad i = 1, 2 \tag{8.29}$$

or, assuming that $U_i \neq c$, $\partial_{yy}\widetilde{\psi}_i - k^2\widetilde{\psi}_i = 0$. (This is just Laplace's equation, coming from $\nabla^2\psi' = \zeta'$, with $\zeta' = 0$ everywhere except at the interface.) Solutions of this that decay away on either side of the interface are

$$y > 0: \qquad \widetilde{\psi}_1 = \Psi_1 e^{-ky}, \tag{8.30a}$$

$$y < 0: \qquad \widetilde{\psi}_2 = \Psi_2 e^{ky}, \tag{8.30b}$$

where Ψ_1 and Ψ_2 are constants. The boundary condition (8.22) gives

$$(U_1 - c)(-k)\Psi_1 = (U_2 - c)(k)\Psi_2, \tag{8.31}$$

and (8.28) gives

$$\frac{\Psi_1}{(U_1 - c)} = \frac{\Psi_2}{(U_2 - c)}. \tag{8.32}$$

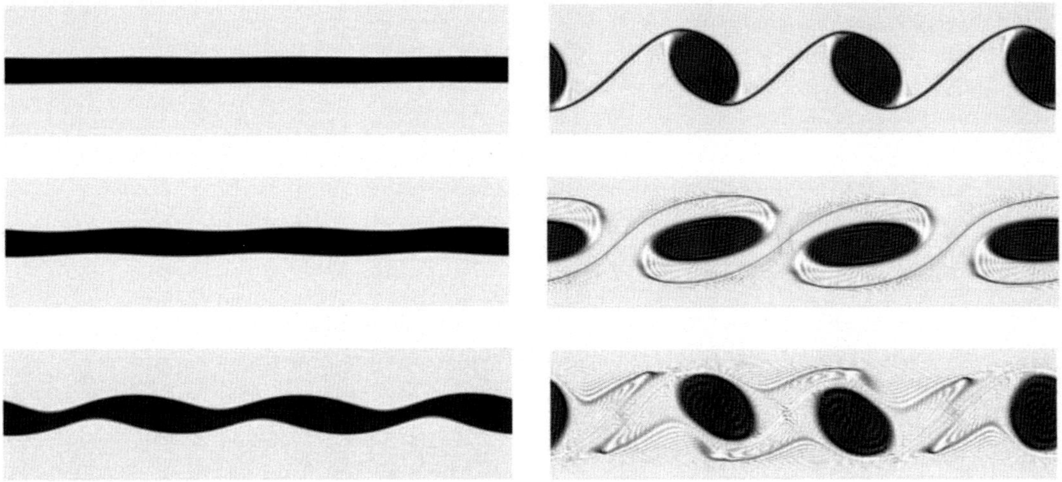

Fig. 8.2: A sequence of plots of the vorticity, at equal time intervals, from a numerical solution of the nonlinear vorticity equation (8.12), with initial conditions of an unstable shear flow, as in the figure below. Time increases first down the left column and then down the right column. The instability first grows exponentially, as in the left column, and then nonlinearities cause vortices to form, as in the right column.

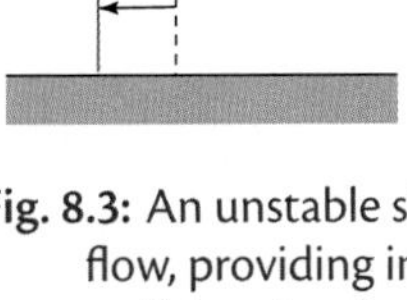

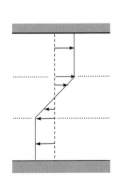

Fig. 8.3: An unstable shear flow, providing initial conditions for the evolution shown in Fig. 8.2.

The last two equations combine to give $(U_1 - c)^2 = -(U_2 - c)^2$, which, supposing that $U = U_1 = -U_2$ gives $c^2 = -U^2$. Thus, since U is purely real, $c = \pm iU$, and the disturbance grows exponentially as $\exp(kU_1 t)$, just as we obtained in Section 8.1, with all wavelengths being unstable, An example of a numerically calculated instability is given in Fig. 8.2, although here the initial vorticity is not just confined to a sheet but spread out over a finite width, as in Fig. 8.3, resulting in a preferred instability scale at about nondimensional wavenumber 3 (i.e., with 3 wavelengths spanning the domain).

8.3 Necessary Conditions for Instability

8.3.1 Rayleigh's Criterion

For simple profiles it may be possible to calculate, or even intuit, the instability properties, but for continuous profiles of $U(y)$ this is often impossible and it would be nice to have some general guidelines as to when a profile might be unstable. To this end, we derive a *necessary* condition for instability, or *sufficient* conditions for stability, that will at least tell us if a flow *might* be unstable.

Consider the vorticity equation, linearized about a parallel shear flow (cf. (8.14) with a β-term),

$$\frac{\partial \zeta}{\partial t} + U\frac{\partial \zeta}{\partial x} + v\left(\frac{\partial Z}{\partial y} + \beta\right) = 0, \tag{8.33}$$

where $Z = -\partial U/\partial y$ is the vorticity of the mean state and we have dropped the primes on the perturbation quantities. If we multiply by ζ and divide by $\beta + \partial_y Z$ we obtain

$$\frac{\partial}{\partial t}\left(\frac{\zeta^2}{Q_y}\right) + \frac{U}{Q_y}\frac{\partial \zeta^2}{\partial x} + 2v\zeta = 0, \tag{8.34}$$

where $Q_y = \partial_y Q = \beta + \partial_y Z$ is the gradient of the absolute vorticity, here equal to the gradient of the potential vorticity (hence the symbol Q). Integrating the above equation with respect to x gives

$$\frac{\partial}{\partial t}\int\left(\frac{\zeta^2}{Q_y}\right)\mathrm{d}x = -2\int v\zeta\,\mathrm{d}x. \tag{8.35}$$

Now, using $\nabla\cdot\boldsymbol{u} = 0$, the vorticity flux may be written as

$$v\zeta = -\frac{\partial}{\partial y}(uv) + \frac{1}{2}\frac{\partial}{\partial x}(v^2 - u^2). \tag{8.36}$$

The classical stability criterion for shear flow was written-down by Rayleigh (1880) and extended to a beta-plane by Kuo (1949) and then to a baroclinic flow (Section 8.7) by Charney & Stern (1962) and Pedlosky (1964).

That is, *the flux of vorticity is the divergence of some quantity.* Its integral therefore vanishes provided there are no contributions from the boundary, and integrating (8.35) with respect to y gives

$$\frac{\mathrm{d}}{\mathrm{d}t}\int\left(\frac{\zeta^2}{Q_y}\right)\mathrm{d}x\,\mathrm{d}y = 0. \tag{8.37}$$

If there is to be an instability ζ must grow, but the integral is identically zero. These two conditions can only be simultaneously satisfied if Q_y, or equivalently $\beta - U_{yy}$, is zero somewhere in the domain. Thus,

> *A necessary condition for instability is that the expression*
> $$\beta - U_{yy}$$
> *change sign somewhere in the domain.*

Equivalently, a sufficient criterion for stability is that $\beta - U_{yy}$ does not vanish in the domain interior. This condition is known as Rayleigh's inflection-point criterion, or when $\beta \neq 0$, the Rayleigh–Kuo inflection point criterion. The quantity ζ^2/Q_y is an example of a *wave-activity density* — a wave activity being a conserved quantity, quadratic in the amplitude of the wave. Such quantities play an important role in instabilities, and they crop up again in Chapter 9.

Interestingly, the β-effect can be either stabilizing or destabilizing: it can stabilize the middle two profiles of Fig. 8.5, because if it is large enough $\beta - U_{yy}$ will be one-signed. However, the β-effect will destabilize a westward point jet, $U(y) = -(1 - |y|)$ (the negative of the jet in Fig. 8.4), because $\beta - U_{yy}$ is negative at $y = 0$ and positive elsewhere. An eastward point jet is stable, with or without β. Finally, we emphasize that Rayleigh's criterion is a necessary condition for instability, not a sufficient one. In some circumstances a flow may satisfy the criterion for instability yet still be stable to infinitesimal perturbations, and this is found to be the case in the last row of Fig. 8.5.

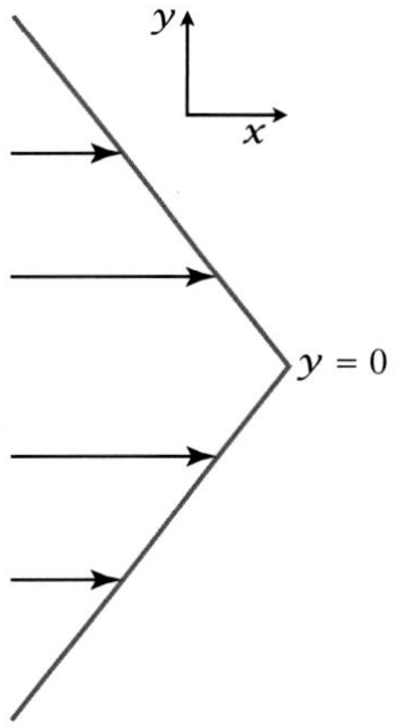

Fig. 8.4: Velocity profile of a point jet, in which vorticity is concentrated at a point. Although the vorticity is discontinuous, a small perturbation gives rise only to *edge waves* centred at $y = 0$, and the jet is stable

8.4 Baroclinic Instability

Baroclinic instability is an instability that occurs in stably stratified, rotating fluids, and it is ubiquitous in the Earth's atmosphere and oceans, and almost certainly occurs in other planetary atmospheres. It gives rise to weather, and so is perhaps the form of hydrodynamic instability that affects us most, and that certainly we talk about most.

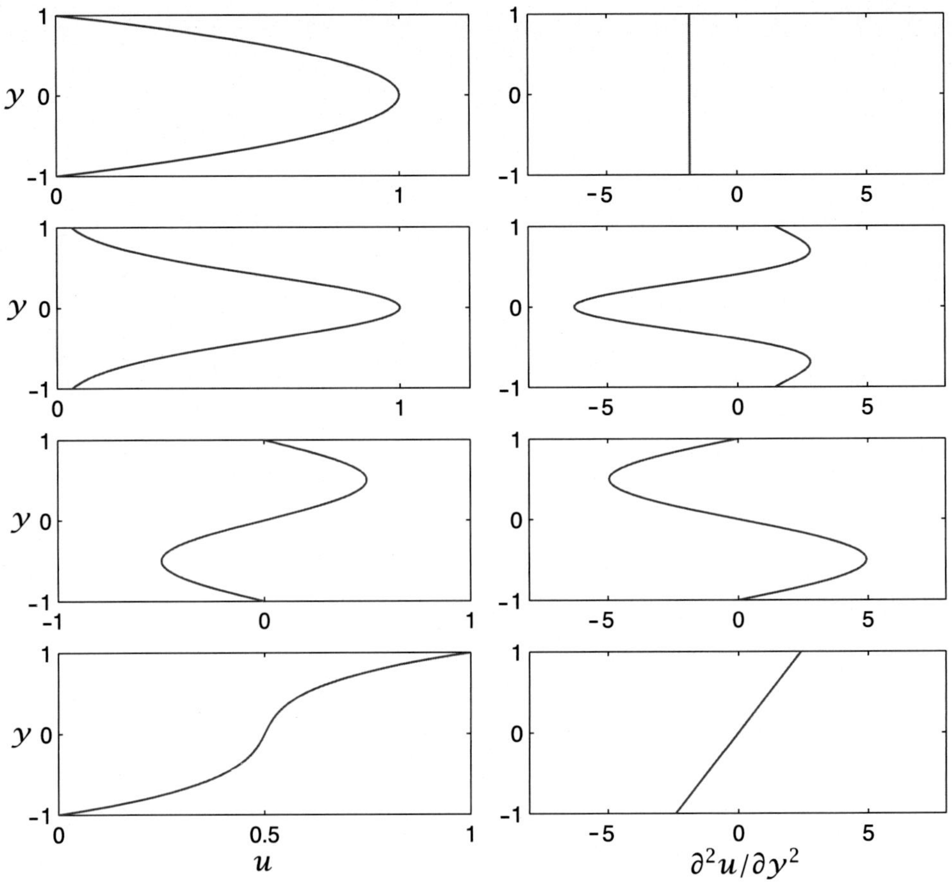

Fig. 8.5: Example parallel velocity profiles (left column) and their second derivatives (right column). From the top: Poiseuille flow ($u = 1 - y^2$); a Gaussian jet; a sinusoidal profile; a polynomial profile. By Rayleigh's criterion, the top profile is stable, whereas the lower three are potentially unstable. The bottom profile is in fact stable (although we do not demonstrate that here). If the β-effect were present and large enough it would stabilize the middle two profiles.

8.4.1 A Physical Picture

We first draw a picture of baroclinic instability as a form of 'sloping convection' in which the fluid, although statically stable, is able to release available potential energy when parcels move along a sloping path. To this end, let us first ask: what is the basic state that is baroclinically unstable? In a stably stratified fluid potential density decreases with height; we can also easily imagine a state in which the basic state temperature decreases, and the potential density increases, polewards. (We couch most of our discussion in terms of the Boussinesq equations and drop the qualifier 'potential' from density.) Can we construct a steady solution from these two conditions? The answer is yes, provided the fluid is also rotating; rotation is necessary because the meridional temperature gradient generally implies a meridional pressure gradient; there is nothing to balance this in the absence of rotation, and a fluid parcel would therefore accelerate. In a rotating fluid this pressure gradient can be balanced by the Coriolis force and a steady solution can be maintained even in the absence of viscosity. Consider a stably stratified Boussinesq fluid in geostrophic and hydrostatic balance on an f-plane, with buoyancy decreasing uniformly

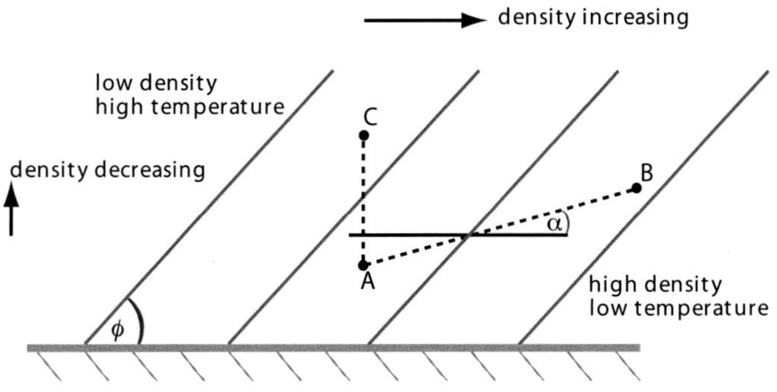

Fig. 8.6: A steady basic state giving rise to baroclinic instability. Potential density decreases upwards and equatorwards, and the associated horizontal pressure gradient is balanced by the Coriolis force. Parcel 'A' is heavier than 'C', and so statically stable, but it is lighter than 'B'. Hence, if 'A' and 'B' are interchanged there is a release of potential energy.

polewards. Then $fu = -\partial\phi/\partial y$ and $\partial\phi/\partial z = b$, where $b = -g\delta\rho/\rho_0$ is the buoyancy. These together give the thermal wind relation, $\partial u/\partial z = \partial b/\partial y$. If there is no variation of these fields in the zonal direction, then, for *any* variation of b with y, this is a steady solution to the primitive equations of motion, with $v = w = 0$.

The density structure corresponding to a uniform increase of density in the meridional direction is illustrated in Fig. 8.6. Is this structure stable to perturbations? The answer is no, although the perturbations must be a little special. Suppose the particle at 'A' is displaced upwards; then, since the fluid is (by assumption) stably stratified it will be denser than its surroundings and hence experience a restoring force, and similarly if displaced downwards. Suppose, however, we interchange the two parcels at positions 'A' and 'B'. Parcel 'A' finds itself surrounded by parcels of higher density than itself, and it is therefore buoyant; it is also higher than where it started. Parcel 'B' is negatively buoyant, and at a lower altitude than where is started. Thus, overall, the centre of gravity of the fluid has been lowered, and so its overall potential energy lowered. This loss in potential energy (PE) of the basic state must be accompanied by a gain in kinetic energy of the perturbation. Thus, the perturbation amplifies and converts potential energy into kinetic energy.

The loss of potential energy is easily calculated. Since

$$PE = \int \rho g \, dz, \tag{8.38}$$

the change in potential energy due to the interchange is

$$\Delta PE = g(\rho_A z_A + \rho_B z_B - \rho_A z_B - \rho_B z_A) = g(z_A - z_B)(\rho_A - \rho_B) = g\,\Delta\rho\,\Delta z. \tag{8.39}$$

If both $\rho_B > \rho_A$ and $z_B > z_A$ then the initial potential energy is larger than the final one, energy is released and the state is unstable. If the slope of the isopycnals is ϕ [so that $\phi = -(\partial_y\rho)/(\partial_z\rho)$] and the slope of the displacements is α, then for a displacement of horizontal distance L the change in potential energy is given by

$$\Delta PE = g\Delta\rho\Delta z = g\left(L\frac{\partial\rho}{\partial y} + L\alpha\frac{\partial\rho}{\partial z}\right)\alpha L = gL^2\alpha\frac{\partial\rho}{\partial y}\left(1 - \frac{\alpha}{\phi}\right), \tag{8.40}$$

if α and ϕ are small. If $0 < \alpha < \phi$ then energy is released by the perturbation, and it is maximized when $\alpha = \phi/2$. For the atmosphere the actual slope of the isotherms is about 10^{-3}, so that the slope and potential parcel trajectories are indeed shallow.

Although intuitively appealing, the thermodynamic arguments presented in this section pay no attention to satisfying the dynamical constraints of the equations of motion, and we now turn our attention to that.

8.4.2 Linearized Quasi-Geostrophic Equations

To explore the dynamics of baroclinic instability we use the quasi-geostrophic equations, specifically a potential vorticity equation for the fluid interior and a buoyancy or temperature equation at two vertical boundaries, one representing the ground and the other the tropopause. (The tropopause is the boundary between the troposphere and stratosphere at about 10 km; it is not a true rigid surface, but the higher static stability of the stratosphere does inhibit vertical motion.) For a Boussinesq fluid, the potential vorticity equation is

$$\begin{aligned} \frac{\partial q}{\partial t} + \boldsymbol{u} \cdot \nabla q = 0, \qquad 0 < z < H, \\ q = \nabla^2 \psi + \beta y + \frac{\partial}{\partial z}\left(F \frac{\partial \psi}{\partial z}\right), \end{aligned} \tag{8.41}$$

where $F = f_0^2/N^2$, and the buoyancy equation, with $w = 0$, is

$$\begin{aligned} \frac{\partial b}{\partial t} + \boldsymbol{u} \cdot \nabla b = 0, \qquad z = 0, H, \\ b = f_0 \frac{\partial \psi}{\partial z}. \end{aligned} \tag{8.42}$$

A solution of these equations is a purely zonal flow, $\boldsymbol{u} = U(y,z)\hat{\mathbf{i}}$ with a corresponding temperature field given by thermal wind balance. The potential vorticity of this basic state is

$$Q = \beta y - \frac{\partial U}{\partial y} + \frac{\partial}{\partial z} F \frac{\partial \Psi}{\partial z} = \beta y + \frac{\partial^2 \Psi}{\partial y^2} + \frac{\partial}{\partial z} F \frac{\partial \Psi}{\partial z}, \tag{8.43}$$

where Ψ is the streamfunction of the basic state, related to U by $U = -\partial\Psi/\partial y$. Linearizing (8.41) about this zonal flow gives the potential vorticity equation for the interior,

$$\frac{\partial q'}{\partial t} + U \frac{\partial q'}{\partial x} + v' \frac{\partial Q}{\partial y} = 0, \qquad 0 < z < H, \tag{8.44}$$

where $q' = \nabla^2 \psi' + \partial_z (F \partial_z \psi')$ and $v' = \partial_x \psi'$. Similarly, the linearized buoyancy equation at the boundary is

$$\frac{\partial b'}{\partial t} + U \frac{\partial b'}{\partial x} + v' \frac{\partial B}{\partial y} = 0, \qquad z = 0, H, \tag{8.45}$$

where $b' = f_0 \partial_z \psi'$ and $\partial_y B = \partial_y(f_0 \partial_z \Psi) = -f_0 \partial U/\partial z$.

Just as for the barotropic problem, a standard way of proceeding is to seek normal-mode solutions. Since the coefficients of (8.44) and (8.45) are functions of y and z, but not of x, we seek solutions of the form

$$\psi'(x, y, z, t) = \mathrm{Re}\, \widetilde{\psi}(y, z) e^{ik(x-ct)}, \tag{8.46}$$

and similarly for the derived quantities u', v', b' and q'. In particular,

$$\widetilde{q} = \frac{\partial^2 \widetilde{\psi}}{\partial y^2} + \frac{\partial}{\partial z} F \frac{\partial \widetilde{\psi}}{\partial z} - k^2 \widetilde{\psi}. \tag{8.47}$$

Using (8.47) and (8.46) in (8.44) and (8.45) gives, with subscripts y and z denoting derivatives,

$$(U - c)\left(\widetilde{\psi}_{yy} + (F\widetilde{\psi}_z)_z - k^2 \widetilde{\psi}\right) + Q_y \widetilde{\psi} = 0 \qquad 0 < z < H, \tag{8.48a}$$

$$(U - c)\widetilde{\psi}_z - U_z \widetilde{\psi} = 0 \qquad z = 0,\ H. \tag{8.48b}$$

These equations are analogous to Rayleigh's equations for parallel shear flow, and illustrate the similarity between the baroclinic instability problem and that of a parallel shear flow.

8.5 THE EADY PROBLEM

We now proceed to explicitly calculate the stability properties of a particular configuration that has become known as the *Eady problem.* This was one of the first two mathematical descriptions of baroclinic instability, the other being the *Charney problem.* The Charney problem is in some respects more complete (for example in allowing a β-effect), but the Eady problem displays the instability in a more transparent form. The β-effect can be incorporated relatively simply in the two-layer model of the next section.

Eric Eady (1915–1966) and Jule Charney (1917–1981) were the two great pioneers of the mathematical theory of baroclinic instability. They worked independently in the late 1940s, each formulating and solving a problem that became the (largely unsupervised) PhD thesis of its respective author, Eady at Imperial College London and Charney at UCLA. Their achievements in this problem rank among the most significant in all the physical sciences in the twentieth century.

To begin, let us make the following simplifying assumptions:

(i) The motion is on the f-plane ($\beta = 0$). This assumption, although not particularly realistic for the Earth's atmosphere, greatly simplifies the analysis.

(ii) The fluid is uniformly stratified; that is, N^2 is a constant. This is a decent approximation for the atmosphere below the tropopause, but less so for the ocean where stratification varies considerably, being much larger in the upper ocean.

(iii) The basic state has uniform shear; that is, $U_0(z) = \Lambda z = Uz/H$, where Λ is the (constant) shear and U is the zonal velocity at $z = H$, where H is the domain depth. This profile is more appropriate for the atmosphere than the ocean — below the thermocline the ocean is relatively quiescent and the shear is small.

(iv) The motion is contained between two rigid, flat horizontal surfaces. In the atmosphere this corresponds to the ground and a 'lid' at a constant-height tropopause.

Although, apart from *(i)*, these assumptions are more appropriate for the atmosphere than the ocean, the same qualitative nature of baroclinic instability carries through to the ocean.

8.5.1 The Linearized Problem

With a basic state streamfunction of $\Psi = -\Lambda z y$, the basic state potential vorticity, Q, is

$$Q = \nabla^2\Psi + \frac{H^2}{L_d^2}\frac{\partial}{\partial z}\left(\frac{\partial \Psi}{\partial z}\right) = 0. \tag{8.49}$$

The fact that $Q = 0$ makes the Eady problem a special case, albeit an illuminating one. The linearized potential vorticity equation is

$$\left(\frac{\partial}{\partial t} + \Lambda z\frac{\partial}{\partial x}\right)\left(\nabla^2\psi' + \frac{H^2}{L_d^2}\frac{\partial^2\psi'}{\partial z^2}\right) = 0. \tag{8.50}$$

This equation has no x-dependent coefficients and in a periodic channel we may seek solutions of the form $\psi'(x, y, z, t) = \mathrm{Re}\,\widetilde{\psi}(y, z)\mathrm{e}^{\mathrm{i}k(x-ct)}$, yielding

$$(\Lambda z - c)\left(\frac{\partial^2\widetilde{\psi}}{\partial y^2} + \frac{H^2}{L_d^2}\frac{\partial^2\widetilde{\psi}}{\partial z^2} - k^2\widetilde{\psi}\right) = 0. \tag{8.51}$$

This equation is (8.48a) applied to the Eady problem.

Boundary conditions

There are two sets of boundary conditions to satisfy, the vertical boundary conditions at $z = 0$ and $z = H$ and the lateral boundary conditions. In the horizontal plane we may either consider the flow to be in a channel, periodic in x and confined between two meridional walls, or, with a slightly greater degree of idealization but with little change to the essential dynamics, we may suppose that the domain is doubly-periodic. Either case is dealt with easily enough by the choice of geometric basis function; we choose a channel of width L and impose $\psi = 0$ at $y = +L/2$ and $y = -L/2$ and, to satisfy this, seek solutions of the form $\Psi = \Phi(z)\sin ly$ or, using (8.46)

$$\psi'(x, y, z, t) = \mathrm{Re}\,\Phi(z)\sin ly\mathrm{e}^{\mathrm{i}k(x-ct)}, \tag{8.52}$$

where $l = n\pi/L$, with n being a positive integer.

The vertical boundary conditions are that $w = 0$ at $z = 0$ and $z = H$. We follow the procedure of Section 8.4.2 and from (8.45) we obtain

$$\left(\frac{\partial}{\partial t} + \Lambda z\frac{\partial}{\partial x}\right)\frac{\partial\psi'}{\partial z} - \Lambda\frac{\partial\psi'}{\partial x} = 0, \qquad \text{at } z = 0, H. \tag{8.53}$$

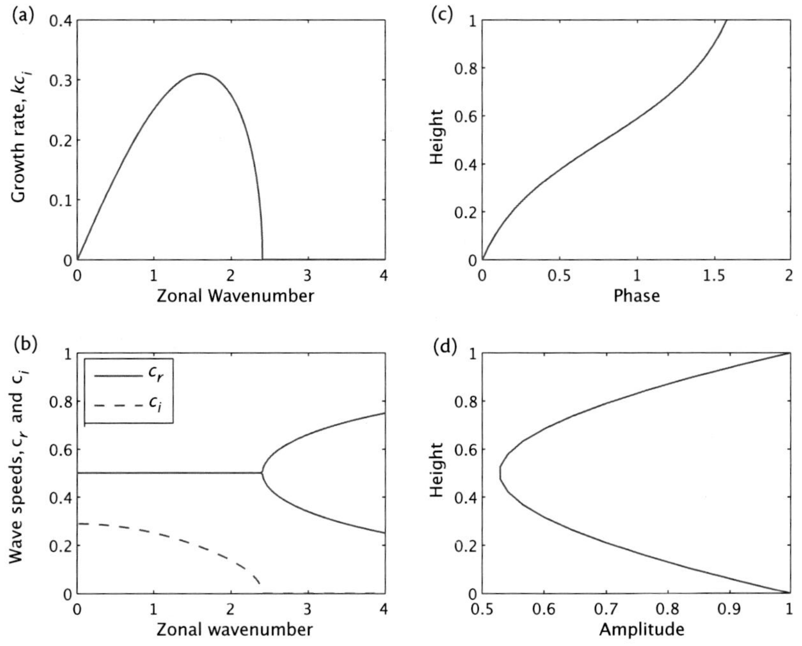

Fig. 8.7: Solution of the Eady problem, in nondimensional units. (a) The growth rate, kc_i as a function of scaled wavenumber μ, from (8.62) with $\Lambda = H = 1$ and for the largest meridional mode. (b) The real (solid) and imaginary (dashed) wave speeds of those modes, as a function of horizontal wavenumber. (c) The phase of the single most unstable mode as a function of height. (d) The amplitude of that mode as a function of height. To obtain dimensional values, multiply the growth rate by $\Lambda H/L_d$ and the wavenumber by $1/L_d$.

Solutions

Substituting (8.52) into (8.51) gives the interior potential vorticity equation

$$(\Lambda z - c)\left[\frac{H^2}{L_d^2}\frac{\partial^2 \Phi}{\partial z^2} - (k^2 + l^2)\Phi\right] = 0, \tag{8.54}$$

and substituting (8.52) into (8.53) gives, at $z = 0$ and $z = H$,

$$c\frac{\mathrm{d}\Phi}{\mathrm{d}z} + \Lambda\Phi = 0 \qquad \text{and} \qquad (c - \Lambda H)\frac{\mathrm{d}\Phi}{\mathrm{d}z} + \Lambda\Phi = 0. \tag{8.55a,b}$$

These are equivalent to (8.48b) applied to the Eady problem. If $\Lambda z \neq c$ then (8.54) becomes

$$H^2\frac{\mathrm{d}^2\Phi}{\mathrm{d}z^2} - \mu^2\Phi = 0, \tag{8.56}$$

where $\mu^2 = L_d^2(k^2 + l^2)$. The nondimensional parameter μ is a horizontal wavenumber, scaled by the inverse of the Rossby radius of deformation. Solutions of (8.56) are

$$\Phi(z) = A\cosh\mu\hat{z} + B\sinh\mu\hat{z}, \tag{8.57}$$

where $\hat{z} = z/H$; thus, μ determines the vertical structure of the solution. The boundary conditions (8.55) are satisfied if

$$A\,[\Lambda H] + B\,[\mu c] = 0,$$
$$A\,[(c - \Lambda H)\mu\sinh\mu + \Lambda H\cosh\mu] + B\,[(c - \Lambda H)\mu\cosh\mu + \Lambda H\sinh\mu] = 0. \tag{8.58}$$

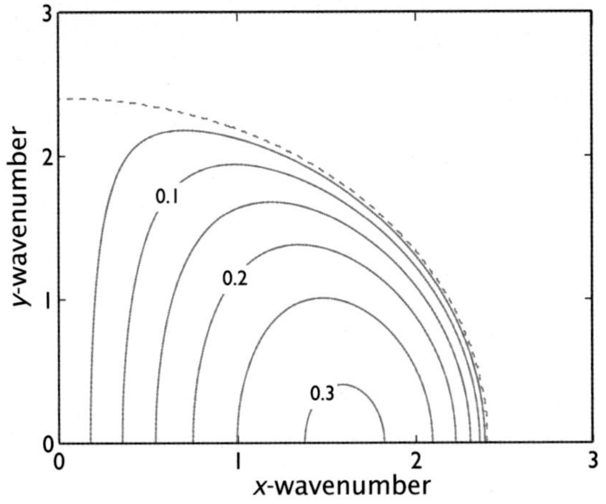

Fig. 8.8: Contours of the growth rate, σ, in the Eady problem, in the k–l plane using (8.62), nondimensionalized as in Fig. 8.7. The growth rate peaks near the deformation scale, and for any given zonal wavenumber the most unstable wavenumber is that with the largest meridional scale.

Equations (8.58) are two coupled homogeneous equations in the two unknowns A and B. Non-trivial solutions only exist if the determinant of their coefficients (the terms in square brackets) vanishes, and this leads to

$$c^2 - Uc + U^2(\mu^{-1}\coth\mu - \mu^{-2}) = 0, \tag{8.59}$$

where $U \equiv \Lambda H$ and $\coth\mu = \cosh\mu/\sinh\mu$. The solution of (8.59) is

$$c = \frac{U}{2} \pm \frac{U}{\mu}\left[\left(\frac{\mu}{2} - \coth\frac{\mu}{2}\right)\left(\frac{\mu}{2} - \tanh\frac{\mu}{2}\right)\right]^{1/2}. \tag{8.60}$$

The waves, being proportional to $\exp(-\mathrm{i}kct)$, will grow exponentially if c has an imaginary part. Since $\mu/2 > \tanh(\mu/2)$ for all μ, for an instability we require that

$$\frac{\mu}{2} < \coth\frac{\mu}{2}, \tag{8.61}$$

which is satisfied when $\mu < \mu_c$ where $\mu_c = 2.399$. The growth rates of the instabilities themselves are given by the imaginary part of (8.60), multiplied by the x-wavenumber; that is

$$\sigma = kc_i = k\frac{U}{\mu}\left[\left(\coth\frac{\mu}{2} - \frac{\mu}{2}\right)\left(\frac{\mu}{2} - \tanh\frac{\mu}{2}\right)\right]^{1/2}. \tag{8.62}$$

These solutions suggest a natural nondimensionalization: scale length by L_d, height by H and time by $L_d/U = L_d/(H\Lambda)$. The growth rate scales as the inverse of the time scaling and so by U/L_d. The timescale is also usefully written as

$$T_E = \frac{L_d}{U} = \frac{NH}{f_0 U} = \frac{N}{f_0 \Lambda} = \frac{1}{Fr f_0} = \frac{\sqrt{Ri}}{f_0}, \tag{8.63}$$

where $Fr = U/(NH)$ and $Ri = N^2/\Lambda^2$ are the Froude and Richardson numbers for this problem.

From (8.62) we can (with a little work) determine that the maximum growth rate occurs when $\mu = \mu_m = 1.61$. For any given x-wavenumber, the most unstable wavenumber has the largest meridional scale, which here is $n = 1$, and we may further consider a wide channel so that $l^2 \ll k^2$. The maximum growth rate, σ_E, is then given by

$$\sigma_E = \frac{0.31U}{L_d} = \frac{0.31\Lambda H}{L_d} = \frac{0.31\Lambda f}{N}, \tag{8.64}$$

and this is known as the *Eady growth rate*. We have removed the subscript 0 from the Coriolis parameter here. Although f is taken as constant in the quasi-geostrophic derivation, we might wish to calculate the Eady growth rate at various locations around the globe, in which case we should use the local value of the Coriolis parameter and deformation radius. Evidently, the growth rate is proportional to the shear times the Prandtl ratio, f/N. The associated phase speed is the real part of c and is given by $c_r = 0.5U$.

For small l the unstable x-wavenumbers and corresponding wavelengths occur for

$$k < k_c = \frac{\mu_c}{L_d} = \frac{2.4}{L_d}, \qquad \lambda > \lambda_c = \frac{2\pi L_d}{\mu_c} = 2.6L_d. \tag{8.65a,b}$$

The wavenumber and wavelength at which the instability is greatest are:

$$k_m = \frac{1.6}{L_d}, \qquad \lambda_m = \frac{2\pi L_d}{\mu_m} = 3.9L_d. \tag{8.66a,b}$$

These properties can be seen in the left-hand panels of Fig. 8.7 and in Fig. 8.8.

Given c, we may use (8.58) to determine the vertical structure of the Eady wave and this is, to within an arbitrary constant factor,

$$\Phi(\hat{z}) = \cosh \mu\hat{z} - \frac{U}{\mu c}\sinh \mu\hat{z} = \left(\cosh \mu\hat{z} - \frac{Uc_r \sinh \mu\hat{z}}{\mu|c^2|} + \frac{\mathrm{i}Uc_i \sinh \mu\hat{z}}{\mu|c^2|}\right). \tag{8.67}$$

The wave therefore has a phase, $\theta(z)$, given by

$$\theta(\hat{z}) = \tan^{-1}\left(\frac{Uc_i \sinh \mu\hat{z}}{\mu|c^2| \cosh \mu\hat{z} - Uc_r \sinh \mu\hat{z}}\right). \tag{8.68}$$

The phase and amplitude of the Eady waves are plotted in the right panels of Fig. 8.7, and their overall structure in Fig. 8.9, where we see the unstable wave tilting into the shear.

8.5.2 Atmospheric and Oceanic Parameters

To get a qualitative sense of the nature of the instability we choose some typical parameters, as follows:

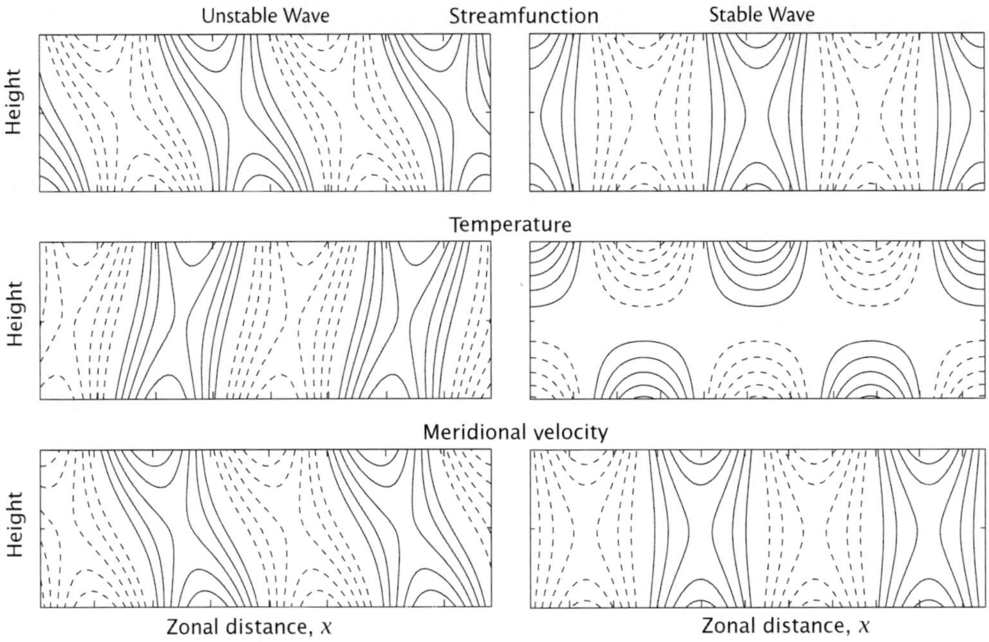

Fig. 8.9: Left column: vertical structure of the most unstable Eady mode. Top: contours of streamfunction. Middle: temperature, $\partial\psi/\partial z$. Bottom: meridional velocity, $\partial\psi/\partial x$. Negative contours are dashed, and two complete wavelengths are present in the horizontal direction. Polewards flowing (positive v) air is generally warmer than equatorwards flowing air. Right column: the same, but now for a wave just beyond the short-wave cut-off.

For the atmosphere

Let us choose

$$H \sim 10\,\text{km}, \qquad U \sim 10\,\text{m}\,\text{s}^{-1}, \qquad N \sim 10^{-2}\,\text{s}^{-1}.$$

We then obtain:

$$\text{deformation radius:} \qquad L_d = \frac{NH}{f} \approx \frac{10^{-2}\,10^4}{10^{-4}} \approx 1000\,\text{km},$$

$$\text{scale of max. instability:} \qquad L_{\max} \approx 3.9 L_d \approx 4000\,\text{km},$$

$$\text{growth rate:} \qquad \sigma \approx 0.3\frac{U}{L_d} \approx \frac{0.3\times 10}{10^6}\,\text{s}^{-1} \approx 0.26\,\text{day}^{-1}.$$

For the ocean

For the main thermocline in the ocean let us choose

$$H \sim 1\,\text{km}, \qquad U \approx 0.1\,\text{m}\,\text{s}^{-1}, \qquad N \sim 10^{-2}\,\text{s}^{-1}.$$

We then obtain:

$$\text{deformation radius:} \qquad L_d = \frac{NH}{f} \approx \frac{10^{-2}\times 1000}{10^{-4}} = 100\,\text{km},$$

$$\text{scale of max. instability:} \qquad L_{\max} \approx 3.9\,L_d \approx 400\,\text{km},$$

$$\text{growth rate:} \qquad \sigma \approx 0.3\frac{U}{L_d} \approx \frac{0.3 \times 0.1}{10^5}\,\text{s}^{-1} \approx 0.026\,\text{day}^{-1}.$$

In the ocean, the Eady problem is not quantitatively applicable because of the non-uniformity of the stratification. Nevertheless, the above estimates give a qualitative sense of the scale and growth rate of the instability relative to the corresponding values in the atmosphere. A summary of the main points of the Eady problem is given in the shaded box on the next page.

8.6 Two-Level Baroclinic Instability

This two-level instability problem is often called the 'Phillips problem', after Norman Phillips who formulated it in 1954. He also made a great many other contributions to dynamical meteorology and was one of the pioneers of numerical weather prediction.

The eigenfunctions displaying the largest growth rates in the Eady problem have a relatively simple vertical structure. This suggests that an even simpler mathematical model of baroclinic instability might be constructed in which the vertical structure is a priori restricted to a very simple form, namely the two-layer or two-level quasi-geostrophic (QG) model of Section 5.6. One notable advantage over the Eady model is that it is possible to include the β-effect in a simple way.

8.6.1 Posing the Problem

For two layers or two levels of equal thickness, we write the potential vorticity equations in the dimensional form,

$$\frac{\mathrm{D}}{\mathrm{D}t}\left[\zeta_i + \beta y + \frac{k_d^2}{2}(\psi_j - \psi_i)\right] = 0, \qquad i = 1,2, \quad j = 3-i, \tag{8.69}$$

where, using two-level notation for definiteness,

$$\frac{k_d^2}{2} = \left(\frac{2f_0}{NH}\right)^2 \quad \rightarrow \quad k_d = \frac{\sqrt{8}}{L_d}, \tag{8.70}$$

where H is the total depth of the domain, as in the Eady problem. The basic state we choose is

$$\Psi_1 = -U_1 y, \qquad \Psi_2 = -U_2 y = +U_1 y. \tag{8.71}$$

It is possible to choose $U_2 = -U_1$ without loss of generality because there is no topography and the system is Galilean invariant. The basic state potential vorticity gradient is then given by

$$Q_1 = \beta y + k_d^2 U y, \qquad Q_2 = \beta y - k_d^2 U y, \tag{8.72}$$

where $U = U_1$. (Note that U differs by a constant multiplicative factor from the U in the Eady problem.) Even in the absence of β there is a non-zero potential vorticity gradient. Why should this be different from the Eady problem? — after all, the shear is uniform in both problems. The difference arises from the vertical boundary conditions. In the two-layered formulation the temperature gradient at the boundary is *absorbed into the definition of the potential vorticity in the interior.* This results in a

Some Results in Baroclinic Instability

Eady Problem

- The length scales of the instability are characterized by the deformation scale. The most unstable scale has a wavelength a few times larger than the deformation radius L_d, where $L_d = NH/f$.
- The growth rate of the instability is approximately

$$\sigma_E \approx \frac{0.3U}{L_d} = 0.3\Lambda\frac{f}{N}. \tag{B.1}$$

That is, it is proportional to the shear scaled by the Prandtl ratio f/N. The value σ_E is known as the *Eady growth rate.*

- There is a *short-wave cutoff* beyond which (i.e., at higher wavenumber than) there is no instability. This occurs near the deformation radius.
- The two-layer (Phillips) problem with zero beta captures many of the results of the Eady model, including the short-wave cutoff.

Effects of beta

- Potential vorticity may now change sign because of the relation between the surface temperature gradient and the interior potential vorticity gradient. Thus, in the continuously-stratified problem, unstable modes do not need to extend the full depth of the atmosphere.
- There is a long-wave cut-off to the main instability branch. At scales larger than this the instabilities are slowly growing, and absent in the two-layer (Phillips) problem.
- In the continuously-stratified problem there is no short-wave cut-off. These modes are slowly growing and shallow, and in the two-layer model they are absent.
- In the two-layer model with beta, there is a minimum shear, Λ_c, for instability given by

$$\Lambda_c = \frac{\beta H}{2}\frac{N^2}{f^2}. \tag{B.2}$$

This shear does not arise in the continuous problem, but it may be a useful criterion for the onset of rapidly growing deep modes.

non-zero interior potential vorticity gradient at the two levels adjacent to the boundary (the only layers in the two-layer problem), but with isothermal boundary conditions. In the Eady problem we have a zero interior gradient of potential vorticity but a temperature gradient at the top and bottom boundary.

The linearized potential vorticity equation is, for each layer,

$$\frac{\partial q_i'}{\partial t} + U_i\frac{\partial q_i'}{\partial x} + v_i'\frac{\partial Q_i}{\partial y} = 0, \qquad i = 1, 2, \tag{8.73}$$

or, more explicitly,

$$\left(\frac{\partial}{\partial t}+U\frac{\partial}{\partial x}\right)\left[\nabla^2\psi_1'+\frac{k_d^2}{2}(\psi_2'-\psi_1')\right]+\frac{\partial\psi_1'}{\partial x}(\beta+k_d^2U)=0, \qquad (8.74a)$$

$$\left(\frac{\partial}{\partial t}-U\frac{\partial}{\partial x}\right)\left[\nabla^2\psi_2'+\frac{k_d^2}{2}(\psi_1'-\psi_2')\right]+\frac{\partial\psi_2'}{\partial x}(\beta-k_d^2U)=0. \qquad (8.74b)$$

For simplicity let us set the problem in a square, doubly periodic domain, and so seek solutions of the form,

$$\psi_i' = \mathrm{Re}\,\widetilde{\psi}_i e^{i(kx+ly-\omega t)} = \mathrm{Re}\,\widetilde{\psi}_i e^{ik(x-ct)}e^{ily}, \qquad i=1,2. \qquad (8.75)$$

Here, k and l are the x- and y-wavenumbers, and $(k,l)=(2\pi/L)(m,n)$, where L is the size of the domain, and m and n are integers. The constant $\widetilde{\psi}_i$ is the complex amplitude.

8.6.2 The Solution

Substituting (8.75) into (8.74) we obtain

$$[ik(U-c)]\left[-K^2\widetilde{\psi}_1+k_d^2(\widetilde{\psi}_2-\widetilde{\psi}_1)/2\right]+ik\widetilde{\psi}_1(\beta+k_d^2U)=0, \qquad (8.76a)$$

$$[-ik(U+c)]\left[-K^2\widetilde{\psi}_2+k_d^2(\widetilde{\psi}_1-\widetilde{\psi}_2)/2\right]+ik\widetilde{\psi}_2(\beta-k_d^2U)=0, \qquad (8.76b)$$

where $K^2=k^2+l^2$. A little algebra turns these equations into a quadratic equation for c, and solving that gives

$$c=-\frac{\beta}{K^2+k_d^2}\left\{1+\frac{k_d^2}{2K^2}\pm\frac{k_d^2}{2K^2}\left[1+\frac{4K^4(K^4-k_d^4)}{k_\beta^4k_d^4}\right]^{1/2}\right\}, \qquad (8.77)$$

where $K^4=(k^2+l^2)^2$ and $k_\beta=\sqrt{\beta/U}$ (its inverse is known as the Kuo scale). We may nondimensionalize this equation using the deformation radius L_d as the length scale and the shear velocity U as the velocity scale. Then, denoting nondimensional parameters with hats, we have

$$k=\frac{\widehat{k}}{L_d}, \qquad c=\hat{c}U, \qquad t=\frac{L_d}{U}\hat{t}, \qquad (8.78)$$

and the nondimensional form of (8.77) is just

$$\hat{c}=-\frac{\widehat{k}_\beta^2}{\widehat{K}^2+\widehat{k}_d^2}\left\{1+\frac{\widehat{k}_d^2}{2\widehat{K}^2}\pm\frac{\widehat{k}_d^2}{2\widehat{K}^2}\left[1+\frac{4\widehat{K}^4(\widehat{K}^4-\widehat{k}_d^4)}{\widehat{k}_\beta^4\widehat{k}_d^4}\right]^{1/2}\right\}, \qquad (8.79)$$

where $\widehat{k}_\beta=k_\beta L_d$ and $\widehat{k}_d=\sqrt{8}$, as in (8.70). The nondimensional parameter

$$\gamma=\frac{1}{4}\widehat{k}_\beta^2=\frac{\beta L_d^2}{4U}, \qquad (8.80)$$

is often useful as a measure of the importance of β; it is proportional to the ratio of the deformation radius to the Kuo scale $\sqrt{U/\beta}$, squared. Let us look at two special cases first, before considering the general solution to these equations.

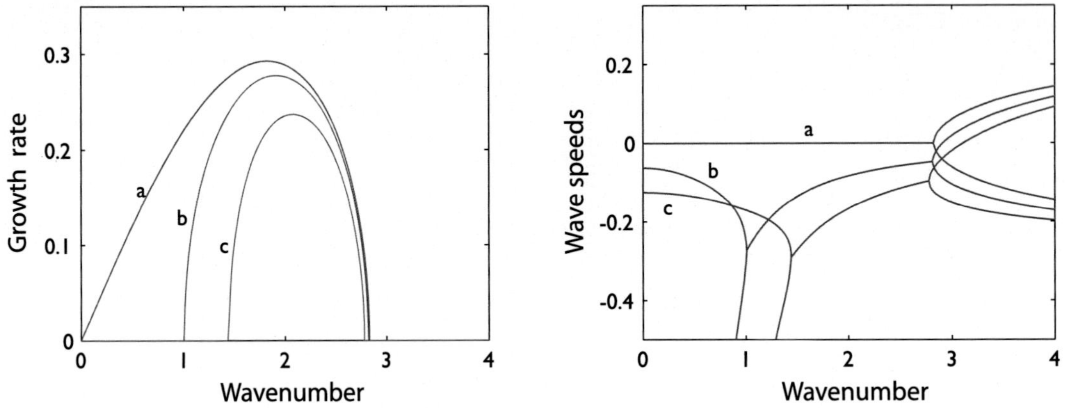

Fig. 8.10: Growth rates and wave speeds for the two-layer baroclinic instability problem, from (8.79), with three (nondimensional) values of β as labelled: a, $\gamma = 0$ ($\widehat{k}_\beta = 0$); b, $\gamma = 0.5$ ($\widehat{k}_\beta = \sqrt{2}$); c, $\gamma = 1$ ($\widehat{k}_\beta = 2$). As β increases, so does the low-wavenumber cut-off to instability, but the high-wavenumber cut-off is little changed. The solutions are obtained from (8.79), with $\widehat{k}_d = \sqrt{8}$ and $U_1 = -U_2 = 1/4$.

I. *Zero shear, non-zero β*
If there is no shear (i.e., $U = 0$) then the phase speeds are purely real and are given by

$$c = -\frac{\beta}{K^2} \qquad \text{and} \qquad c = -\frac{\beta}{K^2 + k_d^2}. \tag{8.81}$$

The first of these is the dispersion relationship for Rossby waves in a purely barotropic flow, and corresponds to the eigenfunction $\widetilde{\psi}_1 = \widetilde{\psi}_2$. The second solution corresponds to the baroclinic eigenfunction $\widetilde{\psi}_1 + \widetilde{\psi}_2 = 0$.

II. *Zero β, non-zero shear*
If $\beta = 0$, then (8.76) yields, after a little algebra,

$$c = \pm U \left(\frac{K^2 - k_d^2}{K^2 + k_d^2} \right)^{1/2} \qquad \text{or} \qquad \sigma = Uk \left(\frac{k_d^2 - K^2}{K^2 + k_d^2} \right)^{1/2}, \tag{8.82}$$

where $\sigma = -i\omega$ is the growth rate. These expressions are similar to those in the Eady problem. We note the following:

- There is an instability for *all* values of U.
- There is a high-wavenumber cut-off, at a scale proportional to the radius of deformation. For the two-layer model, if $K > k_d = 2.82/L_d$ there is no growth. For the Eady problem, the high wavenumber cut-off occurs at $2.4/L_d$.
- There is no low wavenumber cut-off.
- For any given k, the highest growth rate occurs for $l = 0$. In the two-layer model, from (8.82), for $l = 0$ the maximum growth rate occurs when $k = 0.634k_d = 1.79/L_d$. For the Eady problem, the maximum growth rate occurs at $1.61/L_d$.

Solution with non-zero β

Using (8.79), the growth rate and wave speeds as a function of wavenumber are plotted in Fig. 8.10. We observe that there still appears to be a

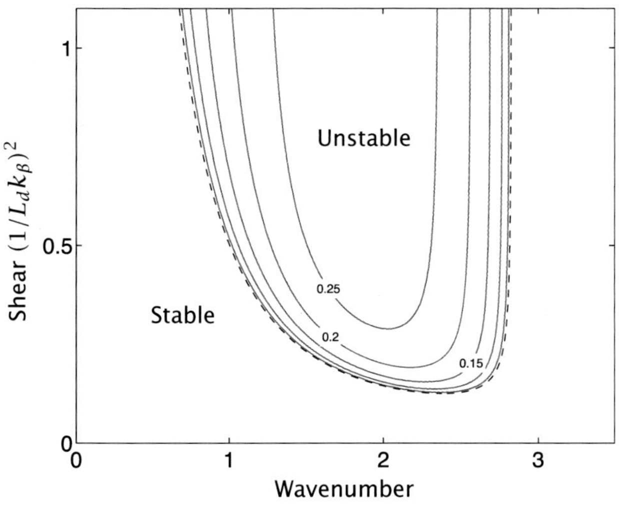

Fig. 8.11: Contours of growth rate in the two-layer baroclinic instability problem. The dashed line is the neutral stability curve, (8.84), and the other curves are contours of growth rates obtained from (8.79). The wavenumber is scaled by $1/L_d$ (i.e., by $k_d/\sqrt{8}$) and growth rates are scaled by the inverse of the Eady time scale (i.e., by U/L_d). Thus, for $L_d = 1000$ km and $U = 10\,\mathrm{m\,s^{-1}}$, a nondimensional growth rate of 0.25 corresponds to a dimensional growth rate of $0.25 \times 10^{-5}\,\mathrm{s^{-1}}$ $= 0.216\,\mathrm{day^{-1}}$.

high-wavenumber cut-off and, for $\beta = 0$, there is a low-wavenumber cut-off. A little analysis elucidates the origin of these features:

The neutral curve

For instability, there must be an imaginary component to the phase speed in (8.77); that is, we require

$$k_\beta^4 k_d^4 + 4K^4(K^4 - k_d^4) < 0. \tag{8.83}$$

This is a quadratic equation in K^4 for the value of K, K_c say, at which the growth rate is zero. Solving, we find

$$K_c^4 = \frac{1}{2}k_d^4\left(1 \pm \sqrt{1 - k_\beta^4/k_d^4}\right), \tag{8.84}$$

and this is plotted in Fig. 8.11.

Minimum shear for instability

From (8.83), instability arises when $\beta^2 {k_d}^4/U^2 < 4K^4({k_d}^4 - K^4)$. The maximum value of the right-hand side of this expression arises when $K^4 = \widehat{k}_d^4/2$; thus, instability arises only when

$$\frac{\beta^2 k_d^4}{U^2} < 4\frac{k_d^4}{2}\frac{k_d^4}{2} \qquad \text{or} \qquad \kappa_\beta < k_d. \tag{8.85}$$

That is, *instability only arises if the deformation radius is sufficiently smaller than the Kuo scale,* $\sqrt{U/\beta}$. The critical velocity difference required for instability is then

$$U_1 - U_2 > U_c = \frac{2\beta}{k_d^2} = \frac{1}{4}\beta L_d^2, \tag{8.86}$$

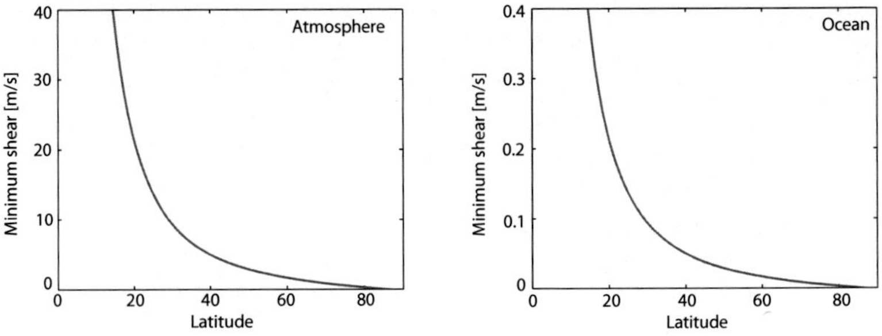

Fig. 8.12: The minimum shear (the velocity difference $U_1 - U_2$) required for baroclinic instability in a two-layer model, calculated using (8.86), i.e., $U_{\min} = \beta L_d^2/4$ where $\beta = 2\Omega a^{-1}\cos\vartheta$ and $L_d = NH/f$, with $f = 2\Omega\sin\vartheta$. The left panel uses $H = 10$ km and $N = 10^{-2}\,\mathrm{s}^{-1}$, and the right panel uses parameters representative of the main thermocline, $H = 1$ km and $N = 10^{-2}\,\mathrm{s}^{-1}$. The results are not quantitatively accurate, but the implications that the minimum shear is much less for the ocean, and that in both the atmosphere and the ocean the shear increases rapidly at low latitudes, are robust.

recalling our notation that $U_1 - U_2 = 2U$ and that $k_d^2 = 8/L_d^2$, as in (8.70). The critical shear for instability is

$$\Lambda_c = \frac{\beta H}{2}\frac{N^2}{f^2}, \tag{8.87a,b}$$

where the shear Λ is defined by $(U_1 - U_2)/(0.5H)$, where H is the total depth of the domain. Note the relationship of the minimum shear to the basic state potential vorticity gradient in the respective layers. In the upper and lower layers the potential vorticity gradients are given by, respectively,

$$\frac{\partial Q_1}{\partial y} = \beta + k_d^2 U, \qquad \frac{\partial Q_2}{\partial y} = \beta - k_d^2 U. \tag{8.88a,b}$$

Thus, the requirement for instability is exactly that which causes the potential vorticity gradient to change sign somewhere in the domain, in this case becoming negative in the lower layer.

Figure 8.12 sketches how the critical shear varies — very approximately — with latitude in the atmosphere and ocean because of the variation of the Coriolis parameter. (In any given quasi-geostrophic calculation f is held constant, but if we wish to see how the critical shear varies with latitude we vary f accordingly.)

High-wavenumber cut-off

Instability can only arise when, from (8.83),

$$4K^4(k_d^4 - K^4) > k_\beta^4 k_d^4, \tag{8.89}$$

so that a necessary condition for instability is

$$k_d^2 > K^2. \tag{8.90}$$

Thus, waves shorter than the deformation radius are always stable, no matter what the value of β.

Low-wavenumber cut-off

Suppose that $K \ll k_d$. Then (8.83) simplifies to $k_\beta^4 < 4K^4$. That is, for instability we require

$$K^2 > \frac{1}{2}k_\beta^2 = \frac{\beta}{2U}. \tag{8.91}$$

Thus, using (8.90) and (8.91) the unstable waves lie approximately in the interval $\beta/(\sqrt{2}U) < K < k_d$.

8.7 ♦ Necessary Conditions for Baroclinic Instability

As in the two-dimensional case discussed in Section 8.3, we can derive a necessary condition for instability, or a sufficient condition for stability, for small perturbations. The procedure is entirely analogous to that of Section 8.3 but now we must take into account boundary conditions at the top and bottom and this complicates the algebra. Trusting readers may skip to the results.

Starting with the linear potential vorticity equation, (8.44), we multiply by q' and divide by Q_y to give (dropping the primes on perturbed quantities to avoid clutter)

$$\frac{\partial}{\partial t}\frac{q^2}{Q_y} + \frac{U}{Q_y}\frac{\partial q^2}{\partial x} + 2vq = 0, \qquad 0 < z < H, \tag{8.92}$$

where, as before, Q is the background potential vorticity and Q_y its meridional gradient. The above equation holds in the fluid interior, and at the top and bottom boundary we use (8.45) to similarly obtain

$$\frac{\partial b^2}{\partial t} + \frac{U}{B_y}\frac{\partial b^2}{\partial x} + 2vb = 0, \qquad z = 0, H, \tag{8.93}$$

where B is buoyancy at the boundary and B_y its meridional gradient.

Now, the poleward eddy flux of potential vorticity is

$$\overline{vq} = -\frac{\partial}{\partial y}\overline{uv} + \frac{\partial}{\partial z}\left(\frac{f_0}{N^2}\overline{vb}\right), \tag{8.94}$$

and integrating this expression with respect to both y and z gives

$$\int_A \overline{vq}\,\mathrm{d}y\,\mathrm{d}z = \left[\frac{f_0}{N^2}\overline{vb}\right]_0^H, \tag{8.95}$$

assuming that the meridional boundaries are at quiescent latitudes. Integrating (8.92) over y and z and using (8.95) gives

$$\frac{\mathrm{d}}{\mathrm{d}t}\iint \frac{1}{2}\frac{\overline{q^2}}{Q_y}\,\mathrm{d}y\,\mathrm{d}z = -\left[\frac{f_0}{N^2}\overline{v'b'}\right]_0^H. \tag{8.96}$$

Using (8.93) to eliminate $\overline{vb}$ finally gives

$$\frac{\mathrm{d}}{\mathrm{d}t}\left\{\iint \frac{1}{2}\frac{\overline{q'^2}}{Q_y}\,\mathrm{d}y\,\mathrm{d}z - \int\left[\frac{1}{2}\frac{f_0}{N^2}\frac{\overline{b^2}}{B_y}\right]_0^H \mathrm{d}y\right\} = 0. \tag{8.97}$$

The perturbation can only grow if the expression within curly brackets has both positive and negative terms, because (8.97) tells us that the expression is a constant. Thus, If the expression is positive or negative definite the perturbation cannot grow and the basic state is stable. Stability thus depends on the meridional gradient of potential vorticity in the interior, and the meridional gradient of buoyancy at the boundary. If Q_y changes sign in the interior, or B_y changes sign at the boundary, we have the potential for instability. If these are both one signed, then various possibilities exist, and using the thermal wind relation ($f_0 \partial U/\partial z = -\partial B/\partial y$) we obtain the following:

I. *A stable case:*

$$\frac{\partial Q}{\partial y} > 0 \text{ and } \left.\frac{\partial U}{\partial z}\right|_{z=0} < 0 \text{ and } \left.\frac{\partial U}{\partial z}\right|_{z=H} > 0 \Rightarrow \text{stability.} \tag{8.98}$$

Stability also ensues if all inequalities are switched.

II. *Instability via interior–surface interactions:*

$$\frac{\partial Q}{\partial y} > 0 \text{ and } \left.\frac{\partial U}{\partial z}\right|_{z=0} > 0 \text{ or } \left.\frac{\partial U}{\partial z}\right|_{z=H} < 0 \Rightarrow \text{potential instability.} \tag{8.99}$$

The condition $\partial Q/\partial y > 0$ and $(\partial U/\partial z)_{z=0} > 0$ (or $f_0^{-1}\partial B/\partial y < 0$) is the most common criterion for instability that is met in the atmosphere. Stability is then determined by the interior potential vorticity gradient and the surface buoyancy gradient.

III. *Instability via interaction between edge waves at top and bottom:*

$$\left.\frac{\partial U}{\partial z}\right|_{z=0} > 0 \text{ and } \left.\frac{\partial U}{\partial z}\right|_{z=H} > 0 \Rightarrow \text{potential instability.} \tag{8.100}$$

(And similarly, with both inequalities switched.) Such an instability may occur where the troposphere acts like a lid, as for example in the Eady problem. If $Q_y = 0$ and there is no lid at $z = H$ (e.g., the Eady problem with no lid) then the instability disappears.

Readers should explore the consequences of these conditions for themselves, and in particular how they are consistent with the instability results from the Eady problem and the two-level problem calculated directly. The necessary conditions above are in some ways more powerful than the explicit calculations for they do not rely on the instability being of normal mode form, although they do not supply the actual growth rates.

Notes and References

The book by Drazin & Reid (1981) discusses many classical problems in hydrodynamic stability, although its treatment of baroclinic instability is rather brief. The continuously stratified baroclinic instability problem with non-zero beta, namely the Charney problem, is discussed by Pedlosky (1987a).

Problems

8.1 Using values of physical parameters obtained from the literature, estimate the baroclinic growth rates on Venus, Mars and Jupiter, using the Eady growth rate and/or the growth rate determined from the two-level model. (Make sensible choices as to the physical parameters on these planets, but there is no need to do a detailed calculation.) Given these results, discuss whether baroclinic instability is likely to be a significant source of atmospheric motion on those planets.

8.2 Show, using the two-layer model (or otherwise) that the presence of β reduces the efficiency of baroclinic instability. For example, show that it makes the meridional velocity slightly out of phase with the temperature.

8.3 Following the same procedure used in Sections 8.3 and 8.7, obtain the necessary conditions for instability in the two-level quasi-geostrophic model in the case with uniform shear. Show that these conditions are consistent with the conditions for instability calculated directly with a normal-mode approach.

8.4 Suppose that the equator to pole temperature gradient on Earth and all other parameters remain fixed, except that Earth's rotation is slowed. At what value of rotation will baroclinic instability cease, and why? (You may use the Eady model, or the two-layer model with beta, or another of your choice.) Suppose now that the rotation rate is increased. Will baroclinic instability ever cease to be important, and why? Briefly discuss and justify your answers in both cases.

8.5 *Stability conditions in the continuously stratified QG model*
Consider the modified Eady problem (Boussinesq, uniform stratification, flow contained between two flat horizontal surfaces at 0 and H), but instead of a uniform shear suppose that the basic state is given by

$$U = -U_0 \cos(\pi m z/H), \tag{P8.1}$$

where $U_0 > 0$, and allow β to be non-zero.

(*a*) Show there is a critical shear, and that this diminishes as m increases.
(*b*) Compare the results to those of the original Eady model and two-layer model with β.

Sketch of solution to (a).

The basic state has no temperature gradient at the boundary, and so stability is assured if $\partial Q/\partial y$ in the interior is positive everywhere. The basic state potential vorticity is $Q = \nabla^2\Psi + (f_0^2/N^2)\partial^2\Psi/\partial z^2 + \beta y$, so that

$$\frac{dQ}{dy} = \beta - k_d^2\pi^2 m^2 U_0 \cos(\pi p z/H), \tag{P8.2}$$

where $k_d^2 = f_0^2 H^2/N^2$ and $U = -\partial\Psi/\partial y$. Thus, stability is assured if $U_0 < (\beta/k_d^2\pi^2 m^2)$.

CHAPTER

9

Waves and Mean-Flows

LINEAR DYNAMICS is mainly concerned with waves and instabilities that live on a pre-determined background flow. But the real world isn't quite like that. Rather, the mean state is the *result* of the combined effects of thermal and mechanical forcing (by radiation from the sun and, for the ocean, the winds) *plus the action of the waves and instabilities themselves.* In this chapter we explore the geophysical fluid dynamics underlying such wave–mean-flow interactions. We try to keep our discussion as elementary as possible by staying within the comfortable bounds of the quasi-geostrophic approximation and considering only zonal averages. Nevertheless the subject is often regarded as an advanced one, and all the sections in this chapter may be considered to be implicitly marked with a diamond, ✦.

9.1 QUASI-GEOSTROPHIC WAVE–MEAN-FLOW INTERACTION

9.1.1 Preliminaries

The Jacobian operator for any two quantities a and b is given by $J(a,b) = \partial_x a\, \partial_y b - \partial_y a\, \partial_x b$. When a is the streamfunction the Jacobian gives the horizontal advection of b, namely $\boldsymbol{u} \cdot \nabla b$.

To fix our dynamical system and notation, we write down the Boussinesq quasi-geostrophic potential vorticity equation

$$\frac{\partial q}{\partial t} + J(\psi, q) = D, \tag{9.1}$$

where $J(\psi, q) = \partial\psi/\partial x\, \partial q/\partial y - \partial\psi/\partial y\, \partial q/\partial x$ and D represents any non-conservative terms. The potential vorticity in a Boussinesq system is

$$q = \beta y + \zeta + \frac{\partial}{\partial z}\left(\frac{f_0}{N^2} b\right), \tag{9.2}$$

where ζ is the relative vorticity and b is the buoyancy perturbation from a background state characterized by N^2. (Nearly all the derivations in this chapter could be done in pressure coordinates with minor modifications.)

We refer to lines of constant b as isentropes. In terms of the streamfunction, the variables are

$$\zeta = \nabla^2 \psi, \qquad b = f_0 \frac{\partial \psi}{\partial z}, \qquad q = \beta y + \left[\nabla^2 + \frac{\partial}{\partial z}\left(\frac{f_0^2}{N^2}\frac{\partial}{\partial z}\right)\right]\psi, \tag{9.3}$$

where $\nabla^2 \equiv (\partial_x^2 + \partial_y^2)$. The potential vorticity equation holds in the fluid interior; the boundary conditions on (9.3) are provided by the thermodynamic equation

$$\frac{\partial b}{\partial t} + J(\psi, b) + wN^2 = S, \tag{9.4}$$

where S represents heating terms. The vertical velocity at the boundary, w, is zero in the absence of topography and Ekman friction so that the boundary condition is just

$$\frac{\partial b}{\partial t} + J(\psi, b) = S. \tag{9.5}$$

Equations (9.1) and (9.5) are the evolution equations for the system, and if both D and S are zero they conserve both the total energy, $\widehat{E}$ and the total enstrophy, $\widehat{Z}$:

$$\begin{aligned} \frac{\mathrm{d}\widehat{E}}{\mathrm{d}t} &= 0, \qquad \widehat{E} = \frac{1}{2}\int_V (\nabla\psi)^2 + \frac{f_0^2}{N^2}\left(\frac{\partial \psi}{\partial z}\right)^2 \mathrm{d}V, \\ \frac{\mathrm{d}\widehat{Z}}{\mathrm{d}t} &= 0, \qquad \widehat{Z} = \frac{1}{2}\int_V q^2 \,\mathrm{d}V, \end{aligned} \tag{9.6}$$

where V is a volume bounded by surfaces at which the normal velocity is zero, or that has periodic boundary conditions. The enstrophy is also conserved layerwise; that is, the horizontal integral of q^2 is conserved at every level.

9.1.2 Potential Vorticity Flux in the Linear Equations

Let us decompose the fields into a mean (to be denoted with an overbar) plus a perturbation (denoted with a prime), and let us suppose the perturbation fields are of small amplitude. (In linear problems, such as those considered in Chapter 8, we decomposed the flow into a 'basic state' plus a perturbation, with the basic state fixed in time. Our approach here is similar, but soon we will allow the mean state to evolve.) The linearized quasi-geostrophic potential vorticity equation is then

$$\frac{\partial q'}{\partial t} + \overline{u}\frac{\partial q'}{\partial x} + u'\frac{\partial \overline{q}}{\partial x} + \overline{v}\frac{\partial q'}{\partial y} + v'\frac{\partial \overline{q}}{\partial y} = D', \tag{9.7}$$

where D' represents eddy forcing and dissipation and, in terms of streamfunction,

$$(u'(x,y,z,t), v'(x,y,z,t)) = \left(-\frac{\partial \psi'}{\partial y}, \frac{\partial \psi'}{\partial x}\right), \tag{9.8a}$$

$$q'(x,y,z,t) = \nabla^2\psi' + \frac{\partial}{\partial z}\left(\frac{f_0^2}{N^2}\frac{\partial \psi'}{\partial z}\right). \tag{9.8b}$$

If the mean is a zonal mean then $\partial\overline{q}/\partial x = 0$ and $\overline{v} = 0$ (because v is purely geostrophic) and (9.7) simplifies to

$$\frac{\partial q'}{\partial t} + \overline{u}\frac{\partial q'}{\partial x} + v'\frac{\partial \overline{q}}{\partial y} = D', \tag{9.9}$$

where

$$\overline{q} = \beta y - \frac{\partial \overline{u}}{\partial y} + \frac{\partial}{\partial z}\left(\frac{f_0}{N^2}\overline{b}\right), \qquad \frac{\partial \overline{q}}{\partial y} = \beta - \frac{\partial^2 \overline{u}}{\partial y^2} - \frac{\partial}{\partial z}\left(\frac{f_0^2}{N^2}\frac{\partial \overline{u}}{\partial z}\right), \tag{9.10a,b}$$

having used the thermal wind relation,

$$f_0\frac{\partial \overline{u}}{\partial z} = -\frac{\partial \overline{b}}{\partial y}. \tag{9.11}$$

Multiplying (9.9) by q' and zonally averaging gives the enstrophy equation:

$$\frac{1}{2}\frac{\partial}{\partial t}\overline{q'^2} = -\overline{v'q'}\frac{\partial \overline{q}}{\partial y} + \overline{D'q'}. \tag{9.12}$$

The quantity $\overline{v'q'}$ is the meridional flux of potential vorticity; this is downgradient (by definition) when the first term on the right-hand side is positive (i.e., $\overline{v'q'}\partial\overline{q}/\partial y < 0$), and it then acts to increase the variance of the perturbation. (This occurs, for example, when the flux is diffusive so that $\overline{v'q'} = -\kappa\partial\overline{q}/\partial y$, where κ may vary but is everywhere positive.) This argument may be inverted: for unforced, inviscid flow ($D = 0$), if the waves are growing, as for example in the canonical models of baroclinic instability discussed in Chapter 8, then *the potential vorticity flux is downgradient*.

If the second term on the right-hand side of (9.12) is negative, as it will be if D' is a dissipative process (e.g., if $D' = A\nabla^2 q'$ or if $D' = -rq'$, where A and r are positive) then a statistical balance can be achieved between enstrophy production via downgradient transport, and dissipation. If the waves are steady (by which we mean statistically steady, neither growing nor decaying in amplitude) and conservative (i.e., $D' = 0$) then we must have

$$\overline{v'q'} = 0. \tag{9.13}$$

Similar results follow for the buoyancy at the boundary; we start by linearizing the thermodynamic equation (9.5) to give

$$\frac{\partial b'}{\partial t} + \overline{u}\frac{\partial b'}{\partial x} + v'\frac{\partial \overline{b}}{\partial y} = S', \tag{9.14}$$

where S' is a diabatic source term. Multiplying (9.14) by b' and averaging gives

$$\frac{1}{2}\frac{\partial}{\partial t}\overline{b'^2} = -\overline{v'b'}\frac{\partial \overline{b}}{\partial y} + \overline{S'b'}. \tag{9.15}$$

If the flow is adiabatic ($S' = 0$) then growing waves have a downgradient flux of buoyancy at the boundary. In the Eady problem there is no interior gradient of basic-state potential vorticity and all the terms in (9.12) are zero, but the perturbation grows at the boundary. If the waves are steady and adiabatic then, analogously to (9.13), $\overline{v'b'} = 0$. In models with discrete vertical layers or a finite number of levels it is common practice to absorb the boundary conditions into the definition of potential vorticity at top and bottom, as in the two level model of Section 5.6.

9.1.3 Wave–Mean-Flow Interaction

In linear problems we usually suppose that the mean-flow is fixed and that the zonal mean terms, $\overline{u}$ and $\overline{q}$ in (9.9), are functions only of y and z. However, in reality we might expect that the mean-flow would change because of momentum and heat flux convergences arising from the eddy–eddy interactions. To calculate these changes we begin with the potential vorticity equation (9.1) and, in the usual way, express the variables as a zonal mean plus an eddy term and obtain

$$\frac{\partial \overline{q}}{\partial t} + \overline{v}\frac{\partial \overline{q}}{\partial y} + \frac{\partial}{\partial y}(\overline{v'q'}) = \overline{D}. \tag{9.16}$$

Now, $\overline{v} = 0$ (since the flow is geostrophic) and the mean-flow thus evolves according to

$$\frac{\partial \overline{q}}{\partial t} + \frac{\partial}{\partial y}\overline{v'q'} = \overline{D}. \tag{9.17}$$

Similarly, at the boundary the mean buoyancy evolution equation is

$$\frac{\partial \overline{b}}{\partial t} + \frac{\partial}{\partial y}\overline{v'b'} = \overline{S}. \tag{9.18}$$

To obtain $\overline{u}$ from $\overline{q}$ and $\overline{b}$ we use thermal wind balance, (9.11), to define a streamfunction Ψ. That is, since $f_0 \partial \overline{u}/\partial z = -\partial \overline{b}/\partial y$, then

$$\left(\overline{u}, \frac{1}{f_0}\overline{b}\right) = \left(-\frac{\partial \Psi}{\partial y}, \frac{\partial \Psi}{\partial z}\right), \tag{9.19}$$

whence, using (9.10a), the zonal mean potential vorticity is

$$\overline{q}(y, z, t) - \beta y = \frac{\partial}{\partial z}\left(\frac{f_0^2}{N^2}\frac{\partial \Psi}{\partial z}\right) + \frac{\partial^2 \Psi}{\partial y^2}. \tag{9.20}$$

If $\overline{q}$ is known in the interior from (9.18), and $\overline{b}$ (i.e., $f_0 \partial \Psi/\partial z$) is known at the boundaries, then $\overline{u}$ and $\overline{b}$ in the interior may be obtained using (9.20) and (9.19b). The equations are also summarized in the shaded box on page 176.

To close the system we suppose that the eddy terms themselves evolve according to (9.9) and (9.14). If in those equations we were to include the eddy–eddy interaction terms we would simply recover the full system, so in neglecting those terms we have constructed an eddy–mean-flow system, commonly called a *wave–mean-flow* system because by eliminating the nonlinear terms in the perturbation equation the eddies will often be wavelike. It is important to realise that such systems do differ from linear ones in which we regard the mean flow as fixed; we have gone one step further and allowed the mean flow to evolve because of the effects of eddies, but we do not allow the eddies to interact with themselves.

We now consider some more properties of the waves themselves — how they propagate and what they conserve — beginning with a discussion of the potential vorticity flux and its relative, the so-called Eliassen–Palm flux.

9.2 POTENTIAL VORTICITY FLUX

The eddy flux of potential vorticity may be expressed in terms of vorticity and buoyancy fluxes as

$$v'q' = v'\zeta' + f_0 v' \frac{\partial}{\partial z}\left(\frac{b'}{N^2}\right). \tag{9.21}$$

The second term on the right-hand side can be written as

$$\begin{aligned} f_0 v' \frac{\partial}{\partial z}\left(\frac{b'}{N^2}\right) &= f_0 \frac{\partial}{\partial z}\left(\frac{v'b'}{N^2}\right) - f_0 \frac{\partial v'}{\partial z}\frac{b'}{N^2} \\ &= f_0 \frac{\partial}{\partial z}\left(\frac{v'b'}{N^2}\right) - f_0 \frac{\partial}{\partial x}\left(\frac{\partial \psi'}{\partial z}\right)\frac{b'}{N^2} \\ &= f_0 \frac{\partial}{\partial z}\left(\frac{v'b'}{N^2}\right) - \frac{f_0^2}{2N^2}\frac{\partial}{\partial x}\left(\frac{\partial \psi'}{\partial z}\right)^2, \end{aligned} \tag{9.22}$$

using $b' = f_0 \partial \psi'/\partial z$.

Similarly, the flux of relative vorticity can be written

$$v'\zeta' = -\frac{\partial}{\partial y}(u'v') + \frac{1}{2}\frac{\partial}{\partial x}(v'^2 - u'^2), \tag{9.23}$$

and using (9.22) and (9.23), (9.21) becomes

$$v'q' = -\frac{\partial}{\partial y}(u'v') + \frac{\partial}{\partial z}\left(\frac{f_0}{N^2}v'b'\right) + \frac{1}{2}\frac{\partial}{\partial x}\left((v'^2 - u'^2) - \frac{b'^2}{N^2}\right). \tag{9.24}$$

Thus the meridional potential vorticity flux, in the quasi-geostrophic approximation, can be written as the divergence of a vector: $v'q' = \nabla \cdot \mathcal{E}$ where

$$\mathcal{E} \equiv \frac{1}{2}\left((v'^2 - u'^2) - \frac{b'^2}{N^2}\right)\hat{\mathbf{i}} - (u'v')\hat{\mathbf{j}} + \left(\frac{f_0}{N^2}v'b'\right)\hat{\mathbf{k}}. \tag{9.25}$$

A particularly useful form of this arises after zonally averaging, for then (9.24) becomes

$$\overline{v'q'} = -\frac{\partial}{\partial y}\overline{u'v'} + \frac{\partial}{\partial z}\left(\frac{f_0}{N^2}\overline{v'b'}\right). \tag{9.26}$$

The vector defined by

$$\mathcal{F} \equiv -\overline{u'v'}\,\hat{\mathbf{j}} + \frac{f_0}{N^2}\overline{v'b'}\,\hat{\mathbf{k}} \tag{9.27}$$

is the wave activity flux, often called the (quasi-geostrophic) Eliassen–Palm (EP) flux, and its divergence, given by (9.26), gives the poleward flux of potential vorticity:

$$\overline{v'q'} = \nabla_x \cdot \mathcal{F}, \tag{9.28}$$

where $\nabla_x \cdot \equiv (\partial/\partial y, \partial/\partial z)\cdot$ is the divergence in the meridional-vertical plane, at constant x. Unless the meaning is unclear, the subscript x will be dropped.

9.2.1 The Eliassen–Palm Relation

The Eliassen–Palm flux and Eliassen–Palm relation are named for quantities that appeared in Eliassen & Palm (1961). That paper was mainly concerned with the transfer of energy in mountain waves rather than with matters related to potential vorticity flux and large-scale flow. 'Wave activity flux' is a descriptive alternative name for the EP flux.

On dividing by $\partial\overline{q}/\partial y$ and using (9.28), the enstrophy equation (9.12) becomes

$$\frac{\partial \mathcal{P}}{\partial t} + \nabla \cdot \mathcal{F} = \mathcal{D}, \tag{9.29}$$

where

$$\mathcal{P} = \frac{\overline{q'^2}}{2\partial\overline{q}/\partial y}, \qquad \mathcal{D} = \frac{\overline{D'q'}}{\partial\overline{q}/\partial y}, \tag{9.30a,b}$$

and $\mathcal{F}$ is given by (9.27). Equation (9.29) is known as the *Eliassen–Palm relation*, and it is a conservation law (when $\mathcal{D} = 0$) for the *pseudomomentum* $\mathcal{P}$. The conservation law is exact (in the linear approximation) if the mean-flow is constant in time; it is a good approximation if $\partial\overline{q}/\partial y$ varies slowly compared to the variation of $\overline{q'^2}$.

If we integrate (9.29) over a meridional area A bounded by walls where the eddy activity vanishes, and if $\mathcal{D} = 0$, we obtain

$$\frac{\mathrm{d}}{\mathrm{d}t}\int_A \mathcal{P}\,\mathrm{d}A = 0. \tag{9.31}$$

The integral is a 'wave activity' — a quantity that is quadratic in the amplitude of the perturbation and that is conserved in the absence of forcing and dissipation. The quantity $\mathcal{P}$ is an example of a 'wave activity density', generically denoted $\mathcal{A}$; other kinds of wave activity density exist — the pseudoenergy for example, but we do not consider them here. If there is no ambiguity we drop the word density and also refer to $\mathcal{A}$ and $\mathcal{P}$ as wave activities. Note that neither the perturbation energy nor the perturbation enstrophy are wave activities of the linearized equations, because there can be an exchange of energy or enstrophy between mean and perturbation — indeed, this is how a perturbation grows in baroclinic or barotropic instability! This is already evident from (9.12), or in general take (9.7) with $D' = 0$ and multiply by q' to give the enstrophy equation,

$$\frac{1}{2}\frac{\partial q'^2}{\partial t} + \frac{1}{2}\overline{\boldsymbol{u}} \cdot \nabla q'^2 + \boldsymbol{u}'q' \cdot \nabla\overline{q} = 0, \tag{9.32}$$

Quasi-Geostrophic Wave–Mean-Flow Interaction

The inviscid and unforced Boussinesq quasi-geostrophic set of wave–mean-flow equations is

$$\frac{\partial q'}{\partial t} + \overline{u}\frac{\partial q'}{\partial x} + v'\frac{\partial \overline{q}}{\partial y} = 0, \tag{WM.1a}$$

$$\frac{\partial \overline{q}}{\partial t} + \frac{\partial}{\partial y}\overline{v'q'} = 0, \tag{WM.1b}$$

along with similar equations as needed for buoyancy at the boundary (see main text). The eddy terms are

$$q' = \left[\nabla^2 + \frac{\partial}{\partial z}\left(\frac{f_0^2}{N^2}\frac{\partial}{\partial z}\right)\right]\psi', \qquad (u', v') = \left(-\frac{\partial \psi'}{\partial y}, \frac{\partial \psi'}{\partial x}\right). \tag{WM.2a,b}$$

The mean-flow terms are

$$\overline{q}(y,t) = \beta y - \frac{\partial \overline{u}}{\partial y} + \frac{\partial}{\partial z}\left(\frac{f_0}{N^2}\overline{b}\right), \tag{WM.3}$$

and

$$\frac{\partial \overline{q}}{\partial y} = \beta - \frac{\partial^2 \overline{u}}{\partial y^2} + \frac{\partial}{\partial z}\left(\frac{f_0}{N^2}\frac{\partial \overline{b}}{\partial y}\right) = \beta - \frac{\partial^2 \overline{u}}{\partial y^2} - \frac{\partial}{\partial z}\left(\frac{f_0^2}{N^2}\frac{\partial \overline{u}}{\partial z}\right), \tag{WM.4}$$

using thermal wind. To solve for the mean-flow we may define a streamfunction Ψ such that

$$\left(\overline{u}, \frac{1}{f_0}\overline{b}\right) = \left(-\frac{\partial \Psi}{\partial y}, \frac{\partial \Psi}{\partial z}\right), \tag{WM.5}$$

whence

$$\overline{q}(y,t) - \beta y = \frac{\partial}{\partial z}\left(\frac{f_0^2}{N^2}\frac{\partial \Psi}{\partial z}\right) + \frac{\partial^2 \Psi}{\partial y^2}. \tag{WM.6}$$

Given $\overline{q}$ from (WM.1b) we solve (WM.6) to give Ψ and thence $\overline{u}$ and $\overline{b}$. Equivalently, we may derive a single equation for the zonal wind by differentiating (WM.1b) with respect to y and, using (WM.4), we obtain

$$\left[\frac{\partial^2}{\partial y^2} + \frac{\partial}{\partial z}\left(\frac{f_0^2}{N^2}\frac{\partial}{\partial z}\right)\right]\frac{\partial \overline{u}}{\partial t} = \frac{\partial^2}{\partial y^2}\overline{v'q'}. \tag{WM.7}$$

The evolution of the mean-flow may also be written in TEM form as described on page 186.

where here the overbar is an average (although it need not be a zonal average). Integrating this over a volume V gives

$$\frac{\mathrm{d}\hat{Z}'}{\mathrm{d}t} \equiv \frac{\mathrm{d}}{\mathrm{d}t}\int_V \frac{1}{2}q'^2\,\mathrm{d}V = -\int_V \boldsymbol{u}'q' \cdot \nabla\overline{q}\,\mathrm{d}V. \tag{9.33}$$

The right-hand side does not, in general, vanish and so $\hat{Z}'$ is not in general conserved.

9.2.2 ♦ The Group Velocity Property for Rossby Waves

The vector $\mathcal{F}$ describes how the wave activity propagates. We stated in Chapter 6 that in a wave most physical quantities of interest propagate at the group velocity, not the phase velocity. We now show that this is true for the wave activity for Rossby waves; in particular we show that $\mathcal{F} = \mathbf{c}_g \mathcal{P}$, where $\mathbf{c}_g$ is the group velocity and $\mathcal{P}$ the pseudomomentum.

The Boussinesq quasi-geostrophic equation on the β-plane, linearized around a constant zonal flow and with constant static stability, is

$$\frac{\partial q'}{\partial t} + U\frac{\partial q'}{\partial x} + v'\frac{\partial \overline{q}}{\partial y} = 0, \tag{9.34}$$

where $q' = [\nabla^2 + (f_0^2/N^2)\partial^2/\partial z^2]\psi'$ and, if U is constant, $\partial\overline{q}/\partial y = \beta$. Thus we have

$$\left(\frac{\partial}{\partial t} + U\frac{\partial}{\partial x}\right)\left[\nabla^2\psi' + \frac{\partial}{\partial z}\left(\frac{f_0^2}{N^2}\frac{\partial \psi'}{\partial z}\right)\right] + \beta\frac{\partial \psi'}{\partial x} = 0. \tag{9.35}$$

Seeking solutions of the form

$$\psi' = \operatorname{Re}\widetilde{\psi}\mathrm{e}^{\mathrm{i}(kx+ly+mz-\omega t)}, \tag{9.36}$$

we find the dispersion relation,

$$\omega = Uk - \frac{\beta k}{\kappa^2}, \tag{9.37}$$

where $\kappa^2 = (k^2 + l^2 + m^2 f_0^2/N^2)$, and the group velocity components:

$$c_g^y = \frac{2\beta kl}{\kappa^4}, \qquad c_g^z = \frac{2\beta km f_0^2/N^2}{\kappa^4}. \tag{9.38}$$

Also, if $u' = \operatorname{Re}\widetilde{u}\exp[\mathrm{i}(kx + ly + mz - \omega t)]$, and similarly for the other fields, then

$$\begin{aligned} \widetilde{u} &= -\operatorname{Re}\,\mathrm{i}l\widetilde{\psi}, & \widetilde{v} &= \operatorname{Re}\,\mathrm{i}k\widetilde{\psi}, \\ \widetilde{b} &= \operatorname{Re}\,\mathrm{i}mf_0\widetilde{\psi}, & \widetilde{q} &= -\operatorname{Re}\kappa^2\widetilde{\psi}. \end{aligned} \tag{9.39}$$

The wave activity density is then

$$\mathcal{P} = \frac{1}{2}\frac{\overline{q'^2}}{\beta} = \frac{\kappa^4}{4\beta}|\widetilde{\psi}^2|, \tag{9.40}$$

where the additional factor of 2 in the denominator arises from the averaging. Using (9.39) the EP flux, (9.27), is

$$\mathcal{F}^y = -\overline{u'v'} = \frac{1}{2}kl|\widetilde{\psi}^2|, \qquad \mathcal{F}^z = \frac{f_0}{N^2}\overline{v'b'} = \frac{f_0^2}{2N^2}km|\widetilde{\psi}^2|. \tag{9.41}$$

Using (9.38), (9.40) and (9.41) we obtain

$$\mathcal{F} = (\mathcal{F}^y, \mathcal{F}^z) = \mathbf{c}_g\mathcal{P}. \tag{9.42}$$

Equation (9.29) may then be written

$$\frac{\partial \mathcal{P}}{\partial t} + \nabla \cdot (\mathcal{P} \mathbf{c}_g) = 0, \tag{9.43}$$

completing our demonstration. The equation essentially tells us how the disturbance associated with Rossby wave propagates, and is of practical as well as fundamental importance.

9.3 THE TRANSFORMED EULERIAN MEAN

The so-called *transformed Eulerian mean*, or TEM, is a transformation of the equations of motion that provides a useful framework for discussing eddy effects under a wide range of conditions. It is useful because, as we shall see, it is equivalent to a natural form of averaging the equations that serves to eliminate eddy fluxes in the thermodynamic equation and collect them together, in a simple form, in the momentum equation, and in so doing it highlights the role of potential vorticity fluxes. The TEM also provides a useful separation between diabatic and adiabatic effects or between advective and diffusive fluxes and a pleasing simplification of the equations. However, the TEM does have some of its own difficulties, for example in the implementation of proper boundary conditions on the velocity.

9.3.1 Quasi-Geostrophic Form

The Boussinesq zonally-averaged Eulerian mean equations for the zonally-averaged zonal velocity and buoyancy may be written as

$$\frac{\partial \overline{u}}{\partial t} + \overline{v}\frac{\partial \overline{u}}{\partial y} + \overline{w}\frac{\partial \overline{u}}{\partial z} - f\overline{v} = -\frac{\partial}{\partial y}\overline{u'v'} - \frac{\partial}{\partial z}\overline{u'w'} + \overline{F}, \tag{9.44a}$$

$$\frac{\partial \overline{b}}{\partial t} + \overline{v}\frac{\partial \overline{b}}{\partial y} + \overline{w}\frac{\partial \overline{b}}{\partial z} = -\frac{\partial}{\partial y}\overline{v'b'} - \frac{\partial}{\partial z}\overline{w'b'} + \overline{S}, \tag{9.44b}$$

where $\overline{F}$ and $\overline{S}$ represent frictional and heating terms, respectively, and the meridional velocity, $\overline{v}$, is purely ageostrophic. Using quasi-geostrophic scaling we neglect the vertical eddy flux divergences and all ageostrophic velocities except when multiplied by f_0 or N^2, which we take to be equal to $\partial\overline{b}/\partial z$. The above equations then become

$$\frac{\partial \overline{u}}{\partial t} = f_0\overline{v} - \frac{\partial}{\partial y}\overline{u'v'} + \overline{F}, \tag{9.45a}$$

$$\frac{\partial \overline{b}}{\partial t} = -N^2\overline{w} - \frac{\partial}{\partial y}\overline{v'b'} + \overline{S}. \tag{9.45b}$$

These two equations are connected by the thermal wind relation, $f_0\partial\overline{u}/\partial z = -\partial\overline{b}/\partial y$, which itself is a combination of the geostrophic v-momentum equation ($f_0\overline{u} = -\partial\overline{\phi}/\partial y$) and hydrostasy ($\partial\overline{\phi}/\partial z = \overline{b}$). One less than ideal aspect of (9.45) is that in the extratropics the dominant balance is usually between the first two terms on the right-hand sides of each equation, even in time-dependent cases. Thus, the Coriolis force closely balances the divergence of the eddy momentum fluxes, and the advection of the mean stratification (N^2w, or 'adiabatic cooling') often balances the divergence of eddy heat flux, with heating being a small residual. This may lead to an underestimation of the importance of diabatic heating, as this is ultimately responsible for the mean meridional circulation. Furthermore, the link between $\overline{u}$ and $\overline{b}$ via thermal wind dynamically couples buoyancy and momentum, and obscures the understanding of how the eddy fluxes influence these fields — is it through the eddy heat fluxes or momentum fluxes, or some combination?

To address this issue we combine the terms N^2w and the eddy flux in (9.45b) into a single total or *residual* (so recognizing the cancellation between the mean and eddy terms) heat transport term that in a steady state is balanced by the diabatic term $\overline{S}$. To do this, we first note that because $\overline{v}$ and $\overline{w}$ are related by mass conservation we can define a mean meridional streamfunction ψ_m such that

$$(\overline{v}, \overline{w}) = \left(-\frac{\partial \psi_m}{\partial z}, \frac{\partial \psi_m}{\partial y}\right). \qquad (9.46)$$

The velocities then satisfy $\partial\overline{v}/\partial y + \partial\overline{w}/\partial z = 0$ automatically. If we define a *residual streamfunction* by

$$\psi^* \equiv \psi_m + \frac{1}{N^2}\overline{v'b'}, \qquad (9.47a)$$

the components of the *residual mean meridional circulation* are then given by

$$(\overline{v}^*, \overline{w}^*) = \left(-\frac{\partial \psi^*}{\partial z}, \frac{\partial \psi^*}{\partial y}\right), \qquad (9.47b)$$

and

$$\overline{v}^* = \overline{v} - \frac{\partial}{\partial z}\left(\frac{1}{N^2}\overline{v'b'}\right), \qquad \overline{w}^* = \overline{w} + \frac{\partial}{\partial y}\left(\frac{1}{N^2}\overline{v'b'}\right). \qquad (9.48)$$

Note that by construction, the residual overturning circulation satisfies

$$\frac{\partial \overline{v}^*}{\partial y} + \frac{\partial \overline{w}^*}{\partial z} = 0. \qquad (9.49)$$

Substituting (9.48) into (9.45a) and (9.45b) the zonal momentum and buoyancy equations then take the simple forms

$$\frac{\partial \overline{u}}{\partial t} = f_0\overline{v}^* + \overline{v'q'} + \overline{F}, \qquad \frac{\partial \overline{b}}{\partial t} = -N^2\overline{w}^* + \overline{S}, \qquad (9.50a,b)$$

There are no explicit eddy fluxes in the buoyancy budget; the only eddy term is the flux of potential vorticity, and this is the divergence of the Eliassen–Palm flux; that is $\overline{v'q'} = \nabla_x \cdot \mathcal{F}$.

which are known as the (quasi-geostrophic) *transformed Eulerian mean equations,* or TEM equations. The potential vorticity flux, $\overline{v'q'}$, is given in terms of the heat and vorticity fluxes by (9.26), and is equal to the divergence of the Eliassen–Palm flux as in (9.28).

The TEM equations make it apparent that we may consider the potential vorticity fluxes, rather than the separate contributions of the vorticity and heat fluxes, to force the circulation. If we know the potential vorticity flux as well as $\overline{F}$ and $\overline{S}$, then (9.49) and (9.50), along with thermal wind balance, (9.11), form a complete set. The meridional overturning circulation is obtained by eliminating time derivatives from (9.50) using thermal wind to give

$$f_0^2 \frac{\partial^2 \psi^*}{\partial z^2} + N^2 \frac{\partial^2 \psi^*}{\partial y^2} = f_0 \frac{\partial}{\partial z} \overline{v'q'} + f_0 \frac{\partial \overline{F}}{\partial z} + \frac{\partial \overline{S}}{\partial y}. \tag{9.51}$$

Thus, the residual or net overturning circulation is driven by the (vertical derivative of the) potential vorticity fluxes and the diabatic terms — 'driven' in the sense that if we know those terms we can calculate the overturning circulation, although of course the fluxes themselves ultimately depend on the circulation. Note that this equation applies at every instant, even if the equations are not in a steady state.

Use of the equations in TEM form is particularly useful when the eddy potential vorticity flux arises from wave activity, for example from Rossby waves. The potential vorticity flux is the convergence of the EP flux $\boldsymbol{\mathcal{F}}$, as in (9.28), and if the eddies satisfy a dispersion relation the components of the EP flux are equal to the group velocity multiplied by the wave activity density $\mathcal{P}$, as in (9.42). Thus, knowing the group velocity tells us a great deal about how momentum is transported by waves.

Connection to potential vorticity and wave–mean-flow interaction

If we cross-differentiate (9.50) then, after using the residual mass continuity equation (9.49), we recover the zonally-averaged potential vorticity equation, namely

$$\frac{\partial \overline{q}}{\partial t} = -\frac{\partial}{\partial y} \overline{v'q'} - \frac{\partial \overline{F}}{\partial y}, \quad \text{where} \quad \overline{q}(y,t) = \frac{\partial}{\partial z}\left(\frac{f_0}{N^2}\overline{b}\right) - \frac{\partial \overline{u}}{\partial y}. \tag{9.52a,b}$$

These equations are equivalent to (9.17) and (9.20) (since we may fold βy into $\overline{q}$ as it is constant in time).

The corresponding equation for the evolution of eddy potential vorticity is, in its inviscid form,

$$\left(\frac{\partial}{\partial t} + \overline{u}(y,t)\frac{\partial}{\partial x}\right) q' + v' \frac{\partial \overline{q}}{\partial y} = 0, \tag{9.53}$$

as in (9.7). Equations (9.52) and (9.53) are a closed set of quasi-linear equations, and we have recovered the wave–mean-flow system described in Section 9.1.3.

9.3.2 The TEM in the Shallow Water Equations

The residual circulation has an illuminating interpretation if we think of the fluid as comprising one or more layers of shallow water. In a single layer system the momentum and mass conservation equation can be written, in the usual way, as

$$\frac{\partial u}{\partial t} + \boldsymbol{u} \cdot \nabla u - f v = F, \qquad \frac{\partial h}{\partial t} + \nabla \cdot (h\boldsymbol{u}) = S, \tag{9.54a,b}$$

where S is a thickness source term. With quasi-geostrophic scaling, so that variations in Coriolis parameter and layer thickness are small, zonally averaging gives

$$\frac{\partial \overline{u}}{\partial t} - f_0 \overline{v} = \overline{v'\zeta'} + \overline{F}, \qquad \frac{\partial \overline{h}}{\partial t} + H\frac{\partial \overline{v}}{\partial y} = -\frac{\partial}{\partial y}\overline{v'h'} + \overline{S}, \tag{9.55a,b}$$

where the meridional velocity is purely ageostrophic. By analogy with (9.48), we define the residual circulation by

$$\overline{v}^* \equiv \overline{v} + \frac{1}{H}\overline{v'h'}, \tag{9.56}$$

where H is the reference thickness of the layer. Using (9.56) in (9.55) gives

$$\frac{\partial \overline{u}}{\partial t} - f_0 \overline{v}^* = \overline{v'q'} + \overline{F}, \qquad \frac{\partial \overline{h}}{\partial t} + H\frac{\partial \overline{v}^*}{\partial y} = \overline{S}, \tag{9.57a,b}$$

where

$$\overline{v'q'} = \overline{v'\zeta'} - \frac{f_0}{H}\overline{v'h'}, \tag{9.58}$$

is the meridional potential vorticity flux in a shallow water system. From (9.56) we see that the residual velocity is a measure of the *total meridional thickness flux*, eddy plus mean, in an isentropic layer. This is often a more useful quantity than the Eulerian velocity $\overline{v}$ because it is generally the former, not the latter, that is constrained by the external forcing. We have effectively used a thickness-weighted mean in (9.54b); to see this explicitly, define the thickness-weighted mean by

$$\overline{v}_* \equiv \frac{\overline{hv}}{\overline{h}}. \tag{9.59}$$

From (9.59) we have

$$\overline{v}_* = \overline{v} + \frac{1}{\overline{h}}\overline{v'h'}, \tag{9.60}$$

then the zonal average of (9.54b) is just

$$\frac{\partial \overline{h}}{\partial t} + \frac{\partial}{\partial y}(\overline{h}\overline{v}_*) = \overline{S}, \tag{9.61}$$

which is the same as (9.57b) if we take $H = \overline{h}$.

The value of this approach becomes more apparent when we consider multiple layers of fluid, or equivalently if we express the continuous system in isentropic coordinates, in which the thickness of isentropic layer of fluid is used as one of the state variables. The residual velocity in the TEM approach is the same as that which arises from a thickness-weighted average, and this velocity represents the actual average flow of fluid parcels more truthfully than does a conventional Eulerian average at a fixed height. The interested reader may pursue this topic in the references given at the end of the chapter.

9.4 THE NON-ACCELERATION RESULT

In this section we derive an important result in wave–mean-flow dynamics, the so-called non-acceleration condition. Under certain conditions, to be made precise below, we can show that waves have no net effect on the mean-flow, an important and somewhat counter-intuitive result.

9.4.1 A Derivation from the Potential Vorticity Equation

Consider how the potential vorticity fluxes affect the mean fields. The unforced and inviscid zonally-averaged potential vorticity equation is

$$\frac{\partial \overline{q}}{\partial t} + \frac{\partial \overline{v'q'}}{\partial y} = 0. \tag{9.62}$$

Now, in quasi-geostrophic theory the geostrophically balanced velocity and buoyancy can be determined from the potential vorticity via an elliptic equation, as in (9.20), namely

$$\overline{q} - \beta y = \frac{\partial^2 \overline{\psi}}{\partial y^2} + \frac{\partial}{\partial z}\left(\frac{f_0^2}{N^2}\frac{\partial \overline{\psi}}{\partial z}\right), \tag{9.63}$$

where $\overline{\psi}$ is such that $(\overline{u}, \overline{b}/f_0) = (-\partial\overline{\psi}/\partial y, \partial\overline{\psi}/\partial z)$. Differentiating (9.62) with respect to y we obtain

$$\left[\frac{\partial^2}{\partial y^2} + \frac{\partial}{\partial z}\left(\frac{f_0^2}{N^2}\frac{\partial}{\partial z}\right)\right]\frac{\partial \overline{u}}{\partial t} = (\nabla\cdot\boldsymbol{\mathcal{F}})_{yy}, \tag{9.64}$$

where $\nabla\cdot\boldsymbol{\mathcal{F}} = \overline{v'q'}$ is the divergence of the EP flux (in the y–z plane, i.e., $\nabla_x\cdot\boldsymbol{\mathcal{F}}$). This is determined using the wave activity equation for pseudomomentum which, from (9.29), is

$$\frac{\partial \mathcal{P}}{\partial t} + \nabla\cdot\boldsymbol{\mathcal{F}} = \mathcal{D}, \tag{9.65}$$

where $\mathcal{P}$ is the pseudomomentum. If the waves are statistically steady (i.e., $\partial\mathcal{P}/\partial t = 0$) and have no dissipation ($\mathcal{D} = 0$) then evidently $\nabla\cdot\boldsymbol{\mathcal{F}} = 0$. If there is no acceleration at the boundaries then the solution of (9.64) is

$$\frac{\partial \overline{u}}{\partial t} = 0. \tag{9.66}$$

This is a *non-acceleration result.* That is to say, under certain conditions the tendency of the mean fields, and in particular of the zonally-averaged zonal flow, are independent of the waves. To be explicit, those conditions are:

(i) The waves are steady (so that, using the wave activity equation $\mathcal{P}$ does not vary).

(ii) The waves are conservative; that is, $\mathcal{D} = 0$ in (9.29). Given this and item *(i)*, the Eliassen–Palm relation implies that $\nabla \cdot \boldsymbol{\mathcal{F}} = 0$ and the potential vorticity flux is zero.

(iii) The waves are of small amplitude (all of our analysis has neglected terms that are cubic in perturbation amplitude).

(iv) The waves do not affect the boundary conditions (so there are no boundary contributions to the acceleration).

Given the way we have derived it, the result does not seem too surprising; however, it can be powerful and counter-intuitive, for it means that steady waves (i.e., those whose amplitude does not vary) do not affect the zonal flow. However, they *do* affect the Eulerian meridional overturning circulation, and the relative vorticity flux may also be non-zero. In fact, the non-acceleration theorem is telling us that the changes in the vorticity flux are exactly compensated for by changes in the meridional circulation, and there is no net effect on the zonally-averaged zonal flow. It is *irreversibility*, often manifested by the breaking of waves, that leads to permanent changes in the mean-flow.

The derivation of this result by way of the momentum equation, which one might expect to be more natural, is rather awkward because one must consider momentum and buoyancy fluxes separately. Nevertheless, we *can* proceed that way if we use the transformed Eulerian mean, as follows.

9.4.2 Using TEM to Give the Non-Acceleration Result

We may use the TEM formalism to obtain the non-acceleration result. We begin with a purely two-dimensional case as here the derivation is very simple.

A two-dimensional case

Consider two-dimensional incompressible flow on the β-plane, for which there is no buoyancy flux. The linearized vorticity equation is

$$\frac{\partial \zeta'}{\partial t} + \overline{u}\frac{\partial \zeta'}{\partial x} + v'\frac{\partial \overline{\zeta}}{\partial y} = D', \tag{9.67}$$

from which we derive, analogously to (9.29), the Eliassen–Palm relation

$$\frac{\partial \mathcal{P}}{\partial t} + \frac{\partial \mathcal{F}}{\partial y} = \mathcal{D}, \tag{9.68}$$

where $\mathcal{F} = -\overline{u'v'}$, $\mathcal{D}$ represents non-conservative forces, and

$$\mathcal{P} = \frac{\overline{\zeta'^2}}{2\partial_y\overline{\zeta}}. \tag{9.69}$$

Now consider the x-momentum equation, namely

$$\frac{\partial u}{\partial t} = -\frac{\partial u^2}{\partial x} - \frac{\partial uv}{\partial y} - \frac{\partial \phi}{\partial x} + fv. \tag{9.70}$$

Zonally averaging, noting that $\overline{v} = 0$, gives

$$\frac{\partial \overline{u}}{\partial t} = -\frac{\partial \overline{uv}}{\partial y} = \overline{v'\zeta'} = \frac{\partial \mathcal{F}}{\partial y}. \tag{9.71}$$

Finally, combining (9.68) and (9.71) gives

$$\frac{\partial}{\partial t}(\overline{u} + \mathcal{P}) = \mathcal{D}. \tag{9.72}$$

In the absence of non-conservative terms (i.e., if $\mathcal{D} = 0$) the quantity $\overline{u} + \mathcal{P}$ is constant. Further, if the waves are steady and conservative then $\mathcal{P}$ is constant and, therefore, so is $\overline{u}$. This is the non-acceleration result.

The stratified case

Beginning with (9.50a), and using the relation $\overline{v'q'} = \nabla \cdot \boldsymbol{\mathcal{F}}$, the unforced zonally-averaged zonal momentum equation becomes

$$\frac{\partial \overline{u}}{\partial t} - f_0\overline{v}^* = \nabla \cdot \boldsymbol{\mathcal{F}}. \tag{9.73}$$

Using the Eliassen–Palm relation, (9.29) the above expression may be written as

$$\frac{\partial}{\partial t}(\overline{u} + \mathcal{P}) - f_0\overline{v}^* = \mathcal{D}, \tag{9.74}$$

again illustrating how $\mathcal{P}$ is related to the momentum of the flow (and why it is called the pseudomomentum). If the waves are steady ($\partial \mathcal{P}/\partial t = 0$) and conservative ($\mathcal{D} = 0$), then $\partial \overline{u}/\partial t - f_0\overline{v}^* = 0$. However, under these same conditions the residual circulation is also zero, as we now show.

The residual meridional circulation $(\overline{v}^*, \overline{w}^*)$ arises via the necessity to keep the temperature and velocity fields in thermal wind balance and is determined by an elliptic equation, namely (9.51). If the waves are steady and adiabatic then, since $\overline{v'q'} = 0$ in the fluid interior, the right-hand side of the equation is zero and (9.51) becomes

$$f_0^2\frac{\partial^2 \psi^*}{\partial z^2} + N^2\frac{\partial^2 \psi^*}{\partial y^2} = 0. \tag{9.75}$$

If $\psi^* = 0$ at the boundaries, then the unique solution of this is $\psi^* = 0$ everywhere. At the meridional boundaries we may certainly suppose that

ψ^* vanishes if these are quiescent latitudes. At the horizontal boundaries the buoyancy flux will vanish if the waves there are steady, because from (9.15) we have

$$\overline{v'b'}\frac{\partial \overline{b}}{\partial y} = -\frac{1}{2}\frac{\partial}{\partial t}\overline{b'^2} = 0. \tag{9.76}$$

Now, the residual streamfunction is related to the Eulerian streamfunction by (9.47a), and if the eddy buoyancy flux and the Eulerian streamfunction are zero at the boundary, then so is the residual streamfunction. Under these circumstances, then, the residual meridional circulation vanishes in the interior and, from (9.73), the mean-flow is steady, and this is the non-acceleration result.

Compare (9.73) with the momentum equation in conventional Eulerian form, namely

$$\frac{\partial \overline{u}}{\partial t} - f_0\overline{v} = \overline{v'\zeta'}. \tag{9.77}$$

There is no reason that the vorticity flux should vanish when waves are present, even if they are steady. Indeed, such a flux is (under non-acceleration conditions) precisely compensated by the meridional circulation $f_0\overline{v}$, something that is hard to infer or intuit directly from (9.77), and even when non-acceleration conditions do not apply there is significant cancellation between the Coriolis and eddy terms. The difficulty boils down to the fact that, in contrast to $\overline{v'q'}$, $\overline{v'\zeta'}$ is not the flux of a wave activity.

Various results regarding the TEM and non-acceleration are summarized in the shaded box on the next page.

9.4.3 ♦ The EP Flux and Form Drag

It may seem a little magical that the zonal flow is driven by the Eliassen–Palm flux via (9.73). The poleward vorticity flux is clearly related to the momentum flux convergence, but why should a poleward buoyancy flux affect the momentum? The TEM form of the momentum equation may be written as

$$\frac{\partial \overline{u}}{\partial t} = \frac{\partial}{\partial z}\left(\frac{f_0}{N^2}\overline{v'b'}\right) + F_m, \tag{9.78}$$

where $F_m = \overline{v'\zeta'} + f_0\overline{v}^*$ represents forces from the momentum flux and Coriolis force. The first term on the right-hand side certainly does not look like a force; however, it turns out to be directly proportional to the *form drag* between isentropic layers.

Some readers may find (9.79) and (9.80) obvious, some will find them mysterious. They may be derived with a little effort, or the reader may consult the references given at the end of the chapter, or take them on trust. The purpose of this section is simply to make it plausible that a buoyancy flux affects the momentum.

The pressure drag, or the 'form drag', τ_d, at an interface between two isentropic layers of fluid is given by the product of the interface displacement, η', and the pressure gradient:

$$\tau_d = -\overline{\eta'\frac{\partial \phi'}{\partial x}}, \tag{9.79}$$

and the interface displacement is just proportional to the buoyancy perturbation,

$$\eta' = -b'/N^2, \tag{9.80}$$

TEM, Residual Velocities and Non-Acceleration

For a Boussinesq quasi-geostrophic system, the TEM form of the unforced momentum equation and the thermodynamic equation are:

$$\frac{\partial \overline{u}}{\partial t} - f_0 \overline{v}^* = \nabla \cdot \boldsymbol{\mathcal{F}}, \qquad \frac{\partial \overline{b}}{\partial t} + \overline{w}^* N^2 = \overline{S}, \tag{T.1}$$

where $N^2 = \partial \overline{b}_0/\partial z$, $\overline{S}$ represents diabatic effects, $\boldsymbol{\mathcal{F}}$ is the Eliassen–Palm (EP) flux and its divergence is the potential vorticity flux; thus, $\nabla \cdot \boldsymbol{\mathcal{F}} = \nabla_x \cdot \boldsymbol{\mathcal{F}} = \overline{v'q'}$. The residual velocities are

$$\overline{v}^* = \overline{v} - \frac{\partial}{\partial z}\left(\frac{1}{N^2}\overline{v'b'}\right), \qquad \overline{w}^* = \overline{w} + \frac{\partial}{\partial y}\left(\frac{1}{N^2}\overline{v'b'}\right). \tag{T.2}$$

We may define a meridional overturning streamfunction such that $(\overline{v}^*, \overline{w}^*) = (-\partial\psi^*/\partial z, \partial\psi^*/\partial y)$, and using thermal wind to eliminate time-derivatives in (T.1) we obtain

$$f_0^2 \frac{\partial^2 \psi^*}{\partial z^2} + N^2 \frac{\partial^2 \psi^*}{\partial y^2} = f_0 \frac{\partial}{\partial z}\overline{v'q'} + \frac{\partial \overline{S}}{\partial y}. \tag{T.3}$$

The manipulations (given in the main text) that lead to the above equations may seem formal, in that they simply transform the momentum and thermodynamic equations from one form to another. However, the resulting equations have two potential advantages over the untransfomed ones:

(i) The residual meridional velocity is approximately equal to the average thickness-weighted velocity between two neighbouring isentropic surfaces, and so is a measure of the total (Eulerian mean plus eddy) meridional transport of thickness or buoyancy.

(ii) The EP flux is directly related to certain conservation properties of waves. The divergence of the EP flux is the meridional flux of potential vorticity:

$$\boldsymbol{\mathcal{F}} = -(\overline{u'v'})\,\hat{\mathbf{j}} + \left(\frac{f_0}{N^2}\overline{v'b'}\right)\hat{\mathbf{k}}, \qquad \nabla \cdot \boldsymbol{\mathcal{F}} = \overline{v'q'}, \tag{T.4}$$

and the *pseudomomentum*, $\mathcal{P}$, satisfies

$$\frac{\partial \mathcal{P}}{\partial t} + \nabla \cdot \boldsymbol{\mathcal{F}} = \mathcal{D}, \qquad \text{where} \quad \mathcal{P} = \frac{\overline{q'^2}}{2\partial\overline{q}/\partial y}, \quad \mathcal{D} = \frac{\overline{D'q'}}{\partial\overline{q}/\partial y}. \tag{T.5}$$

The quantity $\mathcal{P}$ is a *wave activity density*, and $\mathcal{D}$ is its dissipation. For nearly plane waves, $\mathcal{P}$ and $\boldsymbol{\mathcal{F}}$ are connected by the *group velocity property*,

$$\boldsymbol{\mathcal{F}} = (\mathcal{F}^y, \mathcal{F}^z) = \mathbf{c}_g \mathcal{P}, \tag{T.6}$$

where $\mathbf{c}_g$ is the group velocity of the waves. If the waves are steady ($\partial\mathcal{P}/\partial t = 0$) and dissipationless ($\mathcal{D} = 0$) then $\nabla \cdot \boldsymbol{\mathcal{F}} = 0$ and using (T.1) and (T.3) there is no wave-induced acceleration of the mean-flow; this is the 'non-acceleration' result. Commonly there is enstrophy dissipation, or wave-breaking, and $\nabla \cdot \boldsymbol{\mathcal{F}} < 0$; such *wave drag* leads to flow deceleration and/or a poleward residual meridional velocity, discussed more in Chapter 12.

and using (9.79), (9.80) and geostrophic balance gives

$$\tau_d = \frac{f_0}{N^2}\overline{v'b'}. \tag{9.81}$$

Thus, the vertical component of the EP flux (i.e., the meridional buoyancy flux) is in fact a real stress acting on a fluid layer and equal to the momentum flux caused by the wavy interface. One might say that the vertical component of the EP flux is a force in drag, masquerading as a buoyancy flux.

Notes and References

The TEM and the non-acceleration corollary were introduced by Boyd (1976) and Andrews & McIntyre (1976). A precursor is the paper of Riehl & Fultz (1957), who noted the shortcomings of zonal averaging in uncovering the meaning of indirect cells in laboratory experiments, and by extension the atmosphere. An advanced text on wave–mean-flow is that of Bühler (2009).

CHAPTER

10 Turbulence

Horace Lamb (1849–1934) was a British applied mathematician who published the classic text *Hydrodynamics* in 1895, probably the oldest book on fluid mechanics still in print. Werner Heisenberg (1901–1976) was a German physicist most famous for the matrix formulation of quantum mechanics and the uncertainty principle. His doctoral thesis of 1923 was, however, concerned with turbulence, and he returned to the subject in the late 1940s.

AN APOCRYPHAL STORY that has been attributed to both Horace Lamb and Werner Heisenberg goes as follows. 'When I die and go to Heaven,' they are each said to have predicted, 'I would like to ask my Maker to explain two things, namely turbulence and quantum electrodynamics. About the latter I am hopeful of getting an answer.' Aside from the confidence of these two men as to where they were headed, the story speaks to the inherent difficulty of turbulence. But they may have been more likely to get an answer had their lives been more dissolute, for it is said that turbulence is the invention of the Devil, put on Earth to torment us.

Putting aside these metaphysical issues, in this chapter we will give a introduction to three concrete aspects of turbulence: (i) turbulent diffusion; (ii) the classical spectral scaling theory of turbulence in two and three dimensions; and (iii) the theory of geostrophic turbulence. Before all that we'll describe what the 'problem of turbulence' actually is.

10.1 THE PROBLEM OF TURBULENCE

What is turbulence? Roughly speaking, turbulence is high Reynolds number fluid flow with both spatial and temporal disorder, and a couple of beautiful sketches of what Leonardo da Vinci called *turbolenza* are shown in Fig. 10.1.

Traditionally, turbulent flow has often been thought of as occurring at small scales but in fact a turbulent flow has, as a consequence of being so disordered, a *range* of scales from large to small. The weather itself is an example of a turbulent flow — the great storms sweeping across the midlatitudes contain many scales of motion within them and, as we know from experience, are very hard to predict. Still, turbulence in general and weather in particular *do* have predictable aspects — we know that next winter will be colder than this summer, and that any given month in

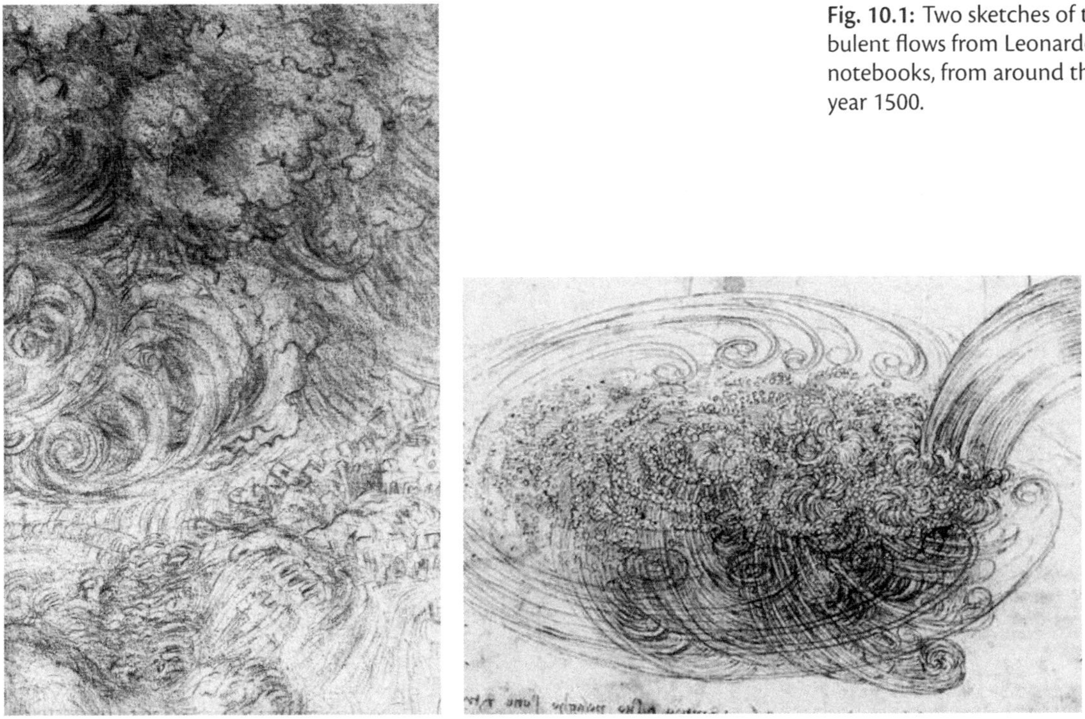

Fig. 10.1: Two sketches of turbulent flows from Leonardo's notebooks, from around the year 1500.

Spain will be warmer than the same month in the UK, and we know that if a storm is approaching from the west it is likely to be windier and rainier than normal, even if we do not know exactly when or where. We might like to be able to predict the *average* flow over a wide area or over a period of a time without necessarily predicting every detail. However, the details may be important — if not in themselves but because they have an effect on the large scale by transporting and mixing properties of the fluid; thus, a turbulent fluid will become well-mixed much more quickly than a laminar fluid. If we drop some ink into a glass of water then we can speed up the mixing by stirring the water, creating turbulent eddies that mix the ink into the water. However, to go beyond this picture and to properly understand the effect of the small scales on the large ones is very difficult because of the 'closure problem', as we now see.

Leonardo da Vinci (1452–1519) is now most famous for his paintings, in particular Mona Lisa and The Last Supper, but he was also a sculptor, architect, inventor and scientist — the embodiment of a 'renaissance man'. He made a number of drawings of eddying flow and coined the word turbolenza *to describe it.*

10.1.1 The Closure Problem

Let us suppose that a flow has a mean component and a fluctuating component, so that the velocity is given by

$$\boldsymbol{v} = \overline{\boldsymbol{v}} + \boldsymbol{v}'. \tag{10.1}$$

Here $\overline{\boldsymbol{v}}$ is the mean velocity field, and $\boldsymbol{v}'$ is the deviation from that mean. The mean might be a time average, in which case $\overline{\boldsymbol{v}}$ is a function only of space and not of time, or it might be a time mean over a finite period (e.g., a season if we are dealing with the weather), or it might be some form of ensemble mean. The average of the deviation is, by definition, zero; that is $\overline{\boldsymbol{v}'} = 0$. We would like to predict the evolution of the mean flow, $\overline{\boldsymbol{v}}$, without predicting the evolution of the eddying flow and to do this we might substitute (10.1) into the momentum equation and try to obtain a closed equation for the mean quantity $\overline{\boldsymbol{v}}$. To keep the algebra simple, and to avoid dealing with the full Navier–Stokes equations, let us carry out this program for a model nonlinear system that obeys

$$\frac{du}{dt} + uu + ru = 0, \tag{10.2}$$

where r is a constant. The average of this equation is:

$$\frac{d\overline{u}}{dt} + \overline{uu} + r\overline{u} = 0. \tag{10.3}$$

The value of the term $\overline{uu}$ (i.e., $\overline{u^2}$) is not deducible simply by knowing $\overline{u}$, since it involves correlations between eddy quantities, namely $\overline{u'u'}$. That is, $\overline{uu} = \overline{u}\,\overline{u} + \overline{u'u'} \neq \overline{u}\,\overline{u}$. We can go to the next order to try (vainly!) to obtain an equation for $\overline{u}\,\overline{u}$. First multiply (10.2) by u to obtain an equation for u^2, and then average it to yield

$$\frac{1}{2}\frac{d\overline{u^2}}{dt} + \overline{uuu} + r\overline{u^2} = 0. \tag{10.4}$$

This equation contains the undetermined cubic term $\overline{uuu}$. An equation determining this would contain a quartic term, and so on in an unclosed hierarchy. Many methods of closing the hierarchy make assumptions about the relationship of $(n+1)$th order terms to nth order terms, for example by supposing that

$$\overline{uuuu} = \alpha\overline{uu}\,\overline{uu} + \beta\overline{uuu}, \tag{10.5}$$

where α and β are parameters to be determined, one may hope, by a theory. If we know that the variables are distributed normally then such closures can be made exact, but this is not generally true in turbulence and all closures that have been proposed so far are, at best, approximations.

This same closure problem arises in the Navier–Stokes equations. If density is constant (as we shall assume in this chapter) the x-momentum equation for an averaged flow is

$$\frac{\partial\overline{u}}{\partial t} + (\overline{\boldsymbol{v}}\cdot\nabla)\overline{u} = -\frac{\partial\overline{\phi}}{\partial x} - \nabla\cdot\overline{\boldsymbol{v}'u'}. \tag{10.6}$$

Written out in full in Cartesian coordinates, the last term is

$$\nabla\cdot\overline{\boldsymbol{v}'u'} = \frac{\partial}{\partial x}\overline{u'u'} + \frac{\partial}{\partial y}\overline{u'v'} + \frac{\partial}{\partial z}\overline{u'w'}. \tag{10.7}$$

These terms, and the similar ones in the y- and z-momentum equations, represent the effects of eddies on the mean flow and are known as *Reynolds stress* terms. The 'closure problem' of turbulence may be thought of as finding a representation of the Reynolds stresses in terms of mean flow quantities. Nobody has been able to usefully close the system without introducing physical assumptions not directly deducible from the equations of motion themselves. Indeed it is not clear that in general a useful closed-form solution even exists.

10.2 TURBULENT DIFFUSION

The most widely used recipe to address the closure problem is by way of turbulent diffusion, or eddy diffusion. The idea comes by way of an analogy with molecular diffusion, and is roughly as follows. Suppose that a fluid carries with it a tracer, φ, that satisfies an equation like

$$\frac{D\varphi}{Dt} = \kappa\nabla^2\varphi, \tag{10.8}$$

where κ is a molecular diffusivity. For simplicity we suppose the flow is two-dimensional and incompressible, and that the flow and the tracer have both a mean and a fluctuating component. The mean component of (10.8) may then be written as

$$\frac{\partial\overline{\varphi}}{\partial t} + \frac{\partial\overline{u}\overline{\varphi}}{\partial x} + \frac{\partial\overline{v}\overline{\varphi}}{\partial y} = -\frac{\partial\overline{u'\varphi'}}{\partial x} - \frac{\partial\overline{v'\varphi'}}{\partial y} + \kappa\nabla^2\overline{\varphi}, \tag{10.9}$$

Now, consider a fluctuating parcel of fluid that, on average, carries its value of φ with it a certain distance ℓ, a 'mixing length', before mixing with its surroundings. If there is a mean gradient of φ in the direction of movement (the y-direction, say) then the value of φ' is given by

$$\varphi' = -\ell\frac{\partial\varphi}{\partial y}. \tag{10.10}$$

If the dominant eddies have a typical speed v' then the eddy transport is given by

$$\overline{v'\varphi'} = -K\frac{\partial\overline{\varphi}}{\partial y}, \qquad \text{where} \qquad K = \overline{v'\ell}. \tag{10.11}$$

In this expression, K is an *eddy diffusivity*, the product of the velocity and length scale of the dominant eddies in the system. If we assume that a similar process occurs in the x-direction then (10.1) becomes

$$\frac{\partial\overline{\varphi}}{\partial t} + \frac{\partial\overline{u}\overline{\varphi}}{\partial x} + \frac{\partial\overline{v}\overline{\varphi}}{\partial y} = K\nabla^2\overline{\varphi} + \kappa\nabla^2\overline{\varphi}. \tag{10.12}$$

To suppose that turbulence acts like an enhanced diffusion is probably the most widely used parameterization of turbulence in practical situations. It is easily implemented and has a rational basis. On the other hand, there are many circumstances when it does not work, and momentum in large-scale atmospheric flows is typically not diffused downgradient, or diffused at all.

In most turbulent flows the eddy diffusivity is much larger than the molecular diffusivity κ because the mixing length is orders of magnitude larger than the corresponding molecular mixing length, which is the average distance that a molecule goes before interacting with another molecule. Thus, we commonly neglect the last term on the right-hand side in (10.12). However, the presence of a molecular viscosity is important in that it allows mixing to take place in the first instance; the turbulence amplifies the molecular mixing enormously, but that mixing must be present.

Equation (10.12) is a practical recipe, and no more, for treating the enhanced transport associated with a turbulent flow. It says that if we are unable to explicitly model the small scales of a turbulent flow, perhaps because we don't know what is happening at those scales, then we might be able to approximately simulate the effects of the small scales using a turbulent diffusion. The idea is rather ad hoc, because we don't have a good theory for the magnitude and structure of the eddy diffusion coefficient K, but it is often better than doing nothing.

The theory of turbulent diffusion stems from work by G. I. Taylor and L. Prandtl, two great figures in fluid dynamics in the early twentieth century.

10.2.1 Homogenization and Lack of Extrema

Now consider a tracer that is advected and diffused. The diffusion might be molecular or, if the effects of turbulence on a tracer are indeed diffusive, there might be an eddy diffusion. An important consequence of this is that, in the absence of additional forcing, there can be no extreme values of the tracer in the interior of the fluid and the diffusion acts to *homogenize* values of the tracer in broad regions.

Consider a tracer that obeys the equation

$$\frac{\mathrm{D}\varphi}{\mathrm{D}t} = \nabla \cdot (\kappa \nabla \varphi), \tag{10.13}$$

where $\kappa > 0$ and the advecting velocity is divergence-free. Given an extremum, there will then be a surrounding surface (in three dimensions), or a surrounding contour (in two), connecting constant values of φ. For definiteness consider three-dimensional incompressible flow which in the steady state flow satisfies

$$\nabla \cdot (\boldsymbol{v}\varphi) = \nabla \cdot (\kappa \nabla \varphi). \tag{10.14}$$

Integrating the left-hand side over the volume, V, enclosed by an isosurface, A, of φ, and applying the divergence theorem, gives

$$\iiint_V \nabla \cdot (\boldsymbol{v}\varphi)\,\mathrm{d}V = \iint_A (\boldsymbol{v}\varphi) \cdot \boldsymbol{n}\,\mathrm{d}A = \varphi \iint_A \boldsymbol{v} \cdot \boldsymbol{n}\,\mathrm{d}A = \varphi \iiint_V \nabla \cdot \boldsymbol{v}\,\mathrm{d}V = 0, \tag{10.15}$$

where $\boldsymbol{n}$ is a unit vector normal to the bounding surface. But the integral of the right-hand side of (10.14) over the same area is non-zero; that is

$$\iiint_V \nabla \cdot (\kappa \nabla \varphi)\,\mathrm{d}V = \iint_A \kappa \nabla \varphi \cdot \boldsymbol{n}\,\mathrm{d}A \neq 0, \tag{10.16}$$

if the integral surrounds an extremum. This is a contradiction for steady flow. Hence, there can be no isolated extrema of a conserved quantity in the interior of a fluid, if there is any diffusion at all. The result is kinematic, in that φ can be any tracer at all, active or passive.

Interpretation and consequences

The physical essence of the result is that the integrated effects of diffusion are non-zero surrounding an extremum, and cannot be balanced by

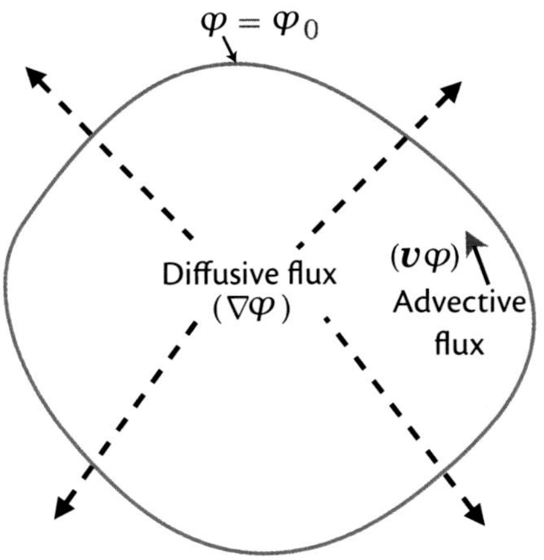

Fig. 10.2: If an extremum of a tracer φ exists in the fluid interior, then diffusion will provide a downgradient tracer flux. But over an area bounded by an isosurface of φ the integrated advective flux is zero. Thus, the diffusion cannot be balanced by advection and so in a steady state no extrema can exist.

advection (Fig. 10.2). Thus, if the initial conditions contain an extremum, diffusion will smooth away the extremum until it no longer exists. Thus, in a steady state, there can be no extrema in the interior of a fluid. One consequence of this is that fluid properties tend to become homogenized in regions of closed contours. Suppose, for example, that the flow is steady and circulates around some point. If that point is initially an extremum, diffusion will effectively *expel* gradients of φ to the enclosing circulation contours, forming a plateau of φ-values.

This result only applies to diffusive steady flows, or to turbulent flows in which the turbulent transport can be parameterized by an eddy diffusivity. This is not always the case and momentum, in particular, is not diffused by turbulence and local maxima of momentum do form on Earth and other planets. However, where such maxima do form the result implies that the flow is not steady, and this is important when studying *super-rotation*, when an atmosphere rotates faster than the planet beneath it: a steady atmosphere cannot super-rotate. However, when turbulence does act as a diffusion mechanism the consequences can be striking, and it appears that potential vorticity is indeed homogenized over large swaths of the ocean, a likely consequence of turbulent diffusion.

10.3 SPECTRAL THEORIES OF TURBULENCE

If we can't close the turbulent equations exactly just what *can* we say about a turbulent flow? One theory that, although it too is not exact, has proven remarkably resilient and that has been shown to agree with experiments in a variety of settings is the spectral theory of Kolmogorov, first proposed in 1941 and often known as K41 theory. The idea is as follows.

Suppose that we stir a fluid at large spatial scale, L_0 say, and thereby we put energy into the fluid at that scale at a rate ε, where ε has units of energy/(time × mass), or $\mathrm{m^2\,s^{-3}}$ or more generically $\mathrm{L^2\,T^{-3}}$. The flow becomes turbulent, creating scales of motion much smaller than L_0. In other words, energy is transferred to smaller scales, and thence to smaller scales still, rather like the ditty in the margin on the following page. Eventually,

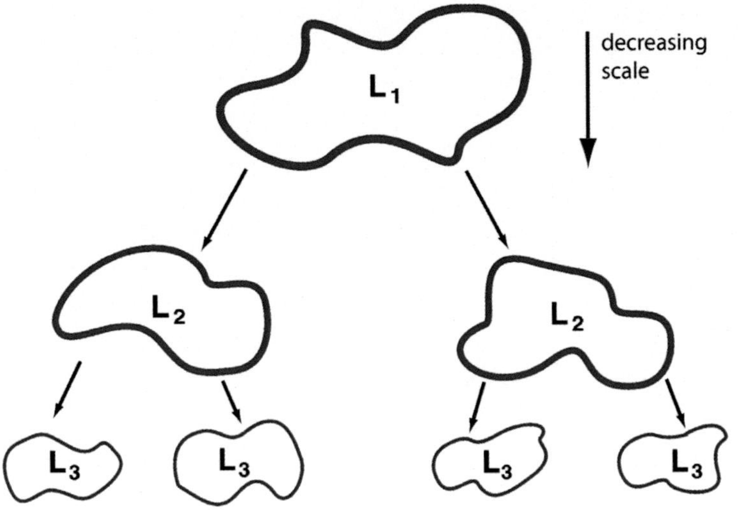

Fig. 10.3: The passage of energy to smaller scales: eddies at large scale break up into ones at smaller scale, thereby transferring energy to smaller scales. (The eddies in reality are embedded within each other.) If the passage occurs between eddies of similar sizes (i.e., if it is spectrally local) the transfer is said to be a cascade.

and no matter how small the molecular viscosity (if not zero), the energy reaches as a scale, L_ν say, at which the Reynolds number, UL_ν/ν, is order one and viscous effects are important. If $\nu = 1 \times 10^{-7}\,\mathrm{m^2\,s^{-1}}$ and $U = 1\,\mathrm{m\,s^{-1}}$ then this viscous scale is given by

Big whorls have little whorls,
that feed on their vorticity;
And little whorls
have lesser whorls,
and so on to viscosity.
L. F. Richardson, 1922 (slightly modified).

$$L_\nu \sim \frac{\nu}{U} = 10^{-7}\,\mathrm{m}. \tag{10.17}$$

This estimate is rather rough and ready, because the velocity itself is likely to be smaller than $1\,\mathrm{m\,s^{-1}}$ at the viscous scale, but nevertheless it is clear that the viscous scale is a small one. We thus have a picture of energy being put into the fluid at large scales, being transferred across scales in an inviscid fashion, and finally being removed by viscosity at very small scales (where the kinetic energy turns into heat). At these intermediate scales the energy transfer from scale to scale must be constant. To describe all this it is useful to deal with Fourier transforms of quantities, so let us have a brief lesson on that.

10.3.1 A Spectral Diversion

Since we are dealing with scales of motion it is convenient to express the velocity field as Fourier components so that we write

$$u(x,y,z,t) = \sum_{\boldsymbol{k}} \tilde{u}(\boldsymbol{k},t)\mathrm{e}^{\mathrm{i}(k^x x + k^y y + k^z z)}, \tag{10.18}$$

where $\tilde{u}$ is the Fourier transformed field of u, with similar identities for v and w, and $\boldsymbol{k} = (k^x, k^y, k^z)$. The sum is a triple sum over all wavenumbers (k^x, k^y, k^z), and in a finite domain these wavenumbers are quantized. Using Parseval's theorem the energy in the fluid is given by

$$\frac{1}{V}\int_V E\,\mathrm{d}V = \frac{1}{2V}\int_V (u^2+v^2+w^2)\,\mathrm{d}V = \frac{1}{2}\sum_{\boldsymbol{k}}\left(|\tilde{u}|^2+|\tilde{v}|^2+|\tilde{w}|^2\right) \equiv \sum_{\boldsymbol{k}} \mathcal{E}_{\boldsymbol{k}}, \tag{10.19}$$

where E is the energy density per unit mass, V is the volume of the domain, and the last equality serves to define the discrete energy spectrum $\mathcal{E}_{\boldsymbol{k}}$. We now assume that the turbulence is isotropic, and that the domain is sufficiently large that the sums in the above equations may be replaced by integrals. We may then write

$$\overline{E} = \frac{1}{V}\widehat{E} = \frac{1}{2V}\int_V \boldsymbol{v}^2 \, \mathrm{d}V = \int \mathcal{E}(k)\, \mathrm{d}k, \tag{10.20}$$

where $\overline{E}$ is the average energy, $\widehat{E}$ is the total energy and $\mathcal{E}(k)$ is the energy spectral density, or the energy spectrum, so that $\mathcal{E}(k)\,\delta k$ is the energy in the small wavenumber interval δk. Because of the assumed isotropy, the energy is a function only of the scalar wavenumber k, where $k^2 = \boldsymbol{k}\cdot\boldsymbol{k} = k^{x2} + k^{y2} + k^{z2}$. The units of $\mathcal{E}(k)$ are L^3/T^2 and the units of $\overline{E}$ are L^2/T^2.

10.3.2 Inertial-Range Theory

We now suppose that the fluid is stirred at large scales and that this energy is transferred to small scales where it is dissipated by viscosity. The key assumption is to suppose that, if the forcing scale is sufficiently larger than the dissipation scale, there exists a range of scales that is intermediate between the large scale and the dissipation scale and where neither forcing nor dissipation are explicitly important to the dynamics. This assumption, known as the *locality hypothesis,* depends on the nonlinear transfer of energy being sufficiently local (in spectral space). This intermediate range is known as the *inertial range,* because the inertial terms and not forcing or dissipation dominate in the momentum balance. If the rate of energy input per unit volume by stirring is equal to ε, then if we are in a steady state there must be a flux of energy from large to small scales that is also equal to ε, and an energy dissipation rate, also ε.

Now, we have no general theory for the energy spectrum of a turbulent fluid, but we might suppose it takes the general form

$$\mathcal{E}(k) = f(\varepsilon, k, k_0, k_\nu), \tag{10.21}$$

where the right-hand side denotes a function of the spectral energy flux or cascade rate ε, the wavenumber k, the forcing wavenumber k_0 and the wavenumber at which dissipation acts, k_ν (and $k_\nu \sim L_\nu^{-1}$). In general, the function f depends on the particular nature of the forcing. Now, the locality hypothesis essentially says that at some scale within the inertial range the flux of energy to smaller scales depends only on processes occurring at or near that scale. That is to say, the energy flux is only a function of $\mathcal{E}$ and k, or equivalently that the energy spectrum can be a function *only* of the energy flux ε and the wavenumber itself. From a physical point of view, as energy cascades to smaller scales the details of the forcing are forgotten but the effects of viscosity are not yet apparent, and the energy spectrum takes the form,

$$\mathcal{E}(k) = g(\varepsilon, k). \tag{10.22}$$

The function g is assumed to be *universal,* the same for every turbulent flow.

The theory described in this section is not an exact theory of turbulence. It relies on assumptions of spectral locality and the near constancy of the energy transfer across scales, and these assumptions are not exactly satisfied. Nevertheless, the theory has been enormously useful and is one of the enduring foundations of the field.

Dimensions and the Kolmogorov Spectrum

Quantity	**Dimension**
Wavenumber, k	1/L
Energy per unit mass, E	$U^2 = L^2/T^2$
Energy spectrum, $\mathcal{E}(k)$	$EL = L^3/T^2$
Energy flux, ε	$E/T = L^2/T^3$

If $\mathcal{E} = g(\varepsilon, k)$ then the only dimensionally consistent relation for the energy spectrum is

$$\mathcal{E} = \mathcal{K}\varepsilon^{2/3}k^{-5/3},$$

where $\mathcal{K}$ is a dimensionless constant.

Let us now use dimensional analysis to give us the form of the function $g(\varepsilon, k)$ (see the shaded box above). In (10.22), the left-hand side has dimensions L^3/T^2; the factor T^{-2} can only be balanced by $\varepsilon^{2/3}$ because k has no time dependence; that is, (10.22), and its dimensions, must take the form

$$\mathcal{E}(k) = \varepsilon^{2/3}g(k), \tag{10.23a}$$

$$\frac{L^3}{T^2} \sim \frac{L^{4/3}}{T^2}g(k), \tag{10.23b}$$

where $g(k)$ is some function. Evidently $g(k)$ must have dimensions $L^{5/3}$, and the functional relationship we must have, if the physical assumptions are right, is

$$\mathcal{E}(k) = \mathcal{K}\varepsilon^{2/3}k^{-5/3}. \tag{10.24}$$

This is the famous 'Kolmogorov -5/3 spectrum', sketched in Fig. 10.4, enshrined as one of the cornerstones of turbulence theory and very well (if not exactly) supported by experiment. The parameter $\mathcal{K}$ is a dimensionless constant, undetermined by this theory; it is known as Kolmogorov's constant and experimentally its value is found to be about 1.5.

The spectral theory of turbulence as presented here was formulated by the Russians A. N. Kologmogorov and A. M. Obukhov in 1941. Prior to that, L. F. Richardson had adumbrated some of its essential qualities in 1920, encapsulated in his ditty on page 194.

An equivalent, and revealing, way to derive this result is to first define an eddy turnover time τ_k, which is the time taken for a parcel with velocity v_k to move a distance $1/k$, v_k being the velocity associated with the (inverse) scale k. On dimensional considerations $v_k = [\mathcal{E}(k)k]^{1/2}$ so that

$$\tau_k = \left[k^3\mathcal{E}(k)\right]^{-1/2}. \tag{10.25}$$

Kolmogorov's assumptions are then equivalent to setting

$$\varepsilon \sim \frac{v_k^2}{\tau_k} = \frac{k\mathcal{E}(k)}{\tau_k}. \tag{10.26}$$

If we demand that ε be a constant then (10.25) and (10.26) yield (10.24).

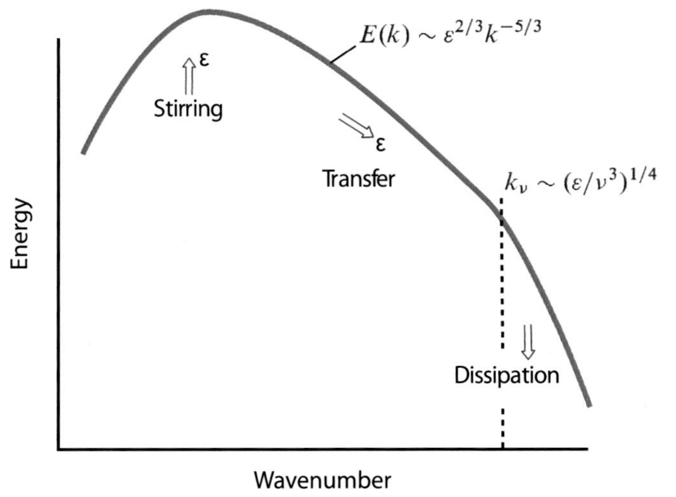

Fig. 10.4: The energy spectrum in three-dimensional turbulence, in the theory of Kolmogorov. Energy is supplied at some rate ε; it is cascaded to small scales, where it is ultimately dissipated by viscosity. There is no systematic energy transfer to scales larger than the forcing scale, so here the energy falls off.

The viscous scale and energy dissipation

At some small length scale we should expect viscosity to become important and the scaling theory we have just set up will fail. What is that scale? In the inertial range friction is unimportant because the time scales on which it acts are too long for it be important and dynamical effects dominate. In the momentum equation the viscous term is $\nu\nabla^2 u$ so that a viscous or dissipation time scale at a scale k^{-1}, τ_k^ν, is

$$\tau_k^\nu \sim \frac{1}{k^2\nu}, \tag{10.27}$$

so that the viscous time scale decreases as k increases and the length scale decreases. The eddy turnover, or the inertial, timescale, τ_k, in the Kolmogorov spectrum is

$$\tau_k = \varepsilon^{-1/3}k^{-2/3}. \tag{10.28}$$

The wavenumber at which dissipation becomes important is then given by equating the above two time scales, yielding the dissipation wavenumber, k_ν and the associated length scale, L_ν,

$$k_\nu \sim \left(\frac{\varepsilon}{\nu^3}\right)^{1/4}, \qquad L_\nu \sim \left(\frac{\nu^3}{\varepsilon}\right)^{1/4}. \tag{10.29a,b}$$

L_ν is called the *Kolmogorov scale*. It is the *only* quantity which can be created from the quantities ν and ε that has the dimensions of length.

An interesting consequence of the theory is this. The rate of energy dissipation by viscous effects is *independent of the magnitude of the coefficient of viscosity.* This might seem strange, since surely there must be less dissipation if the viscosity is smaller? Not so; if we reduce the viscosity, the cascade proceeds to smaller scales but the total dissipation stays the same, for it is equal to the energy input ε. Back in physical space, let us consider the viscous term $\nu\nabla^2\boldsymbol{v}$. If ν decreases the scale at which dissipation takes place also decreases and so the second derivative, ∇^2 becomes correspondingly larger in just such a way as to keep energy dissipation constant.

How big is L_ν in the atmosphere? A crude estimate comes from noting that ε has units of U^3/L, and that at length scales of the order of 100 m in the atmospheric boundary layer (where there might be a three-dimensional energy cascade to small scales) velocity fluctuations are of the order of 1 cm s^{-1}, giving $\varepsilon \approx 10^{-8}\,\mathrm{m^2\,s^{-3}}$. Using (10.29b) we then find the dissipation scale to be of the order of a millimetre or so. In the ocean the dissipation scale is also of the order of millimetres. Various inertial range properties, in both three and two dimensions, are summarized in the shaded box on page 203.

10.4 Two-Dimensional Turbulence

Two-dimensional turbulence behaves in a profoundly different way from three-dimensional turbulence, largely because of the presence of another quadratic invariant, the enstrophy, defined below. In two dimensions, the vorticity equation for incompressible flow is:

$$\frac{\partial \zeta}{\partial t} + \boldsymbol{u} \cdot \nabla \zeta = F + \nu \nabla^2 \zeta, \tag{10.30}$$

where $\boldsymbol{u} = u\boldsymbol{i} + v\boldsymbol{j}$ and $\zeta = \boldsymbol{k} \cdot \nabla \times \boldsymbol{u}$ and F is a stirring term. In terms of a streamfunction, $u = -\partial\psi/\partial y$, $v = \partial\psi/\partial x$, and $\zeta = \nabla^2\psi$, and (10.30) may be written as

$$\frac{\partial \nabla^2 \psi}{\partial t} + J(\psi, \nabla^2 \psi) = F + \nu \nabla^4 \psi. \tag{10.31}$$

We obtain an energy equation by multiplying by $-\psi$ and integrating over the domain, and an enstrophy equation by multiplying by ζ and integrating. When $F = \nu = 0$ we find

$$\widehat{E} = \frac{1}{2}\int_A (u^2 + v^2)\,\mathrm{d}A = \frac{1}{2}\int_A (\nabla\psi)^2\,\mathrm{d}A, \qquad \frac{\mathrm{d}\widehat{E}}{\mathrm{d}t} = 0, \tag{10.32a}$$

$$\widehat{Z} = \frac{1}{2}\int_A \zeta^2\,\mathrm{d}A = \frac{1}{2}\int_A (\nabla^2\psi)^2\,\mathrm{d}A, \qquad \frac{\mathrm{d}\widehat{Z}}{\mathrm{d}t} = 0, \tag{10.32b}$$

where the integral is over a finite area with either no-normal flow or periodic boundary conditions. The quantity $\widehat{E}$ is the energy, and $\widehat{Z}$ is known as the *enstrophy*. The enstrophy invariant arises because the vortex stretching term, so important in three-dimensional turbulence, vanishes identically in two dimensions. In fact, because vorticity is conserved on parcels it is clear that the integral of *any* function of vorticity is zero when integrated over A; that is, from (10.30)

$$\frac{\mathrm{D}f(\zeta)}{\mathrm{D}t} = 0 \qquad \text{and} \qquad \frac{\mathrm{d}}{\mathrm{d}t}\int_A f(\zeta)\,\mathrm{d}A = 0, \tag{10.33}$$

where $f(\zeta)$ is an arbitrary function. Of this infinity of conservation properties, enstrophy conservation (with $f(\zeta) = \zeta^2$) in particular has been found to have enormous consequences to the flow of energy between scales, as we soon discover.

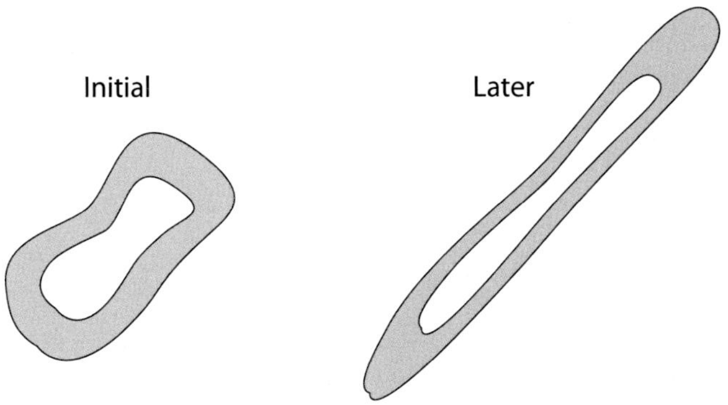

Fig. 10.5: In two-dimensional incompressible flow, a band of fluid is elongated, but its area is preserved. Elongation is followed by folding and more elongation, but still area-preserving in the absence of viscosity. As vorticity is tied to fluid parcels, the values of the vorticity in the shaded area (and in the hole) are maintained; thus, vorticity gradients increase and the enstrophy is thereby, on average, moved to smaller scales until it finally reaches the viscous scale.

10.4.1 Energy and Enstrophy Transfer

In three-dimensional turbulence we posited that energy is cascaded to small scales via vortex stretching. In two dimensions that mechanism is absent, and there is reason to expect energy to be transferred to *larger* scales. This counter-intuitive behaviour arises from the twin integral constraints of energy and enstrophy conservation, and the following two arguments illustrate why this should be so.

I. Vorticity elongation

Consider a band or a patch of vorticity, as in Fig. 10.5, in a nearly inviscid fluid. The vorticity of each element of fluid is conserved as the fluid moves. Now, we should expect that the quasi-random motion of the fluid will act to elongate the band but, as its area must be preserved, the band narrows and so vorticity gradients will increase. This is equivalent to the enstrophy moving to smaller scales. Now, the energy in the fluid is

$$\widehat{E} = -\frac{1}{2}\int \psi\zeta \, \mathrm{d}A, \tag{10.34}$$

where the streamfunction is obtained by solving the Poisson equation $\nabla^2\psi = \zeta$. If the vorticity is locally elongated primarily only in one direction (as it must be to preserve area), the integration involved in solving the Poisson equation leads to the scale of the streamfunction becoming larger in the direction of stretching, but virtually no smaller in the perpendicular direction. Because stretching occurs, on average, in all directions, the overall scale of the streamfunction increases in all directions, and the cascade of enstrophy to small scales is accompanied by a transfer of energy to large scales.

II. An energy–enstrophy conservation argument

A moment's thought will reveal that the distributions of energy and enstrophy in wavenumber space are respectively analogous to the distribution of mass and moment of inertia of a lever, with wavenumber playing the role of distance from the fulcrum. Any rearrangement of mass such that its distribution also becomes wider must be such that the centre of mass moves toward the fulcrum. Thus, analogously, any rearrangement of a flow that preserves both energy and enstrophy, and that causes the distribution to spread out in wavenumber space, will tend to move energy to small wavenumbers and enstrophy to large. To prove this we begin with expressions for the average energy and enstrophy:

$$\overline{E} = \int \mathcal{E}(k)\, \mathrm{d}k, \qquad \overline{Z} = \int \mathcal{Z}(k)\, \mathrm{d}k = \int k^2 \mathcal{E}(k)\, \mathrm{d}k, \tag{10.35}$$

where $\mathcal{E}(k)$ and $\mathcal{Z}(k)$ are the energy and enstrophy spectra. A wavenumber characterizing the spectral location of the energy is the centroid,

$$k_e = \frac{\int k\mathcal{E}(k)\, \mathrm{d}k}{\int \mathcal{E}(k)\, \mathrm{d}k}, \tag{10.36}$$

and, for simplicity, we normalize units so that the denominator is unity. The spreading out of the energy distribution is formalized by setting

$$I \equiv \int (k - k_e)^2 \mathcal{E}(k)\, \mathrm{d}k, \qquad \frac{\mathrm{d}I}{\mathrm{d}t} > 0. \tag{10.37}$$

Here, I measures the width of the energy distribution, and this is assumed to increase. Expanding out the integral gives

$$\begin{aligned} I &= \int k^2 \mathcal{E}(k)\, \mathrm{d}k - 2k_e \int k\mathcal{E}(k)\, \mathrm{d}k + k_e^2 \int \mathcal{E}(k)\, \mathrm{d}k \\ &= \int k^2 \mathcal{E}(k)\, \mathrm{d}k - k_e^2 \int \mathcal{E}(k)\, \mathrm{d}k, \end{aligned} \tag{10.38}$$

where the last equation follows because $k_e = \int k\mathcal{E}(k)\, \mathrm{d}k$ is, from (10.36), the energy-weighted centroid. Because both energy and enstrophy are conserved, (10.38) gives

$$\frac{\mathrm{d}k_e^2}{\mathrm{d}t} = -\frac{1}{\overline{E}} \frac{\mathrm{d}I}{\mathrm{d}t} < 0. \tag{10.39}$$

Thus, the centroid of the distribution moves to smaller wavenumbers and to larger scales (see Fig. 10.6).

An appropriately defined measure of the centre of the enstrophy distribution, on the other hand, moves to higher wavenumbers. The demonstration follows easily if we work with the inverse wavenumber, which is a direct measure of length. Let $q = 1/k$ and assume that the enstrophy distribution spreads out by nonlinear interactions, so that, analogously to (10.37),

$$J \equiv \int (q - q_e)^2 \mathcal{Y}(q)\, \mathrm{d}q, \qquad \frac{\mathrm{d}J}{\mathrm{d}t} > 0, \tag{10.40}$$

where $\mathcal{Y}(q)$ is such that the enstrophy is $\int \mathcal{Y}(q)\mathrm{d}q$ and

$$q_e = \frac{\int q\mathcal{Y}(q)\, \mathrm{d}q}{\int \mathcal{Y}(q)\, \mathrm{d}q}. \tag{10.41}$$

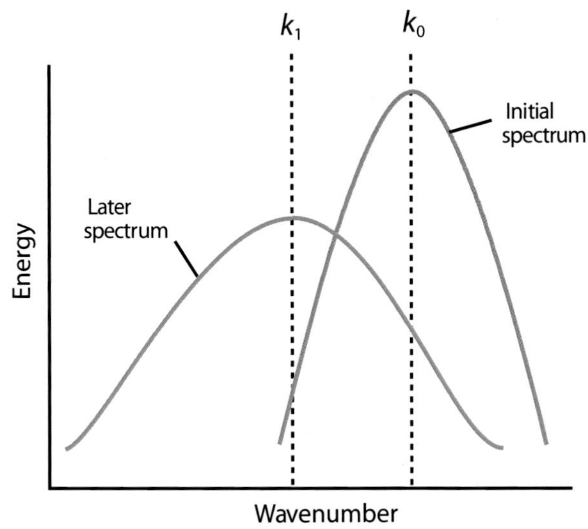

Fig. 10.6: In two-dimensional flow, the centroid of the energy spectrum moves to large scales (smaller wavenumber) provided that the width of the distribution increases — as can be expected in a nonlinear, eddying flow.

Expanding the integrand in (10.40) and using (10.41) gives

$$J = \int q^2 \mathcal{Y}(q)\, dq - q_e^2 \int \mathcal{Y}(q)\, dq. \tag{10.42}$$

But $\int q^2 \mathcal{Y}(q)\, dq$ is conserved, because this is the energy. Thus,

$$\frac{dJ}{dt} = -\frac{d}{dt} q_e^2 \int \mathcal{Y}(q)\, dq, \tag{10.43}$$

whence

$$\frac{dq_e^2}{dt} = -\frac{1}{Z}\frac{dJ}{dt} < 0. \tag{10.44}$$

Thus, the length scale characterizing the enstrophy distribution gets smaller, and the corresponding wavenumber gets larger.

10.4.2 Inertial Ranges in Two-Dimensional Turbulence

The above results suggest that, in a forced-dissipative two-dimensional fluid, energy is transferred to larger scales and enstrophy is transferred to small scales. We thus expect that two inertial ranges may form — an *energy inertial range* carrying energy to larger scales, and an *enstrophy inertial range* carrying enstrophy to small scales (Fig. 10.7). Let us calculate their properties.

The enstrophy inertial range

In the enstrophy inertial range the enstrophy cascade rate η, equal to the rate at which enstrophy is supplied by stirring, is assumed constant. By analogy with (10.26) we may assume that this rate is given by

$$\eta \sim \frac{k^3 \mathcal{E}(k)}{\tau_k}. \tag{10.45}$$

In two-dimensional turbulence energy is generally transferred to large scales whereas enstrophy is transferred to small scales. If these transfers are spectrally local they are referred to as *cascades*. In an upscale energy cascade, meaning a cascade to larger scales, the energy spectrum varies as $k^{-5/3}$ and in an enstrophy spectrum it varies as k^{-3}.

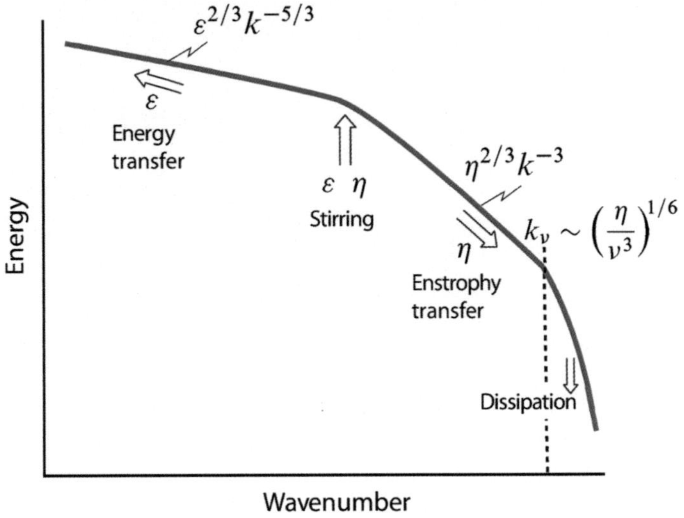

Fig. 10.7: The energy spectrum of two-dimensional turbulence. (Compare with Fig. 10.4.) Energy supplied at some rate ε is transferred to large scales, whereas enstrophy supplied at some rate η is transferred to small scales, where it may be dissipated by viscosity. If the forcing is localized at a scale k_f^{-1} then $\eta \approx k_f^2\varepsilon$.

With τ_k (still) given by (10.25) we obtain

$$\mathcal{E}(k) = \mathcal{K}_\eta \eta^{2/3} k^{-3}, \tag{10.46}$$

where $\mathcal{K}_\eta$ is, we presume, a universal constant, analogous to the Kolmogorov constant of (10.24). This, and various other properties in two- and three-dimensional turbulence, are summarized in the shaded box on the facing page.

The velocity and time at a particular wavenumber then scale as

$$v_k \sim \eta^{1/3}k^{-1}, \qquad t_k \sim l_k/v_k \sim 1/(kv_k) \sim \eta^{-1/3}. \tag{10.47a,b}$$

We may also obtain (10.47) by substituting (10.46) into (10.25). Thus, *the eddy turnover time in the enstrophy range of two-dimensional turbulence is length-scale invariant.* The appropriate viscous scale is given by equating the inertial and viscous terms in (10.30). Using (10.47a) we obtain, analogously to (10.29a), the viscous wavenumber

$$k_\nu \sim \left(\frac{\eta^{1/3}}{\nu}\right)^{1/2}. \tag{10.48}$$

The enstrophy dissipation goes to a finite limit given by

$$\frac{\mathrm{d}}{\mathrm{d}t}\overline{Z} = \nu \int_A \zeta\nabla^2\zeta \,\mathrm{d}A \sim \nu k_\nu^4 v_{k_\nu}^2 \sim \eta, \tag{10.49}$$

using (10.47a) and (10.48). Thus, the enstrophy dissipation in two-dimensional turbulence is (at least according to this theory) independent of the viscosity, just as energy dissipation is independent of the viscosity in three-dimensional turbulence.

Inertial Range Properties in 3D and 2D Turbulence

A few inertial range properties are listed below, omitting nondimensional constants.

	3D energy range	2D enstrophy range	
Energy spectrum	$\varepsilon^{2/3}k^{-5/3}$	$\eta^{2/3}k^{-3}$	(T.1)
Turnover time	$\varepsilon^{-1/3}k^{-2/3}$	$\eta^{-1/3}$	(T.2)
Viscous scale, L_ν	$(\nu^3/\varepsilon)^{1/4}$	$(\nu^3/\eta)^{1/6}$	(T.3)

In these expressions:

ν = viscosity, k = wavenumber, ε = energy cascade rate,
η = enstrophy cascade rate,

Energy inertial range

The energy inertial range of two-dimensional turbulence is quite similar to that of three-dimensional turbulence, except in one major respect: the energy flows from smaller to larger scales! Because the atmosphere and ocean behave in some ways as two-dimensional fluids, this has profound consequences on their behaviour, and is something we return to in the next section. The upscale energy flow is known as the *inverse cascade,* and the associated energy spectrum is, as in the three-dimensional case,

Robert Kraichnan (1928–2008) made a host of significant contributions to the theory of turbulence in the second half of the twentieth century, working independently first in New Hampshire and later in New Mexico. Among other things he is largely responsible for the inertial-range theory of two-dimensional turbulence described here.

$$\mathcal{E}(k) = \mathcal{K}_\varepsilon \varepsilon^{2/3} k^{-5/3}, \tag{10.50}$$

where $\mathcal{K}_\varepsilon$ is a nondimensional constant — sometimes called the Kolmogorov–Kraichnan constant, and not necessarily equal to $\mathcal{K}$ in (10.24) — and ε is the rate of energy transfer to larger scales, where it is removed by boundary-layer friction.

The atmosphere itself is not observed to have an inverse −5/3 spectrum at large scales; indeed there is no well-defined inverse energy cascade in the sense described above. The atmosphere does have an approximate −3 cascade but whether we should attribute this to a classical forward enstrophy cascade is not settled. A numerical simulation of two-dimensional turbulence, illustrating both the energy and enstrophy cascade, is shown in Fig. 10.8.

10.5 ✦ Geostrophic Turbulence

The atmosphere is not two-dimensional but, because of the effects of stratification and rotation, it is 'quasi-two-dimensional' and approximately satisfies the quasi-geostrophic equations. Let us therefore extend our two-dimensional theory to include the effects of stratification.

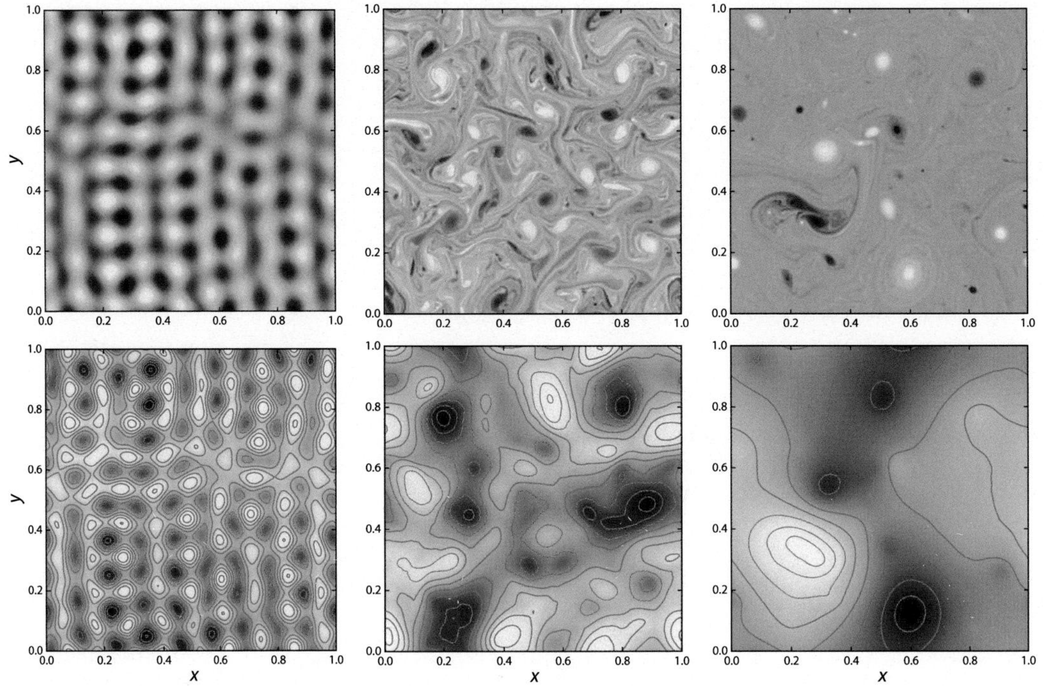

Fig. 10.8: Evolution of vorticity (top) and streamfunction (bottom) in a numerical simulation of two-dimensional turbulence with no forcing. The initial conditions (left) have just a few non-zero Fourier modes (around wavenumber 9) with randomly generated phases. Time proceeds left to right, with like-signed vortices merging (with an example in the top right panel) and enstrophy cascading to smaller scales between the vortices. The scale of the streamfunction grows larger, reflecting the transfer of energy to larger scales. (See Fig. 10.11 to see jets form because of the effects of rotation.)

10.5.1 An Analogue to Two-Dimensional Flow

Consider a quasi-geostrophic atmosphere in which the unforced, inviscid governing equations are

$$\frac{\mathrm{D}q}{\mathrm{D}t} = 0, \qquad q = \nabla^2\psi + Pr^2\frac{\partial^2\psi}{\partial z^2}, \tag{10.51a}$$

where $Pr = f_0/N$ is the *Prandtl ratio* (and Pr/H is the inverse of the deformation radius) and $\mathrm{D}/\mathrm{D}t = \partial/\partial t + \boldsymbol{u}\cdot\nabla$ is the two-dimensional material derivative. The vertical boundary conditions are

$$\frac{\mathrm{D}}{\mathrm{D}t}\left(\frac{\partial\psi}{\partial z}\right) = 0, \qquad \text{at } z = 0, H. \tag{10.51b}$$

These equations are analogous to the equations of motion for purely two-dimensional flow. In particular, with periodic lateral boundary conditions, or conditions of no-normal flow, there are two quadratic invariants of the motion, the energy and the enstrophy, which are obtained by multiplying (10.51a) by $-\psi$ and q and integrating over the domain. The conserved energy is

$$\frac{\mathrm{d}\widehat{E}}{\mathrm{d}t} = 0, \qquad \widehat{E} = \frac{1}{2}\int_V \left[(\nabla\psi)^2 + Pr^2\left(\frac{\partial\psi}{\partial z}\right)^2\right]\mathrm{d}V, \tag{10.52}$$

where the integral is over a *three-dimensional* domain. The enstrophy is conserved at each vertical level, and of course the volume integral is also conserved, namely

$$\frac{d\widehat{Z}}{dt} = 0, \qquad \widehat{Z} = \frac{1}{2}\int_V q^2 \, dV = \frac{1}{2}\int_V \left[\nabla^2\psi + Pr^2\left(\frac{\partial^2\psi}{\partial z^2}\right)\right]^2 dV. \quad (10.53)$$

The analogy with two-dimensional flow is even more transparent if we further rescale the vertical coordinate by $1/Pr$, and so let $z' = z/Pr$. Then the energy and enstrophy invariants are:

$$\widehat{E} = \int (\nabla_3\psi)^2 \, dV, \qquad \widehat{Z} = \int q^2 \, dV = \int (\nabla_3^2\psi)^2 \, dV, \quad (10.54)$$

where $\nabla_3 = \hat{\mathbf{i}}\,\partial/\partial x + \hat{\mathbf{j}}\,\partial/\partial y + \hat{\mathbf{k}}\,\partial/\partial z'$. The invariants then have almost the same form as the two-dimensional invariants, but with a three-dimensional Laplacian operator instead of a two-dimensional one.

Given these invariants, we expect that any dynamical behaviour that occurs in the two-dimensional equations *that depends solely on the energy/enstrophy constraints* should have an analogue in quasi-geostrophic flow. In particular, the transfer of energy to large-scales and enstrophy to small scales will also occur in quasi-geostrophic flow with, in so far as these transfers are effected by a local cascade, corresponding spectra of $k^{-5/3}$ and k^{-3}. However, in the quasi-geostrophic case, it is the *three-dimensional* wavenumber that is relevant, with the vertical component scaled by the Prandtl ratio. As a consequence, the energy cascade to larger horizontal scales is generally accompanied by a cascade to larger vertical scales — a *barotropization* of the flow. Still, the analogy between two-dimensional and quasi-geostrophic cascades should not be taken too far, because in the latter the potential vorticity is advected only by the horizontal flow. Thus, the dynamics of quasi-geostrophic turbulence are *not* in general isotropic in three-dimensional wavenumber. To examine these dynamics more fully we turn to a simpler model, that of two-layer flow.

Geostrophic turbulence is similar to two-dimensional turbulence in that energy is transferred to large scales. The transfer occurs in both the vertical and horizontal directions, and the transfer to large scales leads to the *barotropization* of the flow.

10.5.2 Two-Layer Geostrophic Turbulence

We consider flow in two layers of equal depth, governed by the two-level quasi-geostrophic equations as described in Section 5.6. Taking $\beta = 0$ the equations of motion are

$$\frac{\partial q_i}{\partial t} + J(\psi_i, q_i) = 0, \qquad i = 1, 2, \quad (10.55)$$

where, for any a and b, $J(a,b) = \partial a/\partial x \, \partial b/\partial y - \partial a/\partial y \, \partial b/\partial x$ and

$$q_1 = \nabla^2\psi_1 + \frac{1}{2}k_d^2(\psi_2 - \psi_1), \qquad q_2 = \nabla^2\psi_2 + \frac{1}{2}k_d^2(\psi_1 - \psi_2), \quad (10.56a,b)$$

$$\frac{1}{2}k_d^2 = \frac{4f_0^2}{N^2H^2}, \quad (10.56c)$$

and k_d is inversely proportional to the baroclinic deformation radius.

It turns out to be convenient to rewrite the equations in terms of the sum and difference of the streamfunctions in each layer, and to this end we define the barotropic and baroclinic streamfunctions by

$$\widehat{\psi} \equiv \frac{1}{2}(\psi_1 + \psi_2), \qquad \tau \equiv \frac{1}{2}(\psi_1 - \psi_2), \tag{10.57}$$

so that $\widehat{\psi}$ is the vertically averaged streamfunction and τ is proportional to the buoyancy (since $b = f_0 \partial\psi/\partial z$; think 'tau for temperature'). The potential vorticities for each layer are then

$$q_1 = \nabla^2\widehat{\psi} + (\nabla^2 - k_d^2)\tau, \qquad q_2 = \nabla^2\widehat{\psi} - (\nabla^2 - k_d^2)\tau, \tag{10.58a,b}$$

and the equations of motion may be rewritten as the following evolution equations for $\widehat{\psi}$ and τ:

$$\frac{\partial}{\partial t}\nabla^2\widehat{\psi} + J(\widehat{\psi}, \nabla^2\widehat{\psi}) + J(\tau, (\nabla^2 - k_d^2)\tau) = 0, \tag{10.59a}$$

$$\frac{\partial}{\partial t}(\nabla^2 - k_d^2)\tau + J(\tau, \nabla^2\widehat{\psi}) + J(\widehat{\psi}, (\nabla^2 - k_d^2)\tau) = 0. \tag{10.59b}$$

With these new variables a little algebra reveals that the conservation of energy takes the form

$$\frac{\mathrm{d}\widehat{E}}{\mathrm{d}t} = 0, \qquad \widehat{E} = \frac{1}{2}\int_A \left[(\nabla\widehat{\psi})^2 + (\nabla\tau)^2 + k_d^2\tau^2\right] \mathrm{d}A, \tag{10.60}$$

and the conservation of total enstrophy takes the form

$$\frac{\mathrm{d}\widehat{Z}}{\mathrm{d}t} = 0, \qquad \widehat{Z} = \int_A (\nabla^2\widehat{\psi})^2 + \left[(\nabla^2 - k_d^2)\tau\right]^2 \mathrm{d}A. \tag{10.61}$$

In addition the enstrophy in each layer is conserved. The nonlinear interactions in (10.59) will lead to a transfer of energy between scales, as in two-dimensional turbulence, and to better understand that we note:

(i) $\widehat{\psi}$ and τ are analogous to Fourier modes in the vertical. That is, $\widehat{\psi}$ is a *barotropic mode* with a 'vertical wavenumber' of zero, and τ is a *baroclinic mode* with a vertical wavenumber of one.

(ii) From (10.59) we see there are two distinct kinds of nonlinear interaction, namely

$$(i) \quad (\widehat{\psi}, \widehat{\psi}) \to \widehat{\psi}, \qquad (ii) \quad (\tau, \tau) \to \widehat{\psi} \quad \text{or} \quad (\widehat{\psi}, \tau) \to \tau. \tag{10.62}$$

The first is a *barotropic interaction,* involving only the barotropic mode. The other two are examples of mixed barotropic–baroclinic interactions: two baroclinic modes giving rise to a barotropic model, or a barotropic model interacting with a baroclinic mode. There are no interactions that involve only the baroclinic mode.

(iii) Wherever the Laplacian operator acts on τ in the evolution equations, it is accompanied by $-k_d^2$. That is, it is *as if* the effective wavenumber (squared) of τ is shifted, so that $k^2 \to k^2 + k_d^2$. Similarly, the energy and enstrophy invariants bear the same relation to each other that they do in two-dimensional turbulence, except that the relevant wavenumber for the baroclinic mode is $k^2 + k_d^2$ instead of k^2. Effectively, at any given horizontal wavenumber, the baroclinic modes have a higher 'total' wavenumber than the barotropic ones.

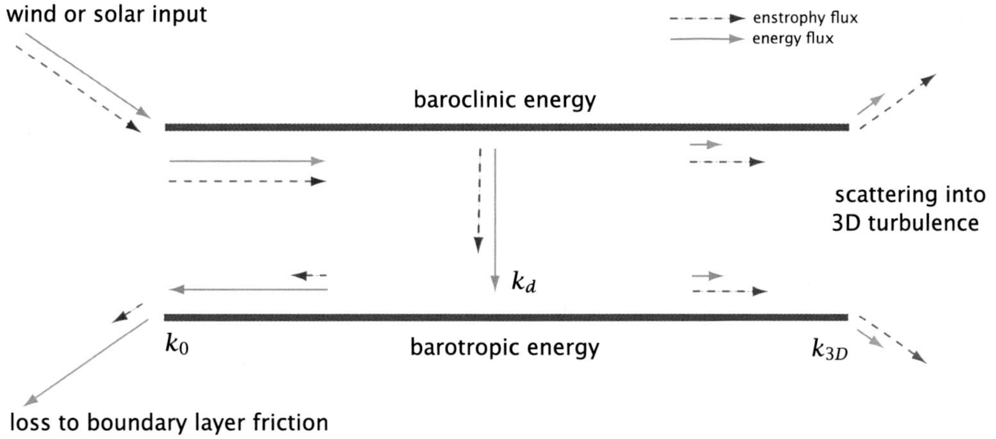

Fig. 10.9: Two-layer baroclinic turbulence. The horizontal axis represents horizontal wavenumber, and the vertical variation is decomposed into two vertical modes — the barotropic and first baroclinic. Energy transfer is shown by solid arrows and enstrophy transfer by dashed arrows. Large-scale forcing provides energy to the baroclinic mode at large scales. Energy is transferred to smaller baroclinic scales, and then into barotropic energy, and finally transferred to larger barotropic scales in an upscale cascade.

Phenomenology of Geostrophic Turbulence

Putting together the above considerations with our experience of two-dimensional turbulence leads to the following picture of geostrophic turbulence, summarized in Fig. 10.9. In general, nonlinear interactions among the various modes will led to a transfer of energy to the largest possible scales, or the smallest possible total wavenumber, but since the effective wavenumber of the baroclinic mode is larger than that of the barotropic mode, energy will also be transferred to the barotropic mode (this is the barotropization process referred to earlier). That is, the energy will be transferred to the largest possible vertical scale as well as the largest horizontal scale. The conversion from baroclinic to barotropic energy tends to occur at the scale of the radius of deformation, since this process is, essentially, baroclinic instability.

The overall energy cycle thus goes as follows. At large horizontal scales we may imagine some source of baroclinic energy, which in the atmosphere might be the differential heating between pole and equator, or in the ocean might be the wind and surface heat fluxes. Baroclinic instability effects a non-local transfer of energy to the deformation scale, where both baroclinic and barotropic modes are excited. Energy is then transferred back to large scales in the barotropic mode (an inverse cascade), and ultimately it is dissipated by bottom friction. At the same time, there is an enstrophy cascade in each layer to smaller and smaller scales, until the wavenumber is large enough (denoted by k_{3D} in Fig. 10.9) that non-geostrophic effects become important and enstrophy is scattered by three-dimensional effects and dissipated.

This cycle of energy occurs in baroclinic eddies in both atmosphere and ocean. Higher baroclinic modes, not captured in the two-layer system, can sometimes be excited (especially in the ocean) but the two-layer system does capture the main features of the real systems. Indeed, in the atmosphere the great mid-latitude weather systems can be succinctly characterized by the phrase 'baroclinic growth and barotropic decay', which is essentially the process illustrated in Fig. 10.9.

10.6 BETA-PLANE TURBULENCE

We now turn to looking at the effects of a variable Coriolis parameter — the beta effect — on geostrophic turbulence. To display the effects in their most austere fashion we restrict attention to two-dimensional turbulence on the beta-plane, and we can write the equation of motion in the various equivalent forms,

$$\frac{\mathrm{D}q}{\mathrm{D}t} = \frac{\mathrm{D}}{\mathrm{D}t}(\zeta + f) = F - D, \quad \text{or} \quad \frac{\partial \zeta}{\partial t} + \boldsymbol{u} \cdot \nabla(\zeta + f) = F - D, \quad (10.63\text{a,b})$$

or

$$\frac{\partial \zeta}{\partial t} + J(\psi, \zeta + \beta y) = F - D \quad \text{or} \quad \frac{\partial \zeta}{\partial t} + J(\psi, \zeta) + \beta v = F - D. \quad (10.63\text{c,d})$$

In these equations $q = \zeta + f$ is the potential vorticity (equal to the absolute vorticity here), $\zeta = \nabla^2 \psi$ is the relative vorticity, $J(\psi, \zeta) = \psi_x \zeta_y - \psi_y \zeta_x$ is the Jacobian, $f = f_0 + \beta y$, and F and D represent the forcing and dissipation.

The form of (10.63d) suggests that the presence of β will lead to the production of zonal flow, since in a statistically steady state we can write

$$\beta v = \text{forcing and nonlinear terms.} \quad (10.64)$$

The forcing is ultimately responsible for the flow, and if we imagine increasing β while keeping the forcing constant then v (the meridional flow) must diminish or the equation simply cannot balance. However, the zonal flow, u, will remain strong and the vorticity equation can then display a three-way balance between nonlinearity, $\boldsymbol{u} \cdot \nabla \zeta$, the beta term, βv, and the forcing, F. Still, the argument is a little heuristic and the precise balance depends on the scale of the motion, so let us look at the dynamics a little more closely.

10.6.1 The Wave–Turbulence Cross-over

Omitting the forcing term we write the equation of motion as

$$\frac{\partial \zeta}{\partial t} + \boldsymbol{u} \cdot \nabla \zeta + \beta v \;\; = 0. \quad (10.65)$$

If $\zeta \sim U/L$ and if $t \sim T$ then the respective terms in this equation scale as

$$\frac{U}{LT} \qquad \frac{U^2}{L^2} \qquad \beta U, \quad (10.66)$$

and for now we do not differentiate between U and V. If the nonlinear term dominates, as (10.66) suggests it will at small scales, then the flow will be turbulent and display all the characteristics of two-dimensional turbulence. If the beta term dominates then Rossby waves will dominate and the flow will be more linear, and this behaviour will occur at large scales. The cross-over scale, denoted L_R, is called the *Rhines scale* and is given by

$$L_R \sim \left(\frac{U}{\beta}\right)^{1/2}. \quad (10.67)$$

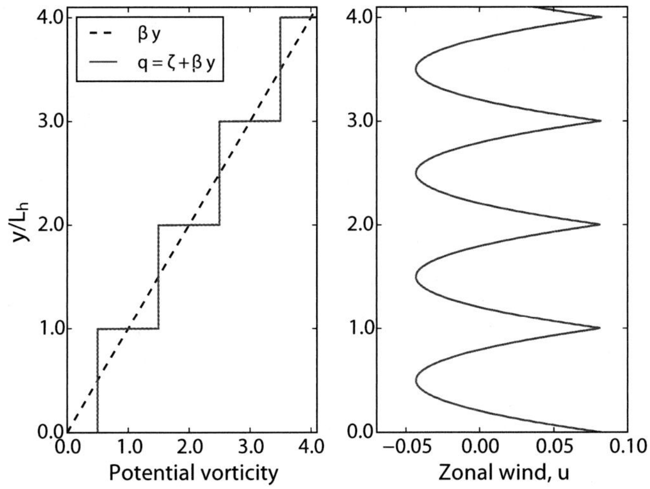

Fig. 10.10: An idealized potential vorticity staircase on a β-plane with $\beta = 1$. The left panel shows homogeneous regions of potential vorticity of meridional extent L_h separated by jumps, as calculated using (10.71). The right-hand panel shows the corresponding zonal flow, calculated using (10.75) with u_0 chosen to be such that the average zonal flow is zero.

At scales larger than L_R Rossby waves are likely to dominate whereas at smaller scales advection and turbulence dominate and the flow is more isotropic. (The U in (10.67) is best interpreted as the root-mean-square velocity at the energy containing scales, not a mean or translational velocity.) At these large scales the Rossby waves interact with the turbulence and produce a zonal flow, as discussed in Problem 10.3 and in Chapter 12. However, here we shall give a complementary argument and show that the tendency of the flow to homogenize potential vorticity leads to jet production on the beta-plane.

10.6.2 Jets and Potential Vorticity Staircases

The gist of the argument is as follows. Suppose that a turbulent flow exists on a background state that has a potential vorticity gradient, such as might be caused by the beta-effect. The turbulent eddies seek to homogenize the potential vorticity but, in general, they are unable to do so completely. Instead, the flow becomes partially homogenized, with regions of homogeneous potential vorticity separated by jumps, as illustrated in Fig. 10.10. Such a structure is called a *potential vorticity staircase,* and the staircase implies the existence of zonal jets.

To go into more detail recall that, as discussed in Section 10.2.1, the gradients of a scalar that is both advected and diffused become smeared out as much as possible and, as a consequence, the scalar may become homogenized. Potential vorticity is such a scalar; however, potential vorticity is not a *passive* scalar and the process of homogenization affects the flow itself, often preventing complete homogenization both because of the need to satisfy boundary conditions and because the process of homogenization may require more energy than the flow contains. To see this, consider freely-evolving flow on a beta-plane obeying the barotropic vorticity equation

$$\frac{\partial \zeta}{\partial t} + J(\psi, \zeta + \beta y) = \nu \nabla^2 \zeta, \tag{10.68}$$

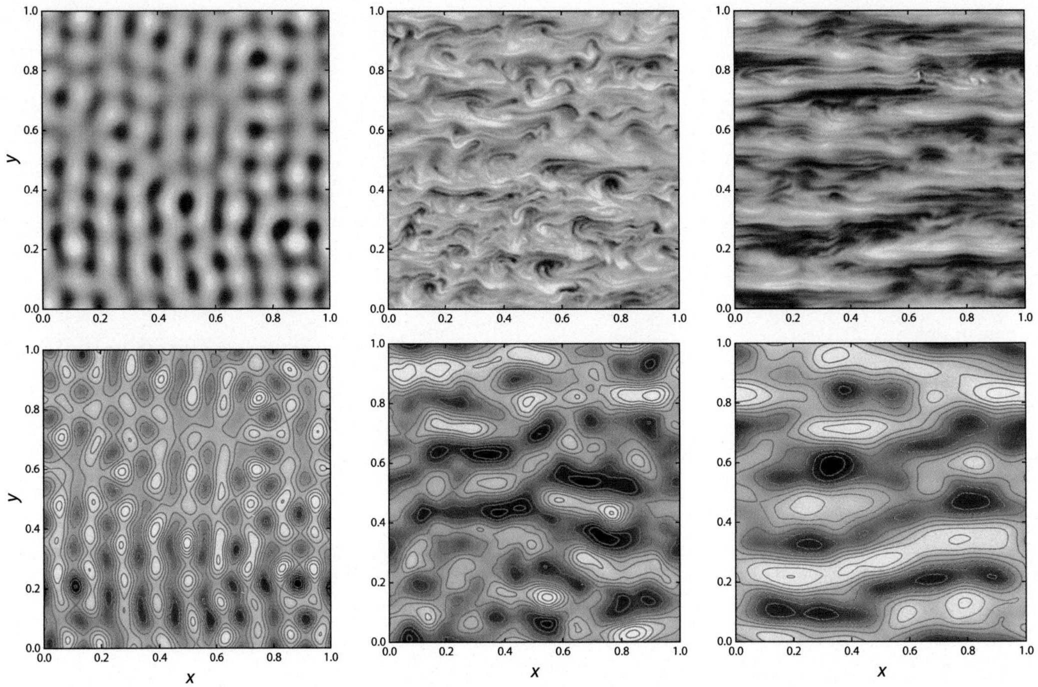

Fig. 10.11: Evolution of vorticity (top) and streamfunction (bottom) in a numerical simulation on the β-plane, obeying (10.65) with the addition of a weak viscous term on the right-hand side. The initial conditions are the same as for Fig. 10.8, and time proceeds from left to right. Compared to Fig. 10.8, vortex formation is inhibited and there is a tendency toward zonal flow

where ν is sufficiently small that energy is well conserved over the timescales of interest. If potential vorticity, $q = \zeta + \beta y$, is to be homogenized over some meridional scale L_h then, in the homogenized region, $\zeta = -\beta y$, taking the value of q to be zero for simplicity. If the flow is predominantly zonal then the zonal flow will have a magnitude $U \sim \beta y^2$ and the energy in the region is, very approximately,

$$\frac{1}{2}\int u^2 \,\mathrm{d}x\,\mathrm{d}y \approx \int (\beta y^2)^2 \,\mathrm{d}x\,\mathrm{d}y, \tag{10.69}$$

giving the estimate $U^2 L_h \sim \beta^2 L_h^5$. Solving the above estimate for L_h suggests that the potential vorticity can become homogenized over the scale

$$L_h \sim \left(\frac{U}{\beta}\right)^{1/2}. \tag{10.70}$$

This is the same as the Rhines scale given in (10.67), as it has to be by dimensional analysis, although now it is supposed that much of the energy lies in the zonal flow. Although the scale is the same, this way of looking at the problem adds something to the physical picture — *an asymmetry between eastward and westward flow,* as we now show.

In the idealized staircase of Fig. 10.10 potential vorticity is piecewise continuous and given by

$$q = \zeta + \beta y = q_0, \qquad 0 < y < L_h, \tag{10.71a}$$

$$q = \zeta + \beta y = q_0 + \beta L_h \equiv q_1, \qquad L_h < y < 2L_h, \quad \text{and so on.} \tag{10.71b}$$

In any one of the homogenized regions the flow is given by solving

$$\frac{\partial u}{\partial y} = \beta y - q_n \quad \text{giving} \quad u = \frac{1}{2}\beta y^2 - q_n y + \text{constant}, \tag{10.72}$$

where $q_n = \beta L_h/2 + n\beta L_h$. The constants in each region may be determined by requiring continuity of u across the regions, and we then obtain

$$u = \frac{1}{2}\beta(y - L_h/2)^2 + u_0, \qquad 0 < y < L_h, \tag{10.73}$$

$$u = \frac{1}{2}\beta(y - 3L_h/2)^2 + u_0, \qquad L_h < y < 2L_h, \tag{10.74}$$

where u_o is a constant. In general we have

$$u = \frac{1}{2}\beta\left(y - (n - 1/2)L_h\right)^2 + u_0, \quad (n-1)L_h < y < nL_h, \quad n = 1, 2, 3 \ldots \tag{10.75}$$

Relative to u_0, the flow is weakly westward in the homogenized regions, whereas in the transition regions the flow has sharp eastward peaks, as shown in Fig. 10.10.

The staircase structure implies that potential vorticity is (obviously) well mixed in the regions of constant potential vorticity, but that there is no mixing across the jumps. The jumps are known as *mixing barriers* and, once formed, are found to be very persistent in numerical simulations. The edge of the stratospheric polar vortex, and the boundaries of ocean gyres, are examples of such barriers. Numerical simulations do show a tendency to produce zonal flow on the beta-plane, as seen in Fig. 10.11. However, the subject continues to evolve and the reader should consult the references below and the recent primary literature to learn more.

Notes and References

The study of turbulence itself goes back to, at least, Leonardo da Vinci, and blossomed in the twentieth century. There are so many relevant papers on the subject that the reader should refer to a book such as Pope (2000) or Davidson (2015) for a discussion of the classical theory and a bibliography. The phenomenology of two-dimensional turbulence was put forward by Kraichnan (1967, 1971), Leith (1968) and Batchelor (1969). Geostrophic turbulence was introduced by Charney (1971) with subsequent work by Salmon (1980) and Rhines (1977). The interaction of Rossby waves and turbulence and the generation of zonal flows on a beta-plane comes from Rhines (1975) and Vallis & Maltrud (1993). A potential vorticity staircase was proposed by Marcus (1993), his motivation being the jets of Jupiter.

Problems

10.1 *Predictability.* The eddy turnover time of three-dimensional turbulence at a wavenumber k is given by $\tau_k = \varepsilon^{-1/3}k^{-2/3}$, and that of two-dimensional turbulence by $\tau_k = \eta^{-1/3}$ where k is the wavenumber and ε and η are constants (the energy and enstrophy cascade rates, respectively). Suppose that in weather prediction the error is confined to small scales and that the time taken for the error to contaminate the next largest scale (in a logarithmic sense) is τ_k, so that the time taken for an error at a small scale k_s to reach the large scales k_l is given by

$$T = \int_{k_s}^{k_l} \tau_k \, dk. \tag{P10.1}$$

If the inertial range extends indefinitely show that this time is infinite for classical two-dimensional turbulence and finite for three-dimensional turbulence, and discuss the implications for weather predictability and weather forecasting. If the atmosphere is two-dimensional down to a scale of 50 km, estimate a limit to weather predictability (e.g., a timescale in days), making sensible (and clearly stated) assumptions about the magnitude of the flow.

10.2 *Heat transport.* Considering large-scale turbulence in the atmosphere, suppose that the mixing length is given by the radius of deformation and a mixing velocity is given by a typical synoptic velocity. Estimate a value for the eddy diffusivity provided by weather systems, and hence estimate the polewards heat transport by these systems. Compare with the values of the heat transport in the atmosphere given by observations, and discuss the realism, or otherwise, of your calculation.

10.3 *Jets.* The Rossby wave frequency is given by $\omega_R = -\beta k/(k^2 + l^2)$, and this is anisotropic in (k, l) space. Suppose that a 'turbulence frequency', in two dimensions, is given by $\omega_t = Uk$, where U is a constant.

(a) By equating the above two expressions, obtain an expression for the line in wavenumber space on which the turbulence frequency and Rossby-wave frequency are the same. Plot the curve; it should look a little like a dumbbell, with the Rossby-wave frequency higher than the turbulence frequency inside the dumbbell.
Partial solution. The crossover x-wavenumber is

$$k^x = (\beta/U)^{1/2} \cos^{3/2}\theta \qquad \text{where} \qquad \theta = \tan^{-1}(k^y/k^x).$$

What is k^y?

(b) Energy is transferred from small scales to large in two-dimensional turbulence; however, it will be unable to excite Rossby waves if they have a higher frequency than the turbulence. Hence argue that the cascade will excite zonal motion, and make a rough estimate for the meridional scale of the zonal flow.

(c) Repeat the above but now with a turbulence frequency given by $\omega_t = \varepsilon^{1/3}k^{2/3}$, which is the inverse of the eddy turnover time in the energy inertial range, given (as in 3D turbulence) by (10.28).

10.4 Suppose that the inverse cascade of two-dimensional turbulence is halted by a linear drag. (That is, the vorticity equation is $D\zeta/Dt = F - r\zeta$, where F is the stirring and r is a constant.)

(a) By equating the drag timescale to that of the turbulence obtain an estimate for the scale at which the cascade is halted, and an estimate for the velocity at that scale.

(b) Show that an estimate of the horizontal eddy diffusivity, $\mathcal{K}$, is then $\mathcal{K} \sim \varepsilon/r^2$, where ε is the energy cascade rate.

Part II

Atmospheres

CHAPTER

11 The Tropical Atmosphere

IN THIS CHAPTER AND THE TWO FOLLOWING we discuss the structure and circulation of planetary atmospheres. This chapter and the next focus mainly on Earth, first on the tropical circulation and then on the midlatitudes and the stratosphere. Then, in Chapter 13, we look at planetary atmospheres a little more generally. We begin with a brief observational overview of Earth's atmosphere as a whole.

11.1 An Observational Overview

Many of the main zonally- and/or time-averaged features of Earth's atmosphere can be seen in Figs. 11.1–11.3. The most prominent features are:

(i) The temperature falls monotonically with height to an altitude of about 16 km (in the tropics) or 8 km at high latitudes, before increasing with height. The lower region is called the *troposphere*, above which lies the *stratosphere*, and the boundary between the two is the *tropopause*.

(ii) The temperature also falls monotonically from equator to pole, at the surface falling from about 300 K to 240 K at the pole. The tropopause temperature is more uniform, varying from about 230 K to 210 K.

(iii) The surface winds are easterly in low latitudes (from about 30° S to 30° N), westerly in mid- and high latitudes, with weak polar easterlies in some seasons.

(iv) The winds increase in height, especially in midlatitudes, with pronounced westerly jets in both hemispheres centred at about 40° latitude in both hemispheres.

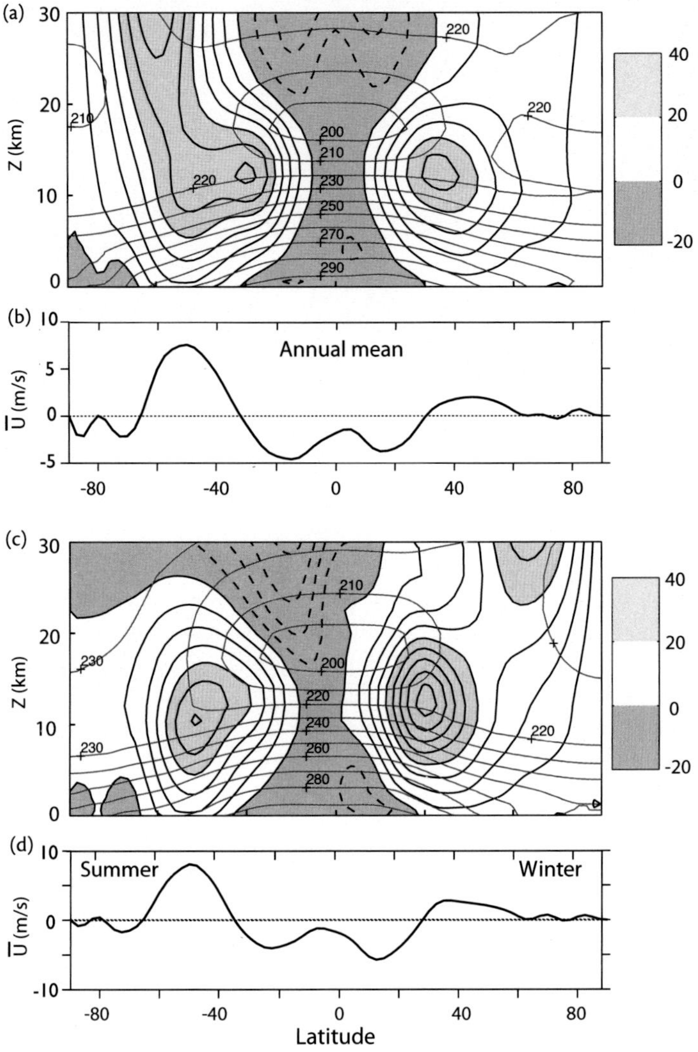

Fig. 11.1: (a) Annual mean, zonally-averaged zonal wind (heavy contours and shading) and the zonally-averaged temperature (red, thinner contours). (b) Annual mean, zonally averaged zonal winds at the surface. (c) and (d) Same as (a) and (b), except for northern hemisphere winter (December–January–February, or DJF).

The wind contours are at intervals of 5 m s^{-1} with dark shading for eastward winds above 20 m s^{-1} and light shading for all westward winds, and the temperature contours are labelled.

(*v*) In the meridional (y–z) plane the circulation in each hemisphere (Fig. 11.3) is characterized by:

(*i*) A 'direct' *Hadley Cell*, with warm air rising near the equator and sinking in the tropics. The winter Hadley Cell is much stronger and has greater latitudinal extent than the summer one, with the warm air rising at low latitudes in the summer hemisphere, sinking in the tropics in the winter hemisphere.

(*ii*) An 'indirect' *Ferrel Cell*, with cool air apparently rising around 60° and sinking in the tropics.

A direct cell is one that is thermally driven, with warm, buoyant fluid rising and cold fluid descending. An indirect cell may be mechanically driven, as we discuss later.

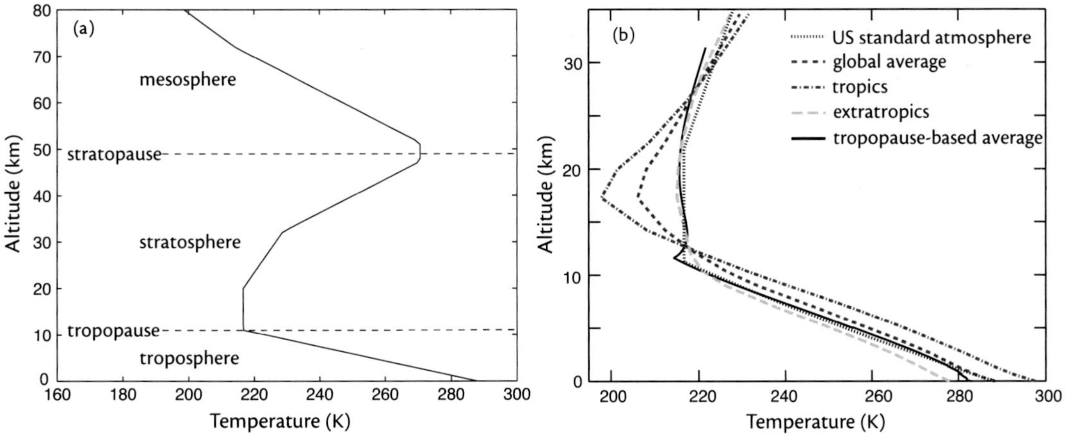

Fig. 11.2: (a) The temperature profile of the so-called US standard atmosphere. (b) Observed, annually averaged profiles of temperature in the atmosphere, where the ordinate is log-pressure. Tropics here is the average from 30° S to 30° N, and extratropics is the average over the rest of the globe. (The 'tropopause-based average' uses the tropopause itself as the origin of the height scale, set to be 11 km, but we shall not discuss it further.)

11.2 An Ideal Hadley Circulation

To makes matters as simple as possible we imagine the Earth to be a sphere with a uniform surface (so no mountains and no oceans), and that there are no seasons (unlike the case in Fig. 11.3. Because of the differential solar heating the air is warmer at low latitudes than at high, so we may reasonably imagine that the warm air rises and moves polewards before cooling and sinking at some high latitude, perhaps near the pole, and returning near the ground. Indeed such a concept was envisioned by George Hadley over 300 hundred years ago. However, observations tell us that the air does not go all the way to the pole; rather, it sinks in the subtropics at about 25–30°. There are two reasons why it must sink, one related to thermodynamic constraints and the other to hydrodynamic instabilities. Both are related to the properties of the air as it moves, conserving its angular momentum.

11.2.1 Zonally-Symmetric Equations of Motion

In Chapter 2 we wrote down the equations of motion on a sphere. If the flow is zonally symmetric (no longitudinal variation) then, with a little manipulation, the zonal momentum equation may be written in the form

$$\frac{\partial u}{\partial t} - (f + \zeta)v + w\frac{\partial u}{\partial z} = 0. \tag{11.1}$$

The variables in this equation are functions of latitude and height (ϑ and z) only, and not longitude, λ, and $\zeta = -(a\cos\vartheta)^{-1}\partial_\vartheta(u\cos\vartheta)$. If the vertical advection is small then a steady solution obeys

$$(f + \zeta)v = 0. \tag{11.2}$$

Presuming that the meridional flow v is non-zero then $f + \zeta = 0$, or equivalently, on the sphere,

$$2\Omega\sin\vartheta = \frac{1}{a}\frac{\partial u}{\partial\vartheta} - \frac{u\tan\vartheta}{a}. \tag{11.3}$$

The Hadley Cell is named for George Hadley (1685–1768), a British meteorologist who put forward perhaps the first scientific model of Earth's overturning circulation, in which the air rose near the equator and sank near the pole.

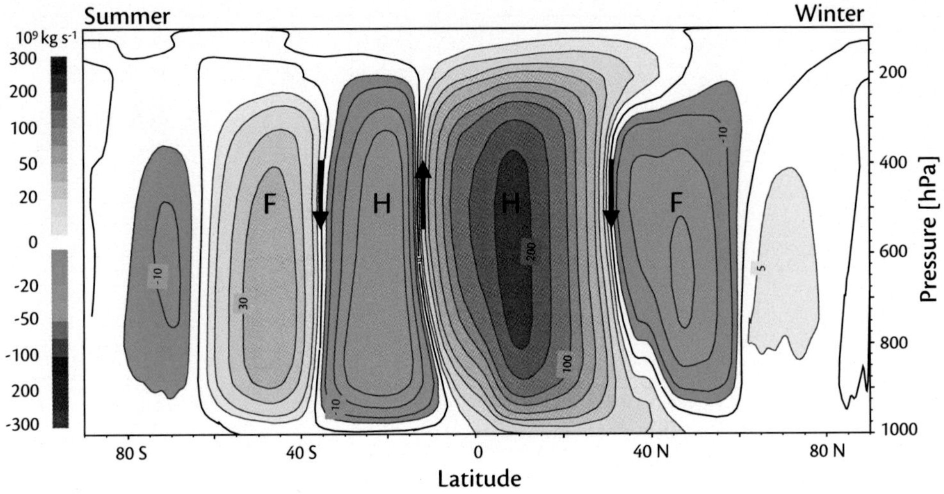

Fig. 11.3: The observed meridional overturning circulation (MOC) of the atmosphere ($\mathrm{kg\,s^{-1}}$) averaged over December–January– February. Note the direct *Hadley Cells* (H) particularly strong in winter with rising motion near the equator, descending motion in the subtropics, and the weaker, indirect, *Ferrel Cells* (F) in midlatitudes. (From an ECMWF reanalysis.)

At the equator we may assume that $u = 0$, because here parcels have risen from the surface where the flow is weak. Equation (11.3) then has a solution of

$$u = \Omega a \frac{\sin^2 \vartheta}{\cos \vartheta}. \tag{11.4}$$

This gives the zonal velocity of the poleward moving air in the upper branch of the (model) Hadley Cell, above the frictional boundary layer. Evidently the zonal velocity increases rapidly with latitude: at 20° and 40° the values of u are about 59 and 256 m $\mathrm{s^{-1}}$respectively (look ahead to Fig. 11.6 or Fig. 11.7), becoming far larger than the observed values at midlatitudes.

Angular momentum conservation

An instructive interpretation of (11.4) comes by way of the conservation of angular momentum, m, of a ring of air at a latitude ϑ, as in Fig. 11.4. The angular momentum per unit pass of a parcel of air with zonal velocity u is

$$m = (u + \Omega a \cos \vartheta) a \cos \vartheta, \tag{11.5}$$

and if $u = 0$ at $\vartheta = 0$ and if m is conserved on a poleward moving parcel, then (11.5) leads directly to (11.4).

The air returning to the equator close to the surface has a small zonal velocity, meaning that there is a large thermal wind, which we may calculate using the thermal wind expression

$$2\Omega \sin \vartheta \frac{\partial u}{\partial z} = -\frac{1}{a}\frac{\partial b}{\partial \vartheta}, \tag{11.6}$$

where $b = g\,\delta\theta/\theta_0$ and $\delta\theta$ is the deviation of potential temperature from a constant reference value θ_0. (Be reminded that θ is potential temperature,

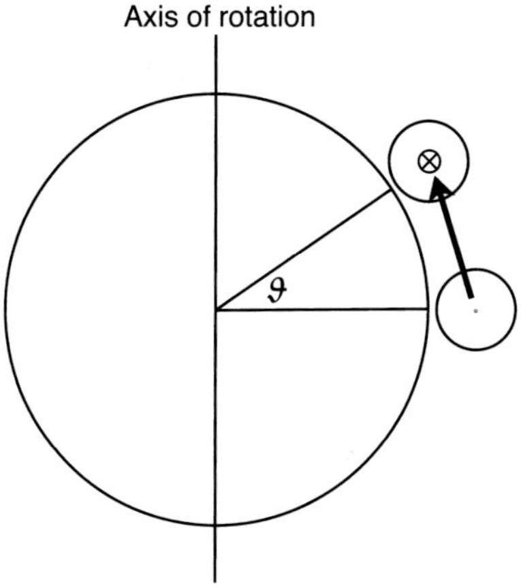

Fig. 11.4: If a ring of air at the equator moves poleward it moves closer to the axis of rotation. If the parcels in the ring conserve their angular momentum their zonal velocity must increase; thus, if $m = (u + \Omega a \cos\vartheta)a\cos\vartheta$ is preserved and $u = 0$ at $\vartheta = 0$ we recover (11.4).

whereas ϑ is latitude.) Vertically integrating from the ground to the height H where the outflow occurs and substituting (11.4) for u yields

$$\frac{1}{a\theta_0}\frac{\partial\theta}{\partial\vartheta} = -\frac{2\Omega^2 a}{gH}\frac{\sin^3\vartheta}{\cos\vartheta}, \tag{11.7}$$

where $\theta = H^{-1}\int_0^H \delta\theta \, dz$ is the vertically averaged potential temperature.

If the latitudinal extent of the Hadley Cell is not too great we can make a small-angle approximation, and replace $\sin\vartheta$ by ϑ and $\cos\vartheta$ by one, and then integrating (11.7) gives

$$\theta = \theta(0) - \frac{\theta_0\Omega^2 y^4}{2gHa^2}, \tag{11.8}$$

where $y = a\vartheta$ and $\theta(0)$ is the potential temperature at the equator. Evidently, the temperature in an angular momentum conserving Hadley Cell

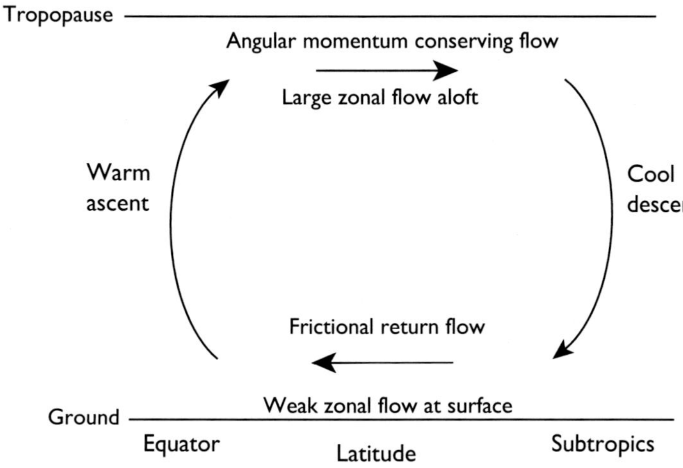

Fig. 11.5: A simple model of the Hadley Cell. Rising air near the equator moves poleward near the tropopause, descending in the subtropics and returning.

The poleward moving air conserves its angular momentum, leading to a shear of the zonal wind that increases away from the equator. By thermal wind the temperature of the air falls as it moves poleward, and to satisfy the thermodynamic budget it sinks in the subtropics.

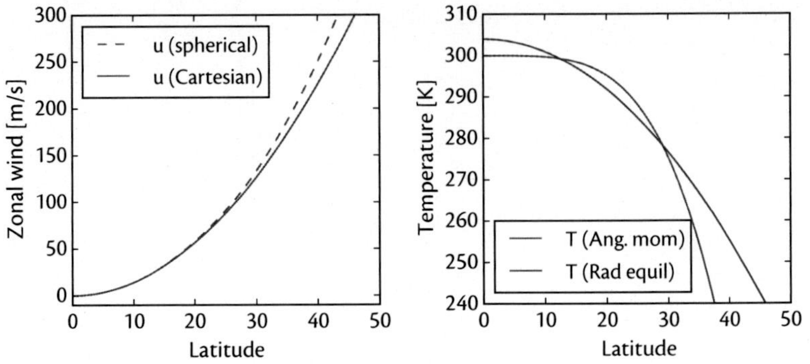

Fig. 11.6: Left: Upper tropospheric zonal wind computed for an angular momentum conserving Hadley Cell. The solid line makes the small-angle approximation. Right: The blue line shows temperature of the angular momentum conserving wind, computed using (11.8). The red curve shows the radiative equilibrium temperature, using (11.9). The Hadley Cell extends to the latitude of the second crossing, at about 30°.

drops rapidly away from the equator — if (11.8) were satisfied the temperature would drop by over 80 K between the equator and 40° latitude, and would drop to absolute zero at around 50°. Plainly, the Hadley Cell cannot extend to the pole and remain angular momentum conserving. What happens? Essentially, the air becomes so cold that it sinks in order to satisfy the thermodynamic constraints, as in Fig. 11.5 and as we now explain.

11.2.2 Thermodynamic Constraints

In the absence of any motion the atmosphere will evolve to a radiative-equilibrium temperature, in which the warming by solar radiation exactly balances the cooling by infra-red radiation. A difficult calculation is needed to determine the true radiative equilibrium temperature, but an approximate one is easier to come by. Essentially, at each latitude the net incoming solar radiation must be balanced by the outgoing infra-red radiation, which is an increasing function of temperature in the upper troposphere. Since the incoming solar radiation is higher at low latitudes then so is the radiative equilibrium temperature, and a simple estimate of the radiative equilibrium potential temperature in the upper troposphere is

$$\theta_E = \theta_{E0} - \Delta\theta\left(\frac{y}{a}\right)^2, \tag{11.9}$$

where θ_{E0} is the equilibrium temperature at the equator and $\Delta\theta$ determines the equator–pole radiative-equilibrium temperature difference.

Now, the Hadley Cell itself carries heat poleward. If its temperature is given by (11.8) then it will not extend polewards beyond the latitude at which its temperature becomes equal to the radiative equilibrium temperature. Although the radiative equilibrium temperature at the equator, θ_{E0}, is a little higher than the actual temperature it is not significantly so and we can take $\theta(0) \approx \theta_{E0}$. Equating (11.8) with (11.9) and solving for y gives the latitudinal extent of the Hadley Cell, y_M, or its equivalent latitude ϑ_M, namely

$$\vartheta_M = \frac{y_M}{a} = \left(\frac{2\Delta\theta g H}{\Omega^2 a^2 \theta_0}\right)^{1/2}, \tag{11.10}$$

The more accurate calculation shown in the right-hand panel of Fig. 11.6 computes the equatorial temperature using the requirement that the circulation in the Hadley Cell does not create energy, so that the areas between the red and blue lines within the Hadley Cell sum to zero. This consideration does not significantly change (11.10).

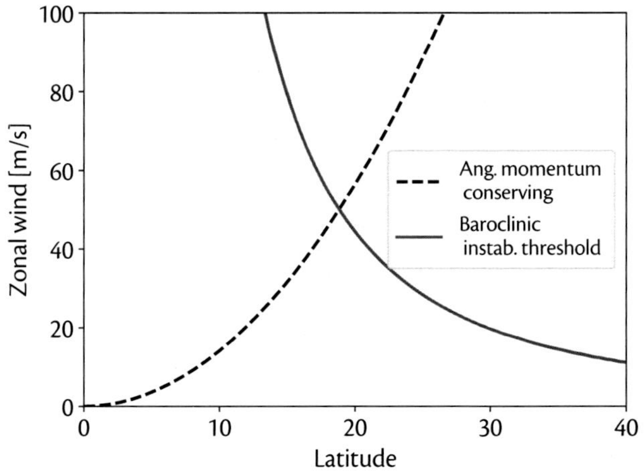

Fig. 11.7: Upper tropospheric zonal winds. The dashed curve shows the approximate angular-momentum conserving wind with $u = 0$ at the equator using (11.12). The solid curve shows the threshold for baroclinic instability of a zonal flow using a two-layer quasi-geostrophic calculation, (11.11). Polewards of about 20° the angular momentum conserving flow is, in this calculation, baroclinically unstable.

with a corresponding latitude $\vartheta_M = y_M/a$. This is the latitude at which the temperature in the outflowing branch of the Hadley Cell equals the radiative equilibrium temperature, at which point the Hadley Cell must terminate. (A more accurate calculation that does not set $\theta(0) = \theta_{E0}$, as shown in Fig. 11.6, makes a small quantitative difference but does not change the scaling.)

The main point here is a qualitative one: an ideal Hadley Cell cannot fill the hemisphere; rather, it descends at some latitude, beyond which the air will take on the radiative equilibrium temperature and have no meridional motion. In reality, the Hadley Cell is also baroclinically unstable, as we are about to see.

11.3 Instability of the Ideal Hadley Circulation

If the air flowing poleward in the Hadley Cell conserves its angular momentum, but the flow at lower levels is smaller, then there will be a large shear that increases with latitude, and eventually this will become baroclinically unstable. A simple model of this uses the two-level quasi-geostrophic equations for which the critical velocity difference between the two layers is given by

$$\Delta U_C \equiv (U_1 - U_2)_C = \frac{1}{4}\beta L_d^2 = \frac{H^2 N^2 a}{8\Omega y^2}, \tag{11.11}$$

where, using the small angle approximation, $\beta = 2\Omega/a$, $f = 2\Omega y/a$ and $L_d = NH/f$ with H being the depth of the domain and N the buoyancy frequency. The velocity difference over the entire depth of the troposphere is roughly twice this value. The angular momentum conserving solution gives a velocity difference over depth of the troposphere of

$$\Delta U_M = \frac{\Omega y^2}{a}, \tag{11.12}$$

and equating twice (11.11) with (11.12) gives a cross-over latitude of

$$\vartheta_C = \frac{y_C}{a} = \left(\frac{NH}{2\Omega a}\right)^{1/2}, \tag{11.13}$$

as shown in Fig. 11.7. For values of parameters typical for Earth this gives a latitude of around 20°.

11.3.1 Discussion

A parameter that all these results depend on is the rotation rate, Ω, and with the value this takes for Earth, along with reasonable values for the stratification and depth of the atmosphere, it is hard for the Hadley Cell to continue polewards of about 30°. On Earth, then, the Hadley Cell is a tropical phenomenon. On more slowly rotating planets the Hadley Cell may extend further polewards, as we discover in Chapter 13, but for the rest of this chapter we focus on the dynamics of Earth's tropics. Finally, we remark that even if the angular momentum conserving solution did not hold and the air were stationary in radiative equilibrium, the horizontal temperature gradient and associated vertical shear would, of themselves, be baroclinically unstable.

11.4 THE TROPICAL CIRCULATION

If the Hadley Cell determines how far the tropics extend, let us now look at the circulation that take place within the tropics. We first look at the quasi-horizontal circulation and then at the phenomenon we most associate with the tropics, moist convection.

In earlier chapters we derived the quasi-geostrophic equations, appropriate when the Rossby number is small and so for midlatitude motion. Is such an approach possible for the tropics? The answer is, 'well, in part'. It *is* possible to derive some reduced sets of equations for the tropics, and that will be the main topic of this section. However, these reduced sets have not proven nearly as useful for the tropics as quasi-geostrophy has been for the midlatitudes, in part because it is harder to make relevant equations that are simple, or simple equations that are relevant. In any case, let us begin with the stratified primitive equations.

11.4.1 Balanced, Adiabatic Flow

We will present a scaling for tropical flow side-by-side with the corresponding scaling for midlatitude flow. We begin with the hydrostatic primitive equations for adiabatic, frictionless flow, using the Boussinesq approximation since the essential dynamics, and no more, are contained there. (A largely equivalent derivation could use pressure coordinates.) As in (5.31b) the equations are

$$\frac{\mathrm{D}\boldsymbol{u}}{\mathrm{D}t} + \boldsymbol{f} \times \boldsymbol{u} = -\nabla_z \phi, \qquad \frac{\partial \phi}{\partial z} = b, \tag{11.14a,b}$$

$$\frac{Db}{Dt} + N^2 w = 0, \qquad \nabla \cdot \boldsymbol{v} = 0. \tag{11.14c,d}$$

We interpret the buoyancy as a perturbation potential temperature, $b = g\,\delta\theta/\theta_0$, and $N^2 = d\tilde{b}/dz$ where $\tilde{b}(z)$ is a reference stratification. Let us suppose that the basic variables scale according to

$$(x, y) \sim L, \quad z \sim H, \quad (u, v) \sim U, \quad w \sim W, \quad t \sim \frac{L}{U}, \\ \phi \sim \Phi, \quad b \sim B, \quad f \sim f_0. \tag{11.15}$$

The quantity B is representative of horizontal variations in buoyancy. Vertical variations scale differently, hence their separate representation in (11.14c). By choosing the time t to scale advectively we are implicitly eliminating gravity waves. The nondimensional numbers that arise are the Rossby, Burger and Richardson numbers,

$$Ro = \frac{U}{f_0 L}, \quad Bu = \left(\frac{L_d}{L}\right)^2 = \left(\frac{NH}{f_0 L}\right)^2, \quad Ri = \left(\frac{NH}{U}\right)^2, \tag{11.16}$$

and evidently

$$Bu = Ri \times Ro^2. \tag{11.17}$$

The Rossby number Ro is generally small in midlatitudes for large-scale flow, but in the tropics it is $\mathcal{O}(1)$ or larger. The Richardson number is usually large in both midlatitudes and tropics except in regions of active convection where N is very small. If we take $N = 10^{-2}\,\mathrm{s}^{-1}$, $H = 10^4\,\mathrm{m}$ and $U = 10\,\mathrm{m\,s}^{-1}$ then $Ri = 100$. In fact, on large-scales the atmosphere is sufficiently stratified that the Richardson number is usually such that $1/(Ro\,Ri)$ is small in both midlatitudes and tropics.

The difference between the tropics and midlatitudes is mainly apparent from the dominant balance in the momentum equation. At small Rossby number we have the familiar geostrophic balance, and with hydrostatic balance we obtain the scaling:

$$\boldsymbol{f} \times \boldsymbol{u} \approx -\nabla_z \phi, \quad \frac{\partial \phi}{\partial z} = b, \quad \Longrightarrow \quad \Phi = f_0 U L, \;\; B = \frac{f_0 U L}{H}. \tag{11.18}$$

In the tropics the advective term, or the advectively-scaled time derivative, balances the pressure gradient meaning that $D\boldsymbol{u}/Dt \sim \nabla_z \phi$ and, since we still have $\partial\phi/\partial z = b$ we find

$$\Phi = U^2, \qquad B = \frac{U^2}{H}. \tag{11.19}$$

If U is of similar magnitude in the tropics and midlatitudes (and in the absence of a dynamical analysis this is an assumption), then *variations of pressure and temperature are smaller in the tropics than in midlatitudes.* This is an important and not-quite obvious result and it is the essence of the *weak temperature gradient approximation,* discussed more in Section 11.4.2.

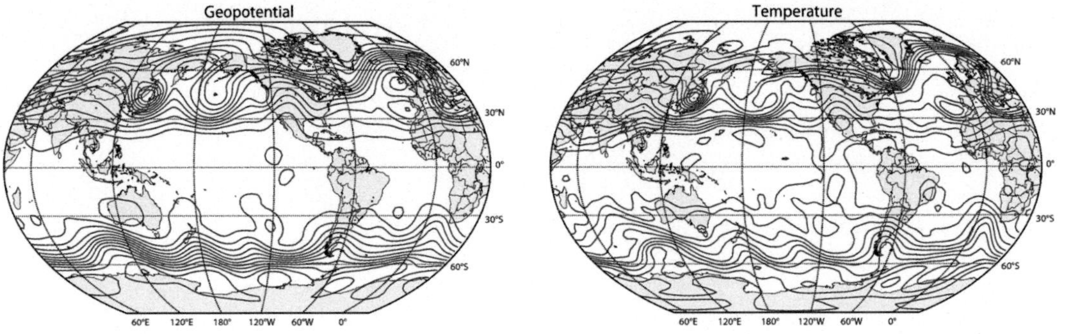

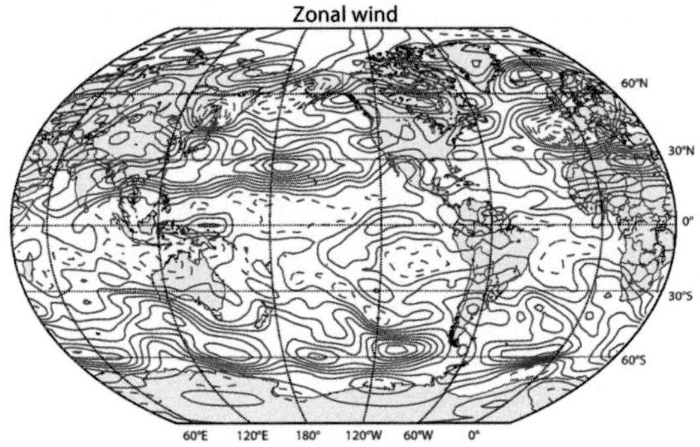

Fig. 11.8: Geopotential height, temperature and zonal wind at 500 hPa on 9 February 2015. The contour interval is uniform in each plot, with the same number of contours for each field, and negative values for u are dashed. Noticeable is the lack of variability of the geopotential and temperature fields in the tropics.

If we were to carry through the derivation in pressure coordinates (see the shaded box on page 42) with geopotential and temperature as the variables we would find

$$\text{Midlatitudes:} \quad \Phi = f_0 U L, \quad T_s = \frac{f_0 U L}{R}, \tag{11.20a}$$

$$\text{Tropics:} \quad \Phi = U^2, \quad T_s = \frac{U^2}{R}, \tag{11.20b}$$

where Φ is now the scaling for geopotential, T_s is the scaling for temperature and R is the ideal gas constant. For $U = 10\,\mathrm{m\,s^{-1}}$, $L = 10^6\,\mathrm{m}$ and, for midlatitudes only, $f_0 = 10^{-4}\,\mathrm{s^{-1}}$, we obtain

$$\text{Midlatitudes:} \quad \Phi \sim 1000\,\mathrm{m^2\,s^{-2}}, \quad T_s \sim 3\,\mathrm{K}, \tag{11.21a}$$

$$\text{Tropics:} \quad \Phi \sim 100\,\mathrm{m^2\,s^{-2}}, \quad T_s \sim 0.3\,\mathrm{K}. \tag{11.21b}$$

The implications of these results are illustrated in Fig. 11.8, which shows a snapshot of observed contours of geopotential height, temperature and zonal wind; the large-scale variability of geopotential and temperature is evidently much smaller in the tropics than midlatitudes.

Vertical velocity

The obvious scaling for the vertical velocity is that suggested by the mass continuity equation, namely $W = UH/L$. However, as we know from

Chapter 5, the vertical velocity may be much less than this estimate in a stratified, rotating fluid, and here we start with the thermodynamic equation for adiabatic flow,

$$\boldsymbol{u} \cdot \nabla b + wN^2 = 0, \qquad \Longrightarrow \qquad W = \frac{UB}{LN^2}. \tag{11.22}$$

Using (11.18) and (11.19) gives

$$\text{midlatitudes:} \quad W = \frac{f_0 U^2}{HN^2} = \frac{f_0 L}{U} \frac{U^2}{N^2 H^2} \frac{UH}{L} = (Ro\, Ri)^{-1} \left(\frac{UH}{L} \right), \tag{11.23a}$$

$$\text{Tropics:} \quad W = \frac{U^3}{LHN^2} = \frac{U^2}{N^2 H^2} \frac{UH}{L} = Ri^{-1} \left(\frac{UH}{L} \right), \tag{11.23b}$$

where $Ri \equiv N^2 H^2 / U^2$ is the Richardson number, with typical values of $\mathcal{O}(10)$–$\mathcal{O}(100)$ for large-scale flow. Thus, again perhaps non-intuitively, the vertical velocity is, for adiabatic flow, *smaller* in the tropics than in midlatitudes, by an order of a midlatitude Rossby number. In midlatitudes the scaling for W is more commonly written as

$$W = \frac{f_0 U^2}{HN^2} = Ro \frac{L^2}{L_d^2} \frac{UH}{L} = \frac{Ro}{Bu} \frac{UH}{L}, \tag{11.23c}$$

so that for scales comparable to the midlatitude deformation radius (i.e., with $Bu \sim 1$) the vertical velocity is an order Rossby-number smaller than mass-continuity scaling suggests.

Vorticity

Cross-differentiating the horizontal momentum equation gives, as in (5.58), the vertical component of the vorticity equation with associated scalings, using (11.23) for W,

$$\frac{\mathrm{D}}{\mathrm{D}t}(\zeta + f) = -(\zeta + f)\left(\frac{\partial u}{\partial x} + \frac{\partial v}{\partial y} \right) + \left(\frac{\partial u}{\partial z} \frac{\partial w}{\partial y} - \frac{\partial v}{\partial z} \frac{\partial w}{\partial x} \right),$$

$$\text{Tropics:} \quad \left(\frac{U}{L} \right)^2 \sim \left(\frac{U}{L} + \frac{1}{Ro} \frac{U}{L} \right) \left(\frac{1}{Ri} \frac{U}{L} \right) \quad \left(\frac{U}{H} \right) \left(\frac{1}{Ri} \frac{UH}{L^2} \right),$$

$$\text{midlatitudes:} \quad \left(\frac{U}{L} \right)^2 \sim \left(\frac{U}{L} + \frac{1}{Ro} \frac{U}{L} \right) \left(\frac{Ro}{Bu} \frac{U}{L} \right) \quad \left(\frac{U}{H} \right) \left(\frac{Ro}{Bu} \frac{UH}{L^2} \right). \tag{11.24a,b,c}$$

In the tropical case with $Ro = \mathcal{O}(1)$ or larger *all* the terms on the right-hand side are much smaller than the terms on the left-hand side, whereas in the midlatitude case vortex stretching by the Coriolis term, $f(\partial u/\partial x + \partial v/\partial y)$, is the same order (because f is big and the divergence is small). Thus, in the tropical case the vorticity equation simplifies severely and at lowest order becomes the two-dimensional vorticity equation,

$$\frac{\mathrm{D}}{\mathrm{D}t}(\zeta + f) = 0. \tag{11.25}$$

The large-scale velocity is, at this order, purely rotational and is given by a streamfunction ψ such that

$$\nabla^2\psi = \zeta, \qquad (u, v) = \left(-\frac{\partial\psi}{\partial u}, \frac{\partial\psi}{\partial x}\right). \tag{11.26}$$

This equation nominally holds independently at each vertical level, although the underlying assumptions take the depth scale to be large.

A few comments

The relative weakness of large-scale horizontal gradients of pressure and temperature in the tropics, for a given velocity field, is a robust result of the scaling analysis and borne out in observations. However, (11.26) is not well satisfied by tropical motion for the following reasons:

- The scaling does not take into account heating, or diabatic sources, which may be expected to be particularly important in tropical regions.
- The smallness of the vertical velocity requires that the Richardson number, N^2H^2/U^2 be large; that is, that stratification be strong. This is only true in regions that are not actively convecting; in convective regions N may be small.
- Equation (11.26) tells us nothing about the vertical structure and so, unlike quasi-geostrophy in midlatitudes, is not sufficiently complete to be a useful prognostic, or even diagnostic, equation for tropical motion.

Let us now look at how diabatic effects might affect large-scale motion.

11.4.2 ♦ Large-Scale Flow with Diabatic Sources

In the tropics, heat sources such as condensational heating are very important and should be explicitly included in the reduced equations. Let us see if and how this is possible, using the shallow water equations for illustration. The underlying physical assumption is that in an air column adiabatic cooling associated with vertical motion balances diabatic heating, and in the shallow water equations this becomes a balance between the heating and the divergence.

Diabatic balanced shallow water equations

On an f-plane the shallow-water equations may be written in vorticity-divergence form as

$$\frac{\partial h}{\partial t} + \nabla\cdot(\boldsymbol{u}h) = Q, \tag{11.27a}$$

$$\frac{\partial \zeta}{\partial t} + \nabla\cdot[\boldsymbol{u}(\zeta + f_0)] = -r\zeta, \tag{11.27b}$$

$$\frac{\partial \delta}{\partial t} + \nabla^2\left(\frac{1}{2}\boldsymbol{u}^2 + gh\right) - \hat{\mathbf{k}}\cdot\nabla\times[\boldsymbol{u}(\zeta + f_0)] = -r\delta, \tag{11.27c}$$

where $\delta = \partial u/\partial x + \partial v/\partial y$ is the divergence, Q is the mass or heating source, r is a frictional coefficient and other notation is standard. The height field is a proxy for both pressure and temperature and we assume that horizontal gradients are weak, by which we mean that the dominant balance in (11.27a) is characterized by the scaling

$$H\Delta = Q_0, \tag{11.28}$$

where H is the mean thickness of the layer, Δ is a scaling for the divergence and Q_0 is the magnitude of the heating. We then choose the velocity scale, U, and the vorticity scale, Z, to be

$$U = \frac{Q_0 L}{H}, \qquad Z = \Delta = \frac{Q_0}{H}. \tag{11.29}$$

Finally, the magnitudes of horizontal deviations in the height field, $\mathcal{H}$, are determined from the divergence equation. Depending on whether rotation is or is not important we deduce

$$\mathcal{H} = \frac{f_0 UL}{g} = \frac{Q_0 f_0 L^2}{gH}, \quad \text{or} \quad \mathcal{H} = \frac{U^2}{g} = \frac{Q_0^2 L^2}{gH^2}. \tag{11.30a,b}$$

The height field may then be separated into a mean and deviation, $h = H + \eta$, where $\eta = \mathcal{H}\widehat{\eta}$. These scalings involve the heating in an essential way and so are fundamentally different from the adiabatic scaling of the previous section.

Using the above scalings (11.27) may be written in nondimensional form as

$$Bu^{-1}\left[\frac{1}{f_0 T}\frac{\partial \widehat{\eta}}{\partial \widehat{t}} + Ro\nabla\cdot(\widehat{\boldsymbol{u}}\widehat{\eta})\right] + \widehat{\delta} = \widehat{Q}, \tag{11.31a}$$

$$\frac{1}{f_0 T}\frac{\partial \widehat{\zeta}}{\partial \widehat{t}} + \nabla\cdot[\widehat{\boldsymbol{u}}(\widehat{\zeta} + \widehat{f}_0)] = -\frac{r}{f_0}\widehat{\zeta}, \tag{11.31b}$$

$$\frac{1}{f_0 T}\frac{\partial \widehat{\delta}}{\partial \widehat{t}} + \nabla^2\left(\frac{1}{2}\widehat{\boldsymbol{u}}^2 + \widehat{\eta}\right) - \hat{\mathbf{k}}\cdot\nabla\times[\widehat{\boldsymbol{u}}(\widehat{\zeta} + \widehat{f}_0)] = -\frac{r}{f_0}\widehat{\delta}, \tag{11.31c}$$

where T is the scaling for time, $Bu = (L_d/L)^2$ with $L_d = \sqrt{gH}/f_0$, and $\widehat{f}_0 = 1$. The Rossby number is given by $Ro = Q_0/f_0 H$ and is not necessarily small.

Reduced equations

Let us now suppose that the mass source determines the divergence in (11.31a). The condition for this is that

$$\max\left(\frac{1}{f_0 T}, Ro\right) Bu^{-1} \ll 1, \tag{11.32}$$

which means that the scale of motion cannot be too large and the time scale cannot be too short. If (11.32) is satisfied then the dimensional height equation, (11.27a), becomes

$$\nabla\cdot\boldsymbol{u} = \frac{Q}{H}. \tag{11.33a}$$

This value of the divergence is used in (11.27b) and (11.27c) which, retaining all terms since none are obviously small, become

$$\frac{\partial \zeta}{\partial t} + \boldsymbol{u} \cdot \nabla(\zeta + f_0) + (\zeta + f)\frac{Q}{H} = -r\zeta, \tag{11.33b}$$

$$g\nabla^2 h = \hat{\mathbf{k}} \cdot \nabla \times [\boldsymbol{u}(\zeta + f_0)] - \frac{1}{H}\frac{\partial Q}{\partial t} - r\delta - \nabla^2 \frac{\boldsymbol{u}^2}{2}. \tag{11.33c}$$

The equation set (11.33) has but one prognostic equation, namely (11.33b), and so is truly balanced and may be thought of as a generalization of (11.25) to the case with non-zero heating. The divergence equation is a nonlinear balance equation except now with a diabatic term on the right-hand side. The divergent flow itself is computed using the height equation, by an assumed balance between adiabatic cooling and diabatic heating. The relationship between velocity and geopotential (or pressure) is the same as in the adiabatic case, because this arises through the momentum equation. Thus, even in the presence of a heating, gradients of geopotential and temperature remain relatively weak, a result that ultimately arises from the smallness of the Coriolis parameter. The importance of the result lies in what it implies about the response of the atmosphere to a localized heating: the equations provide a scaling for the response of the velocity, and suggest that the response may become spread out over a sufficient area to keep the temperature gradients small. This, taken with the scaling arguments of Section 5.3.1, is called the 'weak temperature gradient approximation'.

11.5 EFFECTS OF MOISTURE

In the rest of this chapter we talk about the two things that one immediately notices about the deep tropics — moisture and convection.

Moisture affects nearly every facet of tropical dynamics, and without it the tropics would be a very different place — no towering cumulonimbus clouds for example. However, convection *would* still take place since the basic state set up by the radiative forcing would still be convectively unstable.

11.5.1 Measures of Moisture

There are various measures of the amount of moisture in the atmosphere, so let us summarize them. The *absolute humidity* is the amount of water vapour per unit volume, with units of kg m^{-3}, or informally g m^{-3}. The *mixing ratio, w*, is the ratio of the mass of water vapour, m^v, to that of dry air, m^d, in some volume of air and is thus

$$w \equiv \frac{m^v}{m^d} = \frac{\rho^v}{\rho^d}. \tag{11.34}$$

It is a nondimensional measure but it is often expressed in terms of grams per kilogram. In the atmosphere values range from close to zero to about 2×10^{-2} (20 g kg^{-1}) in the tropics on a humid day.

The *specific humidity, q*, is the ratio of the mass of water vapour to the total mass of air — dry air plus water vapour — and so is

$$q \equiv \frac{m^v}{m^d + m^v} = \frac{w}{1 + w} \quad \text{and} \quad w = \frac{q}{1 - q}. \tag{11.35a,b}$$

The specific humidity is just the mass concentration of water vapour in air. In most circumstances in Earth's atmosphere $m^v \ll m^d$ so that $q \approx w$, usually to an accuracy of about one percent. In most of this chapter we ignore the differences between w and q, but this is not appropriate for all planetary atmospheres.

The partial pressure of water vapour in air, or the vapour pressure, e, is the pressure exerted by water molecules and is proportional to the number of molecules, and so the number of moles, of water vapour in the volume. It is given by

$$e = \frac{n^v}{n^d + n^v} p = \frac{m^v/\mu^v}{m^d/\mu^d + m^v/\mu^v} p, \tag{11.36}$$

where n^v and n^d are the number of moles of water vapour and dry air in the mixture , and μ^v and μ^d are the respective molecular weights, and p is the total pressure. Using (11.34) we can write (11.36) as

$$e = \frac{wp}{w + \epsilon} \quad \text{or} \quad w = \frac{\epsilon e}{p - e}, \tag{11.37}$$

where $\epsilon = \mu^v/\mu^d \approx 0.62$. In terms of q instead of w these expressions are

$$e = \frac{qp}{q + \epsilon(1 - q)} \quad \text{and} \quad q = \frac{\epsilon e}{p - e(1 - \epsilon)}. \tag{11.38}$$

In Earth's atmosphere $w \ll 1$ so that

$$e \approx w\frac{p}{\epsilon} = 1.61wp \qquad \text{and} \qquad q \approx w \approx \epsilon\frac{e}{p}. \tag{11.39}$$

If the mixing ratio of water vapour is $10\,\mathrm{g\,kg^{-1}}$ (a typical tropical value) and $p = 1000\,\mathrm{hPa}$ then $e \approx 16\,\mathrm{hPa}$.

The *relative humidity*, $\mathcal{H}$, is the ratio of the actual vapour pressure to the saturation vapour pressure, e_s, which is the maximum vapour pressure that can occur at a given temperature before condensation occurs. Thus, $\mathcal{H} = e/e_s \approx q/q_s$ where q_s is the specific humidity at saturation — which we now explain.

Saturation and the Clausius–Clapeyron relation

At any given temperature, there is a maximum vapour pressure beyond which condensation normally occurs and this is known as the *saturation vapour pressure,* denoted e_s. This value is given by the *Clausius–Clapeyron* equation, which, to a very good approximation for Earth's atmosphere, is

$$\frac{\mathrm{d}e_s}{\mathrm{d}T} = \frac{Le_s}{R^v T^2}. \tag{11.40}$$

Here, L is the latent heat of condensation, which is the amount of heat released per unit mass when water vapour condenses, R^v is the gas constant for water vapour and T is the temperature. If L is a constant (in

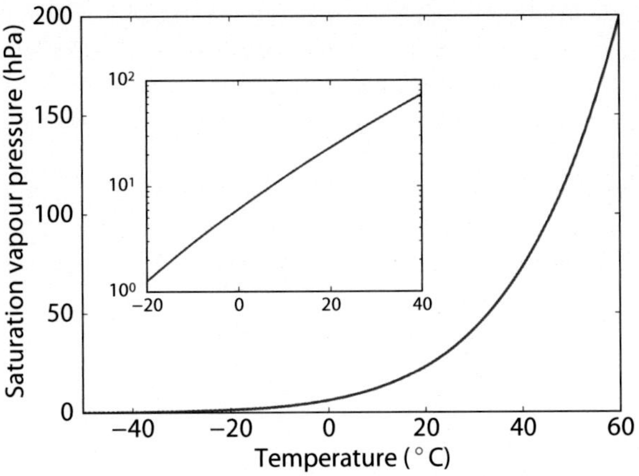

Fig. 11.9: The saturation vapour pressure of water vapour as a function of temperature. The inset plots the same quantities on a log scale over a slightly smaller range.

fact it varies by about 10% as temperature varies from 0°C to 100°C) then (11.40) can be integrated to give

$$e_s = e_0 \exp\left[\frac{L}{R^v}\left(\frac{1}{T_0} - \frac{1}{T}\right)\right], \tag{11.41}$$

where, for water, $T_0 = 273\,\mathrm{K}$, $e_0 = 6.12\,\mathrm{hPa}$, $L = 2.44 \times 10^6\,\mathrm{J\,kg^{-1}}$ and $R^v = 462\,\mathrm{J\,kg^{-1}\,K^{-1}}$. As Fig. 11.9 indicates, the saturation vapour pressure goes up approximately exponentially with temperature. The saturation vapour pressure is a function *only* of temperature; in particular it is not a function of the pressure and its value is independent of the amount of dry air present. Suppose that a parcel of moist air is cooled. The saturation vapour pressure falls according to (11.41) and at some temperature (called the dew point) the actual vapour pressure will become equal to the saturation vapour pressure. The volume is then saturated and any further cooling leads to condensation. This is how rain forms — a parcel of moist air ascends, cools, saturates and condenses.

The amount of water vapour that a volume can hold is independent of the presence or otherwise of dry air in that volume. Rather, it is a function of temperature, and saturation vapour pressure increases approximately exponentially with temperature as in Fig. 11.9.

Why don't other gases cause rain? Why does water, and not nitrogen or oxygen, condense? It is because the other gases are much more volatile than water. For nitrogen, for example, the parameters in (11.41) are such that the saturation vapour pressure at common temperatures is very much higher than the actual pressure. However, on some planets other gases may condense. Thus, temperatures at the winter pole on Mars are sufficiently low that carbon dioxide will condense out, and in the frigid atmosphere of Titan (a moon of Saturn) methane is a condensate.

11.5.2 Distribution of Water Vapour in the Atmosphere

The strong dependence of e_s on temperature means that the absolute amount of water vapour in the atmosphere is strongly tied to temperature, as we see in Fig. 11.10. One could easily imagine that this were a plot of temperature, not moisture and the warm tropics hold a lot of moisture. However, the relative humidity field looks quite different, as we see in Fig. 11.11. What determines its distribution?

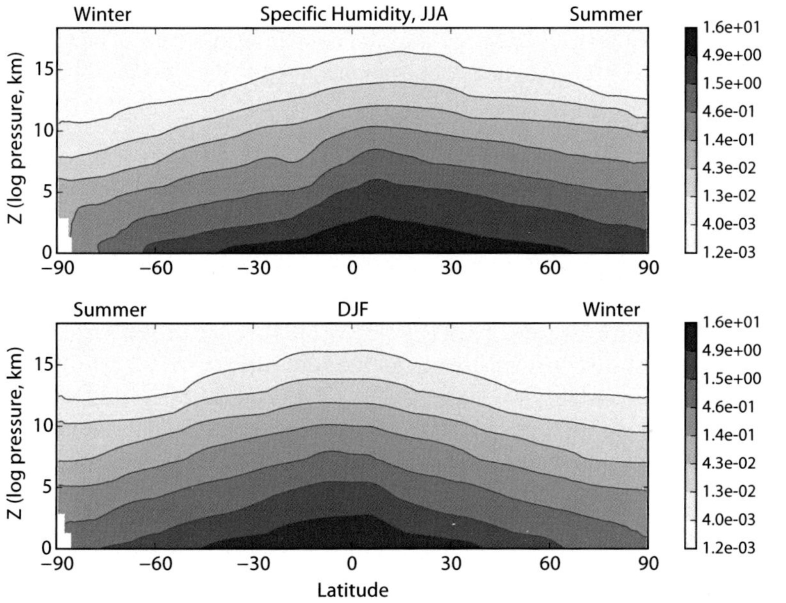

Fig. 11.10: Zonally-averaged specific humidity distribution (g/kg) in the atmosphere, as retrieved from a microwave satellite, for boreal summer in 2008 (top) and boreal winter 2008–2009 (bottom).

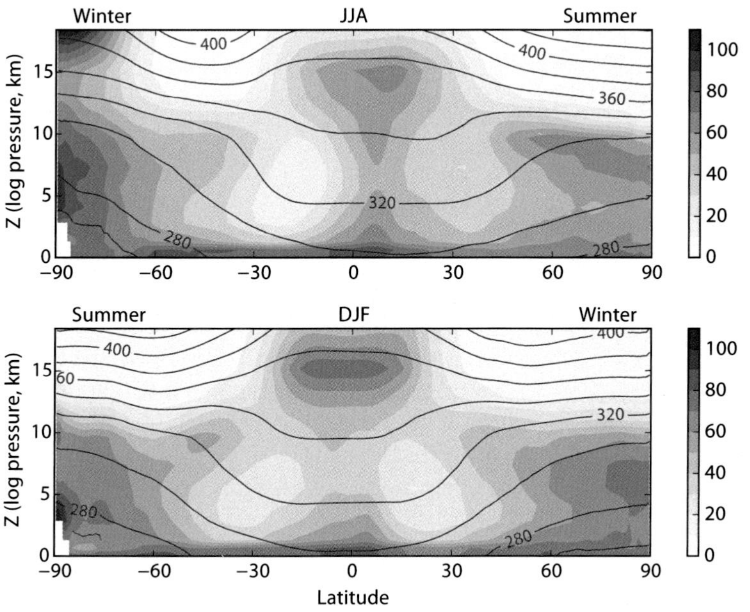

Fig. 11.11: Zonally-averaged relative humidity distribution in the atmosphere (in percent, shading), and isolines of equivalent potential temperature (contours). (Equivalent potential temperature is a modification (generally a small one, except sometimes in the tropics) of potential temperature to account the effects of water vapour.)

The relative humidity of a parcel at a particular location is, to a decent approximation, determined by the temperature at that location and the saturation specific humidity at the position of last saturation. To see this, imagine a moist, but unsaturated, parcel of air moving toward a colder region. At some location the parcel becomes saturated and excess moisture condenses and (we suppose) falls to earth as rain. If the parcel moves to still cooler locations more moisture falls out, but the parcel remains saturated. At some location, $\boldsymbol{x}$ say, we may suppose that the parcel starts to move to a warmer location and the condensation halts. The amount of moisture in the parcel, q, then remains fixed at the amount it had at $\boldsymbol{x}$, namely $q_s(\boldsymbol{x})$. At a new position, $\boldsymbol{x}'$, the relative humidity is thus given by

$$\mathcal{H} = \frac{e(\boldsymbol{x}')}{e_s(\boldsymbol{x}')} \approx \frac{q(\boldsymbol{x}')}{q_s(\boldsymbol{x}')} = \frac{q_s(\boldsymbol{x}')}{q_s(\boldsymbol{x})}. \tag{11.42}$$

This argument neglects the effects of re-evaporation of liquid water, which in some circumstances is quite important, but it does captures the main process.

In the tropics the air typically flows equatorward at low levels, picking up moisture from the ocean and the land surface, before rising in the ascending branch of the Hadley Cell. As it ascends it cools and the air becomes saturated. When the air descends in the subtropics it is moving to a warmer environment and so its relative humidity falls, as given by (11.42). The other features of Fig. 11.11 can be explained by similar reasoning, although the trajectories of parcels in the midlatitudes are more complex because they are affected by baroclinic instability. However, one robust feature can be readily explained and that is the very low relative humidity of the stratosphere. The dryness arises because the tropopause is a local minimum of temperature, so that air passing beyond the tropopause does not condense in the stratosphere and its relative (and absolute) humidity are very small. In fact most air entering the stratosphere does so in equatorial regions, where the tropopause is high and cold and serves as a *cold trap* for the moisture.

11.6 TROPICAL CONVECTION

Convection — often manifested in towering cumulonimbus clouds — is perhaps the most obvious feature of the tropics. Moisture has a large destabilizing effect on the tropics, as we can see by a simple qualitative argument. We saw in Section 3.4 that a column of dry air will be convectively unstable if its lapse rate exceeds (i.e., the temperature falls off more rapidly than) the dry adiabatic lapse rate, g/c_p, which is about 10°C/ km. Now consider a column of air that is stable by this criterion but that contains water vapour (or any other condensible for that matter), and suppose we raise a parcel of air. If the water does not condense then the parcel is stable and it sinks back to its original position. But if the parcel cools sufficiently that it becomes saturated then some water will condense and the parcel will warm. It is then possible that the parcel warms to the extent that it is now more buoyant than its surroundings and it will then be unstable, and continue to rise. Let us use a parcel argument similar to that used in Section 3.4 to quantify this.

11.6.1 Saturated Adiabatic Lapse Rate

The first law of thermodynamics may be written as

$$c_p \, \mathrm{d}T - \alpha \, \mathrm{d}p = \text{đ}Q, \tag{11.43}$$

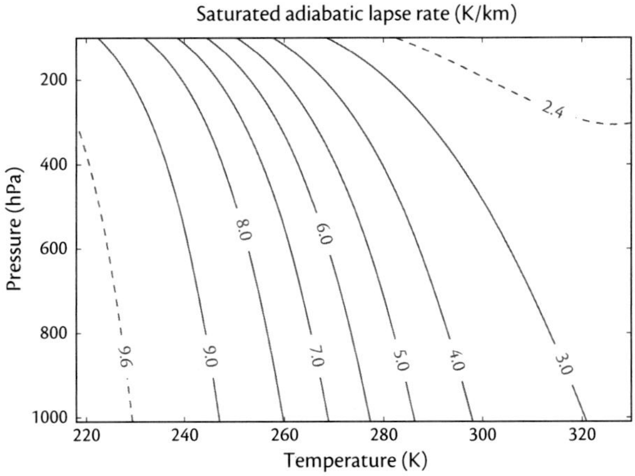

Fig. 11.12: Contours of saturated adiabatic lapse rate (K/km) as a function of pressure and temperature, calculated using (11.49) with $q_s = \epsilon e_s/p$, and e_s given by the solution of the Clausius–Clapeyron equation, (11.41).

where $đQ$ is the heat source. In an adiabatic process, such as may occur in a dry atmosphere, $đQ = 0$ so that $c_p \,\mathrm{d}T = \alpha \,\mathrm{d}p$, and in a hydrostatic atmosphere $\alpha \,\mathrm{d}p = -g \,\mathrm{d}z$. Hence an adiabatic profile is characterized by $c_p \,\mathrm{d}T = -g \,\mathrm{d}z$, giving the dry adiabatic lapse rate $\mathrm{d}T/\mathrm{d}z = -g/c_p$.

In a moist process there exists the possibility of condensation so that (11.43) becomes

$$c_p \,\mathrm{d}T - \alpha \,\mathrm{d}p = -L \,\mathrm{d}q. \tag{11.44}$$

Here $-L \,\mathrm{d}q$ is the amount of the heat released by the condensation of a small amount of water vapour $\mathrm{d}q$, and the negative sign arises because the amount of water vapour in the fluid parcel is then decreasing. For a hydrostatic atmosphere the above equation becomes, treating c_p and, a little less accurately, L, as constants,

$$\mathrm{d}(c_p T + gz + Lq) = 0. \tag{11.45}$$

The quantity in brackets, $\mathcal{M} = c_p T + gz + Lq$, is known as the moist static energy (and $c_p T + gz$ is the dry static energy). A parcel that is moving will conserve its moist static energy even if there is condensation. However, condensation will only occur if $q = q_s$ and, given that fact, we can obtain an expression for the profile of temperature as follows.

As a parcel ascends $\mathrm{d}\mathcal{M}/\mathrm{d}z = 0$ and using (11.45) an ascending parcel then has a lapse rate given by

$$c_p \frac{\mathrm{d}T}{\mathrm{d}z} = -L \frac{\mathrm{d}q}{\mathrm{d}z} - g. \tag{11.46}$$

If the air is moist but not saturated then an ascending parcel will follow the dry adiabatic lapse rate (because then $\mathrm{d}q/\mathrm{d}z = 0$) but if it is saturated (with $q = q_s$) then as the parcel ascends it will cool and some water vapour will condense. Since $q_s \approx \epsilon e_s/p$, and so is a function of temperature and pressure, we have

$$\frac{\mathrm{d}q_s}{\mathrm{d}z} = \left(\frac{\partial q_s}{\partial T}\right)_p \frac{\mathrm{d}T}{\mathrm{d}z} + \left(\frac{\partial q_s}{\partial p}\right)_T \frac{\mathrm{d}p}{\mathrm{d}z} = \left(\frac{\partial q_s}{\partial T}\right)_p \frac{\mathrm{d}T}{\mathrm{d}z} + \left(\frac{q_s}{p}\right)\rho g, \tag{11.47}$$

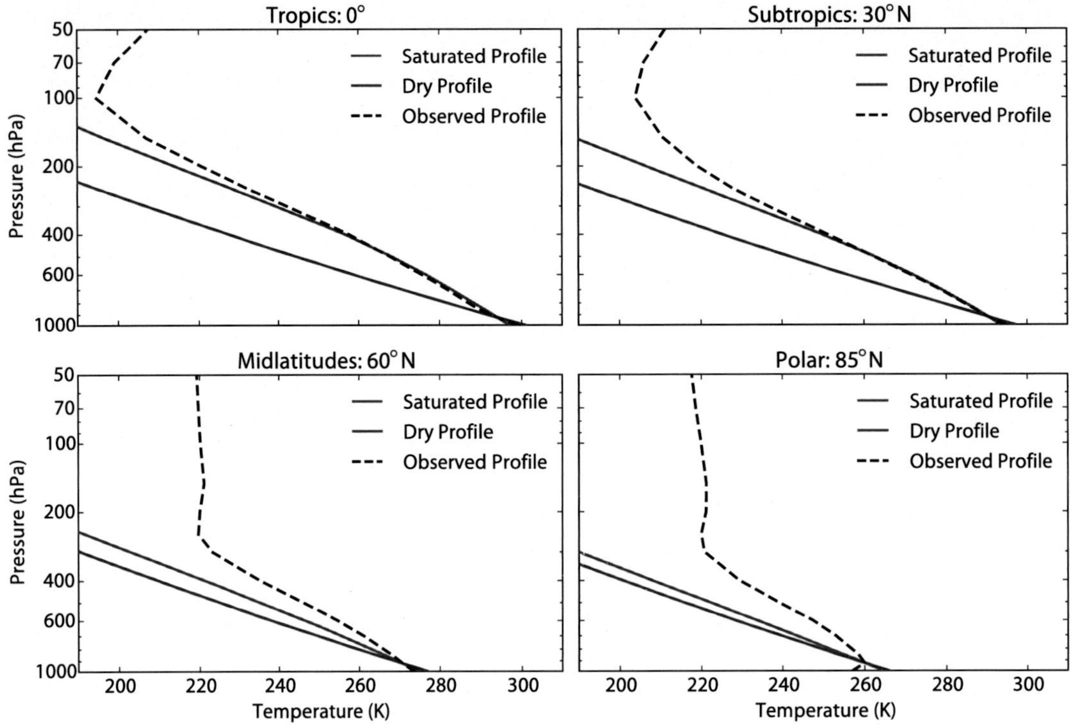

Fig. 11.13: The annually- and zonally-averaged observed temperature profiles at various latitudes, along with profiles constructed by integrating the saturated and dry adiabatic lapse rates, chosen to coincide with the observed temperature at 925 hPa, or about 750 m.

using hydrostasy. Using the Clausius–Clapeyron equation and the ideal gas equation of state gives

$$\frac{dq_s}{dz} = \frac{Lq_s}{R^v T^2}\frac{dT}{dz} + \frac{gq_s}{R^d T}. \tag{11.48}$$

Using (11.46) and (11.48) we obtain an expression for the lapse rate, Γ_s, of an ascending saturated parcel,

$$\Gamma_s = -\left.\frac{dT}{dz}\right|_{ad} = \frac{g}{c_p}\frac{1 + Lq_s/(R^d T)}{1 + L^2 q_s/(c_p R^v T^2)}. \tag{11.49}$$

This profile is called the *saturated adiabatic lapse rate*, plotted in Fig. 11.12. It is 'adiabatic' because we may regard the condensation process as one that is internal to the fluid parcel, with energy being transferred from one form to another during the condensation process without any external energy source. (However, we have not properly accounted for the presence of liquid water in our derivation, so the profile is sometimes called a 'pseudo-adiabatic lapse rate'.) The lapse rate is a function of temperature and pressure because $q_s = \epsilon e_s/p$ and e_s is given by the solution of the Clausius–Clapeyron equation, (11.41). Values of Γ_s are typically around $6\,\mathrm{K\,km^{-1}}$ in the lower atmosphere although since dq_s/dT is an increasing function of T, Γ_s decreases with increasing temperature and can be as low as $3\,\mathrm{K\,km^{-1}}$. The second term in the numerator of (11.49) is usually

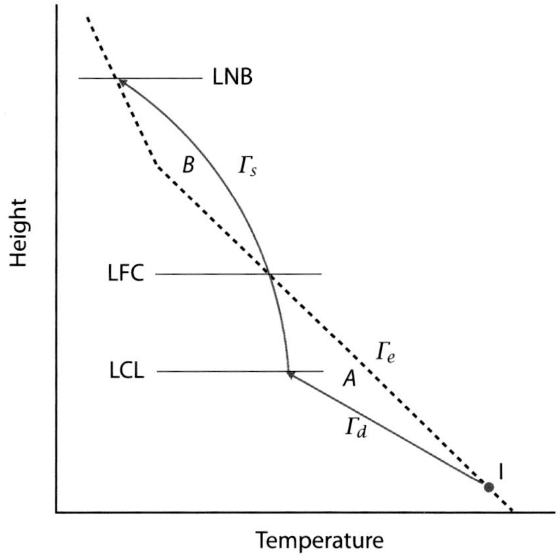

Fig. 11.14: Schematic of a conditional instability in an atmosphere with environmental lapse rate Γ_e (dashed line). A parcel at I is forced to rise, and it does so along the dry adiabat, Γ_d, until it is saturated at the lifting condensation level, LCS. It will then rise along the saturated adiabat, Γ_s, and after reaching the level of free convection (LFC) it is convectively unstable. The parcel continues to rise along the saturated adiabat until it reaches the level of neutral buoyancy (LNB)

quite small (around 0.1, although it can become large at very high temperatures) but the second term in the denominator is positive and order unity. Since g/c_p is the dry adiabatic lapse rate, the saturated adiabatic lapse rate is smaller than the dry adiabatic lapse rate, as can be seen directly from (11.46) since $\mathrm{d}q^v/\mathrm{d}z < 0$.

Using parcel theory, just as in Section 3.4, a saturated column of air will be stable or unstable depending on whether the lapse rate is less than or greater than (11.49); that is

$$\text{Stability}: \quad -\frac{\partial \widetilde{T}}{\partial z} < \Gamma_s, \qquad \text{Instability}: \quad -\frac{\partial \widetilde{T}}{\partial z} > \Gamma_s, \qquad (11.50\text{a,b})$$

where $\widetilde{T}$ is the environmental temperature.

11.6.2 Quasi-Equilibrium and Conditional Instability

The lapse rate of the atmosphere rarely exceeds the saturated adiabatic lapse rate, as we see in Fig. 11.13. In the tropics and subtropics the lapse rate is, on average, very close to the saturated adiabat up to about 300 hPa (about 9 km), whereas in midlatitudes it is considerably less, and so more stable, because of the upward transport of heat by baroclinic eddies, which we come to in the next chapter. In the tropics, however, moist convection is the primary determinant of the lapse rate. In the absence of convection temperature would fall off much more rapidly with height in the lower atmosphere, because most of the heating of the atmosphere occurs close to the surface — the ground is heated by the sun, and the ground in turn heats the lower atmosphere. This heating produces an unstable lapse rate and convection then acts, typically on a timescale of hours, to neutralize that lapse rate. *Convective quasi-equilibrium* is the posited state in which the forcing of a convectively unstable profile by large-scale dynamics and/or radiation is statistically balanced by convection. For quasi-

The maintenance of the tropical lapse rate by convection remains an active topic of research. The degree to which the convection simply acts rather passively to maintain a moist adiabatic lapse rate in response to changes in the large-scale circulation, as well as the degree to which convection actually determines that large-scale circulation, are argued about to this day.

equilibrium to hold the convection should act on a shorter timescale than changes in the large-scale circulation.

Convection only directly determines the lapse rate over the small fraction of the tropics where convection actually occurs, but nevertheless the average lapse rate over the entire tropics is close to moist neutral. This is because the gravity waves emanating from convective regions adjust the tropics to have a weak horizontal temperature gradient, in a process akin to geostrophic adjustment, hence maintaining approximately the same vertical profile even away from regions of active convection.

If the air is not saturated then the critical lapse rate is not the saturated adiabat, but it is the much larger dry adiabat, and it is common for a parcel to be stable if dry but unstable if saturated, and this leads to the notion of *conditional stability,* whereby a parcel is stable to an infinitesimal perturbation but unstable to a finite perturbation. Consider an environment in which the lapse rate lies between the moist and dry rates, as in Fig. 11.14, and consider a parcel near the surface at position I. Suppose the parcel is adiabatically lifted (perhaps mechanically) then its temperature profile follows the dry adiabat until it is saturated at, by definition, the *lifting condensation level* (LCL). However, the parcel is actually negatively buoyant and so would sink unless the parcel is forced to continue rising, but if it does continue to rise it will be along a saturated adiabat and eventually, at the *level of free convection* (LFC) is will become buoyant and convectively unstable. It will continue to rise until its buoyancy no longer exceeds that of the environment, at the *level of neutral buoyancy,* which may well be at or close to the tropopause where the temperature starts to increase again. (The height of the tropopause is not independent of the convection itself, a matter we discuss in Section 13.3.3.)

Evidently, to trigger such an instability a *finite* perturbation is needed (for example a flow over a hill) and this is known as *conditional instability*. Convection itself can in fact provide the trigger for more convection, leading us into the aforementioned quasi-equilibrium state, but the interested reader may pursue that topic elsewhere, perhaps starting with the references below.

Notes and References

The angular momentum conserving model of the Hadley Cell stems from Schneider (1977) and Held & Hou (1980). The notion that the Hadley Cell and the radiative equilibrium temperature distribution itself are baroclinically unstable is implicit in the discussion of Lorenz (1967). The weak temperature and pressure gradients in the tropics are consequences of the scaling of Charney (1963) and the topic was further developed by Sobel *et al.* (2001) and others. Moist thermodynamics and convection are dealt with in more detail in a number of books, for example Emanuel (1994), Bohren & Albrecht (1998) and Ambaum (2010).

Problems

Atmospheric problems are given at the end of Chapter 13.

CHAPTER

12

Midlatitudes and the Stratosphere

WHY DO THE DYNAMICS OF THE MIDLATITUDES differ from their tropical counterparts? Is the difference common to most planets or special to Earth? One difference, at least on Earth, is that the midlatitudes are baroclinically unstable, producing the weather. Even if the Hadley Cell were to terminate of its own accord before becoming baroclinically unstable, the radiative equilibrium temperature in midlatitudes has a meridional gradient that would be unstable. On other, more slowly rotating planets, the Hadley Cell might extend nearly all the way to the pole, in which case the planet may be thought of as entirely tropical! Venus and Titan (a moon of Saturn) are examples of such all-tropical planets, and others likely abound outside our Solar System.

Given this rather general point, in this chapter we will discuss two more specific properties of the Earth's midlatitudes: (i) The predominantly eastward surface winds and the strong eastward winds extending up to the tropopause. (ii) The meridional overturning circulation, or Ferrel Cell. Both features can be seen in Fig. 11.1 and Fig. 11.3, and both are consequences of the general phenomena of baroclinic instability and geostrophic turbulence, moulded by Earth's atmosphere, and they become intertwined in our discussion. None of the dynamics that we discuss in this chapter involves density variation in a truly essential way and readers may simplify the discussion by regarding density as constant.

12.1 JET FORMATION AND SURFACE WINDS

The atmosphere above the surface has a generally eastward flow, with a broad maximum about 10 km above the surface at around 40° in either hemisphere. But if we look a little more at the zonally average wind in Fig. 11.1(a), especially in the Southern Hemisphere, we see hints of there being two jets — one (the subtropical jet) at around 30°, and another somewhat poleward of this. The subtropical jet is associated with a strong meridional temperature gradient at the edge of the Hadley Cell and it is quite baroclinic — that is, there is a noticeable shear in the zonal wind. On the other hand, the midlatitude jet is more barotropic (it has little vertical structure, with less shear than the subtropical jet) and lies above an eastward surface flow. This flow feels the effect of surface friction and so there must be a momentum *convergence* into this region, as is seen in Fig. 12.4. This jet is known as the *eddy-driven jet.*

We encountered eddy-driven jets in our discussion of barotropic turbulence in Section 10.6. However, that case was homogeneous, with no preferred latitude for a particular jet, whereas in the atmosphere there appears to be but one midlatitude jet with a preferred average location, and in the sections that follow we discuss how this jet is maintained.

12.1.1 The Mechanism of Jet Production

For reference later on we establish a useful form of the zonal momentum equation. For two-dimensional, horizontally non-divergent flow we have

$$\frac{\partial u}{\partial t} + \frac{\partial u^2}{\partial x} + \frac{\partial uv}{\partial y} - fv = -\frac{\partial \phi}{\partial x} - D_u, \tag{12.1}$$

where D_u represents the effects of dissipation. We write the variables as the sum of a zonal mean plus a deviation so that $u = \overline{u} + u'$ and $v = \overline{v} + v'$, and for incompressible two dimensional flow $\overline{v} = 0$. The zonal average of (12.1) is then

$$\frac{\partial \overline{u}}{\partial t} + \frac{\partial \overline{u}\,\overline{v}}{\partial y} + \frac{\partial \overline{u'v'}}{\partial y} - f\overline{v} = -D_u, \tag{12.2}$$

but since $\overline{v} = 0$ we have

$$\frac{\partial \overline{u}}{\partial t} + \frac{\partial \overline{u'v'}}{\partial y} = -r\overline{u}, \tag{12.3}$$

where we also represent dissipation as a linear drag, with r being a constant.

We can write the momentum flux in terms of the vorticity flux since, for non-divergent two-dimensional flow,

$$v\zeta = \frac{1}{2}\frac{\partial}{\partial x}\left(v^2 - u^2\right) - \frac{\partial}{\partial y}(uv), \tag{12.4}$$

where $\zeta = \partial v/\partial x - \partial u/\partial y$ is the vorticity. After zonal averaging (12.4) gives

$$\overline{v'\zeta'} = -\frac{\partial \overline{u'v'}}{\partial y}, \tag{12.5}$$

and (12.3) becomes

$$\frac{\partial \overline{u}}{\partial t} = \overline{v'\zeta'} - r\overline{u}. \tag{12.6}$$

If we integrate the vorticity flux between two quiescent latitudes then, from (12.5), the integral vanishes. Thus, from (12.6), the mean wind, $\overline{u}$, must also vanish after integration over latitude and time.

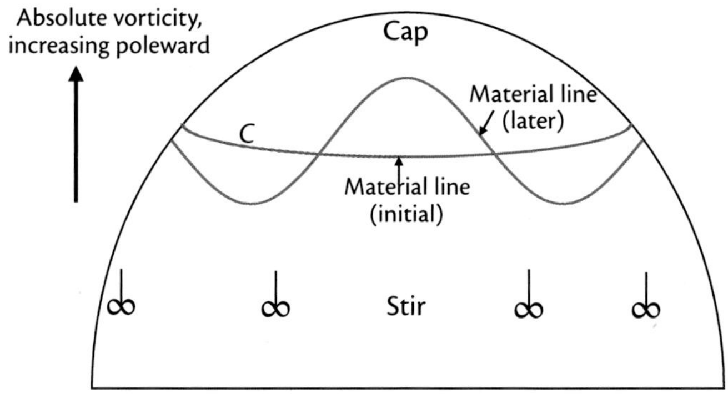

Fig. 12.1: Sketch of the effects of a midlatitude disturbance on the circulation around the latitude line C. If initially the absolute vorticity increases monotonically poleward, then the disturbance will bring fluid with lower absolute vorticity into the cap region. Then, using Stokes theorem, the velocity around the latitude line C will become more westward.

I. The vorticity budget

The argument we first present does not use the momentum equation directly; rather, it uses Kelvin's circulation theorem, and we use spherical coordinates. Suppose that the absolute vorticity normal to the surface , $\zeta + f$, where $f = 2\Omega \sin\vartheta$, increases monotonically poleward. (A sufficient condition for this is that the fluid is at rest.) By Stokes' theorem, the initial circulation, I_i, around a line of latitude circumscribing the polar cap is equal to the integral of the absolute vorticity over the cap. That is,

$$I_i = \int_{\text{cap}} \boldsymbol{\omega}_{ia} \cdot \mathrm{d}\boldsymbol{A} = \oint_C u_{ia}\,\mathrm{d}l = \oint_C (u_i + \Omega a \cos\vartheta)\,\mathrm{d}l, \tag{12.7}$$

where $\boldsymbol{\omega}_{ia}$ and u_{ia} are the initial absolute vorticity and absolute velocity, respectively, u_i is the initial zonal velocity in the Earth's frame of reference, and the line integrals are around the line of latitude. Let us take $u_i = 0$ and suppose there is a disturbance equatorward of the polar cap, and that this results in a distortion of the material line around the latitude circle C (Fig. 12.1).

Since the source of the disturbance is distant from the latitude of interest, if we neglect viscosity the circulation along the material line is conserved, by Kelvin's circulation theorem. Thus, vorticity with a lower value is brought into the region of the polar cap — that is, the region poleward of the latitude line C. Using Stokes' theorem again the circulation around the latitude circle C must therefore fall; that is, denoting later values with a subscript f,

$$I_f = \int_{\text{cap}} \boldsymbol{\omega}_{fa} \cdot \mathrm{d}\boldsymbol{A} < I_i, \tag{12.8}$$

so that

$$\oint_C (u_f + \Omega a \cos\vartheta)\,\mathrm{d}l < \oint_C (u_i + \Omega a \cos\vartheta)\,\mathrm{d}l, \tag{12.9}$$

and thus

$$\overline{u}_f < \overline{u}_i, \tag{12.10}$$

with the overbar indicating a zonal average. Thus, there is a tendency to produce *westward* flow poleward of the disturbance. By a similar argument westward flow is also produced equatorward of the disturbance —

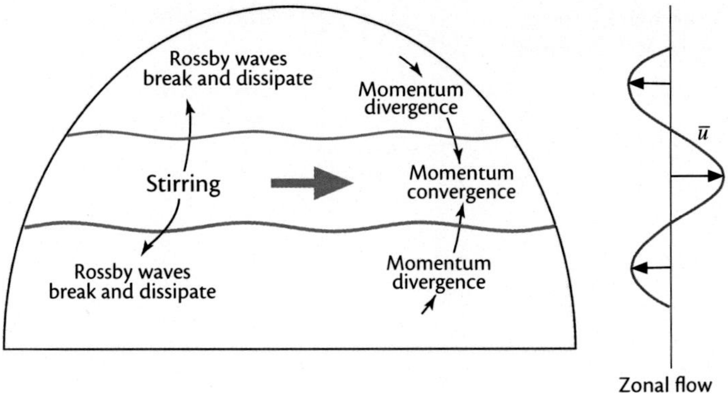

Fig. 12.2: Generation of zonal flow on a rotating sphere. Stirring in midlatitudes (by baroclinic eddies) generates Rossby waves that propagate away. Momentum converges in the region of stirring, producing eastward flow there and weaker westward flow on its flanks.

to see this one may (with care) apply Kelvin's theorem over all of the globe south of the source of the disturbance. Finally, note that the overall situation is the same in the Southern Hemisphere. Thus, on the surface of a rotating sphere, external stirring will produce westward flow *away* from the region of the stirring.

If the disturbance imparts no net angular momentum to the fluid then the integral of $\overline{u}\cos\vartheta$ over the entire hemisphere must be unaltered. But the fluid is accelerating westward away from the disturbance. Therefore, the fluid in the region of the disturbance must accelerate *eastward*, and this is the essence of the production of midlatitude westerlies on Earth, where the stirring is maintained by baroclinic instability.

II. Rossby waves and momentum flux

We have seen that a mean gradient of vorticity is an essential ingredient in the mechanism whereby a mean flow is generated by stirring. Given that, we expect Rossby waves to be excited, and we now show how those waves are intimately related to the momentum flux maintaining the mean flow.

If a stirring is present in midlatitudes then Rossby waves will be generated there before propagating away where they dissipate. To the extent that the waves are quasi-linear, then just away from the source region each wave has the form

$$\psi = \mathrm{Re}\, C e^{\mathrm{i}(kx+ly-\omega t)} = \mathrm{Re}\, C e^{\mathrm{i}(k(x-ct)+ly)}, \tag{12.11}$$

where C is a constant, with dispersion relation (now back to the β-plane)

$$\omega = ck = \overline{u}k - \frac{\beta k}{k^2 + l^2}, \tag{12.12}$$

provided that there is no meridional shear in the zonal flow. The meridional component of the group velocity is given by

$$c_g^y = \frac{\partial \omega}{\partial l} = \frac{2\beta k l}{(k^2 + l^2)^2}\,. \tag{12.13}$$

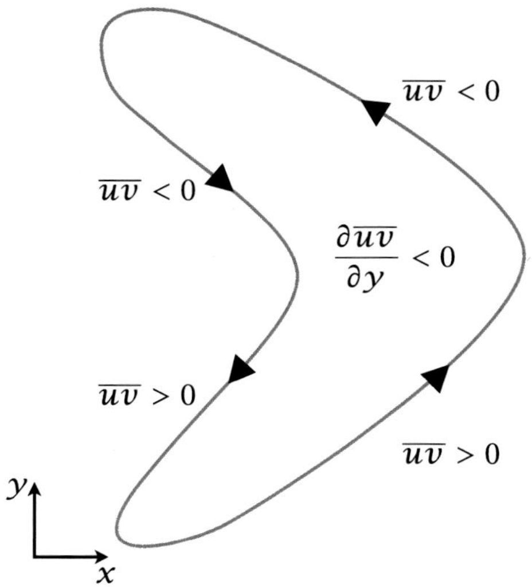

Fig. 12.3: The momentum transport in physical space, caused by the propagation of Rossby waves away from a source in midlatitudes. The ensuing bow-shaped eddies are responsible for a convergence of momentum in the centre of the bow, which in turn accelerates the mean flow eastward. If the arrows were reversed the momentum transport would still have the same sign.

Now, the direction of the group velocity must be away from the source region, because Rossby waves transport energy away from the disturbance. Thus, northward of the source kl is positive and southward of the source kl is negative. That the product kl can be positive or negative arises because for each k there are two possible values of l that satisfy the dispersion relation (12.12), namely

$$l = \pm\left(\frac{\beta}{\overline{u} - c} - k^2\right)^{1/2}, \tag{12.14}$$

assuming that the quantity in parentheses is positive.

The velocity variations associated with the Rossby waves are

$$u' = -\mathrm{Re}\, C\, il\mathrm{e}^{\mathrm{i}(kx+ly-\omega t)}, \qquad v' = \mathrm{Re}\, C\, ik\mathrm{e}^{\mathrm{i}(kx+ly-\omega t)}, \tag{12.15a,b}$$

and the associated momentum flux is

$$\overline{u'v'} = -\frac{1}{2}C^2 kl. \tag{12.16}$$

Thus, given that the sign of kl is determined by the group velocity, northward of the source the momentum flux associated with the Rossby waves is southward (i.e., $\overline{u'v'}$ is negative), and southward of the source the momentum flux is northward (i.e., $\overline{u'v'}$ is positive). That is, the momentum flux associated with the Rossby waves is *toward* the source region. Momentum thus converges in the region of the stirring, producing net eastward flow there and westward flow to either side (see Fig. 12.2).

If we think of this effect in physical space, then if kl is positive lines of constant phase ($kx + ly$ = constant) are tilted north-west/south-east, and the momentum flux associated with such a disturbance is negative (that is, $\overline{u'v'} < 0$). Similarly, if kl is negative then the constant-phase lines are tilted north-east/south-west and the associated momentum flux is positive ($\overline{u'v'} > 0$). The net result is a convergence of momentum flux into the source region; that is, $\partial(u'v')/\partial y < 0$ and from (12.3) the flow accelerates eastward. In physical space the momentum flux is reflected by having eddies that are shaped like a boomerang, or a bow, as in Fig. 12.3.

Although our argument is barotropic, it captures the pattern of the eddy momentum fluxes in Earth's atmosphere, as illustrated in the bottom panel of Fig. 12.4. In the Northern Hemisphere the eddy momentum flux is in fact negative ($\overline{u'v'} < 0$) poleward of about 55°N and positive equatorward of that, thus converging momentum into midlatitudes and producing eastward surface winds. The same pattern (accounting for a change in sign of v) occurs in the Southern Hemisphere.

III. ♦ Group velocity and the Eliassen–Palm flux

Using the Eliassen–Palm (EP) flux (Section 9.2) provides a convenient way of re-expressing the above Rossby-wave argument. In the unforced case, the zonally-averaged momentum equation may be written as

$$\frac{\partial \overline{u}}{\partial t} - f_0 \overline{v}^* = \nabla_x \cdot \boldsymbol{\mathcal{F}}, \tag{12.17}$$

where $\overline{v}^*$ is the residual meridional velocity, $\boldsymbol{\mathcal{F}}$ is the Eliassen–Palm flux and $\nabla_x \cdot$ is the divergence in the meridional plane. In the barotropic case $\overline{v}^* = 0$ and

$$\boldsymbol{\mathcal{F}} = -\mathbf{j}\,\overline{u'v'}. \tag{12.18}$$

If the momentum flux is primarily the result of interacting waves, then the EP flux is equal to the group velocity multiplied by the flux. That is,

$$\mathcal{F}^y = -\overline{u'v'} = c_g^y \mathcal{P}, \tag{12.19}$$

where $\mathcal{P}$ is the pseudomomentum given by

$$\mathcal{P} = \frac{\overline{\zeta'^2}}{2\gamma}, \tag{12.20}$$

where $\gamma = \beta - \overline{u}_{yy}$, and this is usually positive in Earth's atmosphere, in which case $\mathcal{P}$ is also positive. The zonal momentum equation, (12.2) with drag set to zero, and the Eliassen–Palm relation (9.29) become, respectively,

$$\frac{\partial \overline{u}}{\partial t} - \frac{\partial}{\partial y}(c_g^y \mathcal{P}) = 0, \qquad \frac{\partial \mathcal{P}}{\partial t} + \frac{\partial}{\partial y}(c_g^y \mathcal{P}) = 0. \tag{12.21a,b}$$

Evidently, $\overline{u} + \mathcal{P}$ is constant, as we saw in (9.72) if dissipation is zero.

Now suppose that we create a disturbance and then let the fluid evolve freely. The disturbance generates Rossby waves whose group velocity is directed away, and from (12.21b) the wave activity density $\mathcal{P}$ diminishes in the region of the disturbance (and increases elsewhere). However, from (12.21a), the zonal velocity *increases* in the region of the disturbance, and an eastward flow is generated. The argument depends on the sign of γ,

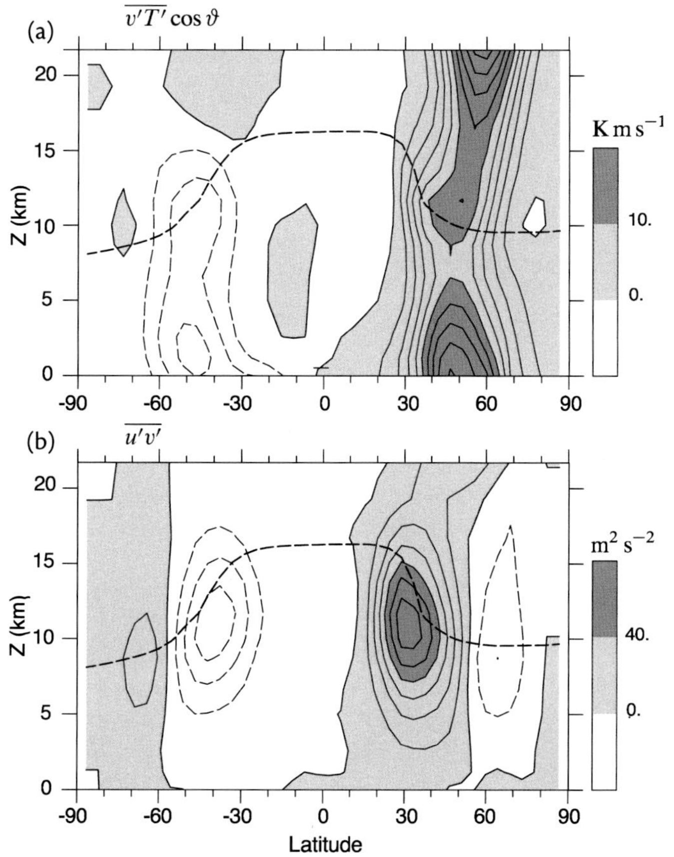

Fig. 12.4: (a) The average meridional eddy heat flux and (b) the eddy momentum flux in the northern hemisphere winter (DJF). The ordinate is log-pressure, with scale height $H = 7.5$ km. Positive fluxes are shaded, and the dashed line marks the thermal tropopause.

The eddy heat flux (contour interval 2 K m s^{-1}) is largely poleward and downgradient in both hemispheres. The eddy momentum flux (contour interval 10 m^2 s^{-2}) is upgradient and *converges* in midlatitudes in the region of the mean jet, leading to eastward surface winds.

and thus on β. If γ were negative (e.g., if Earth's rotation were reversed), then a westward flow would be generated in the region of the disturbance. Unlike the argument of the previous section, we have not had to explicitly use the form of the Rossby wave dispersion relation, but the effect nevertheless clearly depends on β.

IV. ♦ A statistically steady state

Now consider the maintenance of a statistically steady state, and the vorticity flux necessary to produce it. The vorticity equation is

$$\frac{\partial \zeta}{\partial t} + \boldsymbol{u} \cdot \nabla \zeta + v\beta = F_\zeta - D_\zeta, \tag{12.22}$$

where F_ζ and D_ζ are forcing and dissipation of vorticity. The forcing comes from the stirring, and the dissipation might be a linear drag and/or a viscous effect. Linearize (12.22) about a mean zonal flow to give

$$\frac{\partial \zeta'}{\partial t} + \overline{u}\frac{\partial \zeta'}{\partial x} + \gamma v' = F'_\zeta - D'_\zeta, \tag{12.23}$$

where $\gamma = \beta - \partial^2\overline{u}/\partial y^2$ is the meridional gradient of absolute vorticity, which as before is usually positive in both Northern and Southern

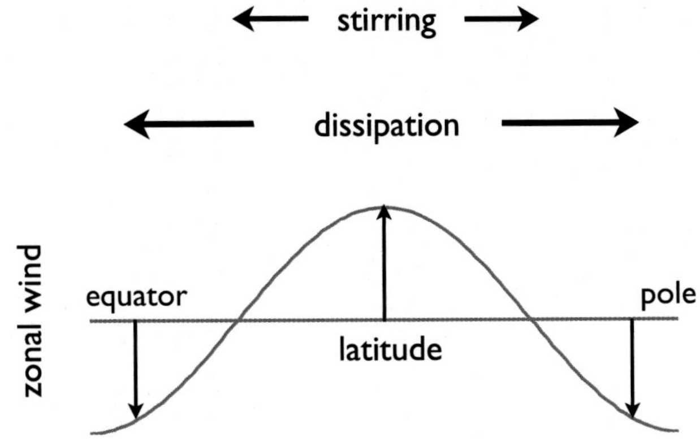

Fig. 12.5: Mean flow generation by a meridionally confined stirring. Rossby waves propagate away from the source region and the distribution of pseudomomentum dissipation is consequently broader than that of pseudomomentum forcing, and the sum of the two leads to the zonal wind distribution shown, with positive (eastward) values in the region of the stirring.

Hemispheres. Now multiply (12.23) by ζ'/γ and zonally average, assuming that $\overline{u}_{yy}$ is small compared to β or varies only slowly in time, to form the forced-dissipative pseudomomentum equation,

$$\frac{\partial \mathcal{P}}{\partial t} + \overline{v'\zeta'} = \frac{1}{\gamma}(\overline{\zeta' F'_\zeta} - \overline{\zeta' D'_\zeta}), \tag{12.24}$$

where $\mathcal{P} = \overline{\zeta'^2}/2\gamma$ is the pseudomomentum.

If we combine (12.24) with the zonal momentum equation, (12.6), we obtain

$$\frac{\partial \overline{u}}{\partial t} + \frac{\partial \mathcal{P}}{\partial t} = -r\overline{u} + \frac{1}{\gamma}\left(\overline{\zeta' F'_\zeta} - \overline{\zeta' D'_\zeta}\right). \tag{12.25}$$

This is just the forced-dissipative version of (9.72). In a statistically steady state we have

$$r\overline{u} = \frac{1}{\gamma}\left(\overline{\zeta' F'_\zeta} - \overline{\zeta' D'_\zeta}\right). \tag{12.26}$$

The terms on the right-hand side represent the stirring and dissipation of vorticity, and integrated over latitude their sum will vanish, or otherwise the pseudomomentum budget cannot be in a steady state. However, let us suppose that forcing is confined to midlatitudes, and let us further suppose that Rossby waves are generated by the forcing and propagate away. In the forcing region, the first term on the right-hand side of (12.26) will be larger than the second, and an eastward mean flow will be generated. Away from the direct influence of the forcing, the dissipation term will dominate and westward mean flows will be generated, as sketched in Fig. 12.5. Thus, *on a β-plane or on the surface of a rotating sphere an eastward mean zonal flow can be maintained by a vorticity stirring that imparts no net momentum to the fluid.*

It is crucial to this argument that the dissipation has a broader latitudinal distribution than the forcing: if all the dissipation occurred in the region of the forcing then from (12.26) no mean flow would be generated. This broadening arises via the action of Rossby waves that are generated in the forcing region and that propagate meridionally before dissipating,

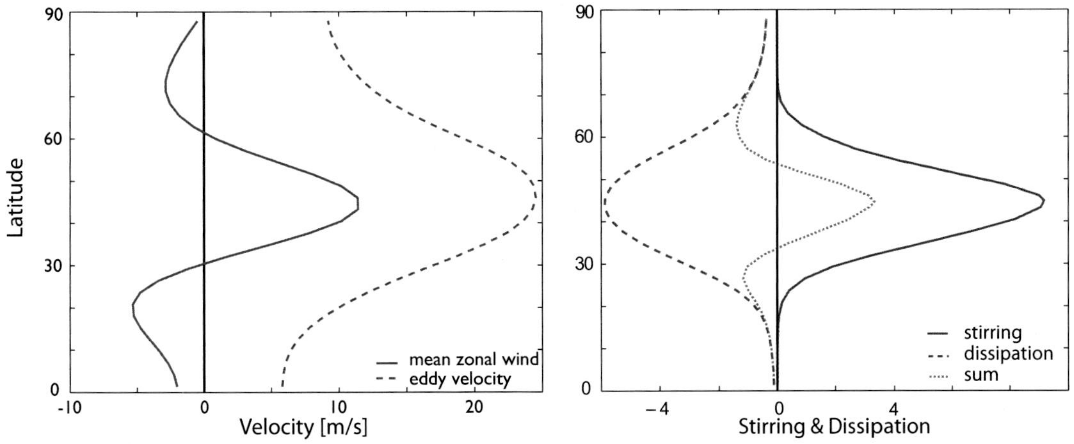

Fig. 12.6: Left: The time and zonally-averaged wind (solid line) and the rms (eddy) speed obtained by an integration of the barotropic vorticity equation (12.27) on the sphere. The fluid is stirred in midlatitudes by a wavemaker that supplies no net momentum. Right: The pseudomomentum stirring (solid line, $\overline{F_\zeta'\zeta'}$), dissipation (red dashed line, $\overline{D_\zeta'\zeta'}$) and their sum (dotted).

Momentum converges in the stirring region leading to an eastward jet with a westward flow to either side. Because Rossby waves propagate away from the stirred region before breaking, the distribution of dissipation is broader than the forcing, resulting in an eastward jet, with westward flow on either side.

as described in the previous subsection, so allowing the generation of a mean flow.

12.1.2 A Numerical Example

We conclude from the above arguments that momentum converges into a rapidly rotating flow that is stirred in a meridionally localized region. To illustrate this, we numerically integrate the barotropic vorticity equation on the sphere, with a meridionally localized stirring term; explicitly, the equation that is integrated is

$$\frac{\partial \zeta}{\partial t} + J(\psi, \zeta) + \beta \frac{\partial \psi}{\partial x} = -r\zeta + \kappa \nabla^4 \zeta + F_\zeta. \tag{12.27}$$

The first term on the right-hand side is a linear drag, parameterizing momentum loss in an Ekman layer. The second term, $\kappa\nabla^4\zeta$, removes small scale noise and has a negligible impact at large scales. The forcing term F_ζ is confined to a zonal strip of about 15° meridional extent, centred at about 45°N, that is statistically zonally uniform and that spatially integrates to zero. Within that region it is a random stirring with spatial decorrelation scale corresponding to about wavenumber 8, thus mimicking weather scales. It provides no net source of vorticity or momentum, but it is a source of pseudomomentum because $\overline{F_\zeta\zeta} > 0$. The results of a numerical integration of (12.27) are illustrated in Fig. 12.6. An eastward jet forms in the vicinity of the forcing, with westward flow on either side.

12.2 The Ferrel Cell

We are now in a position to understand the Ferrel Cell (labelled F in Fig. 11.3) the indirect overturning cell in midlatitudes in which cold air seemingly rises and cold air sinks. It is 'indirect' because cool air apparently rises in high latitudes, moves equatorward and sinks in the subtrop-

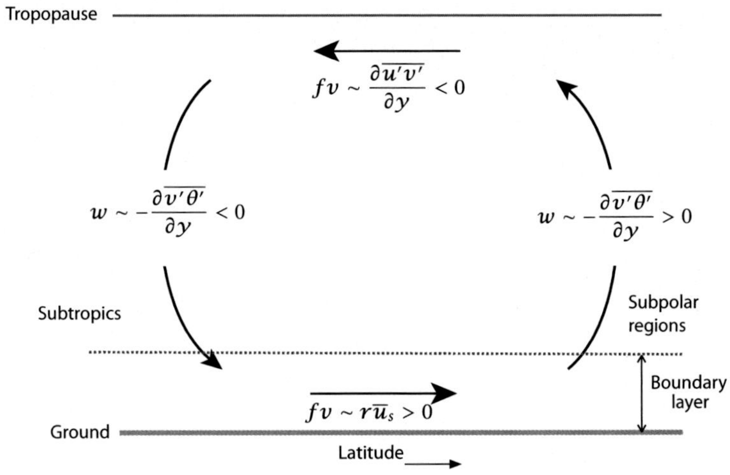

Fig. 12.7: The eddy-driven Ferrel Cell, from an Eulerian point of view. Above the planetary boundary layer the mean flow is largely in balance with the eddy heat and momentum fluxes. The lower branch of the Ferrel Cell is largely confined to the boundary layer, where it is in a frictional–geostrophic balance.

ics. We first discuss the cell from an Eulerian perspective and then from a 'residual' perspective.

Why should the Ferrel circulation exist? The answer, in short, is that it is there to balance the eddy momentum convergence of the midlatitude eddies and it can be thought of as being driven by those eddies. To see this we begin with the zonally-averaged zonal momentum equation,

$$\frac{\partial \overline{u}}{\partial t} - (f + \overline{\zeta})\overline{v} + \overline{w}\frac{\partial \overline{u}}{\partial z} = -\frac{\partial}{\partial y}\overline{u'v'} - \frac{\partial}{\partial z}\overline{u'w'} + \frac{1}{\rho}\frac{\partial \tau}{\partial z}. \tag{12.28}$$

The Ferrel Cell is named for William Ferrel (1817–1891), an American teacher and meteorologist, sometimes regarded as the world's first geophysical fluid dynamicist. He put forward various theories of the overturning circulation in the mid-nineteenth century and was probably the first scientist to appreciate the importance of the conservation of angular momentum (as opposed to linear momentum) in the overturning circulation.

The terms on the right-hand side are eddy fluxes of vertical and horizontal momentum and the frictional force, $\rho^{-1}\partial\tau/\partial z$ where τ is the zonal stress. All the terms involving vertical velocity are relatively small and we will neglect them, and at low Rossby number $|f| \gg \zeta$. For steady flow the equation then becomes

$$-f\overline{v} = -\frac{\partial}{\partial y}(\overline{u'v'}) + \frac{1}{\rho}\frac{\partial \tau}{\partial z}. \tag{12.29}$$

At the surface the stress is (to a decent approximation) proportional to the zonal wind itself and may be represented by a drag, $\tau = \rho_s r\overline{u}_s$, where r is a constant and ρ_s and $\overline{u}_s$ are the density and zonally-averaged zonal wind, respectively, at the surface. The stress dominates the eddy fluxes at the surface but falls away with height so that it is important only in the lowest kilometre or so. Above this frictional layer, the eddy momentum flux convergence is balanced by the Coriolis force on the meridional flow, and in midlatitudes (from about 30° to 70°) the eddy momentum flux divergence is negative in both hemispheres, as seen in Fig. 12.4. The cause of this divergence is Rossby wave propagation away from the region of baroclinic activity, as discussed in Section 12.1.1 and in Fig. 12.3. Using (12.29) we see the zonally-averaged meridional flow in the free atmosphere must be *equatorward*, as illustrated schematically in Fig. 12.7.

The flow cannot be equatorward everywhere, simply by mass continuity, and the return flow occurs largely in the Ekman layer, of depth d

say. Here the eddy balance is between the Coriolis term and the frictional term, and integrating over this layer and taking the density there to be constant gives

$$-fV \approx -r\overline{u}_s, \tag{12.30}$$

where $V = \int_0^d \overline{v}\,dz$ is the meridional transport in the boundary layer of height d, above which the stress vanishes. The surface return flow is poleward (i.e., $V > 0$ in the Northern Hemisphere) producing an eastward Coriolis force and an eastward surface flow. In this picture, then, the midlatitude eastward zonal flow at the surface is a consequence of the poleward flowing surface branch of the Ferrel Cell, this poleward flow being required by mass continuity given the equatorward flow in the upper branch of the cell. Seen this way, the Ferrel Cell is responsible for bringing the midlatitude eddy momentum flux convergence to the surface where it may be balanced by friction, as in Fig. 12.7.

A direct way to see that the surface flow must be eastward, given the eddy momentum flux convergence, is to vertically integrate (12.29) from the surface to the top of the atmosphere. By mass conservation, the Coriolis term vanishes (i.e., $\int_0^\infty f\rho\overline{v}\,dz = 0$) and we obtain

$$\int_0^\infty \frac{\partial}{\partial y}(\overline{u'v'})\rho\,dz = [\tau]_0^\infty = -r\rho_s\overline{u}_s. \tag{12.31}$$

That is, the surface wind is proportional to the vertically integrated eddy momentum flux convergence. Because there *is* a momentum flux convergence, the left-hand side is negative and the surface winds are positive, or eastward.

12.2.1 The Eulerian Meridional Overturning Circulation

We can obtain an explicit equation for the overturning circulation by combining the momentum equation and the thermodynamic equation using thermal wind balance. Neglecting all but the largest terms, the zonally-averaged zonal momentum equation may be written

$$\frac{\partial\overline{u}}{\partial t} - f\overline{v} = M, \tag{12.32a}$$

where $M = -\partial_y(\overline{u'v'}) + \rho^{-1}\partial\tau/\partial z$ contains the main eddy flux and frictional terms. At a similar level of approximation let us write the thermodynamic equation as

$$\frac{\partial\overline{b}}{\partial t} + N^2\overline{w} = J, \tag{12.32b}$$

where $J = Q_b - \partial_y(\overline{v'b'})$ is the sum of the heating, Q_b, and eddy forcing. We are assuming, as in quasi-geostrophic theory, that the mean stratification, N^2, is fixed and $\overline{b}$ represents only the (zonally averaged) deviations from this. Finally, we use the mass continuity equation to define a meridional streamfunction Ψ; that is

$$\frac{\partial\overline{v}}{\partial y} + \frac{\partial\overline{w}}{\partial z} = 0 \quad \text{allows} \quad \overline{w} = \frac{\partial\Psi}{\partial y}, \quad \overline{v} = -\frac{\partial\Psi}{\partial z}. \tag{12.33a,b}$$

We now use the thermal wind relation, $f\partial\overline{u}/\partial z = -\partial\overline{b}/\partial y$, to eliminate time derivatives in (12.32a) and (12.32b), giving

$$f^2\frac{\partial^2\Psi}{\partial z^2} + N^2\frac{\partial^2\Psi}{\partial y^2} = f\frac{\partial M}{\partial z} + \frac{\partial J}{\partial y}, \tag{12.34}$$

where

$$\frac{\partial M}{\partial z} = -\frac{\partial}{\partial z}\left(\frac{\partial(\overline{u'v'})}{\partial y}\right) + \frac{\partial}{\partial z}\left(\frac{1}{\rho}\frac{\partial \tau}{\partial z}\right), \quad \frac{\partial J}{\partial y} = \frac{\partial Q_b}{\partial y} - \frac{\partial^2}{\partial y^2}(\overline{v'b'}). \tag{12.35a,b}$$

Equation (12.34) is a linear equation for the overturning streamfunction, one that holds even if the flow is not in a steady state The equation is not especially accurate, especially in low latitudes where the small Rossby number assumption is poor, but it does capture the essential dynamics producing the Ferrel Cell and even, with less accuracy and different types of terms on the right-hand side, the Hadley Cell. Evidently, the midlatitude overturning circulation is forced by eddy fluxes of heat and momentum, as well as heating and other terms that might appear on the right-hand sides of (12.32a) and (12.32b).

Given the sign convention we have chosen for Ψ, a positive streamfunction corresponds to a clockwise circulation, with (in the Northern Hemisphere) ascending motion at low latitudes and descending motion at high. Furthermore, since the left-hand side is a sum of second derivatives then, roughly speaking, a negative term on the right-hand side produces a positive value of Ψ and a clockwise circulation. Given this, let us compare the Hadley and Ferrel Cells (using Northern Hemisphere notation):

- The horizontal gradient of the thermodynamic forcing — heating at low latitudes, cooling in high latitudes, and $\partial Q_b/\partial y < 0$ — produces a positive streamfunction and drives a direct circulation at all latitudes (i.e., warm air rising at low latitudes, sinking at low), for both Hadley and Ferrel Cells. This term is dominant in low latitudes, producing the Hadley Cell, which is a direct cell.
- The eddy heat flux divergence, shown in Fig. 12.4, is positive in the subtropics (causing cooling) and negative in mid- to high-latitudes, causing heating. This leads to descent in the subtropics and ascent in higher latitudes, as in Fig. 12.7.
- In midlatitudes, the vertical gradient of the horizontal eddy momentum divergence is generally positive with $\partial_z(\overline{u'v'}) > 0$. Using Fig. 12.4 to look at the latitudinal gradient suggests $\partial_z(\partial_y(\overline{u'v'})) < 0$ over most of the midlatitudes. This convergence produces a negative streamfunction, with ascent at high latitudes and descent in the subtropics. Thus, both the eddy momentum flux and the eddy heat flux act in the same fashion to produce an eddy-driven cell, the Ferrel Cell.

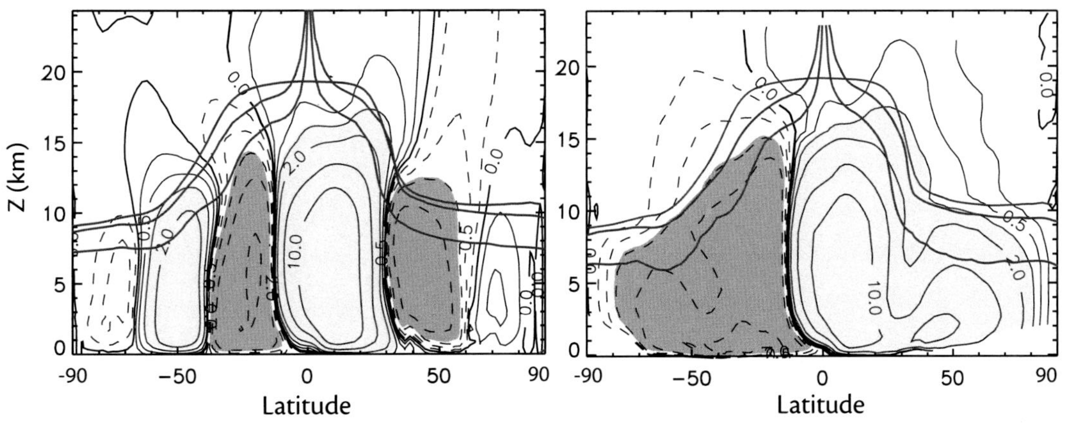

Fig. 12.8: Left: The observed zonally-averaged, Eulerian-mean, streamfunction in Northern Hemisphere winter (DJF, 1994–1997). Negative contours are dashed, and values greater or less than $10^{10}\ \mathrm{kg\,s^{-1}}$ (10 Sv) are shaded, darker for negative values. The circulation is clockwise around the lighter shading. The three thick solid lines indicate various measures of the tropopause. Right: The residual meridional mass streamfunction. (Adapted from Juckes (2001).)

In a steady state we have, from (12.32b),

$$w = \frac{1}{N^2}\left[Q_b - \frac{\partial(\overline{v'b'})}{\partial y}\right]. \tag{12.36a}$$

Similarly, from the momentum equation the horizontal velocity and eddy momentum fluxes are related by, in a steady state,

$$-f\overline{v} = -\frac{\partial(\overline{u'v'})}{\partial y} + \frac{1}{\rho}\frac{\partial \tau}{\partial z}. \tag{12.36b}$$

Figure 12.7 (and Fig. 12.4) shows that both eddy heat and momentum fluxes produce an overturning circulation in the same sense as the observed Ferrel Cell. However, these fluxes are not independent of each other: it is a combination of them, and in particular the potential vorticity flux, that is really responsible for the overturning circulation, as we now see.

12.3 ♦ The Residual Ferrel Cell

A revealing way to describe the meridional overturning is by way of the residual circulation, as discussed in Section 9.3. Written in residual form on the f-plane, as in (9.50), the zonal momentum and buoyancy equations are

$$\frac{\partial \overline{u}}{\partial t} - f_0\overline{v}^* = \overline{v'q'} + F_u, \tag{12.37a}$$

$$\frac{\partial \overline{b}}{\partial t} + N^2\overline{w}^* = Q_b. \tag{12.37b}$$

In these equations F_u is a frictional term, Q_b is the heating term and $\overline{v'q'}$ is the eddy potential vorticity flux. The residual velocities, $\overline{v}^*$ and $\overline{w}^*$, are related to their Eulerian counterparts by

$$\overline{v}^* = \overline{v} - \frac{\partial}{\partial z}\left(\frac{1}{N^2}\overline{v'b'}\right), \qquad \overline{w}^* = \overline{w} + \frac{\partial}{\partial y}\left(\frac{1}{N^2}\overline{v'b'}\right). \tag{12.38}$$

The residual velocities are related to the residual streamfunction by

$$(\overline{v}^*, \overline{w}^*) = \left(-\frac{\partial \psi^*}{\partial z}, \frac{\partial \psi^*}{\partial y}\right), \tag{12.39}$$

where the residual streamfunction is, in turn, related to the Eulerian streamfunction by

$$\psi^* \equiv \psi_m + \frac{1}{N^2}\overline{v'b'}. \tag{12.40}$$

We see from (12.37) that only the eddy flux of potential vorticity is needed to evolve the system, not the momentum flux and buoyancy flux separately. This is as it must be, since under quasi-geostrophic scaling *all* the dynamical fields may be inferred from the potential vorticity alone. The meridional overturning circulation of the residual flow is obtained by using thermal wind balance, $f_0 \partial \overline{u}/\partial z = -\partial \overline{b}/\partial y$, to eliminate time derivatives from (12.37), giving

$$f_0^2 \frac{\partial^2 \psi^*}{\partial z^2} + N^2 \frac{\partial^2 \psi^*}{\partial y^2} = f_0 \frac{\partial}{\partial z}\overline{v'q'} + f_0 \frac{\partial F_u}{\partial z} + \frac{\partial Q_b}{\partial y}. \tag{12.41}$$

The left-hand side has the same form as (12.34), and the diabatic and frictional terms appear in the same way on the right-hand side, but the only eddy fluxes that appear are the potential vorticity fluxes.

In Earth's atmosphere, the residual overturning circulation is a *direct* circulation over the whole hemisphere, as may be in inferred directly from (12.37b). In a statistically steady state the balance is just $\overline{w}^* = Q_b/N^2$, with rising air (positive $\overline{w}^*$) in regions of heating and descending motion in regions of cooling, and the difference between the residual and Eulerian circulations is clearly seen in Fig. 12.8. There is no conventional Ferrel Cell in the residual circulation, although the Hadley Cell and the midlatitude cell can be clearly differentiated: the right-hand panel of Fig. 12.8 shows the Hadley Cell extending to much greater height (about 16 km) versus a midlatitude circulation extending to about 8 km. In midlatitudes, the steady-state balance in the momentum equation, (12.37), in the upper troposphere is $-f_0\overline{v}^* = \overline{v'q'}$, and the potential vorticity flux is negative, particularly in winter, as seen in Fig. 12.9, producing a polewards flow.

12.3.1 Potential Vorticity Fluxes and Surface Winds

We may infer the surface winds from (12.37). If the friction term on the right-hand side has the form $F_u = \rho^{-1}\partial\tau/\partial z$ then integrating the equation over the depth of the atmosphere gives, in a statistically steady state,

$$-\int f_0 \overline{v}^* \rho \,\mathrm{d}z = \int \rho \overline{v'q'} \,\mathrm{d}z - \tau_s, \tag{12.42}$$

where τ_s is the surface stress. The left-hand side vanishes because the residual circulation, like the Eulerian circulation, satisfies a mass conservation equation, as can be seen by cross-differentiating (12.38). If we

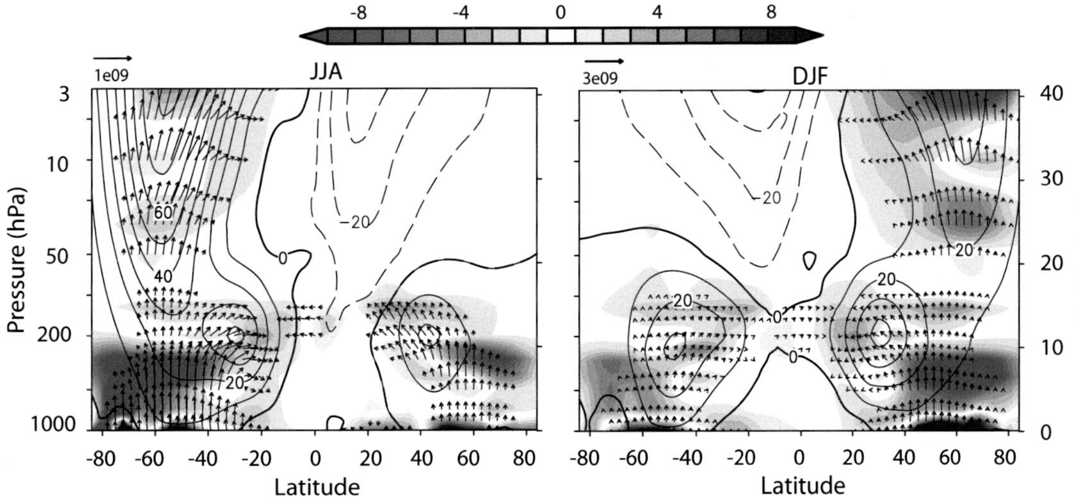

Fig. 12.9: The Eliassen–Palm (EP) flux vectors (arrows), the EP flux divergence (shading, equal to the potential vorticity flux, $\overline{v'q'}$) and the zonally averaged zonal wind (contours) for Northern Hemisphere summer and winter. The tropospheric EP fluxes are of similar magnitudes in the summer and winter hemispheres, but are almost zero in the summer in the stratosphere. The convergence of the EP flux (negative potential vorticity flux) at mid- and high latitudes in upper troposphere and winter stratosphere leads to poleward residual flow. (Figure kindly made by Blanca Ayarzaguena Porras.)

(again) parameterize the surface stress by $\tau_s = \rho_s r\overline{u}$ then (12.42) becomes

$$r\rho_s\overline{u} = \int \rho\overline{v'q'}\,\mathrm{d}z. \tag{12.43}$$

This result is equivalent to that of (12.31), which can be written in the form

$$r\rho_s\overline{u} = \int \rho\overline{v'\zeta'}\,\mathrm{d}z. \tag{12.44}$$

The equivalence arises because the potential vorticity flux is equal to the flux of relative vorticity plus the vertical derivative of a buoyancy flux, and the latter vanishes on vertical integration. Thus, the surface winds are given by either the vertical integral of the vorticity fluxes or the vertical integral of the potential vorticity fluxes. Although the potential vorticity fluxes are negative over much of the atmosphere they are strongly positive near the surface and this produces eastward surface winds in mid-latitudes.

12.4 Stratosphere

The stratosphere is the region of the atmosphere above the troposphere and below the mesosphere; thus, it extends from the tropopause at a height of about 8–16 km, or a pressure of around 200–300 hPa, to the stratopause at about 50 km or 1 hPa (see Fig. 11.2). It is a region of stable stratification — stratosphere means layered sphere and it has much less vertical motion that the troposphere, which is the 'turning sphere'. Above the stratosphere lie the mesosphere (approximately between 50 and 90 km), the thermosphere (from about 90 to 600 km, depending on solar activity) and finally the exosphere, where the atmosphere fades into the endless vacuum of space.

12.4.1 Large-Scale Structure

In the troposphere the stratification is determined by dynamical processes — convection at low latitudes and also baroclinic instability at high latitudes — and the tropopause is the height to which the dynamical activity reaches. In the stratosphere the temperature is determined to a much greater degree by radiative processes and the dynamics are relatively slow. In fact the temperature increases with height, and this is due to a layer of ozone that absorbs solar radiation in the mid-stratosphere between about 20 and 30 km. If there were no ozone there would still be a tropopause and a stratosphere, but the temperature in the stratosphere would increase much less with height.

The radiative-equilibrium temperature for January is illustrated in Fig. 12.10. This temperature is that which would ensue without any stratospheric fluid motion, although we take the distribution of absorbers (such as ozone) to be those present in the actual, moving, atmosphere. There is quite a strong lateral gradient in the winter hemisphere and a weaker reversed temperature in the summer hemisphere, and in fact the part of the stratosphere with the highest radiative equilibrium temperature is the upper-stratosphere summer pole, at around 1 hPa. The actual observed zonally averaged temperature and zonal-wind structure are plotted in Fig. 12.11. From these figures we infer the following:

- The stratosphere is very stably stratified, with a typical lapse rate corresponding to $N \approx 2\times10^{-2}$ s, about twice that of the troposphere on average.
- In the summer the solar absorption at high latitudes leads to a reversed temperature gradient (warmer pole than equator) and, by thermal wind, a negative vertical shear of the zonal wind. Over much of the summer stratosphere the mean zonal winds are negative (westward).
- In winter high latitudes receive very little solar radiation and there is a strong meridional temperature gradient and consequently a strong vertical shear in the zonal wind. Nevertheless, this temperature gradient is significantly weaker than the radiative equilibrium temperature gradient, implying a poleward heat transfer by the fluid motion in the stratosphere.

The absorption of solar radiation by ozone between 20 and 50 km is largely responsible for the increase in temperature with height in the stratosphere, but even without ozone there would still be a tropopause separating a dynamical troposphere from a stratosphere that is closer to being in radiative equilibrium.

Let us now explore some of the causes and consequences of this structure.

12.4.2 The Quasi-Horizontal Circulation

Baroclinic instability

One consequence of the stable stratification in the stratosphere is that baroclinic instability is inhibited. In the stratosphere $N \sim 2 \times 10^{-2}\,\mathrm{s}^{-1}$, and using a height scale of 20 km gives a value of the deformation radius NH/f of about 4000 km, as opposed to the canonical value of 1000 km in the troposphere. Thus, even with the same horizontal temperature gradient as the troposphere, a typical instability scale (of an isolated strato-

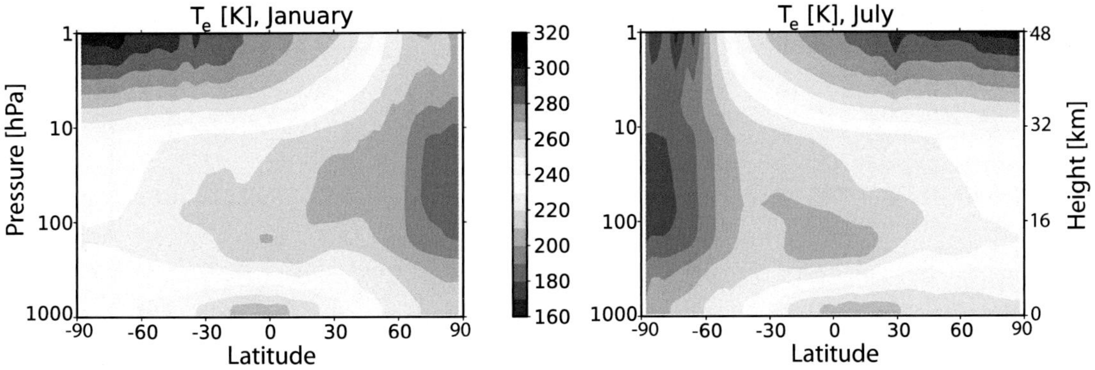

Fig. 12.10: The zonally averaged radiative-equilibrium temperature in January and July; that is, the temperature that would arise in the absence of fluid motion in the stratosphere, but with the actual distribution of radiative absorbers. The stratosphere is highly stratified in the summer hemisphere, with temperature increasing with height, but is fairly uniform with height in the inter hemisphere. Above about 50 km the equilibrium temperature generally diminishes with height.

sphere) would be large, perhaps around wavenumber 2. The stratospheric growth rate would then be much less than in the troposphere: the Eady growth rate is given by $\sigma_E \equiv 0.31\Lambda H/L_d = 0.31U/L_d$, where Λ is the shear, giving a growth rate several times smaller than its tropospheric counterpart. Furthermore, if we think of the stratosphere in isolation then there is no clear reversal of the potential vorticity gradient and no real opportunity for counter-propagating edge waves or Rossby waves to interact in the stratosphere, and the stratosphere is then baroclinically stable.

Of course, if baroclinic instability has a modal form in the vertical then the instability has the same horizontal scale and grows at the same rate in the stratosphere as the tropospheric one — it is the same mode! But in this case the higher lapse rate suppresses the amplitude of the stratospheric instability, and the amplitude of the instability falls significantly in the stratosphere. No matter which way we look at it, *in situ* baroclinic instability is not a significant driver of motion in the stratosphere. Rather, the driving for the stratosphere largely comes from below, from the gravity and (especially) the Rossby waves that propagate up from the troposphere and break in the stratosphere, driving both quasi-horizontal flow and an overturning circulation.

Quasi-two-dimensional flow

The high stratification of the stratosphere inhibits vortex stretching, and the quasi-geostrophic flow becomes quasi-two-dimensional. To see this we first write down the quasi-geostrophic equation in its simplest, constant density, unforced, form, namely

$$\frac{\partial q}{\partial t} + \boldsymbol{u} \cdot \nabla q = 0, \quad \text{where} \quad q = \nabla^2 \psi + \beta y + \frac{\partial}{\partial z}\left(\frac{f_0^2}{N^2}\frac{\partial \psi}{\partial z}\right). \qquad (12.45\text{a,b})$$

Now, if N is sufficiently large the third term on the right-hand side of (12.45b) is small compared to the first. The condition for this is that the deformation radius should be much larger than the horizontal scale L, or

$$\frac{N^2 H^2}{f_0^2} \gg L^2. \qquad (12.46)$$

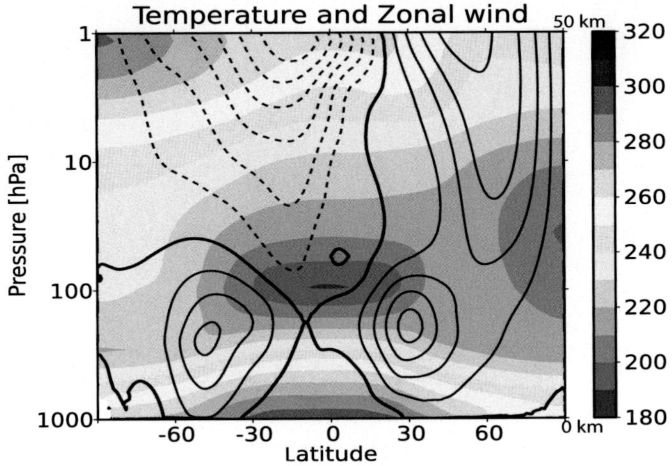

Fig. 12.11: The zonally averaged temperature and zonal wind in January. Colour shading shows the temperature and the contours are the zonal wind, with a contour interval of 10 m s^{-1} and negative values are dashed. In both hemispheres the stratosphere is stably stratified.

The deformation radius in the stratosphere is about 4000 km, so that for all but the largest scales (12.45) reduces to the equation for two-dimensional flow,

$$\frac{\partial q}{\partial t} + \boldsymbol{u} \cdot \nabla q = 0, \qquad \text{where} \qquad q = \zeta = \nabla^2 \psi + \beta y. \tag{12.47}$$

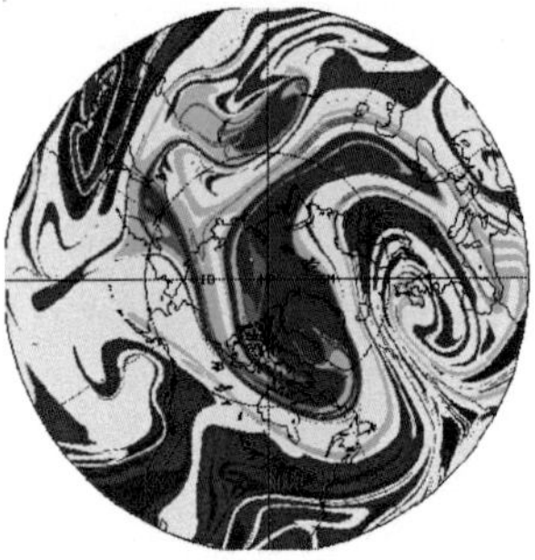

Fig. 12.12: The tracer distribution in the northern hemisphere lower stratosphere on 28 January 1992. The tracer was initialized on 16 January by setting it equal to the potential vorticity field calculated from an observational analysis, and then advected for 12 days by the observed wind fields

The observed stratospheric flow shows the typical features of two-dimensional turbulence. Figure 12.12 shows the lower stratosphere, obtained by using a numerical model to integrate forward a tracer initially set equal to potential vorticity. We see a midlatitude stratosphere characterized by vortices forming and then stretching into filaments and tendrils, and Rossby waves breaking — all the features of an enstrophy cascade. For this reason, the midlatitude region is sometimes known as the *surf zone*. The surf zone does not usually extend all the way to the pole, for in winter dense cold air over the pole forms itself into a cyclonic vortex, even more visible in the colder Southern Hemisphere. The boundary of the vortex, as measured by the value of the potential vorticity or of the tracer, is quite sharp with the value of PV often jumping by a factor of 2 or so, and the vortex is quite persistent — a near-permanent feature of the winter hemisphere.

Within the polar vortex potential vorticity tends to homogenize, and once formed the main communication that the vortex has with the surf zone is via occasional wave breaking at its boundary. It is interesting that, although the potential vorticity gradient is strong at the edge of the vortex, the exchange of properties is weak, implying a failure of notions of diffusion, or at least diffusion with a constant value of diffusivity; the edge of the vortex is a *mixing barrier*. Stable as it is, the polar vortex is nevertheless sometimes disrupted by wave activity from below; this tends to occur when the wave activity itself is quite strong, and when the mean conditions are such as to steer that wave activity polewards. Occasionally, this activity is sufficiently strong so as to cause the vortex to break down,

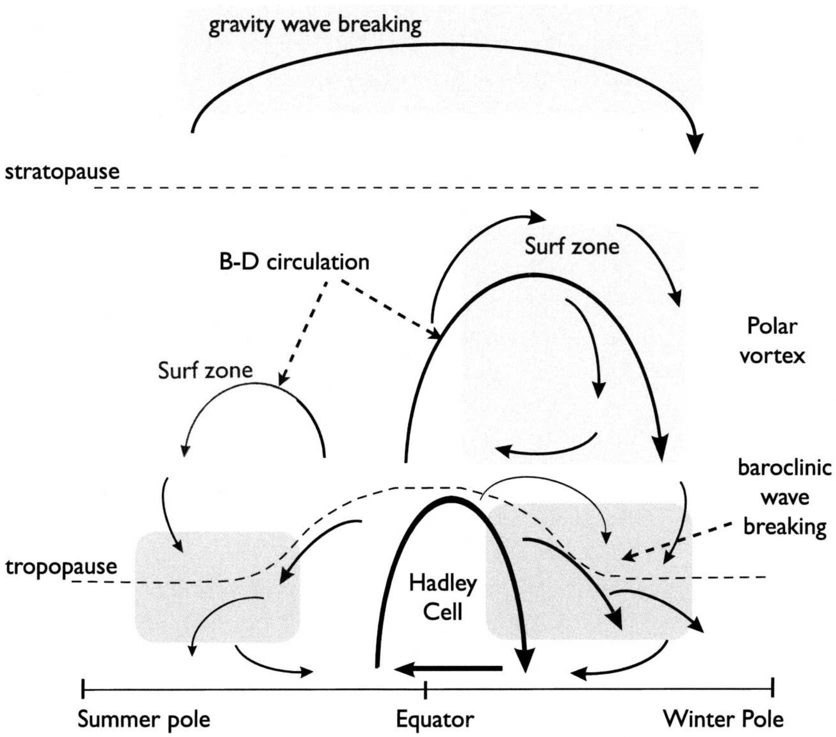

Fig. 12.13: A sketch of the residual mean meridional circulation of the atmosphere. The solid arrows indicate the residual circulation (B-D for Brewer–Dobson) and the shaded areas the main regions of wave breaking. In the surf zone the breaking is mainly that of planetary Rossby waves, and in the troposphere and lower stratosphere the breaking is that of baroclinic eddies. The surf zone and residual flow are much weaker in the summer hemisphere. Only in the Hadley Cell does the residual circulation consist mainly of the Eulerian mean — elsewhere the eddy component dominates.

or to split into two smaller vortices, and so allow warm midlatitude air to reach polar latitudes — an event known as a *stratospheric sudden warming*.

12.4.3 The Stratospheric Overturning Circulation

The stratosphere not only has a quasi-horizontal circulation but it has an overturning circulation known as the *Brewer–Dobson circulation*, sketched in Fig. 12.13. The circulation is weaker than the corresponding circulations (e.g., the Ferrel Cell) in the troposphere because the high stratification of the stratosphere inhibits vertical motion, but nevertheless it is an important feature of the stratosphere.

It is the *residual* meridional overturning circulation (the RMOC) that was identified by Brewer and Dobson and that is important in the stratosphere, because it is this circulation that more nearly represents the Lagrangian circulation of the atmosphere and that carries tracers such as ozone and water vapour, and the equation describing this circulation is just (12.41). The most important terms on the right-hand side are the diabatic term, $\partial Q_b/\partial y$ and the potential vorticity flux term, $f_0\partial_z\overline{v'q'}$. Of these, in the steady-state limit the potential vorticity flux term tends to dominate on the large-scale, because the solar heating is closely balanced by the infra-red cooling and the net heating is small.

In the extra-tropics much of the wave forcing arises from the breaking of Rossby waves that propagate up from the troposphere, as described in Section 6.4.2. When they break high in the stratosphere they try to *decelerate* the flow — that is, they provide a westward acceleration. In midlatitudes this is balanced by the Coriolis force on the residual meridional velocity and in a statistically steady state the zonal momentum equation becomes $-f_0\overline{v}^* \approx \overline{v'q'}$. The potential vorticity flux is negative for breaking Rossby waves (in the Northern Hemisphere), inducing a poleward residual circulation. In terms of (12.41), the term $f_0\partial_z\overline{v'q'}$ is negative, inducing a positive circulation $\psi^* > 0$ and a clockwise circulation, with rising air at low latitudes and sinking at high. The breaking of Rossby waves is, of course, responsible for the surf zone discussed in the previous section.

A. W. Brewer deduced upward motion into the stratosphere at low latitudes based on the water vapour distribution in the late 1940s, while just a few years later G. M. B. Dobson deduced a poleward transport within the stratosphere based on the ozone distribution — the circulation takes ozone from the low latitudes toward the poles. The circulation they discovered is now commonly known as the Brewer–Dobson circulation.

Caveats and complexities

Although the above description does capture the essential wave-breaking mechanism driving the stratospheric overturning circulation, the real atmosphere is, unsurprisingly, considerably more complex. Two caveats in particular are important:

(i) In low latitudes the Coriolis force cannot balance the potential vorticity flux and the nonlinear and unsteady terms in the momentum equation become important. In fact, wavebreaking of Rossby and Kelvin waves in the equatorial stratosphere leads to an oscillation of the zonal wind with a period of about two years, known as the 'quasi-biennial oscillation'.

(ii) The Rossby waves cannot propagate into the stratosphere in summer, because the westward winds provide a barrier to upward wave propagation, as discussed in Section 6.4.2. The stratosphere is then almost devoid of wave activity, as seen in Fig. 12.9, and the corresponding Brewer–Dobson circulation is quite weak.

Notes and References

The generation of a zonal flow by rearrangements of vorticity on a background state with a meridional gradient was noted by Kuo (1951). Dickinson (1969) and Thompson (1971, 1980) calculated the momentum transport by Rossby waves, with Thompson explicitly noting that the zonal momentum flux was in the opposite direction to the group velocity. A discussion of the mechanisms of jet formation and the general circulation in Earth's troposphere is provided by Held (2000), and a discussion of stratospheric dynamics is provided by Haynes (2005). A prescient early discussion of the atmospheric general circulation, heralding in the modern era, is that of Eady (1950).

Problems

Atmospheric problems are given at the end of Chapter 13.

CHAPTER

13

Planetary Atmospheres

EARTH IS BUT ONE PLANET. It is, at least for us humans, the most important and most interesting one, but there are very many others. There are at least seven other planets in the Solar System, all of them with atmospheres, and some of those planets have moons with atmospheres — Titan (orbiting Saturn) and Io (orbiting Jupiter) are two. At the time of writing we have observed about 4000 planets, and over 500 multi-planet solar systems, outside the Solar System but within our galaxy. With a little extrapolation we can estimate there are billions (yes, billions) of planets in our galaxy alone, and there are billions of other galaxies in the Universe. Many of these planets will have atmospheres, and some undoubtedly have oceans. And some, almost certainly, have life.

In this chapter our task is to apply geophysical fluid dynamical principles to these planetary atmospheres and thereby try to understand their circulation. The task is a hard one because the variety of planets is enormous — they differ from each other in their mass and composition, their emitting temperature, their size and rotation, whether they are terrestrial or gas giants (terms we define later) or something else entirely, and in a host of other parameters. There is much greater variety in planetary atmospheres than in the stars they orbit, and there can be no single theory of their circulation, no planetary equivalent of astronomy's main sequence of stars that shows the relation between stellar luminosity and effective temperature. On the other hand, the basic principles we have learned in earlier chapters apply to all planetary atmospheres, so we should not be engaged in describing planets one by one (although Earth is a special case). Indeed, because there are so many planets, we *must* look for general principles where we can, else we are hardly doing science at all.

In the sections that follow we aim to give a coherent but introductory treatment of these atmospheres, to put them into context and see how and where Earth's atmosphere might fit into the set of all planetary atmospheres. We begin with a descriptive taxonomy.

13.1 A Taxonomy of Planets

The formation and evolution of planetary atmospheres is a subject unto itself which we won't delve into, and here we give just a brief descriptive overview of some of the more common types. These types are not all orthogonal and a given planet may belong in two or more categories, and the definitions themselves are of disputed authority and subject to debate, and may well evolve over the years ahead. Readers of this book 20 years hence may well read this section with a knowing smile.

Capitalization by example
The Solar System is the system of bodies — Earth, the other planets, minor planets and so on — orbiting the Sun, as well as the Sun itself. Many planets have earth at their surface, but there is only one Earth and the Moon orbits around it. Earth has only one moon whereas Mars has two, Phobos and Deimos. Other solar systems exist in our galaxy, the Milky Way, and planets in these solar systems orbit around their own suns. There are many galaxies in the Universe and there may even be many universes.

Planets. A planet is defined to be a body that orbits its host star directly and is massive enough to be in hydrostatic equilibrium (effectively meaning it has formed under its own gravitational force and has a spheroidal shape) and to dominate its own orbit, clearing it of other bodies. In so far as this definition is official (i.e., as stated by the International Astronomical Union (IAU) in a statement in 2006) it does not apply to bodies in other solar systems, but the definition may usefully be taken to apply more generally. (Many scientists believe the requirement of clearing the orbit is also too restrictive, in which case dwarf planets, defined below, are also planets.) The planets in the Solar System are Mercury, Venus, Earth, Mars, Jupiter, Saturn, Uranus, and Neptune.

Dwarf planets are bodies that are large enough to form under their own gravity but are not able to clear their orbit of other bodies. Pluto is the most famous example, and Eris (in an orbit beyond Neptune) and Ceres (in the asteroid belt) are two others. More generally, *minor planets* are objects that orbit around a star that need not have formed under their own gravity, including dwarf planets and asteroids but excluding true planets and comets. There are hundreds of thousands of minor planets in the Solar System. The IAU also defines *small Solar System bodies* to be objects, including asteroids and comets, that directly orbit the Sun that are too small to be planets or dwarf planets.

Planetary bodies. This is a general term for objects that have formed under their own gravity, including objects that are not big enough to be true planets (under the definition above) and objects that are not in direct orbit around their host star, but excluding stars themselves. The category thus includes all the Solar System planets, the dwarf planets such as Pluto, large natural satellites such as Titan (in orbit around Saturn and with a thick methane atmosphere), Triton (in orbit around Neptune and with a thin, nitrogen atmosphere), and exoplanets.

Exoplanets are planets in other solar systems. They must be big enough to form under gravity but not so big as to form stars, and so are generally under the gravitational influence of a host star. The definition of exoplanets is generally taken to be looser than that for planets in our own solar system, and might include dwarf planets (if any were to be discovered) but would normally exclude comets and asteroids. A *rogue planet* is a planetary body that orbits the galactic centre and not a particular star.

Terrestrial planets. These are planetary bodies that, like Earth, have an atmosphere with a distinct lower boundary, often a rocky surface but also possibly an ocean or other distinct change of character. Other examples include Mars, Venus and Mercury, although Mercury's atmosphere is extremely thin. Some sources restrict the definition to planets of similar size to Earth, in which case much larger or smaller planets that are otherwise similar might be called *quasi-terrestrial.* On the other hand, the term terrestrial planet is often applied to objects that are not, by the IAU definition, planets, such as Pluto and Titan.

When discussing other planets a 'day', without a qualifying adjective, still refers to 86,400 seconds, and a 'year' refers to 365 days. One may of course refer to such things as 'Venusian day' or a 'Jovian year', in which case the terms apply to the planet in question.

Giant planets. A giant planet may be defined as any planet at least ten times more massive than Earth (although other definitions may differ slightly), including ice giants, gas giants (both defined below) and massive terrestrial planets.

Gas giants. These are giant planets, like Jupiter and Saturn, that are composed mainly of hydrogen and helium and that do not have a sharp interface between atmosphere and solid planet. Jupiter, for example, most likely has an outer layer of molecular hydrogen, an inner layer of metallic hydrogen and a molten rocky core, and is 300 times more massive than Earth. The outer layer contains water and other heavier compounds but, in general, gas giants are more than 90% hydrogen and helium, although not all of it is gaseous: much of the hydrogen may be in liquid form.

Ice giants. Giant planets that are composed of elements heavier than hydrogen and helium are called ice giants, although the name is something of a misnomer. Uranus and Neptune (which are about 15 times more massive than Earth) are both ice giants and have less than 20% hydrogen and helium, the rest being such elements as oxygen, nitrogen and carbonic compounds such as ammonia and methane.

Super-Earths. A super-Earth is a planet with a mass between that of the Earth and that of a giant planet. A super-Earth might in principle be a terrestrial or a gaseous planet, and occasionally they are called 'mini-Neptunes'. There are no super-Earths in the Solar System — the outer planets, Jupiter, Saturn, Uranus and Neptune, are all 'giants'.

Hot Jupiters. These are gas-giant exoplanets that are in close proximity to their host stars and that may be tidally locked (with one side permanently facing their sun), and so can be expected to have very high surface temperatures on one side, low on the other. Their orbital period may be of order tens of Earth days (which is very short compared to Earth), but this may also be their rotation period around their own axis (which would be very long compared to Earth). *Hot Neptunes* is the analogous name for somewhat less massive planets, of Neptune size, that may also be in close orbit around their star.

Brown dwarfs. These are hybrids between small stars and gas-giant planets, and may at some stage have undergone nuclear fusion and have elements much heavier than hydrogen and helium. We won't consider them further in this chapter.

Others. Various other planet types exist or are hypothesized to exist. For example, the unpronounceable *chthonian planets* are gas giants that are losing or have lost their outer layers and, still more exotically, *lava planets* are planetary bodies with a surface covered by molten lava. And so on.

We cannot hope to provide theories for all of these types of planets or the atmospheres they may contain; rather we will first focus on planets with a shallow, ideal-gas atmosphere and consider the effects of a few key parameters. In this context 'shallow' means that the depth of the atmosphere is a small fraction of the planetary radius in which there may be a 'weather layer' similar to the atmosphere of a terrestrial planet. Later in the chapter we will consider the dynamics of the deeper atmosphere immediately beneath the weather layers that might give rise to such things as the jets on Jupiter and Saturn.

13.2 DIMENSIONAL AND NONDIMENSIONAL PARAMETERS

Consider a terrestrial planet with an atmosphere that obeys the primitive equations. It is forced by incoming solar ('shortwave') radiation which is balanced by outgoing infra-red ('longwave') radiation. There is (if the obliquity is low) more incoming radiation near the equator, and in many atmospheres much of the shortwave radiation is absorbed at the surface, and we represent this thermal forcing by a relaxation to a specified temperature that decreases with latitude and height. There is no external forcing in the momentum equation, aside from the effects of gravity in the vertical, but momentum is dissipated by the effects of friction near the surface, and we may represent this by the effects of a linear drag. If we use pressure coordinates then the equations of motion may be written,

$$\frac{\partial \boldsymbol{u}}{\partial t} + \boldsymbol{u} \cdot \nabla \boldsymbol{u} + \omega \frac{\partial \boldsymbol{u}}{\partial p} + \boldsymbol{f} \times \boldsymbol{u} = \nabla \phi - r\boldsymbol{u}, \tag{13.1a}$$

$$\frac{\partial \phi}{\partial p} = -\frac{RT}{p} = -\frac{R\theta}{p_R}\left(\frac{p_R}{p}\right)^{c_v/c_p}, \tag{13.1b}$$

$$\nabla \cdot \boldsymbol{u} + \frac{\partial \omega}{\partial p} = 0, \tag{13.1c}$$

$$\frac{\partial \theta}{\partial t} + \boldsymbol{u} \cdot \nabla \theta + \omega \frac{\partial \theta}{\partial p} = Q_{[\theta]}, \tag{13.1d}$$

where $\boldsymbol{u}$ is the horizontal velocity and the ∇ operator is taken to be horizontal, meaning at constant pressure and we omit viscous and diffusion terms. The term $-r\boldsymbol{u}$ parameterizes surface drag and r is large only very close to the surface, and is negligible in the free atmosphere. The term $Q_{[\theta]}$ represents heating due to radiative forcing, and θ is the potential temperature, $\theta = T(p_R/p)^\kappa$ where p_R is the reference pressure, which we take to be the mean surface pressure, p_s, and $\kappa = R/c_p$.

For the purposes of nondimensionalization the radiative forcing is taken to be given by a relaxation to a radiative-equilibrium temperature field, θ^*, such that

$$Q_{[\theta]} = \frac{\theta^*(\vartheta, p) - \theta(\vartheta, p)}{\tau}, \qquad (13.2)$$

where θ^* is a function of both latitude and height and τ is a relaxation timescale. On Earth, τ is about 10 days, and θ^* might vary by about 60 K from equator to pole and about 50 K from surface to the tropopause, and we denote these values θ_H and θ_V respectively. It is these variations that produce the circulation. For example (although the exact form is not of particular concern here) a possible recipe for θ^* is

$$\theta^* = \overline{\theta}\left[1 + \frac{\Delta_H}{3}(1 - 3\sin^2\vartheta) + \Delta_V Z\right]. \qquad (13.3)$$

Here, $\overline{\theta}$ is the average surface temperature, Δ_H is a nondimensional parameter that determines the equator to pole temperature difference, Δ_V is a similar parameter for the vertical, and $Z = -\log(p/p_s)$. We then have $\theta_H = \overline{\theta}\,\Delta_H$ and $\theta_V = \overline{\theta}\,\Delta_V$.

We nondimensionalize (13.1) by writing

$$(\widehat{u}, \widehat{v}) = \frac{(u, v)}{U}, \quad (\widehat{x}, \widehat{y}) = \frac{(x, y)}{a}, \quad \widehat{\omega} = \frac{\omega a}{U p_s}, \quad \widehat{p} = \frac{p}{p_s}$$
$$\widehat{\theta} = \frac{\theta}{\overline{\theta}}, \quad \hat{t} = \frac{t}{T} \quad \widehat{\phi} = \frac{\phi}{\Phi}. \qquad (13.4)$$

Here, as usual, the hats denote nondimensional quantities and T, U and Φ denote scaling values for time, horizontal velocity, and pressure, and we scale temperature and potential temperature with $\overline{\theta}$. To scale time we use the planetary rotation rate, $T = 1/\Omega$, and to scale velocity we use thermal wind balance, based on the radiative equilibrium temperature difference between equator and pole, as follows. The thermal wind relation for the zonal wind is

$$f\frac{\partial u}{\partial p} = -\frac{R}{pa}\frac{\partial T}{\partial \vartheta}, \qquad (13.5)$$

where R is the gas constant. This suggests the scaling

$$U = \frac{R\theta_H}{\Omega a}. \qquad (13.6)$$

The usual Rossby number is defined as $Ro = U/fL$. If we use (13.6) we can by analogy define the *external Rossby number*, Ro_E, also called the *thermal Rossby number*, by

$$Ro_E = \frac{U}{\Omega a} = \frac{R\theta_H}{\Omega^2 a^2}. \qquad (13.7)$$

Unlike the Rossby number itself, this is an external parameter of the system and not an emergent property of the flow itself. Finally, for the nondimensionalization of the geopotential we use

$$\Phi = (\Omega a) \times U = R\theta_H, \qquad (13.8)$$

which is analogous to the geostrophic scaling estimate $\phi \sim fUL$ encountered in Section 5.3.1.

The external Rossby number is one of the most important nondimensional numbers affecting the behaviour of a planet's atmosphere. Similar to the conventional Rossby number, U/fL, it is a measure of the importance of rotation, but now we take the velocity to be determined by a thermal wind determined by the radiative forcing of the planet, and the length scale to be the radius of the planet itself. The conventional Rossby number may be called the 'internal Rossby number' in this context.

If we substitute (13.4) and (13.6) into (13.1) then a little manipulation leads to the nondimensional equations

$$\frac{\partial \widehat{\boldsymbol{u}}}{\partial \widehat{t}} + Ro_E\, \widehat{\boldsymbol{u}} \cdot \nabla \widehat{\boldsymbol{u}} + Ro_E\, \widehat{\omega} \frac{\partial \widehat{\boldsymbol{u}}}{\partial \widehat{p}} + \widehat{\boldsymbol{f}} \times \widehat{\boldsymbol{u}} = \nabla \widehat{\phi} - Ek\, \widehat{\boldsymbol{u}}, \tag{13.9a}$$

$$\frac{\partial \widehat{\phi}}{\partial \widehat{p}} = -\frac{\widehat{\theta}}{\Delta_H} \left(\frac{1}{\widehat{p}}\right)^{1/\gamma}, \tag{13.9b}$$

$$\nabla \cdot \widehat{\boldsymbol{u}} + \frac{\partial \widehat{\omega}}{\partial \widehat{p}} = 0, \tag{13.9c}$$

$$\frac{\partial \widehat{\theta}}{\partial \widehat{t}} + \widehat{\boldsymbol{u}} \cdot \nabla \widehat{\theta} + \widehat{\omega} \frac{\partial \widehat{\theta}}{\partial \widehat{p}} = \widehat{Q}_{[\widehat{\theta}]}, \tag{13.9d}$$

where Ek is the Ekman number for the system, $Ek = r/\Omega$ and $\gamma = c_p/c_v$. The nondimensional forcing for the thermodynamic equation is

$$\widehat{Q}_{[\theta]} = \frac{\widehat{\theta}^* - \widehat{\theta}}{\widehat{\tau}}, \tag{13.9e}$$

where $\widehat{\tau} = \Omega\tau$, and $\widehat{\theta}^*$ varies by an amount Δ_V in the vertical and Δ_H in the horizontal.

The equation set (13.9) does not capture the rich entirety of types of planetary atmospheres. We have restricted ourselves to terrestrial atmospheres with a particular type of rather simple thermal forcing, one that has no diurnal or seasonal cycle and no condensate such as water vapour. Nevertheless, these equations can describe the essential large-scale behaviour in a large range of planetary atmospheres — the Hadley Cell and the weather dynamics on Earth, the circulation of Venus, even though it is rotating 200 times more slowly than Earth and its atmosphere is 90 times more massive, and perhaps even the jets of Jupiter, depending on how the weather layer connects to the planet's gaseous interior.

13.2.1 External Nondimensional Parameters

Perhaps the most surprising aspect to (13.9) is that there are relatively few nondimensional numbers. They are all 'external' parameters since they depend only on the planetary parameters and not the flow fields themselves, and they are:

The external Rossby number: $$Ro_E = \frac{R\theta_H}{\Omega^2 a^2}, \tag{13.10a}$$

The Ekman number: $$Ek = \frac{r}{\Omega}, \tag{13.10b}$$

The radiative relaxation timescale: $$\widehat{\tau} = \Omega\tau, \tag{13.10c}$$

Thermal forcing parameters: $$(\Delta_H, \Delta_V) = \frac{(\theta_H, \theta_V)}{\overline{\theta}}, \tag{13.10d}$$

A ratio of heat capacities: $$\gamma = \frac{c_p}{c_v}. \tag{13.10e}$$

Some of these can be expected to vary only slightly from planet to planet (for example γ, if the composition does not vary too much) or to have

The external Rossby number is also called the thermal Rossby number, although sometimes the actual temperature between equator and pole is used in the latter's definition rather than the radiative equilibrium temperature.

The thermal Rossby number seems to have been first proposed as a measure of planetary atmospheres by Raymond Hide (1929–2016), who made a great many creative contributions to Earth and planetary science.

values that may not significantly affect the dynamics. Others, like the external Rossby number, vary enormously.

Height coordinates

If we perform the same analysis in height coordinates we end up with functionally the same set of nondimensional parameters, and this is left as an exercise for the reader. The main difference in procedure arises because the hydrostatic equation has a different form, namely $\partial p/\partial z = -\rho g$. This leads to a thermal wind equation of the approximate form $f\partial u/\partial z = -(g/T)\partial T/\partial y$, and thence to an estimate for the zonal wind of

$$U = \frac{gH\theta_H}{\overline{\theta}\Omega a}. \tag{13.11}$$

A natural value for H is the scale height, $R\overline{\theta}/g$. Using this, (13.11) becomes $U = (R\theta_H)/(\Omega a)$, just as in (13.6). The other scalings are essentially the same as the pressure coordinate form.

13.2.2 Derived Nondimensional Numbers

In earlier chapters we came across other scales and nondimensional numbers such as the Rhines scale and the Burger number. In their usual form they involve properties of the flow itself — for example the Rhines scale, L_R, is normally defined to be $L_R = (U/\beta)^{1/2}$. We can now obtain analogues of these numbers and scales using only the external forcing parameters defined above. These parameters are the 'external' versions of these familiar quantities.

To obtain the external Rhines scale we use $\beta \sim \Omega/a$ and $U = R\theta_H/\Omega a$ so that

$$L_R = \left(\frac{U}{\beta}\right)^{1/2} = \left(\frac{R\theta_H}{\Omega^2}\right)^{1/2}. \tag{13.12}$$

The ratio of the radius of a planet to the Rhines scales — the external Rhines number — is

$$R_N = \frac{a}{L_R} = \left(\frac{\Omega^2 a^2}{R\theta_H}\right)^{1/2} = Ro_E^{-1/2}. \tag{13.13}$$

That is, it is the inverse square root of the external Rossby number itself.

The usual deformation radius is $L_d \sim NH/f$ where N is the buoyancy frequency. Now

$$N^2 = \frac{g}{\theta_0}\frac{\partial\theta}{\partial z}, \tag{13.14}$$

so that

$$N^2H^2 \sim gH\frac{\theta_V}{\overline{\theta}} \sim R\theta_V, \tag{13.15}$$

where we have taken H to be a scale height, namely $R\overline{\theta}/g$. The external deformation radius is then given by

$$L_E = \frac{(R\theta_V)^{1/2}}{\Omega}. \tag{13.16}$$

The external Burger number, $Bu_E = L_E^2/L^2$, is then, taking $L = a$,

$$Bu_E = \frac{R\theta_V}{\Omega^2 a^2} = Ro_E \frac{\theta_V}{\theta_H}. \tag{13.17}$$

Unlike the usual Burger number this is a property of the forcing parameters and not the flow itself. This does not mean that the actual Burger number of a planet's circulation is equal to that given by (13.17), but we can expect it to be a function of the nondimensional numbers given in (13.10).

Relation between internal and external nondimensional numbers

There is not necessarily a one-to-one correspondence between the external and internal versions of these numbers, as we can illustrate with two examples. The internal (i.e., the usual) Rossby number is defined as $Ro = U/fL$. The external Rossby number is not necessarily equal to it — in a planet with large friction U might be quite small and the external Rossby number might be a large overestimate of the internal one. More typically, the external Rossby number is an underestimate of a typical internal Rossby number; on Earth the external Rossby number is about 0.07 whereas a canonical value of the internal Rossby number is 0.1 for the very large scale flow, and bigger for smaller-scale flow. The Rhines scale is another example. The internal Rhines scale, $\sqrt{U/\beta}$, depends on the throughput of energy in the system (the energy cascade rate) as well as friction and radiative damping (and, for example, some numerical simulations find that it depends on the ratio of the frictional damping to the thermal damping, the so-called 'planetary Prandtl number'). The external Rhines scale, as defined above, depends on none of those things.

13.2.3 The Role of Gravity

In this section we have a brief diversion into a rather surprising consequence of our nondimensionalization. None of the dimensionless numbers in (13.10) involve gravity or the surface pressure (or the total mass) of the atmosphere. This is puzzling, for surely the behaviour of an atmosphere depends on gravity? In fact, in some circumstances it does not!

The lack of an effect

Suppose we change the value of gravity from g to αg, where α is a constant — for example, if $\alpha = 2$ then we have doubled gravity. The pressure coordinate form of the equations, even in dimensional form, does not have g as a parameter — the factor g only appears in conjunction with z in the geopotential, $\phi = gz$. This suggests that if we increase gravity the height (the z value) of a geopotential surface would fall in order to keep ϕ invariant, but other quantities would remain the same. This conclusion also emerges using the height form (z coordinate) version of the equations where gravity directly appears in the hydrostatic equation $\partial p/\partial z = -\rho g$.

To see this, note that if we change g, the equation set (2.41) remains invariant under the following transformation:

$$\begin{aligned} g \to \alpha g, \qquad p \to \alpha p, \qquad \rho \to \alpha \rho, \qquad (T,\theta) \to (T,\theta), \\ t \to t, \qquad (x,y) \to (x,y), \qquad z \to z/\alpha, \\ (u,v) \to (u,v), \qquad w \to w/\alpha. \end{aligned} \tag{13.18}$$

If we substitute (13.18) into (2.41) then all the factors of α cancel and the equations are unchanged. If the heating terms in the thermodynamic equations also do not depend on gravity then, quite remarkably, and insofar as the atmosphere obeys the primitive equations *the dynamics and circulation of the atmosphere do not depend on gravity.* The atmosphere is simply squashed down, and half as deep, but other aspects, for example the horizontal scales of motion, are unaltered.

Let us check this rather surprising result by calculating the deformation radius, NH/f. The height, H, becomes H/α. The buoyancy frequency squared, $N^2 = g/\theta_0 \partial\theta/\partial z$ becomes $\alpha^2 N^2$ so that NH/f remains the same. It is easily checked that such dynamical quantities as the Eady growth rate or the frequency of a Rossby wave are also unaltered.

The presence of an effect

There are a number of ways in which changing gravity *will* have an effect in the full system; that is, one described by the Navier–Stokes equations on the sphere:

(i) Through nonhydrostatic effects. The vertical momentum equation $Dw/Dt = -\rho^{-1}\partial p/\partial z - g$ is manifestly *not* invariant under the transformation (13.18); the right-hand side scales like α whereas the left-hand side scales like $1/\alpha$. This is consistent with the low-aspect ratio requirement for hydrostasy, which can evidently be seen as a 'high g' requirement, for the ratio of the scale height to the planetary horizontal scale is $(RT/g)/a = RTa/g$, which is small when g is large. Similarly, nonhydrostatic internal waves obeying (7.33) are not invariant under (13.18) but hydrostatic waves obeying (7.44) are. If nonhydrostatic convection affects the large scale then gravity may have a direct effect.

(ii) Through the effects of spherical geometry. If the atmosphere were to get deeper because gravity was reduced, then the ratio of the depth of the atmosphere to the radius of the Earth would change. This effect is not accounted for in the primitive equations (where it is assumed that the depth of the atmosphere is small), but this change will have an effect in the full Navier–Stokes equations in spherical geometry, albeit a small one in most circumstances.

(iii) Through the effects of thermal forcing if that does depend on gravity, for example in radiative transfer if water vapour is present. The amount of water vapour in the atmosphere is largely governed by temperature and not pressure — the warmer the atmosphere, the more water vapour there can be. Now, if gravity were higher, the

temperature would fall off more rapidly with height, so there would be less total water in the atmosphere than in the original case. Since water vapour is a potent greenhouse gas, the high-gravity planet will be cooler than the low-gravity planet. Furthermore, the total amount of water vapour in an atmospheric column will be less at higher gravity so the hydrology cycle will be weaker.

Nonetheless, it is startling to realize that changes in gravity have little direct effect on the dynamics of a dry terrestrial atmosphere especially since a planet, by definition, owes its existence to gravity. If the gravity is changed then the atmosphere may be squashed and its aspect ratio may change, but the dynamics of the fluid remain essentially the same.

13.2.4 The External Rossby Number

Of the parameters in Section 13.2.1, the external Rossby number, $Ro_E = R\theta_H/(\Omega^2 a^2)$, is one that obviously differs from planet to planet. It does so for three main reasons:

(i) The distance from and the strength of the host star — the further a planet is from its host sun, the smaller will be θ_H. This is because $\theta_H = \overline{\theta}\Delta_H$, and Δ_H is essentially a geometric factor associated with the sphericity of a planet. Thus, as the mean temperature $\overline{\theta}$ falls, so do θ_H and Ro_E. A planet like Earth with a mean temperature of 250 K may have a 50 K difference between equator and pole, but a planet with a mean temperature of only 100 K may have a proportionally smaller equator to pole temperature difference.

(ii) The planet's rotation rate and...

(iii) The planet's size. The size and the rotation rate enter only in the combination Ωa, so that a smaller planet may behave similarly to a larger planet with a slower rotation rate. To see this, imagine a soccer ball that is rotating once every 24 hours, like Earth. We can then certainly imagine that the air flow over the ball's surface hardly knows that the ball is rotating. As with the usual Rossby number, U/fL, larger scales feel the rotation more than smaller scales.

With $R = 287\,\mathrm{J\,kg^{-1}\,K^{-1}}$, $a = 6,371\,\mathrm{km}$, $\theta_H = 50\,\mathrm{K}$ and $\Omega = 7.27 \times 10^{-5}\mathrm{rad/s}$, the external Rossby number of Earth is about 0.07. Venus has a similar radius and similar net incoming solar radiation but a rotation rate that is over 200 times less. Its atmosphere is CO_2, with $R \approx 189\,\mathrm{J\,kg^{-1}\,K^{-1}}$, and, taking $\theta_H = 50\,\mathrm{K}$, the external Rossby number for Venus is about 2000! Mars is an intermediate case: its rotation rate is almost the same as Earth but it is much smaller, with a radius of about 3,400 km. Taking $\theta_H = 40\,\mathrm{K}$ gives Ro_E (Mars) = 0.18. Jupiter has a much bigger radius, about 70,000 km, and a rotation period of only 10 hours ($\Omega = 1.76 \times 10^{-4}\mathrm{rad/s}$). It is hard to estimate a proper value for θ_H, since much of the heating for Jupiter's atmosphere is internal, but let us suppose that $\theta_H = 10\,\mathrm{K}$ (most likely a significant overestimate). We then find Ro_E(Jupiter) $\approx 2 \times 10^{-5}$, far smaller than for Earth and a factor of 10^8 smaller than for Venus.

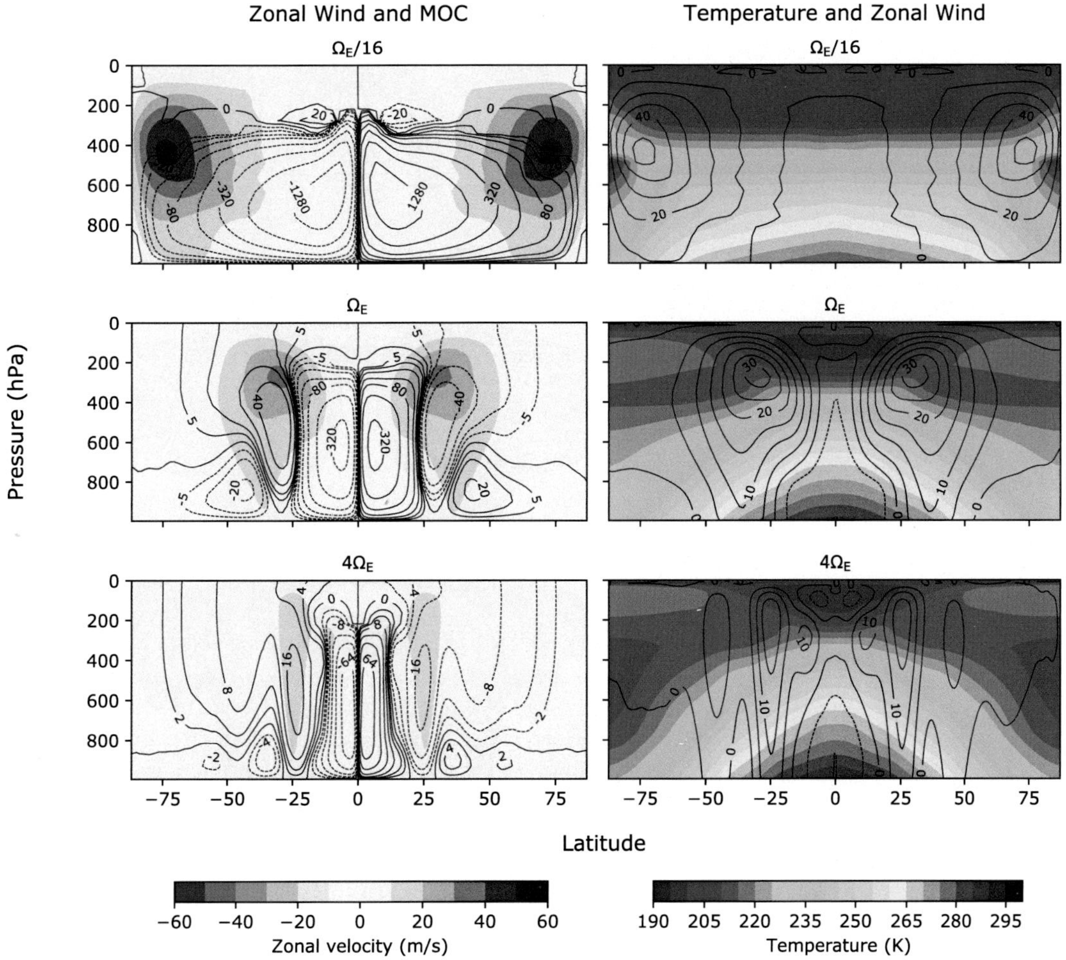

Fig. 13.1: Numerical simulations of a moist, Earth-like planet at varying rotation rates, where Ω_e is equal to the rotation rate of Earth. The external Rossby number ranges from about 20 (top) to 0.004 (bottom). The left column shows zonal-mean zonal wind (colours) and mean meridional mass streamfunction (contours). The right column shows temperature (colours) and zonal mean zonal wind (contours). Note in particular the width of the Hadley Cell decreasing at higher rotation rates. (Simulations and figure by Jake Eager.)

What might the effects of changing the external Rossby number be? We may hypothesize the following effects:

- A Hadley Cell width that increases as the as rotation rate decreases. In the axi-symmetric theory of Section 11.2 the width of the Hadley Cell scales with the square root of the external Rossby number — if H in (11.10) is taken as a scale height the expression is essentially the same as (13.10). If the Hadley Cell terminates because of baroclinic instability, then (11.13) predicts a Hadley Cell width of $(NH/2\Omega a)^{1/2}$. Using (13.15) one may easily show this scales as the fourth root of the external Rossby number. Thus, in any event, the Hadley Cell should expand as Ω falls.
- The number of midlatitude jets will increase as the external Rossby number decreases.

We can illustrate these effects by way of numerical simulations, as we now describe.

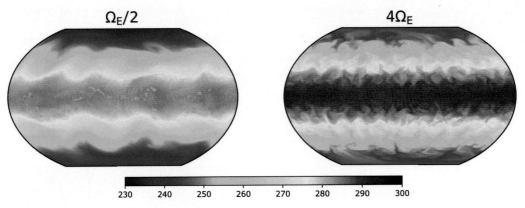

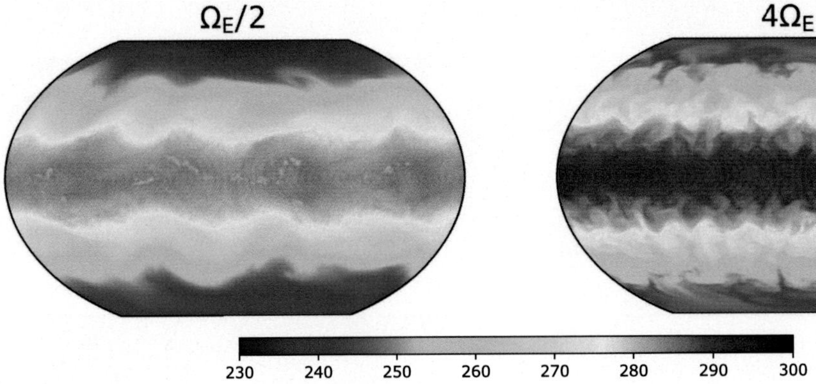

Fig. 13.2: Snapshots of surface temperature (K) in two of the simulations of Fig. 13.1, with rotation rates half that of earth and four times that of Earth. Two effects can be seen. First, in the simulation at lower rotation, the equator to pole temperature gradient is smaller. Second, the eddy scale is much smaller in the simulation at high rotation. (Simulation and figure by Jake Eager, and similar to a figure in Kaspi & Showman 2015.)

13.2.5 Numerical Simulations

Some results of a numerical integration of (13.1), using a thermal forcing similar to that given by (13.3), are shown in Fig. 13.1. The following features are apparent, all consistent with the earlier theoretical explanations.

- The Hadley cells extend further at higher values of the external Rossby number. This property is consistent with arguments in Sections 11.2 and 11.3. Concomitantly, the eastward jets at the Hadley Cell edge move to higher latitude as rotation rate decreases. At the lowest rotation rates the Hadley Cell extends almost to the pole, as do the Hadley Cells on Venus and Titan, both of which are very slowly rotating. (In a sense, the atmospheric circulation of these planets is wholly tropical.) The Hadley Cell is very efficient at transporting heat (because its mixing length is so large) and at low rotation rates the equator to pole temperature gradient is small and no baroclinic instability develops.
- At higher rotation rates a distinct extratropical zone develops, with baroclinic eddies and jets that are distinct from the subtropical jets. The eddies in zones transport heat poleward less efficiently than does the Hadley Cell, and the equator to pole temperature gradient increases with rotation. These jets are driven by an eddy momentum convergence (Section 12.1), which in turn comes from the propagation of Rossby waves away from the baroclinic zones.
- The scale of the eddy motion in midlatitudes decreases as the rotation rate increases, as is apparent in Fig. 13.2. The underlying reason is that, in the absence of changes in stratification, we would expect the deformation radius, NH/f, which is a measure of the linear instability scale, to decrease with increasing rotation.
- At the highest rotation rates, multiple eddy-driven jets can form, rather like the multiple jets discussed in Section 10.6 and, perhaps, similar to the jets seen on Jupiter and Saturn discussed later. Multiple jets arise because the external Rhines scale decreases as the external Rossby number decreases, and so the jet width decreases.

13.3 Radiative Transfer and Vertical Structure

We now look at the *vertical* structure of planetary atmospheres, and this involves a significant diversion into radiation and its interaction with convection. We begin with an elementary introduction to radiative transfer.

13.3.1 Elements of Radiative Transfer

Radiative transfer is a complex subject and in order to make progress we will simplify it in two main ways:

(i) Rather than treating an entire spectrum of wavelengths, we suppose that the radiation exists only in two well-separated bands, namely solar (or shortwave) radiation and infra-red (or longwave) radiation. In each band the absorption is not a function of wavelength, and this is called the grey approximation (or the semi-grey approximation, considering that there are two bands of radiation). We will suppose that the planet only emits longwave radiation, and its host star only emits shortwave radiation.

(ii) We suppose that the radiation travels only vertically, up or down, through a planetary atmosphere. This is the *two-stream* approximation.

Consider a beam of radiation propagating through a thin slab of gas, as in Fig. 13.3. Some of the incoming radiation may be absorbed, some reflected, and the slab may emit radiation of its own. In the three-dimensional problem some radiation may also be scattered into the beam from radiation travelling in other directions. Let us neglect scattering and reflection, a good approximation for longwave radiation. The difference between the outgoing and incoming radiation is then

$$dI = I^{out} - I^{in} = -d\tau I + dE, \tag{13.19}$$

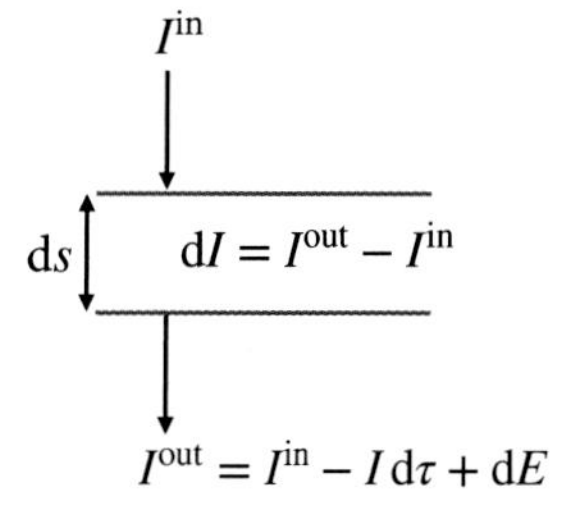

Fig. 13.3: Radiative transfer across a thin slab, with no scattering or reflection and the radiation travelling in the vertical direction only.

where I is the irradiance or radiant intensity (with units of power per unit area, or $W\,m^{-2}$), the term $-d\tau I$ is the absorption and dE is the thermal emission. The quantity $d\tau$ is the nondimensional *optical depth;* it may be written as $d\tau = e_L ds$ where ds is the slab thickness, and e_L is the *emissivity* of the medium. The minus sign on $d\tau$ in (13.19) is appropriate when τ increases in the direction of the beam. The emission of radiation is, in thermal equilibrium, the Planck function, B, such that $dE = B d\tau$. If the atmosphere is grey (as in assumption *(i)* above) then $B = \sigma T^4$ where $\sigma = 5.67 \times 10^{-8}\,W\,m^{-2}\,K^{-4}$ is Stefan's constant. With all this, (13.19) becomes

$$dI = -d\tau(I - B) \quad \text{or} \quad \frac{dI}{d\tau} = -(I - B). \tag{13.20}$$

In planetary atmospheres it is common to choose τ increasing downwards, from 0 at the top of the atmosphere. The downwards (D) and upwards (U) irradiances are then

$$\frac{dD}{d\tau} = B - D, \qquad \frac{dU}{d\tau} = U - B. \tag{13.21a,b}$$

These equations are sometimes known as the *Schwarzchild equations.* The net flux of radiation is $N = U - D$ and the radiative heating is proportional to the net flux divergence, $-\partial N/\partial z$.

The grey assumption is not particularly accurate for many planetary atmospheres but for conceptual or approximate calculations it is often useful to suppose that the atmosphere is grey in the infra-red, in which case (13.21) applies to infra-red radiation. Similar but separate equations (that would normally include reflection, but not thermal emission) are then used for solar radiation. In Earth's atmosphere the absorption of solar radiation in the atmosphere is quite small, and although there is considerable reflection due to clouds most of the net incoming solar radiation is absorbed at the surface.

13.3.2 Radiative Equilibrium

A *radiative equilibrium* state has, by definition, no radiative heating. If the atmosphere is transparent to solar radiation then the condition implies that the vertical divergence of the longwave radiation (denoted with a subscript L) is zero:

$$\frac{\partial(U_L - D_L)}{\partial z} = 0 \quad \text{implying} \quad \frac{\partial(U_L - D_L)}{\partial \tau} = 0. \qquad (13.22a,b)$$

This condition is normally *not* satisfied in the atmosphere because the air moves. But if it were satisfied then (13.21) and (13.22b) form three equations in three unknowns, U_L, D_L and B, and a solution can be found as follows.

Consider an atmosphere with net incoming solar radiation S_{net} and suppose the planet is in radiative equilibrium with the incoming solar radiation balanced by outgoing infra-red radiation. That is, $U_{Lt} \equiv U_L(\tau = 0) = S_{net}$ where U_{Lt} is the net outgoing longwave radiation (OLR) at the top of the atmosphere. The downward infra-red radiation at the top of the atmosphere is zero, so that the boundary conditions on the radiative transfer equations at the top of the atmosphere are

$$D_L = 0, \quad U_L = U_{Lt} \quad \text{at} \quad \tau = 0. \qquad (13.23)$$

A little algebra reveals that a solution of (13.21) and (13.22b) that satisfies (13.23) is

$$D_L = \frac{\tau}{2} U_{Lt}, \qquad U_L = \left(1 + \frac{\tau}{2}\right) U_{Lt}, \qquad B = \left(\frac{1+\tau}{2}\right) U_{Lt}, \quad (13.24a,b,c)$$

as can be easily verified by substitution back into the equations.

It remains to explicitly relate τ to z, and one approximate recipe is to suppose that τ has an exponential profile,

$$\tau(z) = \tau_0 \exp(-z/H_a), \qquad (13.25)$$

where τ_0 is the optical depth at $z = 0$ and H_a is the scale height of the absorber. In the Earth's atmosphere the optical depth is determined by the

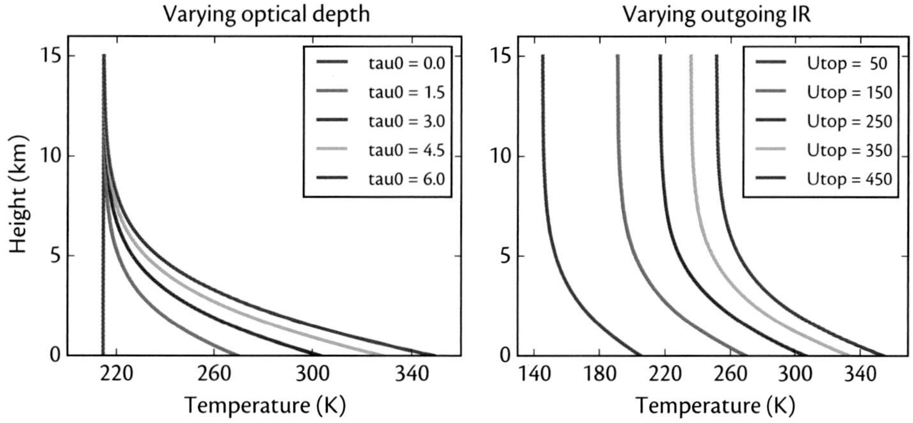

Fig. 13.4: Radiative equilibrium temperature calculated using (13.26), with $H_a = 2$ km. Left: outgoing IR (and incoming solar) radiation is 240 W m^{-2} and surface optical depth varies from 0 to 6. Right: the optical depth is 3.0 and outgoing IR radiation varies from 50 to 450 W m^{-2}

concentrations of water vapour (primarily) and carbon dioxide (secondarily), and τ_0 (the scaled optical depth) typically varies between 2 and 4, depending on the water vapour content of the atmosphere, and $H_a \approx 2$ km, this being a typical scale height for water vapour. Using (13.24c) with $B = \sigma T^4$ the temperature then varies as

$$T^4 = U_{Lt}\left(\frac{1 + \tau_0 e^{-z/H_a}}{2\sigma}\right), \tag{13.26}$$

as illustrated in Fig. 13.4. We note the following aspects of the solution:

(i) Temperature increases rapidly with height near the ground.

(ii) The upper atmosphere, where τ is small, is nearly isothermal.

(iii) The temperature at the top of the atmosphere, T_t is given by

$$\sigma T_t^4 = \frac{U_{Lt}}{2}. \tag{13.27}$$

Thus, if we define the emitting temperature, T_e, to be such that $\sigma T_e^4 = U_{Lt}$, then $T_t = T_e/2^{1/4} < T_e$; that is, the temperature at the top of the atmosphere is *lower* than the emitting temperature.

(iv) Related to the previous point, $B_t/U_{Lt} = 1/2$. That is, the upwards longwave flux at the top of the atmosphere is twice that which would arise if there were a black surface at a temperature T_t. The reason is that there is radiation coming from lower down in the atmosphere where the temperature is higher.

13.3.3 Radiative–Convective Equilibrium

In the radiative equilibrium solution the temperature gradient near the ground varies so rapidly that $-\partial T/\partial z$ may well exceed even the dry adiabatic lapse rate. If so, the radiative equilibrium solution is convectively unstable and convection will be triggered. The convection will then establish a convectively neutral lapse rate, Γ typically given by the dry adi-

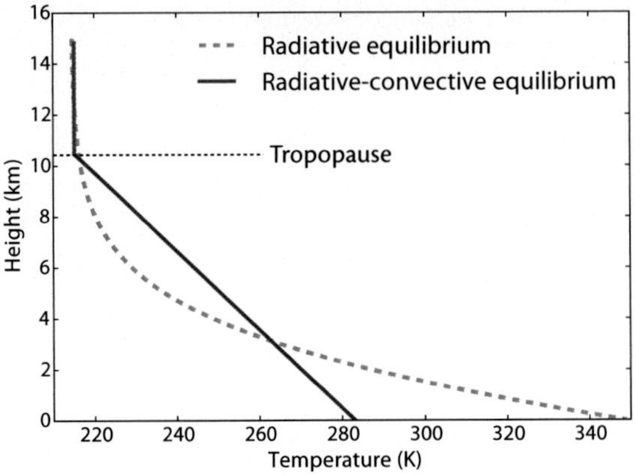

Fig. 13.5: Radiative and radiative–convective equilibrium profiles. The initial radiative equilibrium temperature, dashed blue line, adjusts to a specified profile (red line, in this instance a constant lapse rate) that extends to a finite height, beyond which radiative equilibrium holds. The break in the slope of the radiative–convective temperature profile is the tropopause.

abatic or, if moisture or another condensate is present, the saturated adiabatic lapse rate. The convection cannot proceed to arbitrary height of course — it will cease when the lapse rate is no longer unstable, and at higher altitudes the temperature profile will more-or-less be that of the radiative equilibrium profile, as illustrated in Fig. 13.5. The break in the temperature profile is nothing other than the tropopause, above which lies the stratosphere. The temperature in the real stratosphere is not exactly constant with height, because there may be absorption of solar radiation within it (as by ozone in Earth's stratosphere) and because of the presence of fluid motion. However, it is much more constant than in the troposphere, and a temperature structure similar to that of the red line in Fig. 13.5 is a common feature of many other planets too.

We can obtain a very rough estimate of the average height of the tropopause of a terrestrial planet if we know the optical depth of the atmosphere as well as the lapse rate, Γ. The tropopause temperature, T_T is given by the requirement that the outgoing longwave radiation, U_{Lt} equal the net incoming solar radiation, S_N. For a stratosphere in radiative equilibrium the tropopause temperature, T_T is equal to that at the top of the atmosphere and using (13.27), it must then satisfy

$$\sigma T_T^4 = \frac{S_N}{2} = \frac{\sigma T_e^4}{2} \qquad \text{or} \qquad T_T = \frac{T_e}{2^{1/4}}. \tag{13.28}$$

The surface temperature, T_S, in radiative equilibrium is given by (13.26) evaluated at $z = 0$, giving

$$\sigma T_S^4 = S_N\left(\frac{1+\tau_0}{2}\right) \qquad \text{or} \qquad T_S = T_T(1+\tau_0)^{1/4}. \tag{13.29}$$

The surface temperature in radiative–convective equilibrium is usually considerably lower than this, because convection tends to cool the lower atmosphere and heat the upper atmosphere, but nevertheless (13.29) gives a first estimate. The height of the tropopause, H_T, is then such that $(T_S - T_T)/H_T = \Gamma$ giving

$$H_T = \frac{T_S - T_T}{\Gamma} = \frac{T_T}{\Gamma}\left((1+\tau_0)^{1/4} - 1\right). \tag{13.30}$$

This expression is an overestimate because we have used the radiative equilibrium surface temperature for T_S and to properly account for that requires a complicated calculation (the result of which is the red line in Fig. 13.5). However, (13.30) qualitatively captures three of the main factors determining tropopause height on a terrestrial planet, namely:

(i) The increase of the tropopause height with decreasing lapse rate. With a smaller lapse rate, the tropopause must extend higher in order that the its temperature falls sufficiently so that the outgoing infra-red radiation equals the incoming solar radiation.

(ii) The increase of tropopause height with optical depth. Thus, a planet like Venus with a thick, CO_2 rich atmosphere can be expected to have a higher tropopause than Earth. This effect arises because an increased optical depth raises the surface temperature, but the tropopause temperature, being proportional to the emissions temperature, remains the same. Given a larger value of $T_S - T_T$ the tropopause height must be larger.

(iii) The increase in tropopause height with increased incoming solar radiation. This effect arises because the difference between the surface and tropopause temperatures in radiative equilibrium is proportional to the tropopause temperature itself, as we can see from (13.26) and in the right-hand panel of Fig. 13.5. If we were to move the Earth further from the Sun, or increase its albedo, the tropopause height would fall.

Item *(ii)* has a robust consequence for Earth. As we add greenhouse gases to the atmosphere causing global warming, the optical depth increases, and consequently the tropopause height must increase. Detailed calculations show that the increase will occur at the rate of about 300 m for each degree of warming, and this is one of the most robust predictions about how the atmosphere will change in the future as the planet warms.

We should add some caveats to the above discussion. First, we have treated the atmosphere as if it were a single column. The horizontal transfer of heat may make a significant difference, and on Earth the tropical tropopause is about twice as high as the polar one. Relatedly, it is not always convection that maintains the lapse rate, and in baroclinic zones the stratification may be maintained by baroclinic waves and not convection. However, if we know the lapse rate that the baroclinic eddies produce the above arguments may still be applied. Finally, most planetary atmospheres are not grey, even in the infra-red, and may have clouds, but the complications that these entail is a story for another day.

13.4 Planets in our Solar System

We now look at other planets to see if and how the radiative and dynamical discussions of the previous sections and chapters are reflected in their atmospheric structure. We focus on Venus and Jupiter, since these two planets are both quite different, in different ways, from Earth and many other planets in the universe are likely to lie parametrically between them.

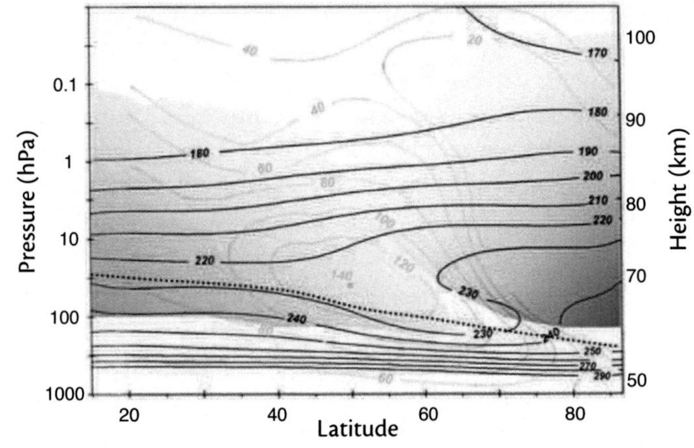

Fig. 13.6: Temperature (black contours, in K) and thermal wind field (green, m/s) as a function of latitude and height for the middle atmosphere of Venus, obtained from an orbital infra-red radiometer during the Pioneer mission. The dotted line shows the approximate cloud-top height, and pink and blue colours indicated radiative heating and cooling, respectively. (Adapted from Titov *et al.* 2013.)

13.4.1 Venus, the Slowly Rotating Hothouse

Venus is a planet that is very similar to Earth in some respects, very dissimilar in others. It is the second planet from the Sun, orbiting it every 225 days. It is thus closer to the Sun (a factor of 0.72) but it is covered by a thick layer of cloud and has a high albedo (about 0.7 vs 0.3 for Earth), with the consequence that its net incoming solar radiation, and hence its emitting temperature, are not very different from those of Earth. Venus is a similar size to Earth, with a radius of about 6,000 km, or 0.95 that of Earth, and a mass of 82% of Earth, and it also has a rocky surface and a gaseous atmosphere. So far, so similar. However, Venus differs in the following striking ways:

A *sidereal day* is the time a planet takes to spin round its own axis, whereas a *solar day* is the time taken for its sun to appear at the same place in the sky. On Earth the two are almost the same (approximately 23 hours 56 minutes and 24 hours, respectively) but on other planets they may be quite different. The solar day of a particular planet is sometimes called a 'sol'.

(i) Venus has a much slower rotation rate, with a sidereal rotation period (i.e., the time it takes to rotate about its own axis, relative to the distant stars) of about 243 Earth days. (A Venusian sol, namely the period taken for the Sun to return to the same point in the sky, is a combination of the sidereal day and the rotation around the sun, and is about 117 Earth days.)

(ii) Venus has a very massive atmosphere, with a surface pressure of about 92 bars (92×10^5 Pa) compared to the 1 bar of Earth.

(iii) Venus's atmosphere is 95% CO_2, which is a potent greenhouse gas. This, combined with the large extent of the atmosphere, means that its optical depth is over 100 times higher than that of Earth.

Venus also has a small obliquity (about 3°) so that seasonal effects are small. Given the above, what can we posit about Venus and how does that compare to observation?

Temperature

The temperature structure of Venus's atmosphere is illustrated in Fig. 13.6 and Fig. 13.7. Three obvious features are:

(i) The very small variation of temperature in the horizontal, both in the upper troposphere and the surface.

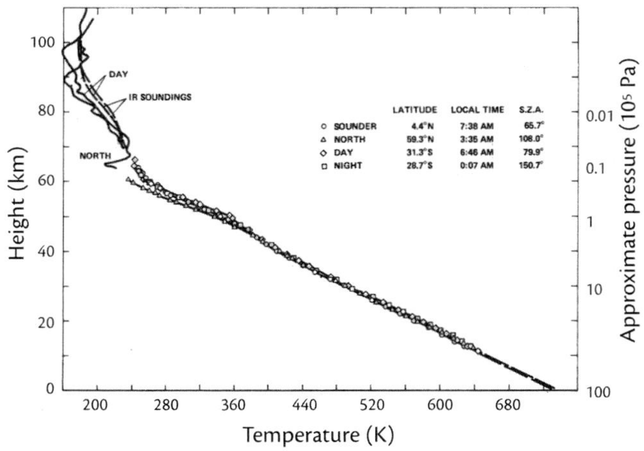

Fig. 13.7: Four temperature profiles on Venus obtained by probes as part of the Pioneer mission. The probes were dropped at various latitudes and times but give remarkably similar results. The lapse rate is close to that of the dry adiabat up to tropopause, at about 60 km. (Adapted from Seiff *et al.* 1979.)

(ii) The almost uniform lapse rate, with a tropopause at about 60 km, some six times higher than that of Earth.

(iii) The very high surface temperature compared to Earth.

The first feature is a consequence of the slow rotation and Hadley Cell extending very far poleward (as discussed below), for the Hadley Cell is a region of almost uniform temperature, as found in Chapter 11. The second and third features are consequences of the high optical depth of the Venusian atmosphere: the tropopause temperature, which is similar to the emitting temperature, is about 240 K, but the surface temperature is much higher, about 700 K, because of the strong greenhouse effect. The dry adiabatic lapse rate, g/c_p varies from about 7.8 K km^{-1} to about 11 K km^{-1} at 50 km ($g = 8.86$ m s^{-2} and c_p varies from about 1130 J kg^{-1} K^{-1} at the surface to about 800 J kg^{-1} K^{-1} at the tropopause, because the temperature varies), and to connect the surface and tropopause with such a lapse rate requires 50 km or more (and in fact the lapse rate is slightly less than, or more stable than, the dry adiabat in the upper troposphere).

A soothing exercise in geometry will reveal that the length of a solar day, *Sol*, and the length of a sidereal day, *Sid*, are related by

$$Sol = \frac{Sid}{1 \pm Sid/OP}$$

where *OP* is the orbital period and the minus or plus sign is taken depending on whether the planet rotates in a prograde or retrograde fashion, respectively. Venus has a sidereal day of 243 Earth days, a sol of 117 Earth days and a year of 225 Earth days.

Winds

As noted in Section 13.2.4, the slow rotation period of Venus leads to an external Rossby number of about 2000 and, given this, we expect that the Hadley Cell will extend much further poleward than on Earth and that its meridional temperature gradient will be much less. These expectations are largely borne out by the observations, as shown in Fig. 13.8, Fig. 13.6 and Fig. 13.7. Observations of the wind are most easily obtained at the cloud layer and these show motion away from the equator, extending to about 50° on either side, which we can interpret as being the edge of the Hadley Cell. We note that the axisymmetric theory of the Hadley Cell extent given in Section 11.2 does not apply without modification to Venus, because the planetary rotation rate of Venus is so slow that geostrophic balance in its usual form is not applicable. Nevertheless, the same general

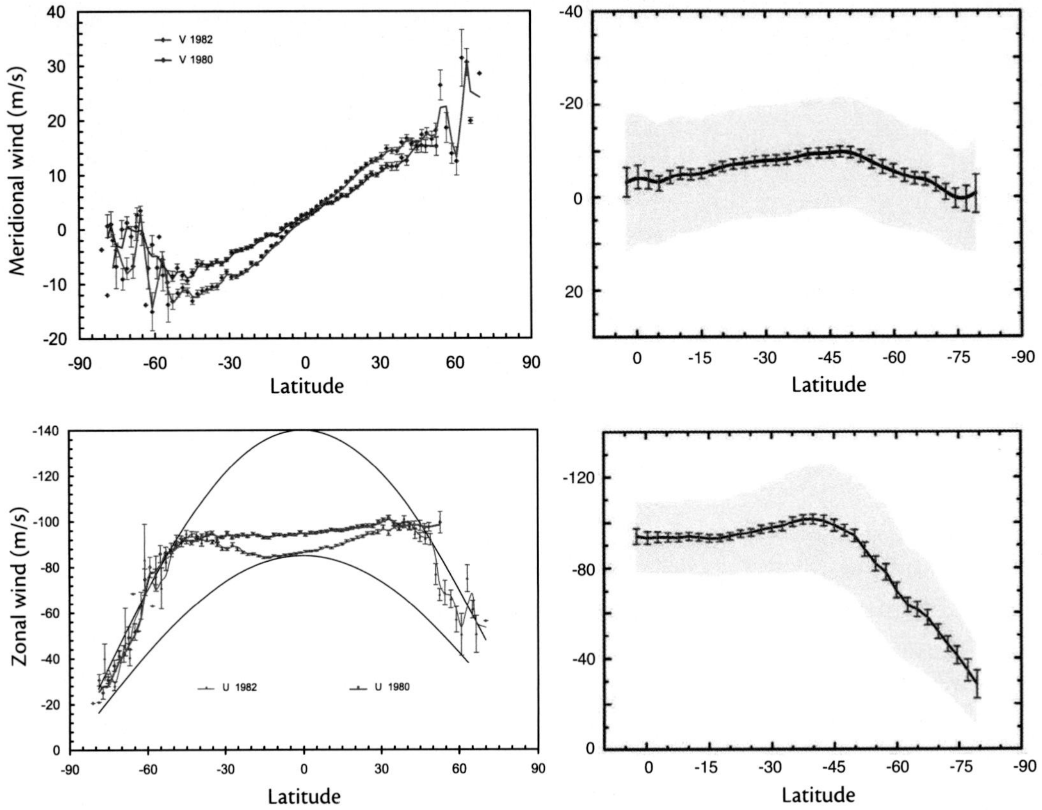

Fig. 13.8: Meridional winds (top) and zonal winds (bottom) on Venus at cloud top height obtained by cloud tracking. The left panels show the wind on the day-side only for 1980 and 1982. The right panels show zonally-averaged winds obtained by the Venus Express orbiter over a period of about 10 Venusian years. The two solid lines in the bottom left panel show constant angular velocity profiles (i.e., $u/\cos\vartheta$ = constant), with two different values at the equator. (Adapted from Limaye (2007), left panels, and Khatuntsev *et al.* (2013), right panels.)

principles apply, with a slower rotation rate leading to a greater poleward extent, and the extended Hadley Cell keeps temperature gradients small.

The zonal winds are notable in two ways:

(i) They show a distinct fall off at latitudes beyond about 50°, that is beyond the inferred Hadley Cell.

(ii) They are large (and negative) at low- and midlatitudes. The negative values just mean the winds are in the same direction as the solid surface, since Venus's rotation is retrograde. The fact that the zonal winds are large at the equator means that the atmosphere is *super-rotating*, with the atmosphere spinning faster than the planet itself.

The fall-off in winds beyond the edge of the Hadley Cell is to be expected, given the Hadley Cell theory of Sections 11.2 and 11.3. The high values of the zonal wind in the Venusian midlatitudes are analogues of the subtropical jets on Earth, rather than the eddy-driven midlatitude jets. Baroclinic instability is very weak on Venus because of the small pole-to-equator temperature gradient and because the radius of deformation, NH/f, is very large, and by and large the zonal mean flow is much stronger than the eddying flow.

The equatorial super-rotation (item *(ii)* above) on Venus is a striking feature with no real analogue on Earth. The velocity of the ground at the equator is about $1.8\,\mathrm{m\,s^{-1}}$, so that a zonal wind of $90\,\mathrm{m\,s^{-1}}$ means the

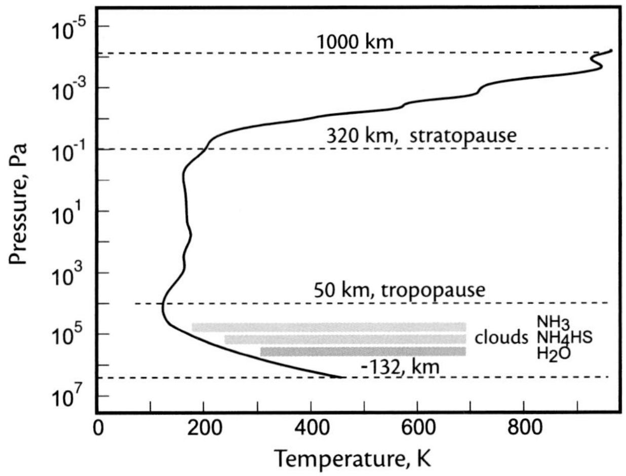

Fig. 13.9: Temperature in Jupiter's upper atmosphere. Heights are measured from the 10^5 Pa (1 bar) level. The approximate altitudes of atmospheric transitions (tropopause and stratopause) are shown, as well as tropospheric cloud layers. The data is from the Galileo probe in 1995 (Seiff *et al.* 1998), and the figure is adapted from Ruslik0 at en.wikipedia.

atmosphere is rotating about *50 times faster than the ground,* and in the same direction. Super-rotation is also apparent in the top two left-hand panels of Fig. 13.1, and is in fact a common feature of numerical simulations at low rotation rates. The remarkable aspect of this phenomenon is that, even though the eddies on Venus are weak, they *must* be involved in the production of super-rotation, for a rather simple reason. If the atmosphere is zonally symmetric then, except for viscous effects, the angular momentum of a ring of air around the axis of rotation, $\overline{m}$, is conserved, as discussed in Section 11.2.1, and no internal maximum (such as super-rotation) can be produced. To see this explicitly, after a little manipulation we find that, if the flow is zonally-symmetric, we can write the zonal momentum equation as

The fact that super-rotation must involve eddying motion is known as Hide's theorem, after Raymond Hide. The mechanism for the super-rotation involves the interaction of equatorial Kelvin waves and off-equatorial Rossby waves giving rise to an eddy momentum convergence at the equator; alas, details are beyond the scope of this book.

$$\frac{\mathrm{D}m}{\mathrm{D}t} = \nu \frac{\partial^2 m}{\partial z^2}, \tag{13.31}$$

where $m = (u + \Omega a \cos\vartheta) a \cos\vartheta$ as in (11.5) and the right-hand side represents viscous effects, keeping only the vertical derivative as it is much larger than the horizontal one. Now, the advective terms certainly cannot create angular momentum, and the viscous terms can only smooth it out, destroying any pre-existing maximum, as discussed in Section 10.2.1. Thus, no maximum of angular momentum, and therefore no super-rotation, can exist in a steady zonally-symmetric flow. If we allow for the possibility of zonal asymmetries then additional terms in the equation arise (such as $\partial_y(\overline{u'v'})$) and an eddy flux of angular momentum can maintain the super-rotation.

13.4.2 Jupiter, the Rapidly Rotating Giant

Jupiter is a rapidly-rotating gas giant, with a sidereal day of just under 10 hours. A temperature profile in the upper 'weather layer' is shown in Fig. 13.9 and a hypothesized radial structure is sketched in Fig. 13.10, based on a combination of observations and theoretical models, although neither of them offer complete or wholly accurate information. Saturn's

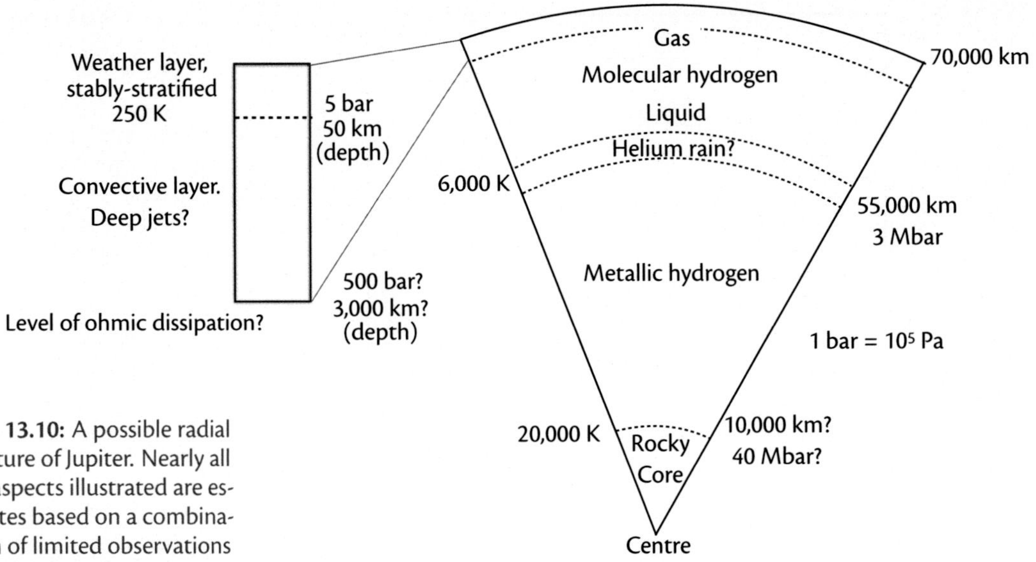

Fig. 13.10: A possible radial structure of Jupiter. Nearly all the aspects illustrated are estimates based on a combination of limited observations and uncertain theoretical models. All the numbers have significant error bars and, for example, the level of significant ohmic dissipation might be much deeper, the presence of helium rain is speculative and the rocky core could be more extensive.

structure may be qualitatively similar, but again we cannot be sure. There are two very obvious differences from Earth:

(i) There is no distinct lower boundary to the atmosphere.

(ii) The planet is *much* bigger, with a radius of about 70,000 km. Over a thousand Earths would fit inside Jupiter (and a thousand Jupiters would fit inside the Sun).

The consequence of the first item is that the circulation of the atmosphere may extend thousands of kilometres into the planetary interior, but we cannot be sure exactly how far. The upper few tens of kilometres of Jupiter's atmosphere are called its 'weather layer', for here we see swirling vortices and jets, somewhat resembling the weather patterns on Earth. The temperature structure (Fig. 13.9) shows a very stably stratified stratosphere lying above a troposphere where the clouds exist, giving us the visual images seen in Fig. 13.11. This weather layer is composed mainly of gaseous hydrogen (H_2) and is perhaps 50 km thick, beneath which is (it is believed) a neutrally stratified layer extending many thousands of kilometres into the planet's interior, and where the hydrogen eventually liquefies because of the immense pressures. This layer is neutrally stratified because it is convective, with the energy source being the heat coming from Jupiter's interior as it very slowly cools. At sufficient pressure the hydrogen starts to behave like a metal, with the layer of metallic hydrogen extending almost to the centre of the planet.

The consequence of the large radius of Jupiter, combined with the planet's short length of day, is that Jupiter is by any measure a 'rapid rotator' — as we calculated in Section 13.2.4, its external Rossby number is about 2×10^{-5}, a few thousand times smaller than that of Earth. The rapid rotation gives rise to zonal jets, as we now discuss.

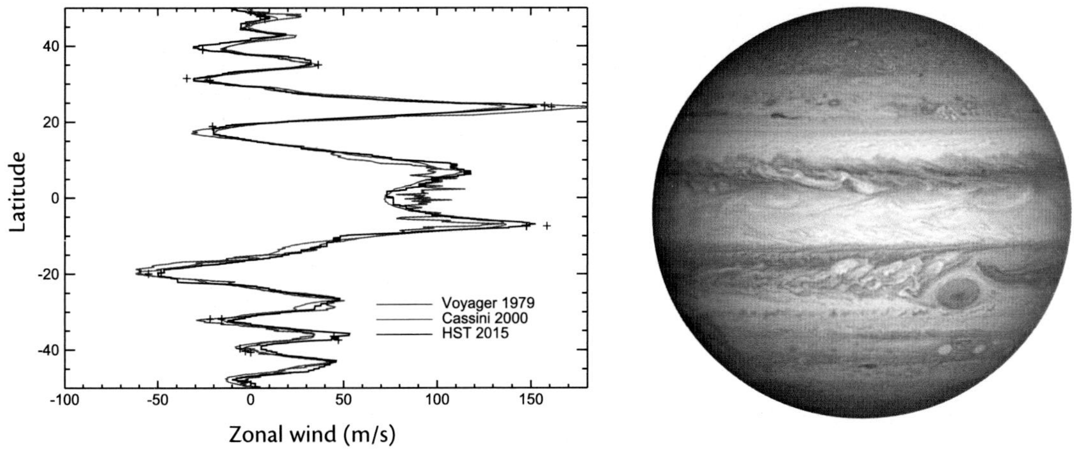

Fig. 13.11: Left: Zonal-average zonal winds on Jupiter at cloud top. The blue line is the Cassini profile in late 2000 (Porco *et al.* 2003), the red line is from Voyager in 1979 (Simon-Miller & Gierasch 2010). The black line is from the Hubble Space Telescope in 2015 (Simon *et al.* 2015, from which the figure is adapted). Right: An image of Jupiter from the Cassini mission.

13.4.3 Jets

Perhaps the most striking visual feature of Jupiter's atmosphere is the strong alternating zonal (i.e., east–west) winds, with a large prograde (i.e., in the direction of rotation) equatorial jet and smaller jets of alternating direction at higher latitudes (see Fig. 13.11). The fact that there are zonal jets is not, in itself, hard to explain, although understanding the actual structure of the jets — their magnitude, width and depth, and why the equatorial jet is prograde is more of a challenge and remains a topic of research.

If there is a source of energy for atmospheric motion then, given Jupiter's rapid rotation rate, zonal jets are an almost inevitable consequence. There are two sources of energy for Jupiter's atmosphere:

(*i*) An average incoming solar radiation of $12.5\,\mathrm{W\,m^{-2}}$, about a third of which is reflected back to space, mainly by clouds. The incoming radiation is much less than that for Earth (where the corresponding value is $340\,\mathrm{W\,m^{-2}}$), but nonetheless important.

(*ii*) An internal source of energy of about $6\,\mathrm{W\,m^{-2}}$. That is, the planet itself has a hot interior and emits radiation, mainly in the infra-red.

If these sources of energy are able to stir the fluid in the weather layer and give rise to eddying motion, then the rapid rotation will organize the flow into jets, just as beta-plane turbulence becomes organized into jets, as we saw in Chapter 10 and again in Chapter 12. Observations tell us that the zonal jets themselves have velocities of about $100\,\mathrm{m\,s^{-1}}$, and that the eddy velocities within the jets are rather less than that. If the jets are primarily a weather-layer phenomenon then an estimate of the meridional jet scale is given by the (internal) Rhines scale

$$L_R = \sqrt{\frac{U}{\beta}}. \tag{13.32}$$

Taking $U = 50\,\mathrm{m\,s^{-1}}$ and $\beta = 2.5\times10^{-12}\,\mathrm{m^{-1}\,s^{-1}}$ we obtain $L_R \approx 4,500\,\mathrm{km}$. The distance between equator and pole is about 110,000 km, suggesting

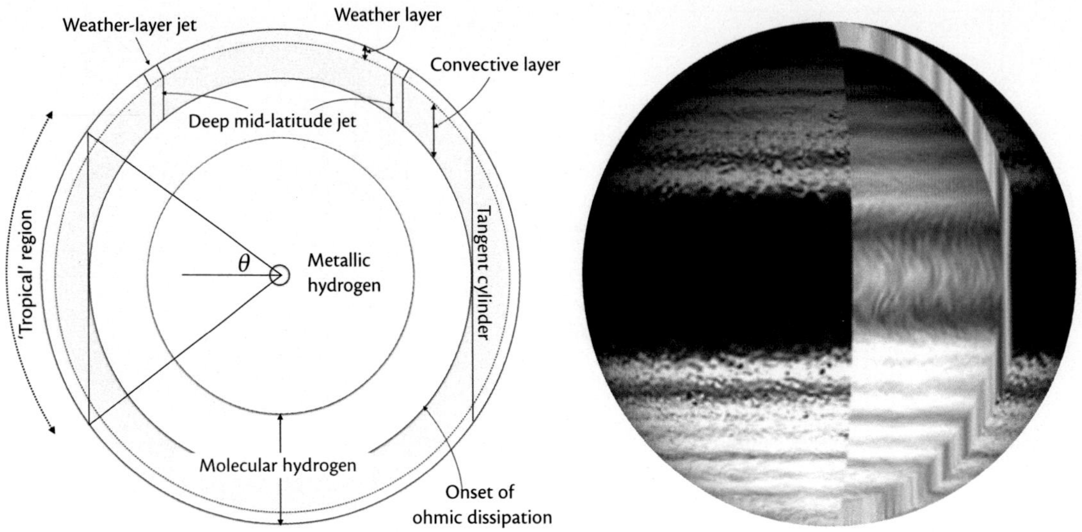

Fig. 13.12: Left: Sketch of the potential structure of Jupiter's atmosphere (not to scale) with the jets in a convective layer between a layer of ohmic dissipation and the very thin weather layer. Right: The zonal velocity on Jupiter obtained from a numerical simulation (from Heimpel *et al.* 2016) of a spherical shell with inner radius equal to 0.9 of the planetary radius, giving $\theta = 23°$. Red colours denote eastward flow and blue colours westward.

there might be about 10 jets between equator and pole. This is a little larger than the number observed (about 6) but the agreement is as good as can be expected given the nature of the scaling argument.

It is, however, quite possible that the jets extend a few thousand kilometres, and perhaps considerably more, into Jupiter's interior, as shown in Fig. 13.12. In this picture, a significant source of energy is the heat emanating from the planetary interior, and the ensuing convection creates a neutrally stratified region extending up to the weather layer, where solar absorption — and possibly water vapour and baroclinic instability — stabilize the fluid to dry convection. How deep the jets go is not known with any certainty. It was once thought that they should go down as far as the metallic core, but other ideas posit that 'ohmic dissipation', which arises because of the finite electrical conductivity of the molecular hydrogen, acts much closer to the surface than the metallic layer and may prevent the jets extending much deeper than a few thousand kilometres, but the exact depth at which this dissipation becomes significant remains uncertain. Assuming this deeper layer (the 'convective layer') in Fig. 13.9 does exist then jets will form within it, as we now discuss.

Jets in a deep atmosphere

If the jets on Jupiter do descend just a few thousand kilometres into the interior then they may be regarded as deep from the point of view of a meteorologist interested in the weather layer, but shallow from the point of a scientist studying planetary interiors!

Consider the schematic in the left panel in Fig. 13.12. We will suppose that the planetary radius is a and that there is a convective layer between some inner shell at a depth d and the outer radius (where the very thin weather layer resides). This convective layer may be naturally divided into three regions, one equatorward of the intersection of the tangent cylinder with the outer radius (and so with latitude less than θ) and denoted the 'tropical' region in Fig. 13.12, and two regions poleward of that, one in either hemisphere. Simple geometry indicates that the angle θ is given by $\theta = \cos^{-1}(a - d)/a$, with $\theta = 15°$ corresponding to $d = 2,400\,\text{km}$. If on the other hand we were to take the inner radius to that of the transition to metallic hydrogen, and so with $d \approx 15{,}000\,\mathrm{km}$, then $\theta \approx 40°$. Let us make a simple model of these convective regions to illustrate how jets form within them, with a super-rotating jet in the tropical region.

We will model the convective region as a layer of shallow water, obeying the potential vorticity equation

$$\frac{\mathrm{D}Q}{\mathrm{D}t} = 0, \qquad Q = \left(\frac{\zeta + 2\Omega}{h}\right). \tag{13.33}$$

Perhaps to a greater extent than elsewhere in the book, here we are making a somewhat speculative model of the phenomenon, not a quantitative or exact theory.

Here Q is the potential vorticity, Ω is the rotation rate of the planet (a constant) and ζ is the vorticity aligned with the planetary rotation (and not with the radial direction which would be conventional in a shallow atmosphere). The quantity h is the thickness of the convecting layer and we write this as $h = H + h'$, where H is the mean shell thickness and h' are small, time-dependent, deviations of that due to fluid motion, and $H \gg h'$. From Fig. 13.13 (and Fig. 13.12) we see that H varies in the y direction, decreasing toward the pole in the region poleward of the intersection with the tangent cylinder, but decreasing toward the equator in the region equatorward of the intersection with the tangent cylinder. It is this variation with mean thickness, and hence the variation of the background potential vorticity, that gives rise to a 'topographic beta effect' and hence to zonal jets. To see this explicitly, we make two more assumptions:

(i) The small Rossby number assumption, that $|2\Omega| \gg |\zeta|$.

(ii) The variations in mean height occur on a larger scale than the variations in vorticity.

The potential vorticity is then given by

$$Q = \left(\frac{\zeta + 2\Omega}{h}\right) \approx \left(\frac{\zeta + 2\Omega}{H}\right), \tag{13.34}$$

and, using the assumptions above, its evolution is given by

$$\frac{\mathrm{D}Q}{\mathrm{D}t} \approx \frac{1}{H}\frac{\mathrm{D}\zeta}{\mathrm{D}t} + 2\Omega\frac{\mathrm{D}}{\mathrm{D}t}\left(\frac{1}{H}\right) = \frac{1}{H}\frac{\mathrm{D}\zeta}{\mathrm{D}t} - \frac{2\Omega}{H^2}\boldsymbol{v}\cdot\nabla H, \tag{13.35}$$

and (13.33) becomes

$$\frac{\mathrm{D}\zeta}{\mathrm{D}t} + \beta^* v = 0 \qquad \text{where} \qquad \beta^* = -\frac{2\Omega}{H}\frac{\partial H}{\partial y}, \tag{13.36a,b}$$

where v is the velocity in the y-direction. We see from Fig. 13.13 that β^* is positive in the region insider the tangent cylinder (the extra-tropics) and negative outside the tangent cylinder, in the tropics.

The planetary rotation organizes the convection below the weather layer into alternating zonal jets because of the topographic beta effect that arises due to the variations in thickness of the convecting layer. The convecting layer effectively forms a lower boundary condition for the weather layer, but the interaction of the two layers is far from understood and the jets in the weather layer might not even have a one-to-one correspondence with those in the convective layer.

Equation (13.36) is very similar to the familiar barotropic vorticity equation on the β-plane — compare it with (6.23) or (10.63). Thus, if the flow is turbulent, we may expect alternating zonal jets to form because of the interaction of Rossby waves with the eddying flow. The intensity of these jets, and the spacing between them, depends on the size of the turbulent flow produced by the convection, and that in turn depends on the heat flux coming up from the planetary interior and the viscosity. The value of β^* is actually similar to the value of β itself, because both are a consequence of the sphericity of the planet.

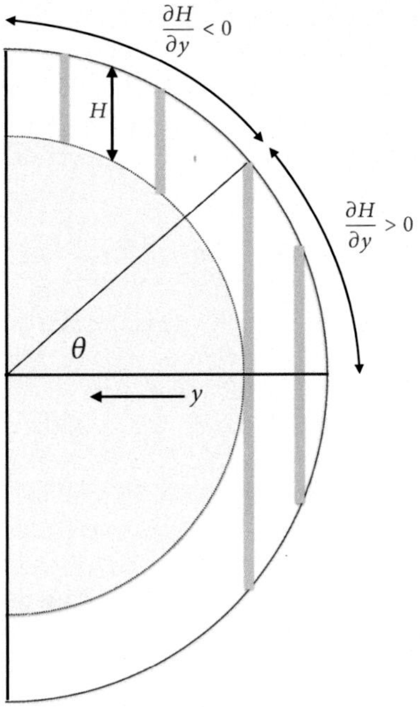

Fig. 13.13: Geometry of columns in a shell. For columns poleward of latitude θ the column height diminishes with latitude and $\partial H/\partial y < 0$, whereas the column height increases with latitude outside the tangent cylinder and $\partial H/\partial y > 0$. The consequence is that the beta parameter, β^*, is positive inside the tangent cylinder (i.e. at higher latitudes) and negative outside the tangent cylinder in the tropical region.

Tropical super-rotation

In the tropical region the zonal flow on Jupiter is *prograde,* that is, in the direction of the planetary rotation, and the atmosphere super-rotates (Fig. 13.11). The analogous jet on Saturn is also prograde. Why should this be so? As discussed above with respect to Venus, the mechanism must involve eddy motion and a plausible explanation is related to the nature of the beta effect in this region. In the tropical region outside the tangent cylinder the beta parameter, β^* given by (13.36b), is negative because the length of the convecting column *increases* with the y-coordinate. (In this region it is given by $H = 2a \sin \vartheta$, where ϑ is latitude and a is the planetary radius.) To simplify matters, let us suppose that β^* in (13.36) is a constant in which case we can write

An alternative explanation for tropical super-rotation on giant planets posits that it is caused by the propagation of Rossby waves away from the equatorial region, similar to the way eastward flow is formed in the midlatitudes on Earth (Section 12.1). This mechanism can operate in the weather layer alone, provided there is a source for the Rossby waves such as moist convection or a heat flux from the interior, but does not preclude the jets going deeper into the interior.

$$\frac{DQ^*}{Dt} = 0, \qquad Q^* = \zeta + \beta^* y. \tag{13.37}$$

If the dynamics are turbulent in the tropical region (for example, if they are stirred by convection) then they will seek to homogenize the potential vorticity, as described in Section 10.6.2, and the zonally-averaged flow is given by

$$\frac{\partial \overline{u}}{\partial y} = A - |\beta^*| y, \qquad \text{or} \qquad \overline{u} = Ay - \frac{1}{2}|\beta^*| y^2 + B, \tag{13.38}$$

where A and B are constants. If $\partial \overline{u}/\partial y = 0$ at $y = 0$ then $A = 0$ and u takes its *maximum* value at $y = 0$; that is, we expect the flow to be prograde, or super-rotating, at the equator.

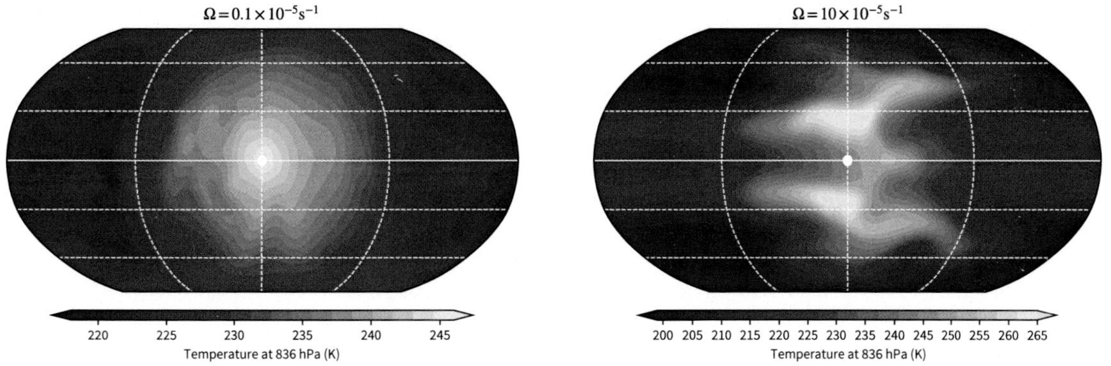

Fig. 13.14: Snapshots of temperature field in the lower troposphere in numerical simulations of two tidally-locked planets, each with the same radius, gravity and atmospheric mass as Earth. The planet on the left has a rotation period of 73 days, and that on the right a period of 18 hours. The substellar point (the position on the planet where the host star is overhead) is stationary, denoted by the small white dot in the centre of each plot. Note the different scales on the colour maps. (Simulation and figure courtesy of James Penn.)

The above argument does not imply that the entire tropical region will necessarily super-rotate. The entire region will only super-rotate if its latitudinal extent is smaller than the Rhines scale. If this is not the case then alternating jets can form within the tropical region, with the westward jets then tending to be sharper than the eastward ones because of the negative beta. (There is in fact some evidence of multiple jets in the tropical region, with distinct eastward maxima at about 7°N and 7°S evident in Fig. 13.11. However, the relatively westward jet between them is no sharper and other mechanisms could be causing these local extrema.)

Finally, we note that the super-rotating jets that are visible in the weather layer are not necessarily coincident with, or the same width as, the jets in the deep layer. although that may be the most likely state. A lack of coincidence could occur if the Rhines scale in the weather layer, and hence the jet width, were significantly different from that in the convective layer, and that would be the case if there were other sources of energy in the weather layer, or if the convective stirring were damped in the weather layer by the stable stratification. The interaction between the weather layer and the deep convective layer is, perhaps needless to say, poorly understood.

13.5 Exoplanets and Tidal Locking

Venus, a slowly rotating terrestrial planet, Jupiter, a rapidly rotating gas giant, as well as Earth itself, span a very large parameter regime that is likely to encompass many planets outside the Solar System. However, there is at least one class of exoplanets that has no analogue in the Solar System, namely tidally-locked planets. When a planet is close to its host star it feels a strong tidal force that drags the planet into a spin-orbit alignment. In this configuration the orbital period and rotation period are equal so that the same face of the planet always points to the star (just as the same face of the Moon always points toward Earth) creating permanent day and night hemispheres.

Figure 13.14 shows the temperature in the lower troposphere in two numerical simulations of tidally-locked Earth-like planets. In both cases the heating is stationary with respect to the planet's surface, centred on the equator and diminishing with both latitude and longitude away from the substellar point (marked with a small white dot). The planet on the left rotates slowly, in about 73 (Earth) days, whereas the planet on the right completes an orbit in 18 hours. For the slowly rotating planet there is efficient advective heat redistribution from day side (the two central quadrants) to night accomplished by a large-scale overturning circulation spanning the entire planet, and the difference in temperature between the day and night side is about 30 K. In the high rotation case the day-night temperature contrast is double that, about 60 K. In this case the circulation is much more localized to the tropics and to the day side of the planet. The heating generates both Kelvin waves and Rossby waves that propagate east and west from the disturbance, respectively, and which are damped by friction forming trapped, quasi-stationary patterns.

Many exoplanets thus far detected are giant planets close to their host star, and are likely to be tidally locked. Such hot Jupiters are relatively easy to detect because when they pass between their host star and us (or behind their host star) there is a detectable change in the solar radiation that we receive. However, most exoplanets in the Universe are almost certainly not tidally-locked hot Jupiters and solar systems with a variety of different planets within them have been discovered. For example, at least five planets of different types are in orbit in the Cancri system and seven terrestrial planets are in orbit in the TRAPPIST-1 system.

The atmosphere becomes hottest in the 'Rossby gyres', which are the cyclonic lobes in both hemispheres west and slightly poleward of the substellar point and which inhibit meridional heat transport. In the more slowly rotating case the deformation radius, NH/f, is larger than the planet itself and there is no such trapping, and although the hotspot is more-or-less coincident with the substellar point the temperature is much more uniform over the entire planetary surface. Because of the weak Coriolis effect the entire planet is in some ways a tropical one obeying the weak temperature gradient approximation discussed in Section 11.4.

Notes and References

A readable introduction to planetary atmospheres can be found in the book by Ingersoll (2013). The book by Sanchez-Lavega (2011) also discusses such topics as atmospheric evolution, chemistry and clouds as well as some dynamics. The book by Pierrehumbert (2010) provides a more complete treatment of radiative and other diabatic processes and discusses planetary climates as a whole. A number of useful review articles on Venus can be found in Bengtsson *et al.* (2013), and there are a large number of review and parameter-surveying articles that discuss various aspects of planetary atmospheres, including Kaspi & Showman (2015), Showman *et al.* (2010), Read *et al.* (2018) and Sanchez-Lavega & Heimpel (2018). Figures 13.1, 13.2 and 13.14 were constructed using the Isca modelling framework (Vallis *et al.* 2018).

Problems

13.1 Suppose that in the model of the Hadley Cell in Section 11.2 the radiative equilibrium temperature fell linearly with latitude, rather than having a quadratic dependence as in (11.9). Obtain a scaling for how the extent of the Hadley Cell varies that is analogous to (11.10).

13.2 (*a*) Consider the angular momentum conserving model of the Hadley Cell in Section 11.2. For a planet otherwise similar to Earth, at what value of the planetary rotation rate will the Hadley Cell extend all the way to the pole (and what then is the length of day compared to Earth)? What is the value of the zonal wind in this case? You may do this problem first in the small angle approximation and then, with more difficulty, try without making that approximation. Are there any planetary bodies in the Solar System for which this is the case? How far poleward can the Hadley Cell be expected to go on Venus, Mars, Titan and Jupiter, assuming it obeys this model?

(*b*) As the rotation rate falls, the latitude at which baroclinic instability occurs also moves poleward. At lower rotation rates, is baroclinic instability less or more likely to be a limiting factor on the Hadley Cell extent, compared to Earth? How would this result differ if the temperature fell linearly with latitude as in Problem 13.1 above?

13.3 Suppose that we are modelling a planet with an atmosphere that obeys the primitive equations with a certain external Rossby number, Ekman number etc., as in Section 13.2.1. Suppose we double the rotation rate of the planet, but that we wish only to change the external Rossby number. What other physical parameters would we need to change to ensure this? Is there an easier way to change only the external Rossby number?

13.4 (*a*) Calculate and plot the saturation vapour pressure of carbon dioxide using an approximate solution to the Clausius–Clapeyron equation such as

$$e_s = e_0 \exp\left[\frac{L}{R^v}\left(\frac{1}{T_0} - \frac{1}{T}\right)\right]. \tag{P13.1}$$

(You will need to look up values of various constants.)

(*b*) Venus has a CO_2 atmosphere with surface pressure of 92 bars (or 92×10^5 Pascals). Is it in danger of raining CO_2?

(*c*) The surface temperature on Mars can drop below 200 K; can the atmosphere sustain gaseous CO_2 and if so how much?

(*d*) On Titan, the surface temperature is about 100 K and the atmosphere contains methane. Approximately how much methane can the atmosphere hold before methane rain occurs?

Part III

Oceans

CHAPTER

14 Wind-Driven Gyres

WE NOW START OUR VOYAGE into that other great fluid covering the Earth, the ocean, and we divide the voyage into three legs. In this first one we look at the essentially horizontal circulation that gives rise to the great gyres in the mid- and high latitudes. In the next chapter we look at the processes giving rise to the vertical structure of the ocean and the meridional overturning circulation, and in the third chapter we look at equatorial circulation and El Niño. Let us first take a brief look at the observations to see what we have to understand.

14.1 AN OBSERVATIONAL OVERVIEW

The aspect of the ocean that most affects the climate is the sea-surface temperature (SST), illustrated in Fig. 14.1. Aside from the to-be-expected latitudinal variation there is significant zonal variation — the western tropical Pacific is particularly warm, and the western Atlantic is warmer than the corresponding latitude in the east. These variations owe their existence to ocean currents, and the vertically averaged currents of the North Atlantic are illustrated in Fig. 14.2. The most striking features are the two main gyres — the clockwise, and anticyclonic, subtropical gyre between about 25° N and 50° N, and the anti-clockwise, and cyclonic, subpolar gyre north of that. We can see that these gyres are intensified in the west; the intensification is most obvious in the subtropical gyre, where the intense northward flowing current is known as the Gulf Stream, but is also present in the subpolar gyre.

The same features are present in *all* of the main basins of the world's ocean, as we see in Fig. 14.3, in both Northern and Southern Hemispheres. The western boundary current of the great subtropical gyre in the North Pacific, flowing northward off the coast of Japan, is known as the Kuroshio, and similar currents flow southward along the west coast of Australia and the west coast of Brazil and Argentina in the Southern

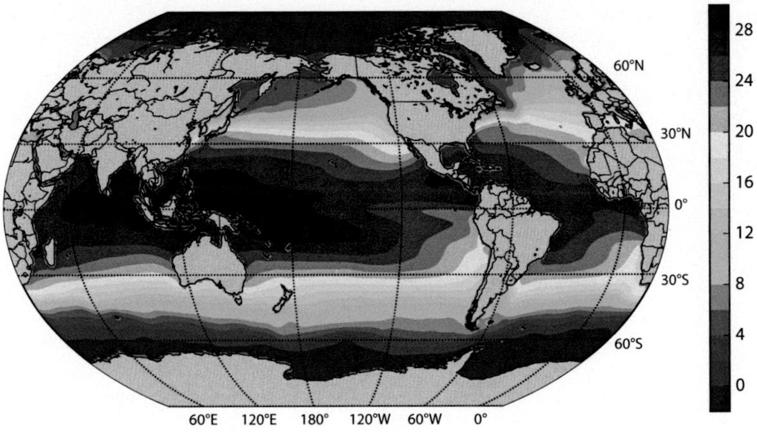

Fig. 14.1: The sea-surface temperature (SST, °C) of the world's ocean, as determined from a great many observations, combined in the World Ocean Circulation Experiment (WOCE).

Hemisphere. The existence of the great gyres, and that they are strongest in the west, has been known for centuries, and our main task in this chapter is to explain that. It turns out to be a much easier task to explain the vertically-integrated flow than the vertical structure of the flow and that is our focus.

14.2 SVERDRUP BALANCE

Harald Sverdrup (1888–1957) was a Norwegian meteorologist/oceanographer who is most famous for the balance that now bears his name, but he also played a leadership role in scientific policy and was the director of Scripps Institution of Oceanography from 1936–1948.

Let us begin by considering an ocean, forced by a wind stress, $\boldsymbol{\tau}_0 = \tau_0^x \hat{\mathbf{i}} + \tau_0^y \hat{\mathbf{j}}$, at the top, that satisfies the equations

$$-fv = -\frac{\partial \phi}{\partial x} + \frac{1}{\rho_0}\frac{\partial \tau^x}{\partial z}, \qquad fu = -\frac{\partial \phi}{\partial y} + \frac{1}{\rho_0}\frac{\partial \tau^y}{\partial z}, \qquad (14.1\text{a,b})$$

and $\tau_x \hat{\mathbf{i}} + \tau_y \hat{\mathbf{j}} = \boldsymbol{\tau}$ is the stress acting on the fluid. Since the ocean's density is very nearly constant we absorb the quantity $1/\rho_0$ into the definition of stress (the quantities $(\tau_x, \tau_y)/\rho_0$ are the 'kinematic stress' but are commonly, if a little loosely, just referred to as the stress). With this new definition of stress we rewrite (14.1) as

$$f(v_g - v) = \frac{\partial \tau^x}{\partial z}, \qquad f(u - u_g) = \frac{\partial \tau^y}{\partial z}, \qquad (14.2)$$

where (u_g, v_g) are the geostrophic velocities given by $f(u_g, v_g) = (-\partial\phi/\partial y, \partial\phi/\partial x)$. The left-hand side is just the ageostrophic velocity, and if we integrate vertically from the top of the ocean to the base of the Ekman layer, where the stress is by definition zero, we obtain

$$fV_a = -\tau_0^x, \qquad fU_a = \tau_0^y, \qquad \text{or} \qquad f\boldsymbol{U}_a = \hat{\mathbf{k}} \times \boldsymbol{\tau}_0, \qquad (14.3)$$

where $U_a = \int_{\mathrm{Ek}} (u - u_g)\, \mathrm{d}z$ is the integral of the ageostrophic velocity over the Ekman layer, and similarly for V_a, and $\boldsymbol{U}_a = U_a \hat{\mathbf{i}} + V_a \hat{\mathbf{j}}$. Evidently the ageostrophic Ekman transport is at right angles to the surface stress, and in the ocean the Ekman layer is of order tens of metres thick.

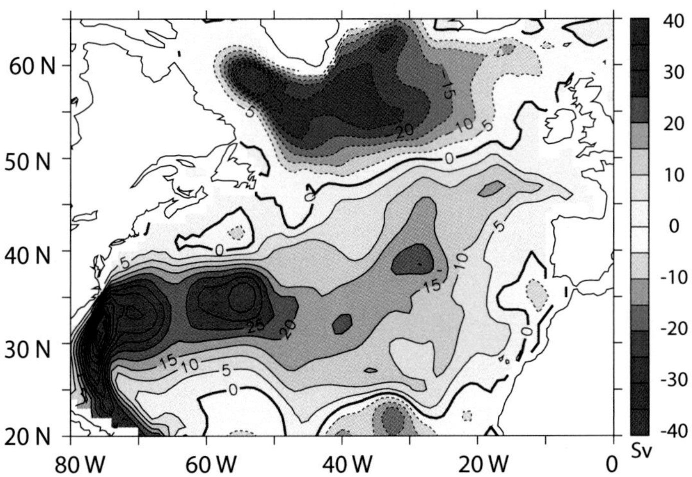

Fig. 14.2: The streamfunction of the vertically averaged flow in the North Atlantic, obtained by constraining a numerical model to observations so giving a 'state estimate'. Red shading indicates clockwise flow, and blue shading anticlockwise. (Courtesy of Rong Zhang using a GFDL model and climatologcal data.)

There is one particularly useful result we can obtain from (14.1). If we cross differentiate and use the mass conservation equation, $\partial u/\partial x + \partial v/\partial y + \partial w/\partial z = 0$, we obtain

$$f\frac{\partial w}{\partial z} + \beta v = \frac{\partial \tau^y}{\partial x} - \frac{\partial \tau^x}{\partial y}. \tag{14.4}$$

Now integrate from the top of the ocean (where $w = 0$) down to some level, z, below the base of the Ekman layer where the stress is zero, to obtain

$$w(z) + \int_z^0 \frac{\beta}{f} v \,\mathrm{d}z' = \frac{1}{f}\left[\frac{\partial \tau_0^y}{\partial x} - \frac{\partial \tau_0^x}{\partial y}\right]. \tag{14.5}$$

If we let the integral go over the entire depth of the ocean, and assume that the vertical velocity and the stress are zero at the ocean bottom, we obtain

$$\int \beta v \,\mathrm{d}z = \frac{\partial \tau_0^y}{\partial x} - \frac{\partial \tau_0^x}{\partial y}. \tag{14.6}$$

This expression is known as the *Sverdrup relation*. It is remarkable because it tells us that, at any location in the ocean, *the vertically integrated meridional velocity is given by the curl of the wind stress at the surface.* Although there are a number of caveats to this statement (as our assumptions are not exactly satisfied), the Sverdrup relation is one of the enduring foundations of physical oceanography.

14.3 Ocean Gyres

The equations of motion that govern the three-dimensional, large-scale flow in the oceans are the planetary-geostrophic equations, discussed in Chapter 5, namely

$$\frac{\mathrm{D}b}{\mathrm{D}t} = \dot{b}, \qquad \nabla_3 \cdot \boldsymbol{v} = 0, \tag{14.7a,b}$$

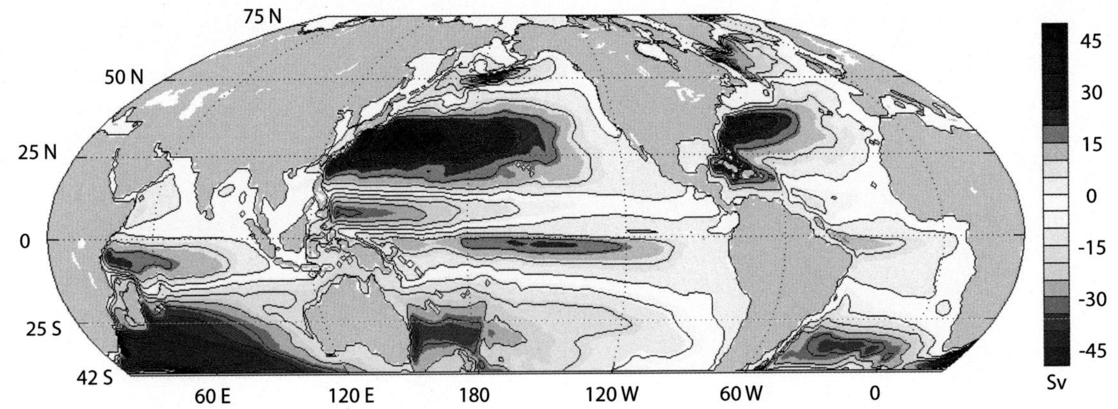

Fig. 14.3: A state estimate of the streamfunction of the vertically integrated flow for the near global ocean. Red shading indicates clockwise flow, and blue shading anticlockwise, but in both hemispheres the subtropical gyres are anticyclonic and the subpolar ones are cyclonic. (Courtesy of Patrick Heimbach, using the ECCO system.)

$$\boldsymbol{f} \times \boldsymbol{u} = -\nabla\phi + \frac{\partial \boldsymbol{\tau}}{\partial z}, \qquad \frac{\partial \phi}{\partial z} = b. \tag{14.8a,b}$$

These equations are, respectively, the thermodynamic equation (14.7a), the mass continuity equation (14.7b), the horizontal momentum equation (14.8a), (i.e., geostrophic balance, plus a stress term), and the vertical momentum equation (14.8b) — that is, hydrostatic balance. The gradient and divergence operators are two dimensional, in the x–y plane, unless noted with a subscript 3. Simple as they may be compared to the full Navier–Stokes equations, the equations are still quite daunting: a prognostic equation for buoyancy is coupled to the advecting velocity via hydrostatic and geostrophic balance, and the resulting problem is quite nonlinear. However, it turns out that thermodynamic effects can effectively be eliminated by the simple device of vertical integration; the resulting equations are linear, and the only external forcing is that due to the wind stress. This device enables us to construct a rather simple but very revealing model of the ocean circulation, as follows.

The model of ocean gyres described here is due to Henry Stommel (1920–1992), one of the most creative physical oceanographers of the twentieth century. He spent most of his career at the Woods Hole Oceanographic Institution (WHOI) and played a major role in making physical oceanography a quantitative scientific endeavour. His forté was the construction and use of models that were both simple and relevant, and the observational and experimental testing of such models.

14.3.1 The Stommel Model

Take the curl of (14.8a) (that is, cross-differentiate its x and y components) and integrate over the depth of the ocean to give

$$\int f\nabla \cdot \boldsymbol{u}\,\mathrm{d}z + \frac{\partial f}{\partial y}\int v\,\mathrm{d}z = \mathrm{curl}_z(\boldsymbol{\tau}_T - \boldsymbol{\tau}_B), \tag{14.9}$$

where the operator curl_z is defined by $\mathrm{curl}_z\boldsymbol{A} \equiv \partial A^y/\partial x - \partial A^x/\partial y = \hat{\mathbf{k}}\cdot\nabla\times\boldsymbol{A}$, and the subscripts T and B are for top and bottom; the stress at the bottom, although small, must be retained to find a solution, as we will discover. Equation (14.9) then becomes

$$\beta V = \mathrm{curl}_z(\boldsymbol{\tau}_T - \boldsymbol{\tau}_B), \tag{14.10}$$

where V is the vertical integral of v over the entire depth of the ocean (and similarly for U later on). Evidently, the thermodynamic fields do not affect the vertically integrated flow.

At the top of the ocean, the stress comes from the wind above. At the bottom we assume that the stress may be parameterized by a linear drag acting on the vertically integrated flow, and it is this assumption that particularly characterizes this model as being due to Stommel. The bottom stress, $\boldsymbol{\tau}_B$, and its curl are then given by

$$\boldsymbol{\tau}_B = r(U\hat{\mathbf{i}} + V\hat{\mathbf{j}}) \qquad \text{and} \qquad \mathrm{curl}_z \boldsymbol{\tau}_B = r\left(\frac{\partial V}{\partial x} - \frac{\partial U}{\partial y}\right) = rZ, \tag{14.11}$$

where Z is the vertically integrated vorticity. Equation (14.10) then becomes

$$\beta V = -rZ + \mathrm{curl}_z \boldsymbol{\tau}_T, \tag{14.12}$$

where $\mathrm{curl}_z \boldsymbol{\tau}_T$ is the wind-stress curl at the top of the ocean. Because the vertical integrated velocity is divergence-free ($\partial U/\partial x + \partial V/\partial y = 0$), we can define a streamfunction ψ such that $U = -\partial\psi/\partial y$ and $V = \partial\psi/\partial x$ and $Z = \nabla^2\psi$. Equation (14.12) may then be written as

$$r\nabla^2\psi + \beta\frac{\partial\psi}{\partial x} = \mathrm{curl}_z \boldsymbol{\tau}_T. \tag{14.13}$$

This equation is often referred to as the *Stommel problem* or the *Stommel model,* and it may be posed in a variety of two-dimensional domains. If we can solve this equation for ψ we have a solution for the vertical integrated flow in our model of the ocean.

14.3.2 Approximate Solution of Stommel Model

Sverdrup balance

We will solve (14.13) perturbatively by supposing that the frictional term is small, meaning there is an approximate balance between wind stress and the β-effect. Friction is small if $|r\zeta| \ll |\beta v|$ or

$$\frac{r}{L} = \frac{f\delta_B}{HL} \ll \beta \tag{14.14}$$

using $r = f\delta_B/H$, and where L is the horizontal scale of the motion, and generally speaking this inequality is well satisfied for large-scale flow. The vorticity equation becomes

$$\beta\overline{v} \approx \mathrm{curl}_z \boldsymbol{\tau}_T, \tag{14.15}$$

which is just the Sverdrup balance, as derived in Section 14.2.

Boundary-layer solution

We set the problem a square domain of side a, and it is then natural to rescale the variables by setting

$$x = a\widehat{x}, \qquad y = a\widehat{y}, \qquad \tau = \tau_0\widehat{\tau}, \qquad \psi = \widehat{\psi}\frac{\overline{\tau}}{\beta}, \tag{14.16}$$

where $\overline{\tau}$ is the amplitude of the wind stress. The hatted variables are nondimensional and, assuming our scaling to be sensible, these are $\mathcal{O}(1)$ quantities in the interior. Equation (14.13) becomes

$$\frac{\partial \widehat{\psi}}{\partial \widehat{x}} + \epsilon_S \nabla^2 \widehat{\psi} = \operatorname{curl}_z \widehat{\boldsymbol{\tau}}_T, \tag{14.17}$$

where $\epsilon_S = (r/a\beta) \ll 1$, in accordance with (14.14). For the rest of this section we drop the hats over nondimensional quantities to avoid clutter. Over the interior of the domain, away from boundaries, the frictional term in (14.17) is small. We can take advantage of this by writing

$$\psi(x, y) = \psi_I(x, y) + \phi(x, y), \tag{14.18}$$

where ψ_I is the interior streamfunction and ϕ is a boundary layer correction. Away from boundaries ψ_I is presumed to dominate the flow, and this satisfies

$$\frac{\partial \psi_I}{\partial x} = \operatorname{curl}_z \boldsymbol{\tau}_T. \tag{14.19}$$

The solution of this equation (called the 'Sverdrup interior') is

$$\psi_I(x, y) = \int_0^x \operatorname{curl}_z \boldsymbol{\tau}(x', y)\, \mathrm{d}x' + g(y), \tag{14.20}$$

where $g(y)$ is an arbitrary function of integration that gives rise to an arbitrary zonal flow. The corresponding velocities are

$$v_I = \operatorname{curl}_z \boldsymbol{\tau}, \qquad u_I = -\frac{\partial}{\partial y} \int_0^x \operatorname{curl}_z \boldsymbol{\tau}_T(x', y)\, \mathrm{d}x' - \frac{\mathrm{d}g(y)}{\mathrm{d}y}. \tag{14.21}$$

The dynamics are most clearly illustrated if we now restrict our attention to a wind-stress curl that is zonally uniform and that vanishes at two latitudes, $y = 0$ and $y = 1$. An example is

$$\tau_T^y = 0, \qquad \tau_T^x = -\cos \pi y, \tag{14.22}$$

for which $\operatorname{curl}_z \boldsymbol{\tau}_T = -\pi \sin(\pi y)$. The Sverdrup (interior) flow may then be written as

$$\psi_I(x, y) = [x - C(y)] \operatorname{curl}_z \boldsymbol{\tau}_T = \pi[C(y) - x] \sin \pi y, \tag{14.23}$$

where $C(y)$ is an arbitrary function of integration $[C(y) = -g(y)/\operatorname{curl}_z \boldsymbol{\tau}]$. If we choose C to be a constant, the zonal flow associated with it is $C \operatorname{curl}_z \boldsymbol{\tau}_T$. We can then satisfy $\psi = 0$ at *either* $x = 0$ (if $C = 0$) *or* $x = 1$ (if $C = 1$). These solutions are illustrated in Fig. 14.4 for the particular stress (14.22).

Regardless of our choice of C we cannot satisfy $\psi = 0$ at both zonal boundaries. We must choose one, and then construct a *boundary layer* solution (i.e., we determine ϕ) to satisfy the other condition. Which choice do we make? On intuitive grounds it seems that we should choose the solution that satisfies $\psi = 0$ at $x = 1$ (the solution on the left in Fig. 14.4),

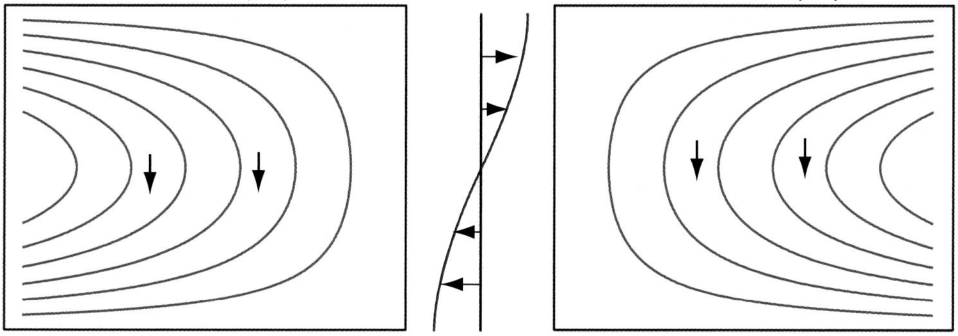

Fig. 14.4: Two possible Sverdrup flows, ψ_I, for the wind stress shown in the centre. Each solution satisfies the no-flow condition at either the eastern or western boundary, and a boundary layer is therefore required at the other boundary. Both flows have the same, equatorward, meridional flow in the interior. Only the flow with the western boundary current is physically realizable, however, because only then can friction produce a curl that opposes that of the wind stress, so allowing the flow to equilibrate.

for the interior flow then goes round in the same direction as the wind: the wind is supplying a clockwise torque, and to achieve an angular momentum balance anticlockwise angular momentum must be supplied by friction. We can imagine that this would be provided by the frictional forces at the western boundary layer if the interior flow is clockwise, but not by friction at an eastern boundary layer when the interior flow is anticlockwise. Note that this argument is not dependent on the sign of the wind-stress curl: if the wind blew the other way a similar argument still implies that a western boundary layer is needed. We will now see if and how the mathematics reflects this intuitive but non-rigorous argument.

Asymptotic matching

Near the walls of the domain the boundary layer correction $\phi(x, y)$ must become important in order that the boundary conditions may be satisfied, and the flow, and in particular $\phi(x, y)$, will vary rapidly with x. To reflect this, let us *stretch* the x-coordinate near this point of failure (i.e., at either $x = 0$ or $x = 1$, but we do not know at which yet) and let

$$x = \epsilon\alpha \qquad \text{or} \qquad x - 1 = \epsilon\alpha. \tag{14.24a,b}$$

Here, α is the stretched coordinate, which has values $\mathcal{O}(1)$ in the boundary layer, and ϵ is a small parameter, as yet undetermined. We then suppose that $\phi = \phi(\alpha, y)$, and using (14.18) in (14.17), we obtain

$$\epsilon_S(\nabla^2\psi_I + \nabla^2\phi) + \frac{\partial \psi_I}{\partial x} + \frac{1}{\epsilon}\frac{\partial \phi}{\partial \alpha} = \operatorname{curl}_z \boldsymbol{\tau}_T, \tag{14.25}$$

where $\phi = \phi(\alpha, y)$ and $\nabla^2\phi = \epsilon^{-2}\partial^2\phi/\partial\alpha^2 + \partial^2\phi/\partial y^2$. Now, by choice, ψ_I exactly satisfies Sverdrup balance, and so (14.25) becomes

$$\epsilon_S\left(\nabla^2\psi_I + \frac{1}{\epsilon^2}\frac{\partial^2\phi}{\partial\alpha^2} + \frac{\partial^2\phi}{\partial y^2}\right) + \frac{1}{\epsilon}\frac{\partial\phi}{\partial\alpha} = 0. \tag{14.26}$$

We now choose ϵ to obtain a physically meaningful solution. An obvious choice is $\epsilon = \epsilon_S$, for then the leading-order balance in (14.26) is

$$\frac{\partial^2\phi}{\partial\alpha^2} + \frac{\partial\phi}{\partial\alpha} = 0, \tag{14.27}$$

the solution of which is

$$\phi = A(y) + B(y)e^{-\alpha}. \tag{14.28}$$

Evidently, ϕ grows exponentially in the negative α direction. If this were allowed, it would violate our assumption that solutions are small in the interior, and we must eliminate this possibility by allowing α to take only positive values in the interior of the domain, and by setting $A(y) = 0$. We therefore choose $x = \epsilon\alpha$ so that $\alpha > 0$ for $x > 0$; the boundary layer is then at $x = 0$, that is, it is a *western boundary,* and it decays eastwards in the direction of increasing α — that is, into the ocean interior. We now choose $C = 1$ in (14.23) to make $\psi_I = 0$ at $x = 1$ in (14.23) and then, for the wind stress (14.22), the interior solution is given by

$$\psi_I = \pi(1 - x)\sin \pi y. \tag{14.29}$$

This alone satisfies the boundary condition at the eastern boundary. The function $B(y)$ is chosen to satisfy the additional condition that

$$\psi = \psi_I + \phi = 0 \qquad \text{at} \quad x = 0, \tag{14.30}$$

and using (14.29) this gives

$$\pi \sin \pi y + B(y) = 0. \tag{14.31}$$

Using this in (14.28), with $A(y) = 0$, then gives the boundary layer solution

$$\phi = -\pi \sin \pi y e^{-x/\epsilon_S}. \tag{14.32}$$

The composite (boundary layer plus interior) solution is the sum of (14.29) and (14.32), namely

$$\psi = (1 - x - e^{-x/\epsilon_S})\pi \sin \pi y. \tag{14.33a}$$

With dimensional variables this is

$$\psi = \frac{\tau_0 \pi}{\beta}\left(1 - \frac{x}{a} - e^{-x/(a\epsilon_S)}\right)\sin\frac{\pi y}{a}. \tag{14.33b}$$

This is a 'single gyre' solution. Two or more gyres can be obtained with a different wind forcing, such as $\tau^x = -\tau_0 \cos(2\pi y)$, as in Fig. 14.5.

It is a relatively straightforward matter to generalize to other wind stresses, provided these also vanish at the two latitudes between which the solution is desired. It is left as a problem to show that in general

$$\psi_I = \int_{x_E}^{x} \mathrm{curl}_z \boldsymbol{\tau}(x', y)\, dx', \tag{14.34}$$

and that the composite solution is

$$\psi = \psi_I - \psi_I(0, y)e^{-x/(x_E \epsilon_S)}. \tag{14.35}$$

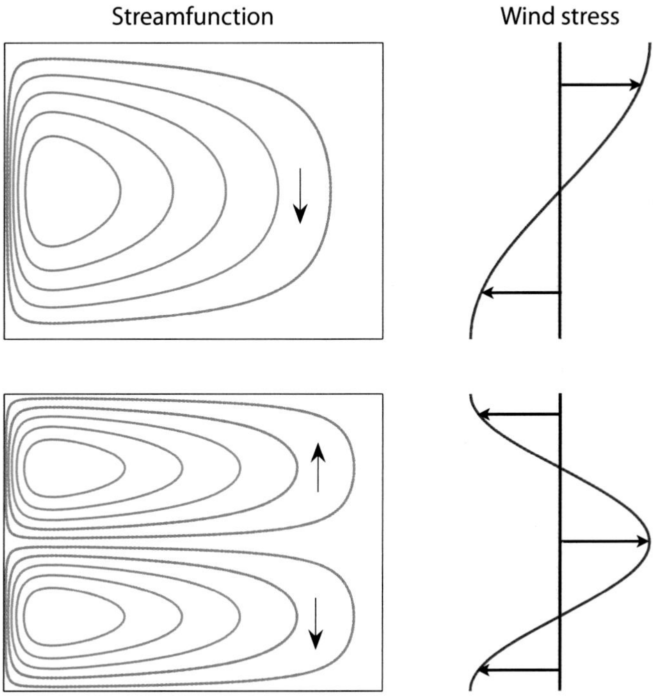

Fig. 14.5: Two solutions of the Stommel model. Upper panel shows the streamfunction of a single-gyre solution, with a wind stress proportional to $-\cos(\pi y/a)$ (in a domain of side a), and the lower panel shows a two-gyre solution, with wind stress proportional to $\cos(2\pi y/a)$. In both cases $\epsilon_S = 0.04$.

14.3.3 The Munk Problem: Using Viscosity Instead of Drag

A natural variation on the Stommel problem is to use a harmonic viscosity, $\nu\nabla^2\zeta$, in place of the drag term $-r\zeta$ in the vorticity equation, the argument being that the wind-driven circulation does not reach all the way to the ocean bottom so that an Ekman drag is not appropriate. This variation is called the 'Munk problem' or 'Munk model'. The problem is to find and understand the solution to the (dimensional) equation

Walter Munk (1917–) is a Viennese-born American physical oceanographer who spent most of his career at Scrippts Institution of Oceanography. He has made important contributions to a host of problems in oceanography, especially in the areas of waves and tides.

$$\beta\frac{\partial\psi}{\partial x} = \mathrm{curl}_z\boldsymbol{\tau}_T + \nu\nabla^2\zeta = \mathrm{curl}_z\boldsymbol{\tau}_T + \nu\nabla^4\psi \qquad (14.36)$$

in a given domain, for example a square of side a. The nondimensional version of this is

$$-\epsilon_M\nabla^4\widehat{\psi} + \frac{\partial\widehat{\psi}}{\partial\widehat{x}} = \mathrm{curl}_z\widehat{\boldsymbol{\tau}}_T, \qquad (14.37)$$

where $\epsilon_M = (\nu/\beta a^3)$.

Because the equation is of higher order we need two boundary conditions at each wall to solve the problem uniquely, and as before for one of them we choose $\psi = 0$ to satisfy the no-normal-flow condition. For the other condition it is common to use a no-slip condition; that is $\psi_n = 0$ where the subscript denotes the normal derivative of the streamfunction, so that, for example, at $x = 0$ and $x = a$ we have $v = 0$. As with the Stommel problem the solution may be found by boundary-layer methods, and

Stommel and Munk Models of the Wind-Driven Circulation

Formulation

- Vertically integrated planetary-geostrophic equations, or a homogeneous fluid with nonlinearity neglected.
- Friction parameterized by a linear drag (Stommel model) or a harmonic Newtonian viscosity (Munk model) or both (Stommel–Munk model).
- Flat bottomed ocean.

Properties

- The transport in the Sverdrup interior is equatorward for an anti-cyclonic wind-stress curl. This transport is exactly balanced by the poleward transport in the western boundary layer.
- There must be a boundary layer to satisfy mass conservation, and this must be a *western* boundary layer if the friction acts to provide a force of opposite sign to the motion itself. As there is a balance between friction and the β-effect, it is a 'frictional boundary layer'. The western location does not depend on the sign of the wind stress, nor on the sign of the Coriolis parameter, but it does depend on the sign of β, and so on the direction of rotation of the Earth.
- In the Stommel model the boundary layer width arises by noting that the terms $r\nabla^2\psi$ and $\beta\partial\psi/\partial x$ are in approximate balance in the western boundary layer, implying boundary-layer scale of $L_S = (r/\beta)$. If r, the inverse frictional time, is chosen to be $1/20\ \text{days}^{-1}$ then $L_S \approx 60\,\text{km}$, similar to the width of the Gulf Stream. Unless the wind has a special form the Sverdrup flow is non-zero on the zonal walls and there must also be boundary layers there, but they are weaker and less visible.
- In the Munk model the balance in western boundary layer is between $\nu\partial^4\psi/\partial x^4$ and $\beta\partial\psi/\partial x$, implying a scale of $L_M = (\nu/\beta)^{1/3}$. There is now a weak boundary layer on the eastern walls, to satisfy the no-slip (or free slip) condition, which is absent in the Stommel model.

after quite some algebra it is found to be

$$\widehat{\psi} = \pi \sin(\pi\widehat{y})\left\{1 - \widehat{x} - \mathrm{e}^{-\widehat{x}/(2\epsilon)}\left[\cos\left(\frac{\sqrt{3}\widehat{x}}{2\epsilon}\right) + \frac{1-2\epsilon}{\sqrt{3}}\sin\left(\frac{\sqrt{3}\widehat{x}}{2\epsilon}\right)\right] + \epsilon \mathrm{e}^{(\widehat{x}-1)/\epsilon}\right\}, \tag{14.38}$$

and this and the Stommel solution are plotted in Fig. 14.6. Evidently, the solution is similar to that of the Stommel problem — the interior is almost identical, being given by the Sverdrup interior flow, and a boundary layer occurs in the west. The boundary layer now brings the alongshore flow to zero, as well as the normal flow, and it has a different thickness to that

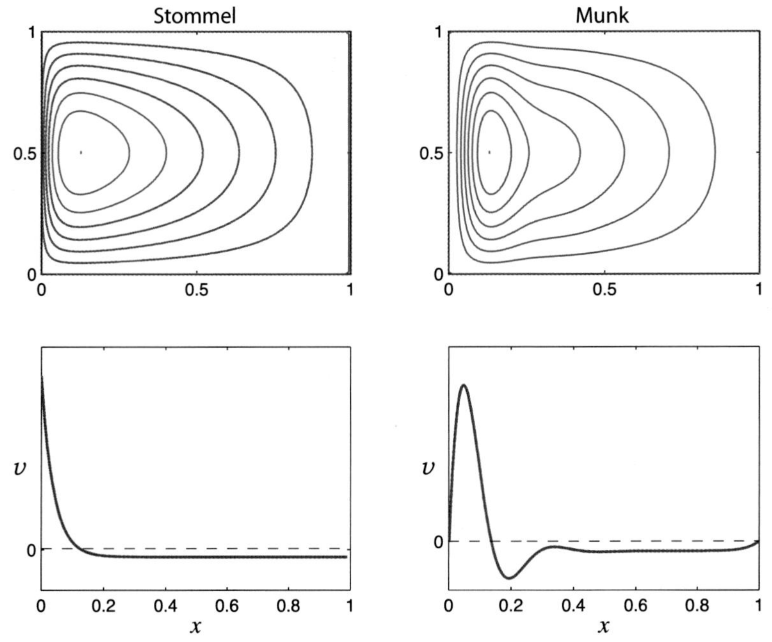

Fig. 14.6: The Stommel and Munk solutions, obtained using (14.33a) and (14.38) respectively, with $\epsilon_S = \epsilon_M^{1/3} = 0.04$, with the wind stress $\tau = -\cos \pi y$, for $x, y \in (0, 1)$. Upper panels are contours of streamfunction in the x–y plane, and the flow is clockwise. The lower panels are plots of meridional velocity, v, as a function of x, in the centre of the domain ($y = 0.5$). The Munk solution can satisfy both no-normal flow and one other boundary condition at each wall, here chosen to be no-slip.

of the Stommel problem.

Even without obtaining the solution, the western boundary layer thickness can be estimated by a scale analysis of (14.36), since in the boundary layer we expect the dominant balance to be between the beta term and the frictional term, namely

$$\beta \frac{\partial \psi}{\partial x} \sim \nu \frac{\partial^4 x}{\partial \psi^4}, \tag{14.39}$$

and the thickness of the boundary layer, L_M, is then given by

$$L_M = \left(\frac{\nu}{\beta} \right)^{1/3}. \tag{14.40}$$

Even though the frictional term in the Munk problem resembles a molecular viscosity the solution should not be regarded as being more realistic than the Stommel model. Rather, the similarity of the two model solutions attests the *qualitative* realism of both models and the relative insensitivity of the solution to the precise form of the friction, but neither model is quantitatively accurate. See the shaded box on the facing page for a summary of the two models.

14.3.4 Application to the North Atlantic

It is all very well to produce an intense western boundary current in a square domain, but another matter to produce a flow field that actually resembles the real ocean. To see if this can be done we solve the Stommel–Munk problem in a domain with realistic geometry, albeit a flat-bottomed

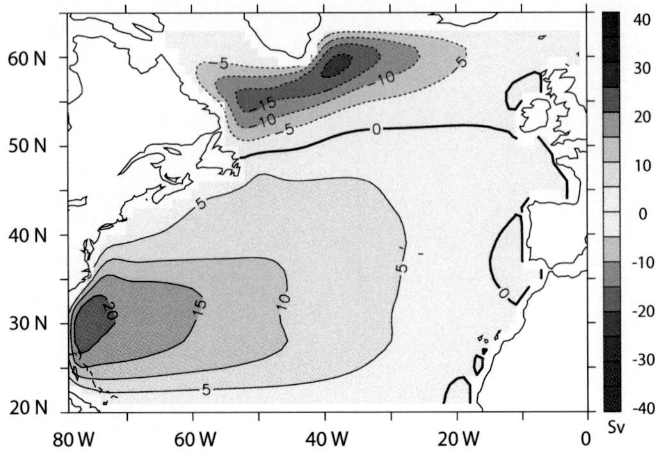

Fig. 14.7: The solution (streamfunction, in Sverdrups) to the Stommel–Munk problem numerically calculated for the North Atlantic, using the observed wind field. The calculation reproduces the observed large-scale patterns, but compared to observations the separation of the Gulf Stream from the coast is a little too far north and the Gulf Stream extension is too diffuse. Compare with Fig. 14.2 and Fig. 14.3.

domain with no topography, using the observed winds to drive the flow. The solution must be obtained numerically, but the equation we solve is equivalent to

$$\beta\frac{\partial\psi}{\partial x} = \operatorname{curl}_z \boldsymbol{\tau}_T - r\zeta + \nu\nabla^2\zeta = \operatorname{curl}_z \boldsymbol{\tau}_T - r\nabla^2\psi + \nu\nabla^4\psi, \tag{14.41}$$

and the solution is illustrated in Fig. 14.7.

The solution captures many of the features of the observed flow — notably the subtropical and subpolar gyres and the Gulf Stream itself — although it is quantitatively deficient in some aspects. For example, only the current in the west is intensified, and not the Gulf Stream extension as it flows eastward, and there is insufficient recirculation in the western part of the gyre. These failings can be ascribed in various ways to the lack of nonlinearity, stratification and bottom topography in the model problem, but overall the model provides a compelling explanation of some of the main features of the ocean circulation.

14.4 ♦ EFFECTS OF NONLINEARITY

Nonlinearity affects the ocean circulation in a variety of ways. One is that the ocean, like the atmosphere, is baroclinically unstable and the resulting baroclinic eddies lead to weather in the ocean, just a they do in the atmosphere. But even if we just consider the depth-integrated or barotropic flow then nonlinearity can be important, as we now explore.

14.4.1 A Nonlinear Wind-Driven Model

Let us suppose that the ocean satisfies the barotropic vorticity equation,

$$J(\psi,\zeta) + \beta\frac{\partial\psi}{\partial x} = \operatorname{curl}_z \boldsymbol{\tau}_T - r\zeta, \tag{14.42}$$

where $J(\psi,\zeta) = \partial_x\psi\partial_y\zeta - \partial_y\psi\partial_x\zeta = \boldsymbol{u}\cdot\nabla\zeta$. This equation is essentially the same as (14.12) except for the addition of the advective term. The addi-

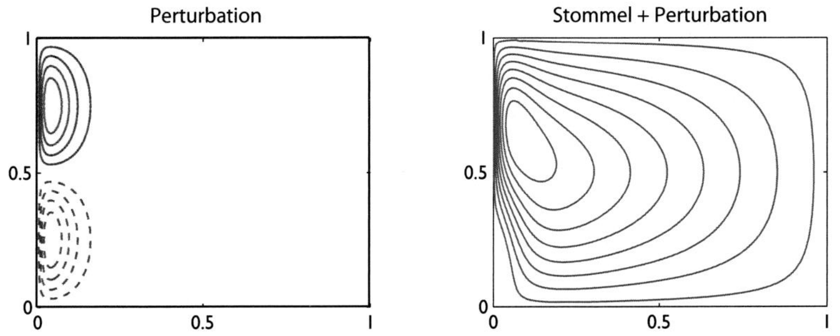

Fig. 14.8: The nonlinear perturbation solution of the Stommel problem, calculated according to (14.43). On the left is the nonlinear perturbation to the Stommel solution and on the right is the total reconstituted solution, with the centre of the gyre pushed poleward. Dashed contours are negative.

tion of such a nonlinear term immediately makes the problem extremely difficult from an analytic point of view, although we can proceed perturbatively if the term is small. In this case we first neglect the nonlinear term and solve the problem in the linear case, as above. If we call this solution ψ_S, with $\zeta_S = \nabla^2\psi_S$, then an approximate nonlinear solution is given by the solution of

$$\beta\frac{\partial\psi}{\partial x} = \mathrm{curl}_z\boldsymbol{\tau}_T - r\zeta - J(\psi_S, \zeta_S). \tag{14.43}$$

The algebra is tedious, but the solution for our canonical wind stress, $\tau^x = -\tau_0\cos(\pi y/a)$, is illustrated in Fig. 14.8. The perturbation is antisymmetric about $\widehat{y} = 1/2$ (where $\widehat{y} = y/a$ as before), being positive for $\widehat{y} > 1/2$ and negative for $\widehat{y} < 1/2$. This tends to move the centre of the gyre polewards, narrowing and intensifying the flow in the poleward half of the western boundary current, whereas the western boundary current equatorwards of $\widehat{y} = 1/2$ is broadened and weakened. The net effect is that the centre of the gyre is pushed poleward — essentially because the western boundary current is advecting the vorticity of the gyre poleward. In the perturbation solution the advection is both by and of the linear Stommel solution; thus, negative vorticity is advected polewards, intensifying the gyre in its poleward half, weakening it in its equatorward half.

A numerical solution

Fully nonlinear solutions show qualitatively similar effects to those seen in the perturbative solutions, as we see in Fig. 14.9, where the solutions of the nonlinear Stommel problem are obtained numerically by Newton's method.

Just as with the perturbative procedure, small values of nonlinearity lead to the poleward advection of the gyre's anticyclonic vorticity in the western boundary current, strengthening and intensifying the boundary current in the northwest corner. A higher level of nonlinearity results in a strong recirculating regime in the upper westward quadrant, and ultimately much of the gyre's transport is confined to this regime. The western boundary current itself becomes less noticeable as nonlinearity increases, the more nonlinear solutions have a much greater degree of

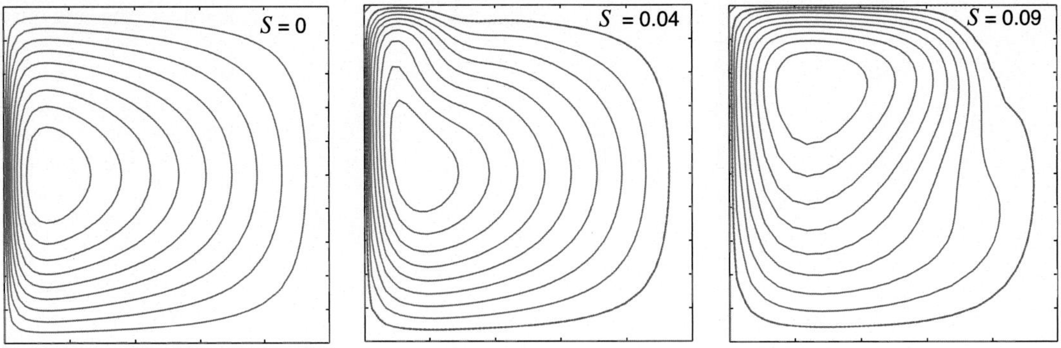

Fig. 14.9: Streamfunctions in solutions of the nonlinear Stommel problem, obtained numerically, for various values of the nonlinearity parameter $S = R_\beta^{1/2}$. As in the perturbation solution, for small values of nonlinearity the centre of the gyre moves polewards, strengthening the boundary current in the north-western quadrant (for a northern-hemisphere solution). As nonlinearity increases, the recirculation of the gyre dominates, and the solutions become increasingly inertial. (Solutions kindly provided by B. Fox-Kemper.)

east-west symmetry than the linear ones. We will find this same effect in the fully nonlinear inertial solutions that we now come to.

14.5 ♦ AN INERTIAL SOLUTION

Rather than attempt to match an inertial boundary layer with an interior Sverdrup flow, we may look for a purely inertial solution that holds basinwide, and such a construction is known as the *Fofonoff model*. That is, we seek solutions to the inviscid, unforced problem,

$$J(\psi, \nabla^2\psi + \beta y) = 0. \tag{14.44}$$

We should not regard this problem as representing even a very idealized wind-driven ocean; rather, we may hope to learn about the properties of purely inertial solutions and this might, in turn, tell us something about the ocean circulation.

The general solution to (14.44) is

$$\nabla^2\psi + \beta y = Q(\psi), \tag{14.45}$$

where $Q(\psi)$ is an arbitrary function of its argument. For simplicity we choose the linear form,

$$Q(\psi) = A\psi + B, \tag{14.46}$$

where $A = \beta/U$ and $B = \beta y_0$, where U and y_0 are arbitrary constants. Thus, (14.45) becomes

$$\left(\nabla^2 - \frac{\beta}{U}\right)\psi = \beta(y_0 - y). \tag{14.47}$$

Nick Fofonoff (1929–2003) was a physical oceanographer at the Woods Hole Oceanographic Institution. As well as his work on ocean circulation he made notable contributions to our understanding of the thermodynamics of seawater.

We further choose $\beta/U > 0$, which we anticipate will provide a westward-flowing interior flow, and which (from our experience in the previous section) is more likely to provide a meaningful solution than an eastward interior, and we use boundary-layer methods to find a solution. A natural scaling for ψ is UL, where L is the domain size, and with this the nondimensional problem is

$$(\epsilon_F\nabla^2 - 1)\widehat{\psi} = \widehat{y}_0 - \widehat{y}, \tag{14.48}$$

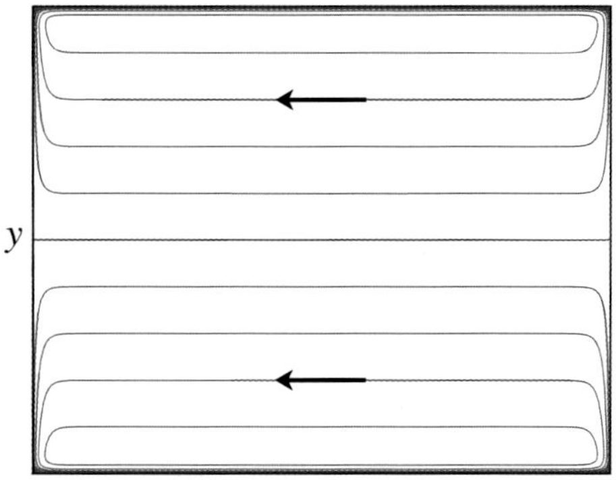

Fig. 14.10: The Fofonoff solution. Plotted are contours (streamlines) of (14.52) in the plane $0 < x < x_E$, $0 < y < y_N$ with $U = 1$, $y_N = 1$, $x_E = y_N = 1$, $y_0 = 0.5$ and $\delta_I = 0.05$. The interior flow is westward everywhere, and $\psi = 0$ at $y = y_0$. In addition, boundary layers of thickness $\delta_I = \sqrt{U/\beta}$ bring the solution to zero at $x = (0, x_E)$ and $y = (0, y_N)$.

where $\epsilon_F = U/(\beta L^2)$ and $\widehat{y} = y/L$. If we take ϵ_F to be small (note that $\epsilon_F = R_\beta$) we can find a solution by boundary-layer methods similar to those used in Section 14.3.2 for the Stommel problem. Thus, we write $\widehat{\psi} = \widehat{\psi}_I + \widehat{\phi}$ where $\widehat{\psi}_I = \widehat{y} - \widehat{y}_0$ (dimensionally, $\psi_I = U(y - y_0)$) and $\widehat{\phi}$ is the boundary layer correction, to be calculated separately for each boundary using

$$\epsilon_F \nabla^2 \widehat{\phi} - \widehat{\phi} = 0 \tag{14.49}$$

and the boundary condition that $\widehat{\phi} + \widehat{\psi}_I = 0$. For example, at the northern boundary, $\widehat{y} = \widehat{y}_N$, the y-derivatives will dominate and (14.49) may be approximated by

$$\epsilon_F \frac{\partial^2 \widehat{\phi}}{\partial \widehat{y}^2} - \widehat{\phi} = 0, \tag{14.50}$$

with solution

$$\widehat{\phi} = B \exp[-(\widehat{y}_N - \widehat{y})/\epsilon_F^{1/2}], \tag{14.51}$$

where $B = \widehat{y}_0 - \widehat{y}_N$, hence satisfying the boundary condition that $\widehat{\phi}(\widehat{x}, \widehat{y}_N) = -\widehat{\psi}_I(\widehat{x}, \widehat{y}_N)$. We follow a similar procedure at the other boundaries to obtain the full solution, and in dimensional form this is

$$\begin{aligned} \psi = U(y - y_0)\left[1 - e^{-x/\delta_I} - e^{-(x_E - x)/\delta_I}\right] \\ + U(y_0 - y_N)e^{-(y_N - y)/\delta_I} + U y_0 e^{-y/\delta_I}, \end{aligned} \tag{14.52}$$

where $\delta_I = (U/\beta)^{1/2}$ is the boundary layer thickness. Evidently, only positive values of U corresponding to a westward interior flow give boundary-layer solutions that decay into the interior.

A typical solution is illustrated in Fig. 14.10. On approaching the western boundary layer, the interior flow bifurcates at $y = y_0$. The western boundary layer, of width δ_I, accelerates away from this point, being constantly fed by the interior flow. The westward return flow occurs in zonal boundary layers at the northern and southern edges, also of width δ_I. Flow along the eastern boundary layers is constantly being decelerated, because it is feeding the interior. If one of the zonal boundaries corresponds to y_0 (e.g., if $y_N = y_0$) there would be no boundary layer along it, since ψ is already zero at $y = y_0$. Rather, there would be westward flow along it, just as in the interior. Indeed, a slippery wall placed at $y = 0.5$ would have no effect on the solution illustrated in Fig. 14.10.

Notes and References

Stommel's eponymous model of ocean gyres was presented in Stommel (1948), with an extension to a harmonic viscosity (instead of a linear drag) given by Munk (1950). Nonlinear extensions were described by Veronis (1966) and the purely inertial Fofonoff solution was described by Fofonoff (1954). Two books, of contrasting length and style, specializing in ocean dynamics and circulation are Samelson (2011) and Olbers *et al.* (2012).

Problems

Oceanic problems are given at the end of Chapter 16.

CHAPTER

15

The Overturning Circulation and Thermocline

IN THE PREVIOUS CHAPTER we studied the horizontal, vertically integrated, flow of the world's oceans. In this chapter we look at the vertical structure of the oceans and the *meridional overturning circulation* (MOC), which is the circulation in the vertical–meridional plane.

15.1 THE OBSERVATIONS

Our main goals in this chapter are to explain two important phenomena:

(i) The structure of the temperature and density of the ocean in the vertical–meridional plane;

(ii) The circulation of the ocean in that same plane.

As one might expect it is much harder to observe the interior of the ocean than the surface ocean, or the atmosphere. Because water is almost opaque to electromagnetic radiation we actually have to drop instruments into the ocean to measure its deep properties. These days measurements come from a combination of moored instruments, hydrographic surveys, floats, gliders and satellites (which mostly measure surface properties). The various measurements are combined in some fashion (often in combination with a numerical model) to give a 'state estimate' of the ocean, and we now have a decent coarse-grained view of the density structure and circulation of the sub-surface ocean, although with far less detail than our view of the atmosphere.

15.1.1 The Thermocline

The density structure of the Atlantic Ocean (and the Pacific is similar) is illustrated in Fig. 15.1. Here we see that the main gradients of density are concentrated in the upper one kilometre or so of the ocean, in the *main thermocline,* which serves to connect the relatively warm surface waters with the much colder abyssal waters. (The main thermocline exists year

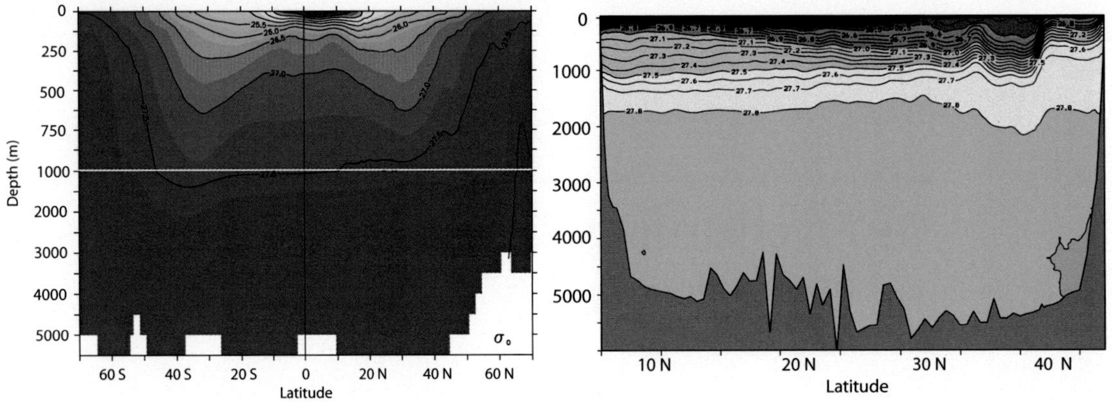

Fig. 15.1: The potential density in the Atlantic ocean. On the left is the climatological zonally-averaged field, plotted with a break in the vertical scale at 1000 m. On the right is a section at 53°W. Both plots show a region of rapid change of density (and temperature) concentrated in the upper kilometre, in the *main thermocline*, below which the density is much more uniform.

round; the seasonal thermocline, which is not visible in these plots, is a much shallower region near the surface over which the temperature gradient varies seasonally.) The thermocline is much weaker at high latitudes, since the near-surface waters are already cold, and it is shallower at low latitudes, as we see in Fig. 15.2. The abyssal temperature at all latitudes is about 2°C, which is similar to the surface temperature at high latitudes, and this is consistent with water at high latitudes sinking, spreading equatorward and filling the abyss.

15.1.2 The Meridional Overturning Circulation

Closely associated with the density structure of the ocean is the meridional overturning circulation, or MOC, and this is illustrated in Fig. 15.3. Focusing on the red, northern cell we see water sinking at high latitudes, spreading south at depth, and upwelling largely in the Southern Ocean; the water in this cell is called *North Atlantic Deep Water*, or NADW. The blue cell shows water sinking at high southern latitudes and spreading north underneath the NADW before rising to mid-depth and returning; this cell contains *Antarctic Bottom Water*, or AABW. The Pacific Ocean has

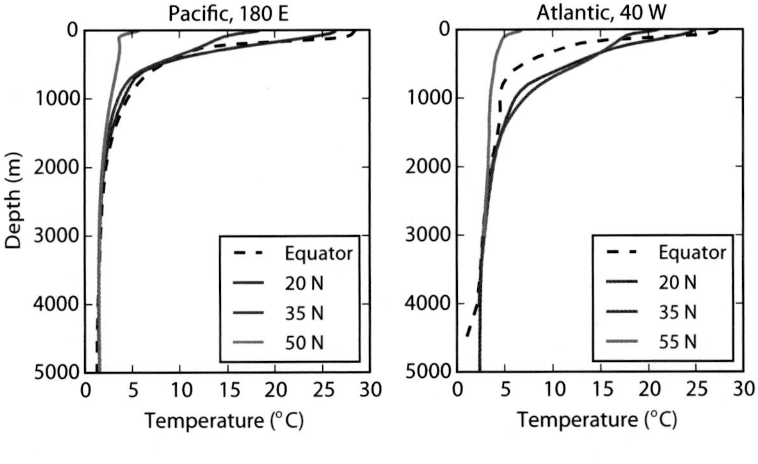

Fig. 15.2: Profiles of observed mean temperature in the North Pacific and North Atlantic at the longitudes and latitudes indicated. Note the shallowness of the equatorial thermoclines (especially in the Atlantic), and the weakness of the subpolar thermoclines.

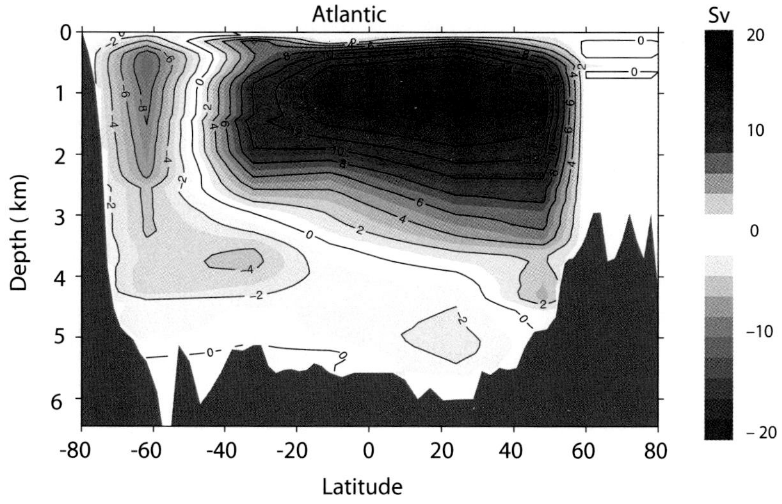

Fig. 15.3: The overturning circulation in the Atlantic Ocean as determined from observations in combination with a simple ('inverse') model. Red shading indicates a clockwise circulation, with water sinking in the North Atlantic and rising in the South, and this cell is predominantly *North Atlantic Deep Water*, or NADW. The blue, anticlockwise deep cell contains *Antarctic Bottom Water* that originates in the Southern Ocean and spreads northwards underneath the NADW. (Figure kindly provided by Loic Juillion using data and a methodology similar to that of Lumpkin & Speer 2007.)

a much weaker overturning circulation to such an extent that the globally-averaged overturning circulation largely reflects that of the Atlantic. The overturning circulation and the thermocline are, as one might expect, intimately linked and to explain one we must explain the other. Let us begin with a phenomenological discussion of the overturning circulation.

15.2 A MIXING-DRIVEN OVERTURNING CIRCULATION

To begin with the simplest case let us consider the circulation in a closed, single hemispheric basin, and suppose that there is a net surface heating at low latitudes and a net cooling at high latitudes that maintains a meridional temperature gradient at the surface. It seems reasonable to imagine that there is a single overturning cell, with water sinking at high latitudes rising at low latitudes before returning to polar regions in the upper ocean, as illustrated schematically in Fig. 15.4 and Fig. 15.5. Is this a reasonable expectation? Can we explain why the water circulates at all?

15.2.1 Why the Water Circulates

Let us suppose that initially all the interior water is at some intermediate temperature, and we will also suppose that the flow in the interior is adiabatic, meaning that to a good approximation the subsurface water conserves its potential temperature as it moves around. Now, given a warm interior, cold surface water at high latitudes will be convectively unstable and will therefore sink, so that very quickly the dense water extends all the way to the ocean floor. By hydrostasy the pressure in the deep ocean is then higher at high latitudes than at low, where the water is warmer, and a pressure gradient then causes water to move equatorward, filling the abyss. Eventually, the *entire* ocean becomes filled with cold dense water of polar origin, except for a very thin layer at the surface, since the ocean surface at lower latitudes is kept at a higher temperature. Once the abyss

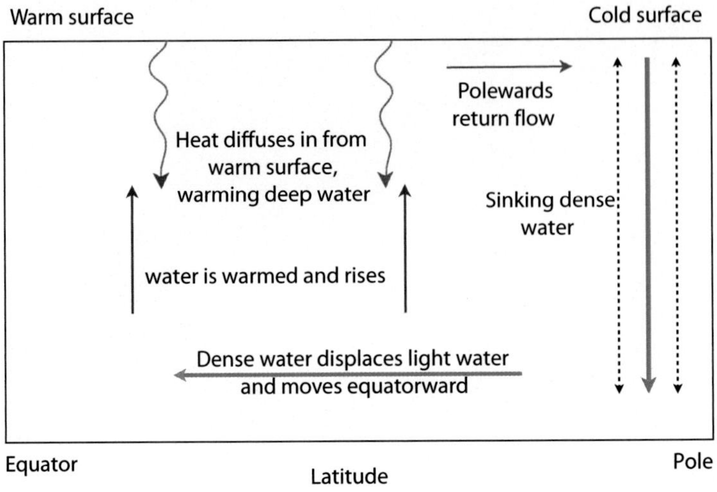

Fig. 15.4: Schematic of a single-celled meridional overturning circulation. Sinking is concentrated at high latitudes and upwelling spread out over lower latitudes.

is filled with dense water the surface polar waters *will no longer be convectively unstable.* The convection will thus cease and the circulation will halt! However, we know from observations that the deep ocean continues to circulate, albeit slowly, with the deep ocean completely overturning and the water being replaced on timescales of a few hundred years. There are two causes of the continued circulation, one being that the ocean mixes and the other being that the wind forcing at the top drives a deep circulation; we consider the effects of mixing first and come back to the wind-driving later in the chapter.

Mixing — either molecular mixing or in reality turbulent mixing, as discussed in Chapter 10 — will cause the higher surface temperatures in lower latitudes to diffuse down into the ocean interior. That is, the interior is slowly *warmed* by heat diffusion from above. This diffusion keeps the deep ocean slightly warmer than the cold polar surface waters, enabling the high-latitude convection and so the circulation itself to persist. The diffusion also extends the vertical temperature gradient down into the interior and we see in Fig. 15.2 how the vertical temperature profile varies with latitude. Except at the highest latitudes where the water is sinking and so almost uniform all the way to the bottom, we see that the temperature gradient is concentrated in the upper kilometre of the ocean, and this region is called the *main thermocline*. Why should the vertical temperature gradient be concentrated in the upper ocean? The upper ocean is the region of the gyres, which certainly creates a temperature gradient, but the underlying reason that the vertical temperature gradient is strongest there is more basic, as we now explore.

15.2.2 A Simple Kinematic Model of the Thermocline

In mid- and low latitudes cold water with polar origins upwells into a region of warmer water where high temperatures are diffusing down, and a simple model of this is the one-dimensional advective–diffusive balance,

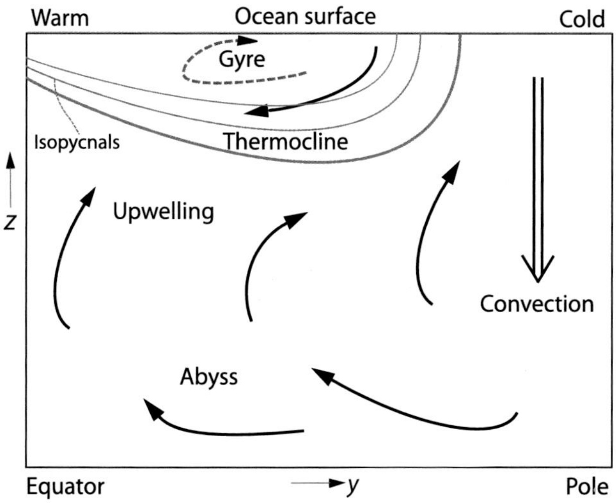

Fig. 15.5: Cartoon of a single-celled meridional overturning circulation, with a wall at the equator. Sinking is concentrated at high latitudes and upwelling spread out over lower latitudes. The thermocline is the boundary between the cold abyssal waters, with polar origins, and the warmer near-surface subtropical water. Wind forcing in the subtropics pushes the warm surface water into the fluid interior, deepening the thermocline as well as circulating as a gyre.

namely

$$w\frac{\partial T}{\partial z} = \kappa\frac{\partial^2 T}{\partial z^2}, \tag{15.1}$$

where w is the vertical velocity (which is positive), κ is a diffusivity and T is temperature. The equation represents a balance between the upwelling of cold water and the downward diffusion of heat. If w and κ are given and are constant, and if T is specified at the top ($T = T_T$ at $z = 0$) and if $T = T_B$ at great depth ($z = -\infty$) then the temperature falls exponentially away from the surface according to

$$T = (T_T - T_B)e^{wz/\kappa} + T_B. \tag{15.2}$$

The scale at which temperature decays away from its surface value is given by

$$\delta = \frac{\kappa}{w}, \tag{15.3}$$

and this is an estimate of the thermocline thickness. It is not a useful a-priori estimate, because the magnitude of w depends on κ. However, it is reasonable to see if the observed ocean is broadly consistent with this expression. The diffusivity κ (which is an eddy diffusivity, maintained by small-scale turbulence) can be measured and is found to have values that range between $10^{-5}\,\mathrm{m^2\,s^{-1}}$ and $10^{-4}\,\mathrm{m^2\,s^{-1}}$ over much of the ocean, with higher values locally in some abyssal and shelf regions.

The vertical velocity is too small to be measured directly, but various estimates based on deep water production suggest a value of about $10^{-7}\,\mathrm{m\,s^{-1}}$. Using this and the smaller value of κ in (15.2) gives an e-folding vertical scale, κ/w, of order a hundred metres, beneath which the stratification is predicted to be very small (i.e., a nearly uniform density). Using the larger value of κ increases the vertical scale to 1000 m, similar to the observed value. Quantitative uncertainties aside, the model has a very robust result, that *the temperature gradient is concentrated in the upper ocean.*

The reason for the up-down asymmetry is that cold water is upwelling and only needs to warm up as it approaches the warm upper surface. If κ were very small there would just be a thin boundary layer at the top of the ocean, and the overturning circulation would be very weak because almost the entire ocean would be as dense as the cold polar surface waters.

The effect of wind

In the above description we considered an ocean without wind, yet we spent the entire previous chapter demonstrating that the wind produces gyres. How do wind effects fit into our picture of the MOC? In the subtropics the wind-stress curl forces water to converge in the subtropical Ekman layer, thereby forcing relatively warm water to downwell and meet the upwelling colder abyssal water at some finite depth. The result of this is that the thermocline is no longer necessarily solely a near-surface phenomenon; rather, the transition from cold abyssal waters to warm subtropical waters can occur at some depth below the surface in an 'internal thermocline'. Between the internal thermocline and the surface waters the gyres circulate, and in general this region will also have a vertical gradient of temperature, so creating the 'ventilated thermocline' — ventilated because the waters feel the effects of, or are ventilated by, the surface. In so far as we can separate the two effects of wind and diffusion, we can say that the strength of the wind influences the *depth* at which the thermocline occurs, whereas the strength of the diffusivity influences the *thickness* of the thermocline, as we discuss further below. However, in practice the regions often merge smoothly together.

15.3 THERMOCLINE DYNAMICS AND THE MOC

Let us now estimate how fast the water in the overturning cell circulates, how deep the thermocline might be, and what parameters these quantities depend upon. The Rossby number of the large-scale circulation is small and the scale of the motion large so the flow obeys the planetary-geostrophic equations. In our standard notation these equations are

$$\boldsymbol{f} \times \boldsymbol{u} = -\nabla\phi, \qquad \frac{\partial \phi}{\partial z} = b, \qquad (15.4\text{a,b})$$

$$\nabla \cdot \boldsymbol{v} = 0, \qquad \frac{\mathrm{D}b}{\mathrm{D}t} = \kappa \frac{\partial^2 b}{\partial z^2}. \qquad (15.4\text{c,d})$$

These equations are, respectively, the horizontal and vertical momentum equations (hydrostatic and geostrophic balance), the mass continuity equation and the buoyancy equation. Let us suppose that these equations hold below an Ekman layer, so that the effects of a wind stress may be included by specifying a vertical velocity, w_E, at the top of the domain. The (eddy) diffusivity, κ, is small and its precise value is uncertain, so a useful practical philosophy is to try to ignore dissipation and viscosity where possible, and to invoke them only if there is no other way out. Let us therefore scale the equations in two ways, with and without diffusion; these scalings will be central to our theory.

15.3.1 A Diffusive Scale

Suppose that the circulation is steady and resembles that of Fig. 15.5, but with no wind forcing. How deep is the diffusive layer in the subtropical gyre? Let us suppose that, as in the kinematic model, the thermodynamic equation reduces to the advective-diffusive balance of (15.1), but we will use the other equations in (15.4) to give an estimate of the vertical velocity. If we take the curl of (i.e., cross-differentiate) the momentum equation (15.4a) and use mass continuity we obtain the linear vorticity equation, $\beta v = f \partial w/\partial z$, and if we take the vertical derivative of the momentum equation and use hydrostasy we obtain thermal wind, $f \partial \boldsymbol{u}/\partial z = \hat{\mathbf{k}} \times \nabla b$. Collecting these equations together we have

$$w \frac{\partial b}{\partial z} = \kappa \frac{\partial^2 b}{\partial z^2}, \qquad \beta v = f \frac{\partial w}{\partial z}, \qquad f \frac{\partial \boldsymbol{u}}{\partial z} = \hat{\mathbf{k}} \times \nabla b, \tag{15.5a,b,c}$$

with corresponding scales

$$\frac{W \Delta b}{\delta} = \frac{\kappa \Delta b}{\delta^2}, \qquad \beta V = \frac{fW}{\delta}, \qquad \frac{fU}{\delta} = \frac{\Delta b}{L}, \tag{15.5d,e,f}$$

where δ is the vertical scale and other scaling values are denoted with capital letters. We suppose that $V \sim U$, where U is the zonal velocity scale, and henceforth we denote both by U and we take L to be the horizontal scale of the motion, which we take as the gyre or basin scale. Typical values for the subtropical gyre are $\Delta b = g \Delta \rho/\rho_0 = g \beta_T \Delta T \sim 10^{-2}\,\mathrm{m\,s^{-2}}$, $L = 5{,}000\,\mathrm{km}$, $f = 10^{-4}\,\mathrm{s^{-1}}$ and $\kappa = 10^{-5}\,\mathrm{m^2\,s^{-2}}$.

Equation (15.5d) is the same as (15.3), as expected, but we can now use (15.5e,f) to obtain an estimate for the vertical velocity, namely

$$W = \frac{\beta \delta^2 \Delta b}{f^2 L}. \tag{15.6}$$

Using this and (15.5d) gives the diffusive vertical scale, and the estimates

$$\delta = \left(\frac{\kappa f^2 L}{\beta \Delta b} \right)^{1/3}, \qquad W = \left(\frac{\kappa^2 \beta \Delta b}{f^2 L} \right)^{1/3}. \tag{15.7a,b}$$

With values of the parameters as above, (15.7) gives $\delta \approx 150\,\mathrm{m}$ and $W \approx 10^{-7}\,\mathrm{m\,s^{-1}}$. With $\kappa = 10^{-4}\,\mathrm{m^2\,s^{-2}}$ (which is larger than observed) then $\delta \approx 700\,\mathrm{m}$ and $W \approx 4.6 \times 10^{-7}\,\mathrm{m\,s^{-1}}$.

15.3.2 An Advective Scale

The values of the vertical velocity above are very small, much smaller than the Ekman pumping velocity at the top of the ocean, which is of order 10^{-6} to $10^{-5}\,\mathrm{m\,s^{-1}}$. This difference suggests that we might ignore the diffusive term in (15.5a) — or, rather, ignore the thermodynamic term completely — and construct an adiabatic scaling estimate for the depth of the wind's influence. Further, in subtropical gyres the Ekman pumping is downward, whereas the diffusive velocity is upward, meaning that at some level, D_a, we expect the vertical velocity to be zero.

The equations of motion are just the thermal wind balance and the linear geostrophic vorticity equation, namely

$$\beta v = f\frac{\partial w}{\partial z}, \qquad \boldsymbol{f} \times \frac{\partial \boldsymbol{u}}{\partial z} = -\nabla b, \tag{15.8}$$

with corresponding scales

$$\beta U = f\frac{W}{D_a}, \qquad \frac{U}{D_a} = \frac{1}{f}\frac{\Delta b}{L}, \tag{15.9}$$

recalling that we take $V \sim U$.

The thermodynamic equation does not enter, but we take the vertical velocity to be that due to Ekman pumping, W_E. From (15.9) we then obtain

$$D_a = W_E^{1/2}\left(\frac{f^2 L}{\beta\,\Delta b}\right)^{1/2}, \tag{15.10}$$

which may be contrasted with the estimate of (15.7a). If we relate U and W_E using mass conservation, $U/L = W_E/D_a$, instead of using (15.8a), then we write L in place of f/β and (15.10) becomes $D_a = (W_E f L^2/\Delta b)^{1/2}$, which is not qualitatively different from (15.10) for large scales.

The important aspect of the above estimate is that the depth of the wind-influenced region increases with the magnitude of the wind stress (because $W_E \propto \mathrm{curl}_z \tau$) and decreases with the meridional temperature gradient. The former dependence is reasonably intuitive, and the latter arises because as the temperature gradient increases the associated thermal wind-shear U/D_a correspondingly increases. But the horizontal transport (the product UD_a) is fixed by mass conservation; the only way that these two can remain consistent is for the vertical scale to decrease. Taking $W_E = 10^{-6}\,\mathrm{m\,s^{-1}}$, and other values as before, gives $D_a \approx 600\,\mathrm{m}$; taking $W_E = 10^{-5}\,\mathrm{m\,s^{-1}}$ (which is unrealistically large over most of the ocean) gives $D_a \approx 2000\,\mathrm{m}$. Such a scaling argument cannot be expected to give more than an estimate of the depth of the wind-influenced region; nevertheless, because D_a is much less than the ocean depth, the estimate does suggest that the wind-driven circulation is predominantly an upper-ocean phenomenon.

The physical picture

What do the vertical scales derived above represent? The wind-influenced scaling, D_a, is the depth to which the directly wind-driven circulation can be expected to penetrate. Thus, over this depth we can expect to see wind-driven gyres and associated phenomena. At greater depths lies the abyssal circulation and this is not wind-driven in the same sense. Now, in general, the water at the base of the wind-driven layer will not have the same thermodynamic properties as the upwelling abyssal water

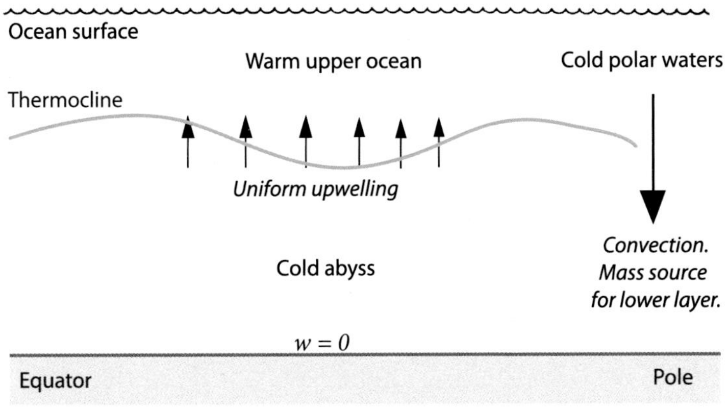

Fig. 15.6: A simple model of the abyssal circulation. Convection at high latitudes provides a localized mass-source to the lower layer, and upwelling through the thermocline provides a more uniform mass sink.

— this being cold and dense, whereas the water in the wind-driven layer is warm and subtropical (look again at Fig. 15.5). The thickness δ characterizes the diffusive transition region between these two water masses and in the limit of very small diffusivity this becomes a *front*. In the diffusive region, no matter how small the diffusivity is in the thermodynamic equation, the diffusive term is important there. In contrast, D_a is the *depth* of the thermocline and this depends on the strength of the wind and not κ. Of course if the diffusion is sufficiently large, the thickness will be as large or larger than the depth, and the two regions will blur into each other, and this may indeed be the case in the real ocean.

15.3.3 A View from Above

In the description above we looked at the circulation in the y–z plane — an 'elevation' in architectural terms. Let us look at the circulation in the x–y plane, namely a plan view, as we did with the wind-driven circulation.

The model of the deep cell described in this section is known as the Stommel–Arons model after its inventors, Henry Stommel and Hank Arons, who put forward the model in 1960.

We imagine a model ocean consisting of two layers of shallow fluid, each obeying the planetary-geostrophic equations of motion. The interface and upper layer represent the thermocline (which to keep matters simple we suppose is motionless), and the lower layer is the abyss. The convection is represented by a localised mass source at high latitudes in the lower layer, and the diffusive upwelling is represented by a transfer of mass from the lower layer to the upper layer, as sketched in Fig. 15.6. In this simple model the lower layer satisfies the mass conservation equation and geostrophic balance in the form

$$\frac{\mathrm{D}h}{\mathrm{D}t} + h\nabla\cdot\boldsymbol{u} = S, \qquad -fv = -g'\frac{\partial h}{\partial x}, \quad fu = -g'\frac{\partial h}{\partial y}, \qquad (15.11\text{a,b,c})$$

where h is the thickness of the lower layer, g' is the reduced gravity and the source term S represents mass transfer between the upper and lower layers. With a little manipulation we combine these equations into the

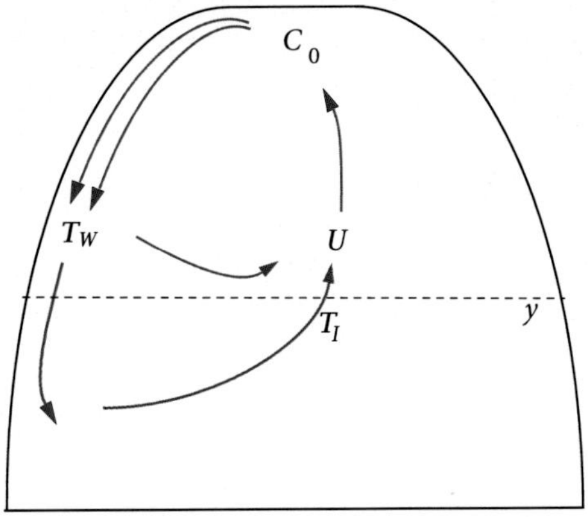

Fig. 15.7: Schematic of the abyssal flow in a Stommel–Arons model of a single sector. Poleward of a latitude y the mass gains in the abyss due to the convective source, C_0, and the interior return flow, T_I, must be balanced by the loss in the western boundary current, T_W and the upwelling, U. The transport in the western boundary current T_W is larger than that of the convective source because some of the flow recirculates. The western boundary current decreases in intensity equatorward, as it loses mass to the polewards interior flow.

planetary-geostrophic potential vorticity equation, namely

$$\frac{\mathrm{D}}{\mathrm{D}t}\left(\frac{f}{h}\right) = -\frac{fS}{h^2}. \tag{15.12}$$

In a steady state, and using (15.11b,c), this equation simplifies to

$$\frac{v}{h}\frac{\partial f}{\partial y} = -\frac{fS}{h^2}, \tag{15.13}$$

which the reader may recognise as a form of the vorticity equation, $\beta v = f\partial w/\partial z$. Now, and considering a situation confined to the Northern Hemisphere, we are supposing that there is an effective point source of mass at high latitudes and a uniform mass sink everywhere else. This means that S is negative nearly everywhere and v is positive — that is, flow is *toward* the mass source in the lower layer! This result seems counter intuitive and at odds with our earlier discussion, in which the flow was away from the source at lower levels, with a shallow return flow.

The reconciliation of these views comes about through an intense western boundary current, similar to the current that closes the circulation in the wind-driven models of the previous chapter, as sketched in Fig. 15.7. In the western boundary current frictional effects are important, allowing the circulation to close. Thus, the model predicts that there should be a deep western boundary current flowing away from the source (and so southwards in the Northern Hemisphere), and a generally polewards return flow in the interior, and these are features of the real ocean.

Following the theoretical prediction by Stommel and Arons a southward flowing deep western boundary current underneath the Gulf Stream was in fact observed by Swallow & Worthington (1961) by tracking neutrally buoyant floats.

A short calculation

We can calculate the strength of the western boundary current as follows. We pose the problem in a Cartesian domain of extent $L_x \times L_y$ and a wall at the equator, $y = 0$. At some latitude y, mass balance in the lower layer must satisfy

$$C_0 + T_I(y) = T_W(y) + U(y), \tag{15.14}$$

as seen in Fig. 15.7. Here, C_0 is the strength of the convective source, which we take as given, $T_I(y)$ is the polewards flow in the interior, in the lower layer, across the latitude line at y, $T_W(y)$ is the equatorial flow in the deep western boundary current at y, and $U(y)$ is the total upwelling polewards of y. The terms on the left-hand side are mass sources to this region and the terms on the right-hand side are losses, and all are in units of $\mathrm{m^3\,s^{-1}}$ (since density is constant, mass balance and volume balance are synonymous). Over the entirety of the domain the source term must balance the upwelling, so that $C_0 = U(0)$, and we assume the upwelling is uniform.

The poleward transport in the interior is given using (15.13),

$$T_I(y) = \int vh\,\mathrm{d}x = \int \frac{fS}{\beta}\,\mathrm{d}x. \tag{15.15}$$

Now, since the upwelling S is uniform, and $\int S\,\mathrm{d}x\,\mathrm{d}y = SL_xL_y = U(0) = C_0$, we have

$$T_I(y) = \frac{fC_0}{\beta L_y} = \frac{C_0 y}{L_y}, \tag{15.16}$$

using $f = \beta y$. It is important to realise that this result is obtained using the potential vorticity equation and not the mass continuity equation.

The upwelling north of latitude y is given by

$$U(y) = SL_x(L_y - y) = C_0\left(1 - \frac{y}{L_y}\right). \tag{15.17}$$

Using (15.16) and (15.17) in (15.14) gives

$$T_W(y) = \frac{2C_0 y}{L_y}. \tag{15.18}$$

This is a remarkable result, for it tells us that the strength of the western boundary current near the source region is *twice* the strength of the source itself! The result arises because some of the flow in the deep layer is *recirculating*, going round and round without upwelling or coming from the source itself. The calculation itself is very approximate, but the fact that there is a deep western boundary current, and that the flow recirculates, transcend its limitations and these are robust predictions.

A final point to note is that we have taken the convective source to have a given magnitude. In reality, the strength of the source must match the strength of the upwelling, this being the strength of the overturning circulation itself. This is a function of the diapycnal diffusivity and the meridional temperature gradient, as described in Sections 15.2 and 15.3.

15.4 An Interhemispheric Overturning Circulation

As attractive as it may be, the theory of the overturning circulation and thermocline described in the preceding sections is only part of the picture.

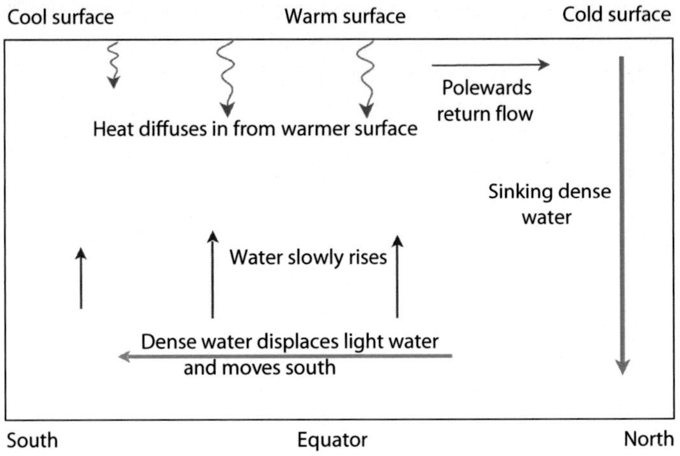

Fig. 15.8: An idealized interhemispheric overturning circulation. Water from the north sinks, because it is the densest water in the system. It displaces any lighter water, filling up the entire basin (i.e., both hemispheres) except for a thermocline near the surface. A circulation is maintained if heat diffuses in from the surface, warming the deep water and enabling it to rise. The strength of the circulation depends on the diffusivity, and if it is zero the circulation eventually halts.

In fact, much of the deep circulation is *interhemispheric:* we can see in Fig. 15.3 that much of the water that sinks in the North Atlantic upwells around 40°S or even further south in the Southern Ocean (although this only became truly apparent at the beginning of twenty-first century). In the rest of the chapter we try to understand why that should be.

15.4.1 A Basic Mechanism

An interhemispheric circulation of itself is of no particular surprise. For simplicity consider a 'shoebox' ocean consisting of a single basin stretching from high northern latitudes to high southern latitudes, and let us suppose that the surface at high latitudes in one hemisphere, say the North, is particularly cold and dense, as in Fig. 15.8. The physical situation then actually differs little from the situation described in Section 15.2. The densest water in the system sinks, and spreads equatorward. However, there is no reason that it should all upwell before it reaches the equator, although if the equatorial regions are warm the upwelling may be strong there because the downward diffusion of heat warms the deep water. Nonetheless, if the diffusion is small the densest water in the system displaces any lighter water and fills up both hemispheres of the basin, except for a thermocline near the surface. The flow away from the convective region occurs, as in the single-hemisphere model, in deep western boundary currents, with upwelling and return flow in the basin interior.

A non-zero circulation depends, as with the mixing-driven circulation, on there being a non-zero diffusivity to warm the deep water and allow it to rise. If the diffusivity were zero, then the entire basin would simply fill with the densest available water (with the exception of an infinitesimally thin layer at the surface) and the circulation would then halt.

15.4.2 A Wind-Driven Interhemispheric Circulation

The mixing-driven circulation described above is not the only mechanism, and is not in fact the main mechanism, whereby deep water ac-

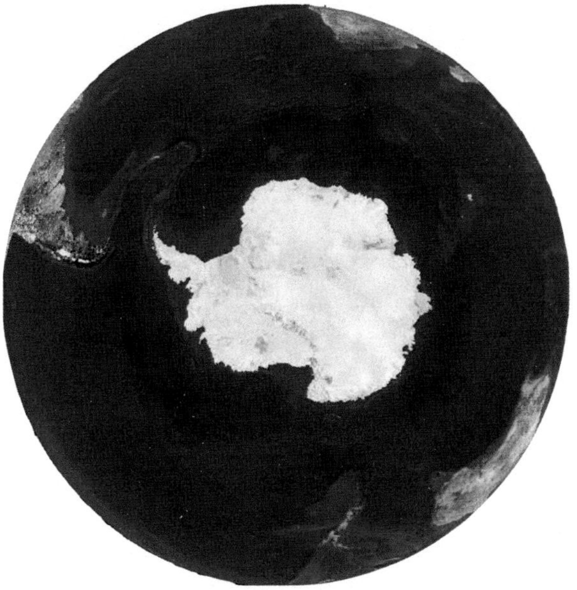

Fig. 15.9: Composite satellite image (courtesy of NASA) of Antarctica and the Southern Ocean, also showing South America (at ten o'clock), South Africa (one o'clock) and Australia (four o'clock). Unlike the Northern Hemisphere basins, the Southern Ocean has no unbroken meridional boundaries and, like the atmosphere, constitutes a zonally-re-entrant channel.

tually circulates. Rather, a significant fraction of the oceanic MOC is *wind-driven,* stemming in large part from the particular geography of the Southern Ocean, as we see in Fig. 15.9. The oceanic relevance of this striking image is simply the following. The Southern Ocean has no unbroken meridional boundaries, and at higher latitudes, between the tip of South America and Antarctica, the ocean forms a zonally-re-entrant channel over which the wind predominantly (and rather strongly) blows eastward — the famous 'roaring forties' and the even stronger 'furious fifties'.

The eastward winds generate a northward flowing Ekman drift in the channel. The strength of the Ekman flow is given using the x-component of the frictional-geostrophic balance,

$$-fv = -\frac{\partial \phi}{\partial x} + \frac{\partial \tau}{\partial z}, \tag{15.19}$$

where, as before, τ is the kinematic stress, namely the actual stress divided by the seawater density. If we integrate this equation around a channel the pressure term vanishes, and if we integrate down in the ocean to a level where the stress effectively vanishes we obtain an estimate for the transport induced by the wind, namely

$$V_w = \frac{\tau_0 L}{f}, \tag{15.20}$$

where V_w is the volumetric transport (in $\mathrm{m^3\,s^{-1}}$) due to the surface wind stress, τ_0, and L is the zonal extent over which wind stress occurs. The wind stress over the Southern Ocean is quite high, about 0.2 Pa (for the actual stress), and if we suppose that L corresponds to the width of

The Southern Ocean shown in Fig. 15.9 has been variously and informally defined as all the ocean south of Australia, just the ocean south of 60°S, or something in between. Its importance lies in the facts that it connects all the ocean basins and that it is partially *re-entrant,* meaning that at the latitudes between South America and the Antarctic Peninsula the flow can circulate around in the zonal direction.

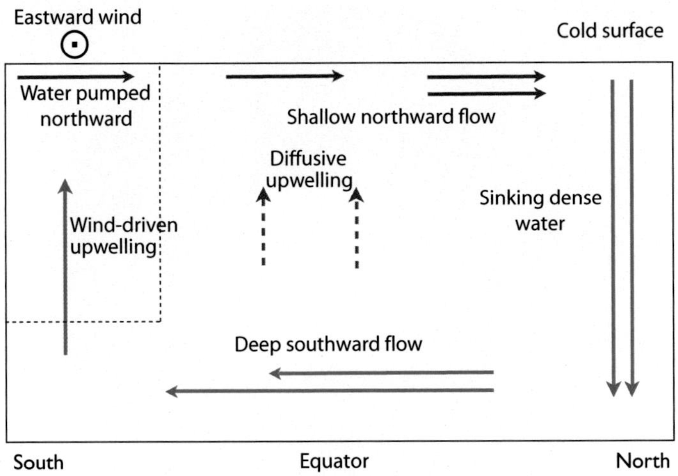

Fig. 15.10: Schematic of an idealized wind- and mixing-driven inter- hemispheric overturning circulation, with a source of cold dense water in the north and a southern channel, marked by the dashed rectangle. The water circulates as shown by the solid arrows, even with no diapyncal diffusivity. If diapycnal diffusivity is non-zero then some water upwells, as in the dashed arrows, enhancing the overturning circulation.

the Atlantic, about 5,000 km, we obtain a transport of about 20 Sv, or $2 \times 10^7\ \mathrm{m^3\,s^{-1}}$ (which should be regarded as a very rough estimate).

What makes this calculation of particular interest to the MOC is that, because of the channel nature of the Southern Ocean, the northwards flowing Ekman transport cannot be returned at the surface. In an enclosed ocean basin the western boundary current can provide any return flow needed to balance the Ekman transport, but this cannot occur in a channel. Thus, *the return flow must occur at depth*, implying an overturning circulation.

If the diffusivity is non-zero then, as we discussed previously, there will also be a diffusively-driven circulation. Using (15.7b), and setting $\beta \sim f/L$ we obtain an estimate for the diffusive component of the volumetric overturning circulation, namely

$$V_d = \left(\frac{\kappa^2 \Delta b L^4}{f}\right)^{1/3}. \tag{15.21}$$

Putting in values for the various parameters gives an estimate of between 1×10^6 to 2×10^7, depending on the values chosen for κ and Δb. The ratio of the two estimates above, namely (15.21) to (15.20), is given by

$$\mathcal{R} = \left(\frac{\kappa^2 \Delta b L f^2}{\tau_0^3}\right)^{1/3}. \tag{15.22}$$

The ratio has large error bars and the reader is invited to substitute representative values. However, $\mathcal{R}$ is neither obviously very large nor very small, implying that wind-driving over the Southern Ocean can play a siginficant role in driving a global overturning circulation in the ocean. Whether the wind-driven or diffusive component is dominant in the real ocean is a question for observations and for comprehensive numerical models, and the reader may refer to the references in the literature, perhaps beginning with those at the end of the chapter.

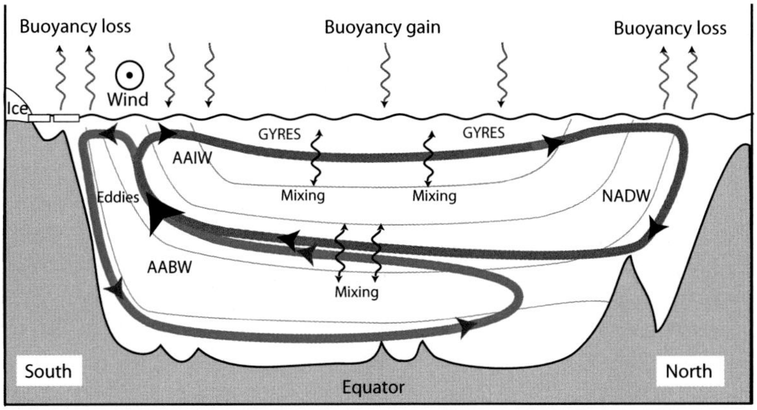

Fig. 15.11: A sketch of the overturning circulation of the ocean, and the main processes that produce it — winds, mixing, baroclinic eddies and surface buoyancy fluxes. The figure shows both an upper interhemispheric cell of NADW and a lower, AABW cell. The figure is most representative of the Atlantic, which is the major contributor to the global average. (Adapted from a figure in Watson *et al.* (2015).)

15.4.3 Dynamics of the Interhemispheric Circulation

Let us now explore the dynamics of the MOC a little more and consider the case with a source of cold dense water in the north and a southern channel, as illustrated in Fig. 15.10, in which we phenomenologically combine our pictures of the channel and the overturning circulation. Consider the case when the diapycnal diffusivity is zero. Cold dense water in the north sinks and fills up the basin, but unlike the case with no wind pumping the circulation does not eventually cease when the basin is full of the densest water, because the winds over the channel continually pump water in the channel northward. On leaving the channel the water does not sink, because it is lighter than the water beneath it that has come from the cold north. Rather, these waters continue a northward journey until they themselves reach high northern latitudes, at which time they are cold and dense and hence they sink, and begin their overturning journey anew.

Antarctic Bottom Water

The picture is still not complete, because at the southern end of the channel the surface waters are actually colder and denser than they are at high northern latitudes. These waters will therefore sink and, we might imagine, fill up the entire basin, and if so there would then be no need for an interhemispheric MOC. However, this reasoning is not wholly correct, because the southward flow at the base of the channel must exactly balance the northward flow at the surface and this requires that some flow from high northern latitudes must be drawn down to the Southern channel. The net result of this is that the circulation forms two cells, as schematically illustrated in Fig. 15.11. The lower cell of Antarctic Bottom Water (AABW) depends for its circulation on the presence of mixing, since it is not pumped by the winds, whereas the upper North Atlantic Deep Water (NADW) cell is driven both by mixing and Southern Ocean winds. In the absence of any mixing the very deep ocean would fill with AABW and then become stagnant, with the upper cell continuing to circulate.

Thus far, we have not paid much attention to the dynamics of the Southern Ocean itself, and how it manages to satisfy the momentum balance and whether such things as baroclinic eddies might form within it. Let us now address that.

15.5 ♦ THE ANTARCTIC CIRCUMPOLAR CURRENT

We now consider a rather idealized model of the Antarctic Circumpolar Current, or ACC. Specifically, we imagine a zonally-re-entrant, flat-bottomed channel of fresh water in the Southern Hemisphere. We suppose the water in the channel is forced by a constant eastward wind at its surface and that the air temperature decreases polewards, and that the surface waters in the channel take on the temperature of the overlying atmosphere and so also decrease polewards. We also suppose that the channel is isolated from the rest of the ocean; that is, it has walls at its southern and northern edges. We will take it that the flow in the channel is in a frictional-geostrophic balance, with the frictional stress terms being important only near the surface (because of the wind forcing) and at the bottom (because of bottom friction), with eddy stresses possible in the interior.

There can be no zonally-averaged meridional geostrophic flow in the channel interior. This follows from geostrophic balance, $fv = \partial\phi/\partial x$: the zonal average of v, $\overline{v}$, then vanishes. The vertically integrated meridional flow must also vanish on average, by mass conservation. A corollary of this result is that Sverdrup balance is not a good foundation on which to build. To see this, suppose that the wind is zonally uniform and the flow is statistically homogeneous in the zonal direction. Sverdrup balance would mean that $\beta[v] = -f\partial\tau/\partial y$, where τ is the zonal wind stress and $\langle v\rangle$ is the vertically integrated meridional velocity. But the zonal average of $\langle v\rangle$ must vanish, so that the local satisfaction of Sverdrup balance can hold *only* in the presence of zonal inhomogeneities, such as continental boundaries or topography. Evidently, the dynamics of the ACC are likely to be different from those of gyres, and may be more like the atmosphere.

15.5.1 Momentum Balance

We noted previously that the eastward wind produces a northward Ekman drift at the surface, and as this water moves north it warms. When it reaches the equatorward edge of the channel it sinks and returns, even though it is warm, in order that mass conservation can be satisfied. This circulation is achieved by a pressure gradient, and the flow returns along the bottom where friction acts. The frictional-geostrophic zonal momentum equation is, in steady state and with no eddy terms,

$$-fv = -\frac{\partial\phi}{\partial x} + \frac{\partial\tau}{\partial z}. \tag{15.23}$$

Averaging zonally (denoted with an overbar) gives $-f\overline{v} = \partial\overline{\tau}/\partial z$, and integrating vertically (denoted with angle brackets) gives

$$-f\langle\overline{v}\rangle = \overline{\tau}(0) - \overline{\tau}(-H), \tag{15.24}$$

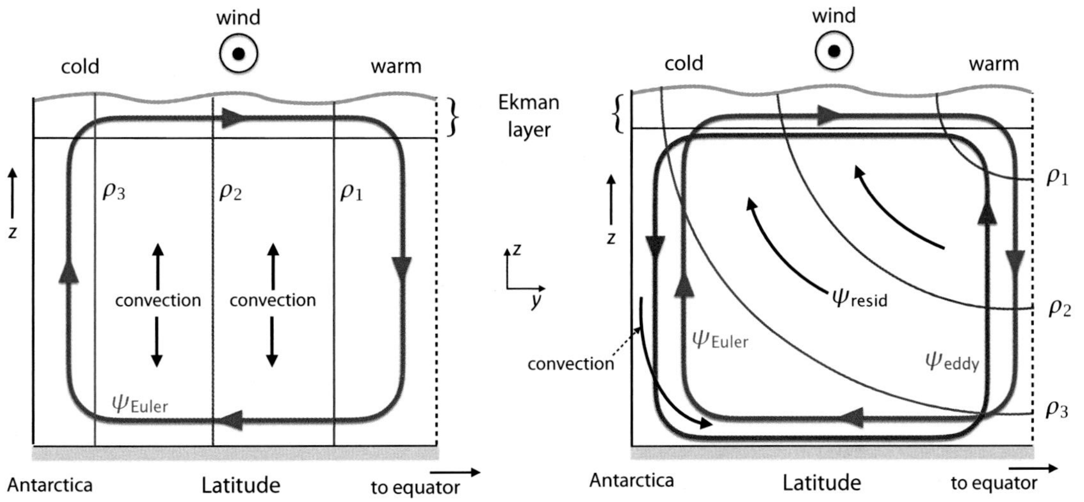

Fig. 15.12: Idealized stratification and overturning circulation in a Southern Hemisphere circumpolar channel, forced by eastward winds and with a surface temperature that increases northwards. The dashed vertical line is a wall in a true channel, and an open boundary in the real ACC.

The left panel shows a channel with no baroclinic eddies and hence vertical isopycnals. The right panel shows the circulation in which baroclinic eddies form, causing the isopycnals to slump, with an Eulerian circulation (dark blue), an eddy circulation (red), and a net, or residual, circulation (black arrows). The clockwise Eulerian circulation is forced by the eastward winds, the eddy circulation opposes it, and the smaller residual circulation is nearly along isopycnals, either recirculating or entering at the northern edge.

where $z = 0$ and $z = -H$ at the top and bottom of the channel. Since $\langle \overline{v} \rangle = 0$ by mass conservation, the wind stress at the top of the ocean must be balanced, at each latitude, by a drag at the bottom. We therefore expect a weak but non-zero eastward zonal flow at the bottom to provide that stress, and the communication between top and bottom is provided by the overturning circulation, as illustrated in the left panel of Fig. 15.12.

Now, once a water parcel leaves the surface it will, except for the rather weak effects of diffusion, keep the potential temperature it had at the surface. Thus, the returning water at the channel bottom has the same potential temperature as that of the warmest part of the channel at the surface, and so as it moves south it will be lighter than the surface water above it, and thus convectively unstable. The column thus convects, producing vertical isopycnals stretching from the bottom to the surface, and so with a meridional temperature gradient at all levels and a corresponding thermal wind shear. The absolute value of the eastward flow in such a situation is set by the requirement that the bottom stress exactly balances the wind stress at the surface. However, such a flow has a large amount of available potential energy and will be baroclinically unstable and provide eddy momentum and buoyancy fluxes, as we now consider.

15.5.2 Eddy Fluxes and the Residual Circulation

Baroclinic instability releases available potential energy (APE), and the result of this is that the isopycnals slump, pushing cold dense water to the abyss and raising the lighter water and so reducing the overall potential energy of the fluid, as illustrated schematically in Fig. 15.12. The slope of the isopycnals is then determined by a balance between the wind forcing (the steepening effect) and baroclinic instability (the slumping effect). A quantitative theory is beyond the scope of this book, but we can go a little way along the path of understanding by using the TEM framework introduced in Chapter 9.

Including the eddy momentum and buoyancy fluxes, we write the f-plane, Eulerian form of the zonally-averaged equations of motion as

$$\frac{\partial \overline{u}}{\partial t} - f_0\overline{v} = \frac{\partial \tau}{\partial z} - \frac{\partial \overline{u'v'}}{\partial y}, \qquad \frac{\partial \overline{b}}{\partial t} + N^2 w = -\frac{\partial \overline{b'v'}}{\partial y} + \mathcal{H}, \tag{15.25a,b}$$

$$\frac{\partial \overline{v}}{\partial y} + \frac{\partial \overline{w}}{\partial z} = 0, \qquad f_0\frac{\partial \overline{u}}{\partial z} = -\frac{\partial \overline{b}}{\partial y}. \tag{15.25c,d}$$

The first two equations above are the zonally-averaged zonal momentum equation and buoyancy equation (where $\mathcal{H}$ represents heating), and the last two equations are the mass continuity and thermal wind equations, respectively. We have neglected the nonlinear terms involving the mean flow but we keep the larger eddy flux terms on the right-hand sides.

The Southern Ocean is the most eddy-active area of the world's ocean. One consequence of these eddies is the production of an eddy circulation that in part balances the Eulerian, wind-driven, circulation, as illustrated in Fig. 15.12.

The velocities, $\overline{v}$ and $\overline{w}$, are related through an Eulerian mean streamfunction,

$$(\overline{v}, \overline{w}) = \left(-\frac{\partial \psi_m}{\partial z}, \frac{\partial \psi_m}{\partial y}\right). \tag{15.26}$$

If N^2 is a function only of z then, as in Section 9.3.1, we can define an eddy streamfunction, ψ_e, and a residual streamfunction, ψ^*, by

$$\psi_e = \frac{1}{N^2}\overline{v'b'}, \qquad \psi^* = \psi_m + \psi_e, \tag{15.27a,b}$$

so that the components of the residual velocities are

$$\overline{v}^* = -\frac{\partial \psi^*}{\partial z} = \overline{v} - \frac{\partial}{\partial z}\left(\frac{1}{N^2}\overline{v'b'}\right), \tag{15.28a}$$

$$\overline{w}^* = \frac{\partial \psi^*}{\partial y} = \overline{w} + \frac{\partial}{\partial y}\left(\frac{1}{N^2}\overline{v'b'}\right). \tag{15.28b}$$

Using these expressions, (15.25a,b) may then be written in the TEM form

$$\frac{\partial \overline{u}}{\partial t} - f_0\overline{v}^* = \overline{v'q'} + \frac{\partial \tau}{\partial z}, \qquad \frac{\partial \overline{b}}{\partial t} + N^2\overline{w}^* = \mathcal{H}, \tag{15.29a,b}$$

where

$$\overline{v'q'} = -\frac{\partial}{\partial y}\overline{u'v'} + \frac{\partial}{\partial z}\left(\frac{f_0}{N^2}\overline{v'b'}\right), \tag{15.30}$$

and just as in our discussion in Section 9.3 the eddy terms in (15.29) only explicitly appear in the momentum equations. If the horizontal scales of the eddies are larger than the deformation radius NH/f then the eddy momentum flux term is smaller than the eddy buoyancy flux and we may ignore it. In this case the steady momentum equation becomes

$$-f_0\overline{v}^* = \frac{\partial \tau}{\partial z} + \frac{\partial}{\partial z}\left(\frac{f_0\overline{v'b'}}{N^2}\right). \tag{15.31}$$

The first term on the right-hand side of (15.31) is the Eulerian contribution to the residual flow. Since τ is large and positive at the surface (because of the wind forcing) and also large and positive at the ocean floor (because of frictional effects) the Eulerian circulation is clockwise, as in the blue circulation in Fig. 15.12 (remember that f is negative). The eddy circulation tends to oppose this, and the residual circulation is much smaller than either the Eulerian circulation (proportional to $\partial\tau/\partial z$) or the contribution from the eddy terms, given by the second term on the right-hand side of (15.31).

The overall sense of the net, residual, circulation can be determined from the steady TEM form of the buoyancy equation, namely $N^2\overline{w}^* = \mathcal{H}$. This equation implies a direct circulation, with water sinking where it is being cooled. To go further we would need to introduce a closure for the eddy buoyancy flux, such as $\overline{v'b'} = -\mathcal{K}\partial\overline{b}/\partial y$ where $\mathcal{K}$ is an eddy diffusivity to be given by a theory of mesoscale turbulence, and (15.31) becomes

The theory and modelling of the overturning circulation of the world's oceans is one of the most challenging and active areas in oceanography today.

$$-f_0\overline{v}^* = \frac{\partial\tau}{\partial z} + f_0\frac{\partial}{\partial z}(\mathcal{K}s_b), \qquad (15.32)$$

where $s_b = -(\partial_y\overline{b})/N^2$ is the slope of the isopycnals (as $N^2 \approx \partial\overline{b}/\partial z$). This way we can begin to have a quantitative theory, but our own story must finish here.

15.5.3 Connection to the World Ocean

The fact that the ACC is strongly eddying does not necessarily change the central role of the Southern Ocean in the global MOC, but there is one adaptation we must make, namely that the circulation in the basins must connect smoothly to the *residual* circulation of the ACC, not the Eulerian circulation, because it is the residual circulation that best represents the flow of water parcels, although in the basin regions the eddy contribution is relatively small. A second point is that the MOC of the real ocean is truly global, meaning that it spans basins and over the course of centuries a water parcel may find itself in the Atlantic, Pacific or Indian Oceans. We leave the reader to explore that topic elsewhere.

Notes and References

Much of the modern theory of the thermocline stems from Robinson & Stommel (1959) and Welander (1959). The ideas of the ventilated and internal thermocline were then developed by many authors with the papers by Luyten *et al.* (1983), Salmon (1990) and Samelson & Vallis (1997) more-or-less describing the theory as it stands today. The ideas of a wind-driven MOC emerged from observations and numerical models by Toggweiler & Samuels (1998) and Döös & Coward (1997). The models of the ACC and MOC that we describe in this chapter draw from Johnson & Bryden (1989), Gnanadesikan (1999), Vallis (2000), Marshall & Radko (2003), Samelson (2004), Wolfe & Cessi (2011) and Nikurashin & Vallis (2011, 2012). Observations of the ocean are discussed in more detail in the books by Talley *et al.* (2011) and Wunsch (2015).

Problems

Oceanic problems are given at the end of Chapter 16.

Chapter

16 Equatorial Oceans and El Niño

EQUATORIAL DYNAMICS DIFFERS from its midlatitude counterpart because the Coriolis parameter is relatively small and the Rossby number large, and balanced and unbalanced dynamics then become intertwined. Yet if we move more than a few degrees away from the equator the Rossby number again becomes quite small, suggesting that familiar ways of investigating the dynamics — Sverdrup balance for example — might yet play a role. Not surprisingly, the equatorial ocean is the home to a multitude of interesting phenomena and in this chapter we discuss just two of the most striking, namely the equatorial undercurrent and El Niño. Let us first see what the observations tell us.

16.1 Observations of the Equatorial Ocean

The most distinctive features of equatorial oceans are illustrated in Fig. 16.1 and the top panel of Fig. 16.2, namely:

(i) A shallow westward flowing surface current, typically confined to the upper 50 m or less, strongest within a few degrees of the equator, although not always symmetric about the equator. Its speed is typically a few tens of centimetres per second.

(ii) A strong coherent eastward undercurrent extending to about 200 m depth, confined to within a few degrees of the equator. Its speed is up to a metre per second, and it is this current that dominates the vertically integrated transport at the equator. Beneath the undercurrent the flow is relatively weak.

(iii) Westward flow on either side of the undercurrent, with eastward countercurrents poleward of this. The Pacific countercurrent is strongest in the Northern Hemisphere, where it reaches the surface.

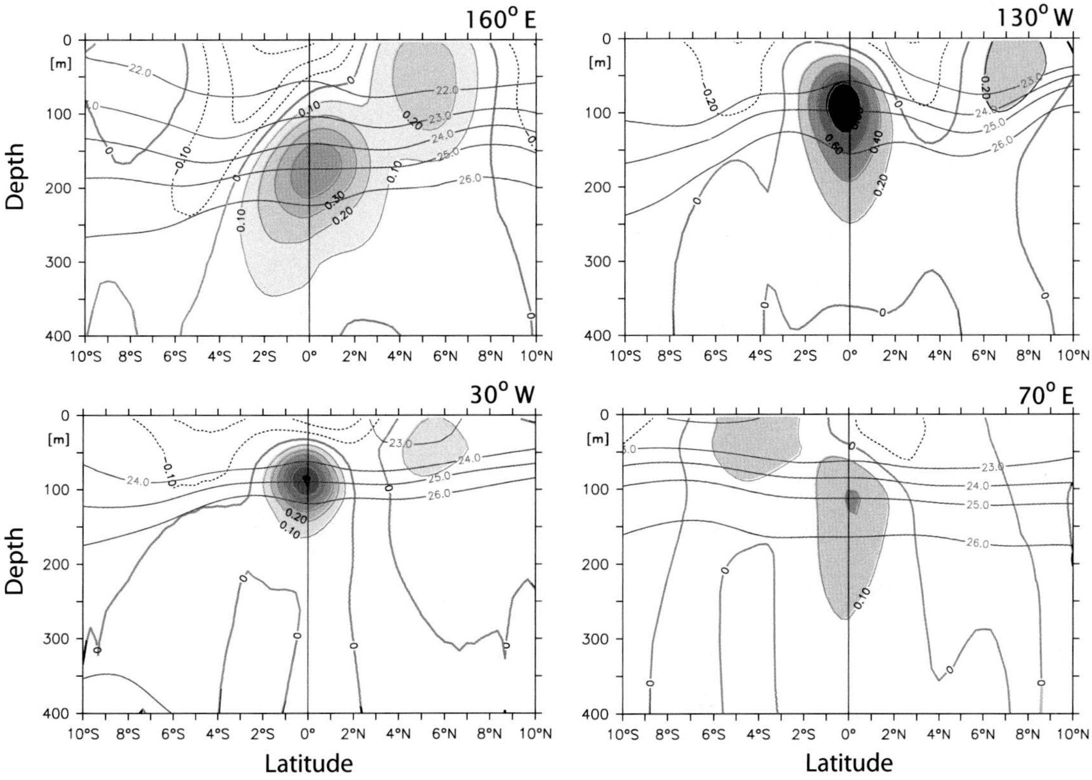

Fig. 16.1: Sections of the mean zonal current (shading and associated contours) at two longitudes in the Pacific (upper panels), in the Atlantic (lower left) and in the Indian Ocean (lower right). The contours are every 20 cm s^{-1} in the upper two panels and every 10 cm s^{-1} in the lower panels. Note the well-defined eastward undercurrent at the equator in all panels, and a weaker eastward countercurrent at about 6° N and/or 6° S. The red, more horizontal, lines are isolines of potential density.

These features are largely common to both the Atlantic and Pacific Oceans and to a somewhat lesser extent in the Indian Ocean. We start our dynamical explorations with the vertically integrated flow.

16.2 Vertically Integrated Flow and Sverdrup Balance

In midlatitudes the large scale currents system may be understood using the planetary geostrophic equations of motion, with Sverdrup balance (Section 14.2) providing a solid foundation on which to build. As we approach lower latitudes the Coriolis parameter, f, decreases and the Rossby number increases and one might expect that dynamics based on geostrophic balance will ultimately fail. However, it is only very close to the equator that the Rossby number exceeds unity: if we take a velocity of 0.5 m $s^{-1}$and a length scale of 500 km then the Rossby number at 5° latitude is 0.08, at 2° it is 0.2 and at 1° it is 0.4. These numbers suggest that until we are virtually at the equator we can use some of the familiar tools from the midlatitude dynamics. Let us first see the extent to which the familiar Sverdrup balance can explain the vertically integrated flow. The horizontal momentum may be written

$$\frac{\partial \boldsymbol{u}}{\partial t} + \boldsymbol{u} \cdot \nabla \boldsymbol{u} + \boldsymbol{f} \times \boldsymbol{u} = -\nabla \phi + \frac{1}{\rho_0} \frac{\partial \boldsymbol{\tau}}{\partial z}, \qquad (16.1)$$

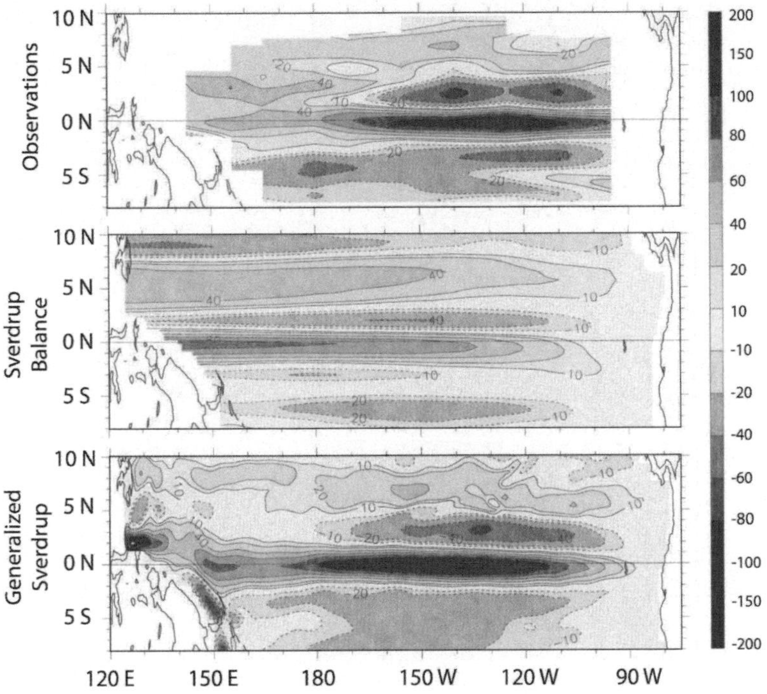

Fig. 16.2: Vertically integrated zonal transport in the Pacific. Red colours indicate eastward flow, blue colours westward. The top panel shows the observed flow, the middle panel shows the flow calculated using Sverdrup balance with the observed wind, and the bottom panel shows the flow calculated with a 'generalized' Sverdrup balance that includes the nonlinear terms in a diagnostic way.

where $\boldsymbol{\tau}$ is the stress on the fluid. As in earlier chapters, we will absorb the constant density, ρ_0, into the stress, so that $\boldsymbol{\tau}/\rho \to \boldsymbol{\tau}$. The mass conservation equation is

$$\frac{\partial u}{\partial x} + \frac{\partial v}{\partial y} + \frac{\partial w}{\partial z} = 0, \tag{16.2}$$

which, on vertical integration over the depth of the ocean, gives

$$\frac{\partial U}{\partial x} + \frac{\partial V}{\partial y} = 0, \tag{16.3}$$

where U and V are the vertically integrated zonal and meridional velocities (e.g., $U = \int u \, \mathrm{d}z$) and we assume the ocean has a flat bottom and a rigid lid at the top. If we assume the flow is steady and integrate (16.1) vertically, then take the curl and use (16.3), we obtain

$$\beta V = \mathrm{curl}_z(\boldsymbol{\tau}_T - \boldsymbol{\tau}_B) + \mathrm{curl}_z \boldsymbol{N}, \tag{16.4}$$

where the subscripts T and B denote top and bottom, $\boldsymbol{N}$ represents all the nonlinear terms and curl_z is defined by $\mathrm{curl}_z \boldsymbol{A} \equiv \partial A^y/\partial x - \partial A^x/\partial y = \hat{\mathbf{k}}\cdot\nabla_3 \times \boldsymbol{A}$. Equations (16.4) and (16.3) are closed equations for the vertically averaged flow.

If we neglect the nonlinear terms and the stress at the bottom (we'll come back to these terms later) then (16.4) becomes

$$\beta V = \mathrm{curl}_z \boldsymbol{\tau}_T. \tag{16.5}$$

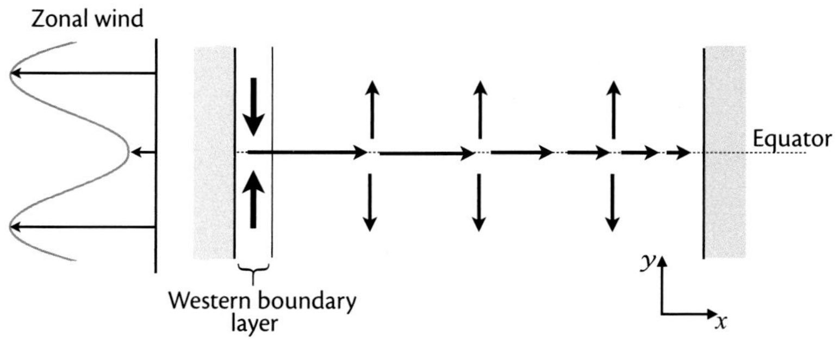

Fig. 16.3: Schema of Sverdrup flow at the equator between two meridional boundaries. The mean winds are all westward, but with a minimum in magnitude at the equator. By Sverdrup balance, (16.5), the wind stress produces the divergent meridional flow shown, which in turn induces an eastward equatorial zonal flow, strongest in the western part of the basin.

This is just Sverdrup balance, familiar from Chapter 14. The zonal transport is obtained by differentiating (16.5) with respect to y, using (16.3) to replace $\partial_y V$ with $\partial_x U$, and then integrating from the eastern boundary (x_E). This procedure gives

$$U = -\frac{1}{\beta}\int_{x_E}^{x} \frac{\partial}{\partial y}\mathrm{curl}_z \boldsymbol{\tau}_T \,\mathrm{d}x' + U(x_E, y). \tag{16.6}$$

We don't integrate from the western boundary because a boundary layer can be expected there, whereas the value of U at the eastern boundary, namely U_E, will be small.

The wind stress is known from observations and we can then use (16.6) to calculate U, which is found to be generally positive (eastward) at the equator. The solution is plotted in the middle panel of Fig. 16.2. There is a good but not perfect agreement with the observations, shown in the top panel. In the western equatorial Pacific the observed eastward flow is quite broad whereas the eastward Sverdrup flow is narrow, flanked on either side by westward flow, and much of this discrepancy can be attributed to the role of the nonlinear and frictional terms, as illustrated in the bottom panel of Fig. 16.2. To obtain the results shown, the the nonlinear terms (which have the form $\mathrm{curl}_z(\int \boldsymbol{u}\cdot\nabla\boldsymbol{u}\,\mathrm{d}z)$ are included in a diagnostic fashion. That is to say, the term $\mathrm{curl}_z\boldsymbol{N}$ is evaluated from observations and included on the right-hand side of (16.4) in order to calculate a 'generalized Sverdrup' flow, which (as one might expect) is in better agreement with the observations. Perhaps the most interesting point is that, even quite close to the equator and even without the nonlinear terms, Sverdrup balance provides a qualitatively correct picture of the vertically averaged flow, with the longitudinal structure of the flow sketched in Fig. 16.3.

16.2.1 Sensitivity of the Sverdrup Flow

Although the calculations of Sverdrup flow do show good agreement with observations, the calculation — and, most likely, the observed flow — is rather sensitive to the precise form of the winds. To illustrate this, suppose that $U(x_E, y) = 0$ and the stress is zonal and uniform, then (16.6) becomes

$$U(x,y) = \frac{1}{\beta}(x - x_E)\frac{\partial^2 \tau_T^x}{\partial y^2}. \tag{16.7}$$

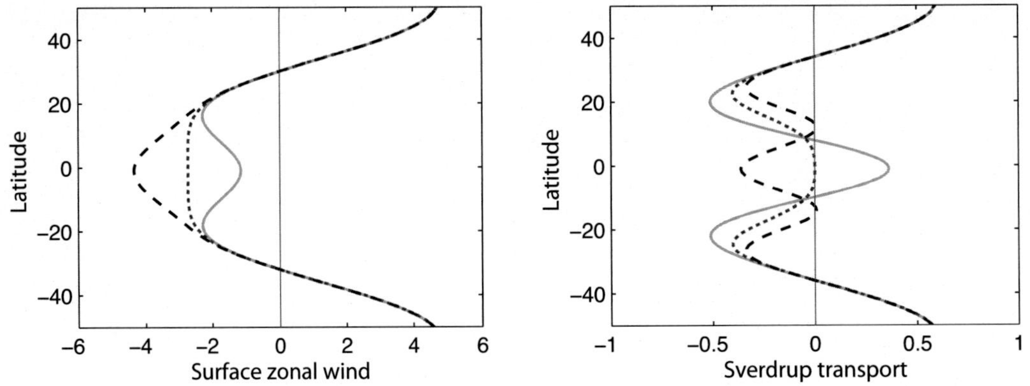

Fig. 16.4: The left panel shows three putative surface zonal (atmospheric) winds, u, all with westward winds in the tropics and with the solid line being the most realistic. The right panel shows the corresponding negative of the second derivative, $-\partial^2 u/\partial y^2$, proportional to the (oceanic) Sverdrup transport, in arbitrary units. The wind represented by solid (blue) line gives an eastward transport at the equator, as is observed, with the others differing markedly.

That is, the depth integrated flow is proportional to the second derivative of the zonal wind stress, and because $x < x_E$ we have $U \propto -\partial^2 \tau_T^x/\partial y^2$. Now, although the zonal wind is generally westward in the tropics there is a minimum in the magnitude of that wind near the equator (that is, there is a local maximum as sketched in the left panel Fig. 16.3) so that $\partial^2 \tau_T^x/\partial y^2$ is negative. Without this local maximum the Sverdrup flow would be westward at the equator.

This sensitivity of the Sverdrup flow to the wind pattern is illustrated in Fig. 16.4. The figure shows three surface zonal wind distributions, with the 'w' shaped solid line having a minimum in the westward flow (i.e., a minimum in the trade winds) at the equator and so being the most realistic. The right-hand panel shows the negative of the second derivative of the winds which is proportional to the zonal Sverdrup flow. Only in the one case (the blue line) does the wind produce an eastward Sverdrup flow. In fact, in the case illustrated with the dashed lines, the small changes in the meridional gradient of the wind between 15° and 20° produce large variations in the Sverdrup transport. Given this sensitivity, the small difference in the latitudinal variation of the Sverdrup flow and the observed flow, illustrated in the top and middle panels of Fig. 16.2, is not surprising and cannot be considered a major failure of the theory. However, the difference in the longitudinal structure of the two fields is indicative of the importance of other terms in the vorticity balance.

Although the Sverdrup flow is rather sensitive to the horizontal derivatives of wind pattern, the undercurrent itself is not, and let us turn our attention to that.

16.3 DYNAMICS OF THE EQUATORIAL UNDERCURRENT

The equatorial undercurrent is perhaps the single most conspicuous feature of the ocean current system at low latitudes and we now describe a model for it. Our model will be a local one, meaning that it is the direct effect of the winds in the equatorial region that drive the current, and although it does provide a simple, compelling explanation for the undercurrent it is an incomplete picture: it does not account for the remote effects of winds in building up a head of pressure that can produce an un-

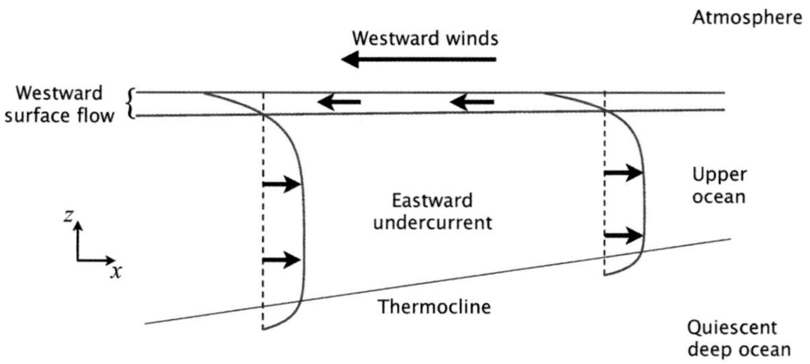

Fig. 16.5: Sketch of an undercurrent, showing a near-surface flow in the direction of the wind, a counter-flowing undercurrent beneath and a quiescent deep ocean below the thermocline

dercurrent even in the absence of equatorial winds. Readers who wish to read about that may consult the references at the end of the chapter.

The essence of the local picture is as follows. The mean winds are westward and provide a stress on the upper ocean, pushing the near-surface waters westward. Given that there is a boundary in the west, the water piles up there so creating an eastward pressure-gradient force. To a large degree the pressure gradient and the wind stress compensate each other leading to a state of no motion. However, the compensation does not hold exactly everywhere. Rather, close to the surface the stress is dominant and a westward surface current results, but below the surface the pressure gradient dominates, resulting in an eastward flowing undercurrent, as in the observations in Fig. 16.1. It is interesting that this description makes no mention of the Coriolis parameter or Sverdrup balance or the wind-stress curl. To understand it better, let us make a simple mathematical model of it.

The undercurrent itself seems to have been first discovered in the Atlantic by J. Y. Buchanan in the 1880s. The discovery of the undercurrent in the Pacific is often credited to Townsend Cromwell (1922–1958) in the early 1950s, and there the current is called the Cromwell Current. Cromwell also provided the first credible theoretical model of the undercurrent. He tragically died in a plane crash in 1958 while en route to an oceanography expedition.

16.3.1 Response of a Homogeneous Layer to a Uniform Zonal Wind

Let us first consider the simple case of the response of a layer of homogeneous fluid obeying the shallow water equations to a steady zonal wind that is uniform in the y-direction. With our usual notation the equations of motion in the presence of momentum and mass forcing are

$$\frac{\mathrm{D}u}{\mathrm{D}t} - fv = -g'\frac{\partial h}{\partial x} + F^x, \tag{16.8a}$$

$$\frac{\mathrm{D}v}{\mathrm{D}t} + fu = -g'\frac{\partial h}{\partial y} + F^y, \tag{16.8b}$$

$$\frac{\mathrm{D}h}{\mathrm{D}t} + h\left(\frac{\partial u}{\partial x} + \frac{\partial v}{\partial y}\right) = 0, \tag{16.8c}$$

where (F^x, F^y) are the zonal and meridional forces on the fluid arising from the wind stress at the surface. The quantity h is the thickness of the fluid layer, which in this case can be taken to be equal to the depth of the thermocline. For steady flow, and neglecting the nonlinear terms, the equations become

$$-fv = -g'\frac{\partial h}{\partial x} + F^x, \quad +fu = -g'\frac{\partial h}{\partial y}, \quad H\left(\frac{\partial u}{\partial x} + \frac{\partial v}{\partial y}\right) = 0. \quad (16.9\text{a,b,c})$$

If we take the y-derivative of (16.9a) and subtract it from the x-derivative of (16.9b), and noting that $\partial F^x/\partial y = 0$, we obtain

$$\beta v = 0. \qquad (16.10)$$

Thus, using the continuity equation (16.9c), we have $\partial u/\partial x = 0$. That is, the zonal velocity is uniform. If there is a zonal boundary at which $u = 0$ then the zonal flow is zero everywhere and the complete solution is

$$u = 0, \qquad v = 0, \qquad g\frac{\partial \eta}{\partial x} = F^x, \qquad \frac{\partial h}{\partial y} = 0. \qquad (16.11)$$

That is to say, the ocean is motionless and the wind stress is balanced by a pressure gradient. If the wind is westward, as it is on the equator, then $\partial h/\partial x < 0$ and the thermocline slopes down and deepens toward the west. Although the real ocean is not as simple as our model of it, the analysis exposes a truth with some generality: *the wind stress is largely opposed by a pressure gradient* rather than inducing a large westward acceleration that is halted by friction.

16.3.2 A Two-Layer Model

Let us now consider a model with some vertical structure, thereby allowing the wind stress to be taken up in the upper ocean, but the deeper ocean feel no stress. As before, the wind pushes the near-surface water westward and creates a zonal pressure gradient. The deeper water feels the pressure-gradient force (because the pressure is hydrostatic) but does not feel the wind stress, and so the deep water flows eastward, as sketched in Fig. 16.5. We can make a simple model of this by considering a homogeneous layer of fluid in which the frictional forces extend only part way through its depth, as follows.

Equations of motion

Consider the one-dimensional equations of motion, with no Coriolis force (for now) and no nonlinear terms, and a linear drag for friction. We divide our homogeneous fluid into an upper layer in which wind-stress is important and a deeper layer in which wind stress is negligible. The momentum equations for each layer are then

$$\frac{\partial u_1}{\partial t} = -\frac{\partial \phi_1}{\partial x} + \frac{\partial \tau^x}{\partial z} - ru_1, \qquad \frac{\partial u_2}{\partial t} = -\frac{\partial \phi_2}{\partial x} - ru_2, \qquad (16.12\text{a,b})$$

where $\partial \tau^x/\partial z$ is the zonal force on the upper layer due to the wind stress. The stress goes to zero at the base of the upper layer and remains zero in the lower layer. A drag is present in both layers, represented by the terms

The Equatorial Undercurrent

What is it?

The equatorial undercurrent is the single most striking feature of the low latitude ocean circulation. It is an eastward flowing subsurface current, mostly confined to depths between about 50 m and 250 m and to latitudes within 2° of the equator, with speeds of up to $1\,\mathrm{m\,s^{-1}}$ (Fig. 16.1). The undercurrent is a permanent feature of the Atlantic and Pacific Oceans, but varies with season in the Indian Ocean because of the monsoon winds.

What are its dynamics?

- The theory that we have described here regards the undercurrent as a direct response to the westward winds at the equator. The winds push water westward and create a balancing eastward pressure gradient force. Below a frictional surface layer the influence of the wind stress is small and the pressure gradient leads to an eastward undercurrent.
  - In the frictional surface layer the flow is away from the equator and there is upwelling at the equator. The circulation is essentially closed in the equatorial region.
  - The dynamics of the simplest models of this ilk are linear, but their quantification relies on the use of somewhat poorly constrained frictional and mixing coefficients.
- In reality, the undercurrent also responds to nonlocal effects, and in particular to off-equatorial winds that build up a pressure gradient almost independently of what happens at the equator. A subsurface current moves inertially from higher latitudes into the equatorial region. A pressure head is created in the western equatorial basin, which then pushes the undercurrent along.

$-ru_1$ and $-ru_2$ where r is a constant, and the terms $\partial\phi_1/\partial x$ and $\partial\phi_2/\partial x$ are the pressure gradient forces in each layer. To obtain a relationship between the pressures in the two layers, note that the hydrostatic equation for a Boussinesq system is

$$\frac{\partial \phi}{\partial z} = b, \tag{16.13}$$

where ϕ is the kinematic pressure and b the buoyancy. If the buoyancy is uniform (so there are no thermodynamic effects) then the horizontal gradient of pressure is the same in the upper layer as in the lower layer, and $\partial\phi_1/\partial x = \partial\phi_2/\partial x$.

Integrating vertically over each layer and looking for steady solutions gives

$$0 = -H_1\frac{\partial \phi_1}{\partial x} + \tau_0^x - rH_1u_1, \qquad 0 = -H_2\frac{\partial \phi_2}{\partial x} - rH_2u_2, \tag{16.14}$$

where τ_0 is the zonal wind stress at the surface, which is negative for a westward wind, and H_1 and H_2 are the layer depths. Dividing the above two equations by H_1 and H_2 respectively and subtracting one from the other, noting that $\partial\phi_1/\partial x = \partial\phi_2/\partial x$, gives us the simple result

$$r\,[u_2 - u_1] = -\frac{\tau_0^x}{H_1}. \tag{16.15}$$

Evidently, the difference between the zonal flow in the lower and upper layers, $u_2 - u_1$, has the opposite sign to the wind. If u_1 is zero, because the pressure gradient in the top layer is balanced by the pressure gradient, then the frictional force on the lower level flow is equal and opposite to the force due to the wind stress, and it is this balance that lies at the heart of this model of the undercurrent. It is also instructive to write the lower layer flow as a deviation from the vertical mean, and a little manipulation enables (16.15) to be written in the equivalent form

$$r\left[u_2 - \left(\frac{u_1 H_1 + u_2 H_2}{H}\right)\right] = -\frac{\tau_0^x}{H}, \tag{16.16}$$

ß where $H = H_1 + H_2$. The quantity in parentheses on the left-hand side is the vertically average flow, and evidently *the deviation of the lower level flow from the vertical average is in the direction opposite to that of the wind stress.* This result is the essence of the undercurrent.

The presence of a Coriolis force is not essential for obtaining an undercurrent, but it has a substantial effect on its dynamics, in particular confining it to near the equator.

Effect of a Coriolis force

Let us now add a Coriolis force to the above model. The equations of motion for the lower layer are

$$-fv_2 = -\frac{\partial\phi}{\partial x} - ru_2, \qquad fu_2 = -\frac{\partial\phi}{\partial y} - rv_2. \tag{16.17a}$$

Let us suppose that the upper layer is stationary, with the pressure gradient balanced by the wind stress, so that the upper layer force balance is given by

$$0 = -\frac{\partial\phi}{\partial x} + \frac{\partial\tau^x}{\partial z}, \qquad 0 = -\frac{\partial\phi}{\partial y} + \frac{\partial\tau^y}{\partial z}. \tag{16.18}$$

The pressure gradients given by (16.18) are, as before, equal to those of the lower layer appearing in (16.17). Integrating (16.18) over the upper layer and substituting into (16.17) gives

$$-fv_2 = -\frac{\tau^x}{H_1} - ru_2, \qquad fu_2 = -\frac{\tau^y}{H_1} - rv_2. \tag{16.19a,b}$$

Solving for u and v then gives

$$u_2 = \frac{-\tau^x r - \tau^y f}{H_1(r^2 + f^2)}, \qquad v_2 = \frac{\tau^x f - \tau^y r}{H_1(r^2 + f^2)}. \tag{16.20a,b}$$

A sample solution is plotted in the left panel of Fig. 16.6 in a case with $\tau^y = 0$ and $f = \beta y$. The zonal flow as given by (16.20a) is then *eastward*, since, as we have discussed, the wind is balanced by an opposing pressure gradient and the deep ocean feels the pressure gradient but not the wind stress, and this effect is unchanged by the presence of a Coriolis force. However, the Coriolis term does have two important effects:

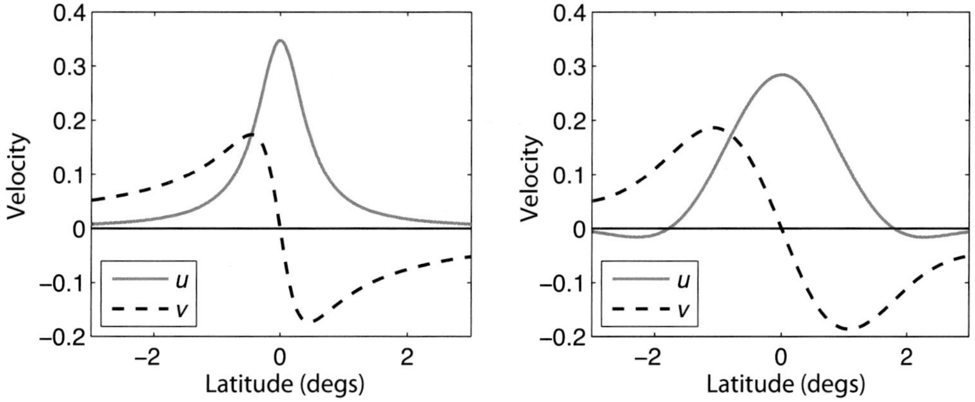

Fig. 16.6: Horizontal profiles of the undercurrent with friction represented by a linear drag (left, as given by (16.20)) and by a harmonic viscosity (right, see text), in dimensional units (m/s for velocity and degrees for latitude).

(i) The undercurrent is concentrated at the equator, decaying quite rapidly with latitude.

(ii) The deep meridional flow is zero at the equator, where $f = 0$, but is toward the equator in both hemispheres and therefore induces equatorial upwelling.

The latitudinal width of the undercurrent is determined by the ratio of β to r. Thus, with $\tau^y = 0$ (16.20a) becomes

$$u_2 = \frac{-\tau^x r}{H_1(r^2 + \beta^2 y^2)}, \tag{16.21}$$

and the width of the undercurrent scales as r/β — more friction gives a broader undercurrent.

Viscosity instead of drag

The frictional parameter r is a little arbitrary and unrealistic — friction does not act as a simple drag in the real ocean. To remedy this we can carry through a similar calculation with a viscosity instead of a drag, and we can also allow a continuous variation in the vertical instead of restricting ourselves to two layers. The equations of motion are similar except that terms like ru are replaced by $\nu\nabla^2 u$. The algebra to obtain a solution is now considerably more complicated, but the underlying mechanism producing the undercurrent is exactly the same and the solution itself is quite similar, as illustrated in the right-hand panel of Fig. 16.6. But we will not purse this topic further in this book; rather, let us turn our attention to that other great equatorial phenomenon, El Niño.

16.4 El Niño and the Southern Oscillation

Without capitalization, el niño is Spanish for male infant, whereas el Niño refers to the Christ Child and El Niño to the oceanic phenomenon.

El Niño! One of the most famous phenomena in the climate sciences, and certainly one with an enormous impact on humankind. El Niño is an anomalous warming of the surface waters in the eastern equatorial Pacific, peaking around Christmas-time, and its appealing name belies

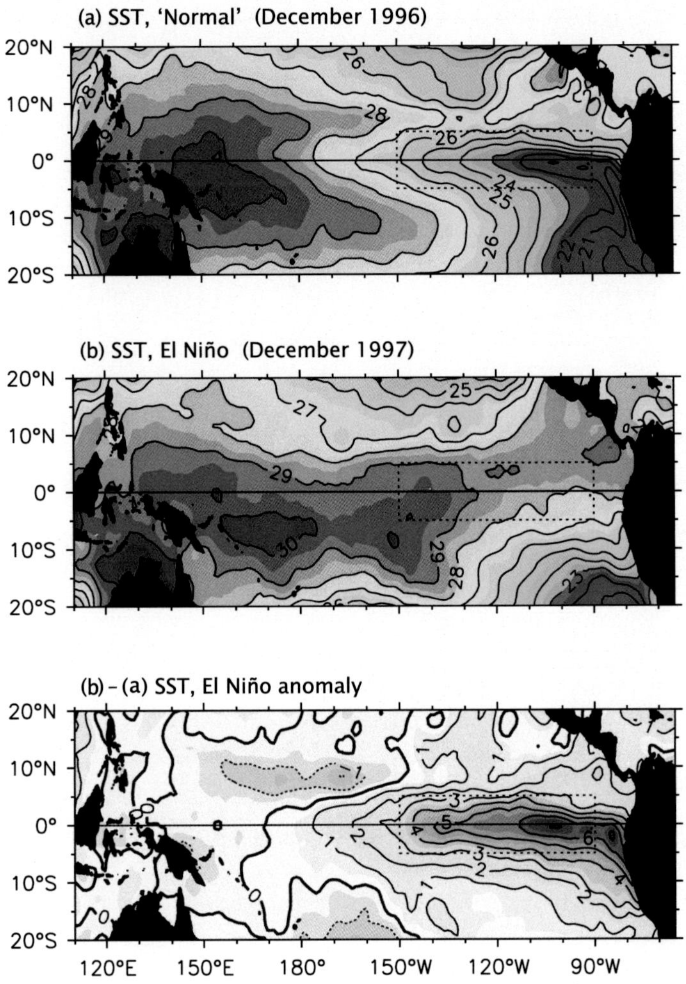

Fig. 16.7: The sea-surface temperature in December of a non-El Niño year (December 1996, top panel), a strong El Niño year (December 1997, middle panel) and their difference (bottom panel). An El Niño year is typically characterized by an anomalously warm tongue of water in the eastern tropical Pacific. The El Niño 3 region is the rectangular region demarcated by thin dotted lines in the eastern equatorial region. (Figure courtesy of A. Wittenberg, using data from NOAA.

its enormous power and global effects, bringing heavy rains to California and Northern Argentina and anomalously dry weather to South East Asia and Northern and Eastern Australia; it also raises the global average surface temperature by about half a degree Celsius. Taken with the associated changes in the atmosphere, in which case the whole phenomenon is known as the El Niño–Southern Oscillation (ENSO), it is the largest and most important source of global climate variability on interannual timescales.

16.4.1 A Descriptive Overview

Every few years the temperature of the surface waters in the eastern tropical Pacific rises quite significantly. The strongest warming takes place between about 5°S to 5°N, and from the west coast of Peru (a longitude of about 80°W) almost to the dateline, at 180°W, as illustrated in Fig. 16.7. The warming is large, with a difference in temperature up to 6°C from an

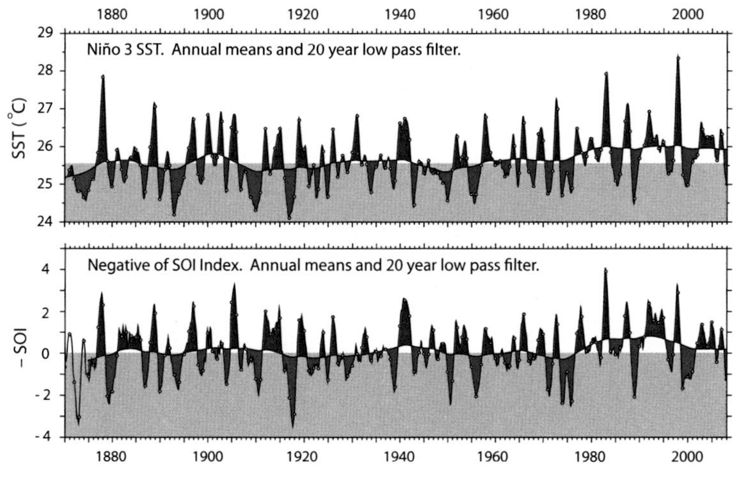

Fig. 16.8: Upper: Time series of the sea-surface temperature (SST) in the Eastern Equatorial Pacific (Niño 3) region. Lower: Negative of the Southern Oscillation Index (SOI), the anomalous pressure difference between Tahiti and Darwin. The pointy curves show the annual means, the dots are values in December, the smoother curves show the variable after application of a 20 year low-pass filter, and the top of the grey shading is the 1876–1975 mean.

El Niño year to a non-El Niño year. The warmings occur rather irregularly, with typical intervals between warmings of 3 to 7 years, as seen in Fig. 16.8, and with particularly large events in 1887–88, 1982–83, 1997–98 and 2015–2016; the development of the last event (at the time of writing) is illustrated in Fig. 16.9.

The warmings have become known as El Niño events or even (abusing the Spanish language) El Niños. The warmings typically last for several months, occasionally up to two years, and appear as an enhancement to the seasonal cycle with high temperatures appearing at a time when the waters are already warming. If definiteness is desired, an event may be said to occur when there is a warming of at least 0.5° C averaged over the eastern tropical Pacific lasting for six months or more. Ocean temperatures tend to fluctuate between warm El Niño years and those years in which the equatorial ocean temperatures are lower in the east and higher in the west, and a *La Niña* is said to occur when it is particularly warm in the west (la niña being Spanish for young girl). We have direct observational evidence — temperature measurements from ships and buoys — of El Niño events for over a century, but the events have almost certainly gone on for a much longer period of time, probably many millennia, judging from proxy records of tree rings and coral growth.

The name El Niño was originally used by fishermen along the coasts of Ecuador and Peru to refer to a warm ocean current that often appears in December (i.e., around Christmas) between Paita and Pascamayo, and lasts for several months. An early, perhaps the first, written account of El Niño is to be found in Carillo (1892), a report by a Navy captain on a warm ocean current know to the fishermen as Corriente del Niño, or Current of the Christ Child. These days, the name is applied only when the warming is particularly strong and to the warming over the whole eastern tropical Pacific.

The overlying atmospheric winds and the surface pressure in the equatorial Pacific tend to covary with the sea-surface temperature (SST) and during El Niño events the equatorial Pacific trade winds become much weaker and may even reverse. A convenient measure of the atmospheric signal is the pressure difference between Darwin (12° S, 130° E) and Tahiti (17° S, 150° W), or similar locations, and the normalized record of this signal is known as the *Southern Oscillation.* The Southern Oscillation and the SST record of El Niño are highly correlated (Fig. 16.8), and, as we mentioned earlier, the combined El Niño–Southern Oscillation phenomenon is denoted ENSO.

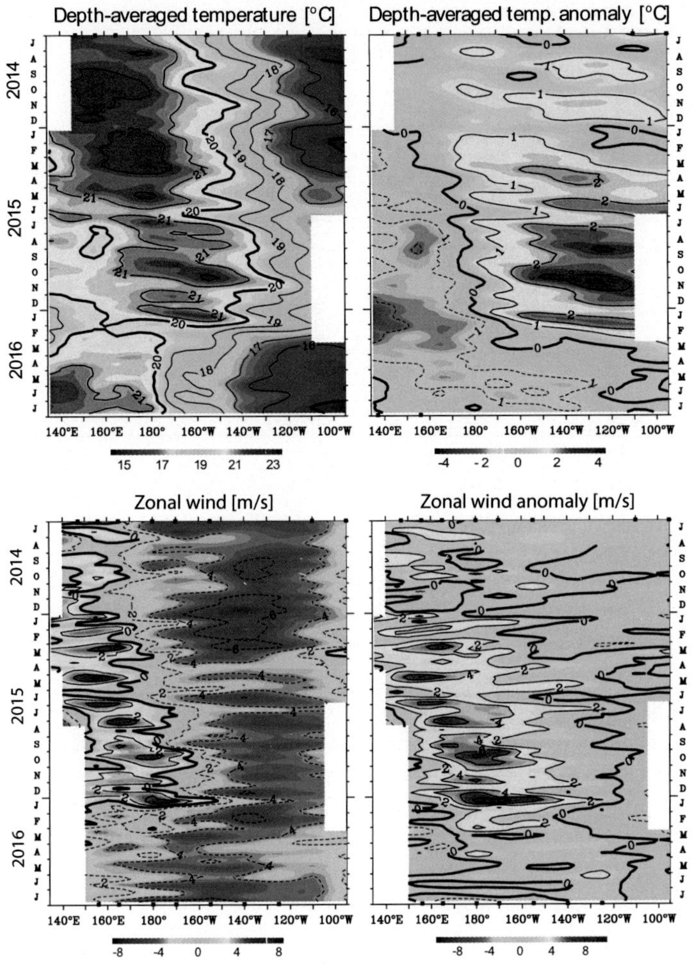

Fig. 16.9: The development of the 2015–2016 ENSO event. The top panels show the evolution of the depth-averaged temperature over the upper 300 m from mid-2014 to mid-2016, and the bottom panels show the surface zonal wind, both over a strip from 2°S to 2°N. An eastward propagation in both fields can be seen over the second half of 2015, ceasing in early 2016 just after the temperature anomalies in the eastern tropical Pacific reach their maximum in the east. (Data comes from the Tropical Atmosphere Ocean (TAO) project, http://www.pmel.noaa.gov/tao/.)

16.4.2 A Qualitative View of the Mechanism

The essential mechanism of El Niño is sketched in Fig. 16.10. First consider the mean state. The trade winds blow predominantly from higher latitudes toward the equator, and from the east to the west. This leads to a current system as described earlier in this chapter, with equatorial divergence of surface waters and upwelling — particularly in the east because here it must also replenish the surface waters moving westward away from the South American continent. The upwelling water here is cold, because much of it comes from below the thermocline, so that the SST of the eastern equatorial Pacific is relatively low, and the surface waters warm as they move westward. Furthermore, the thermocline deepens further west and so the upwelling does not bring as much cold abyssal water to the surface. The result of all this is that the SST is high in the western Pacific, up to about 30°C, and low in the eastern Pacific, about 21°C (see Fig. 16.7, top panel).

The strong zonal gradient of SST affects the atmosphere. The warm

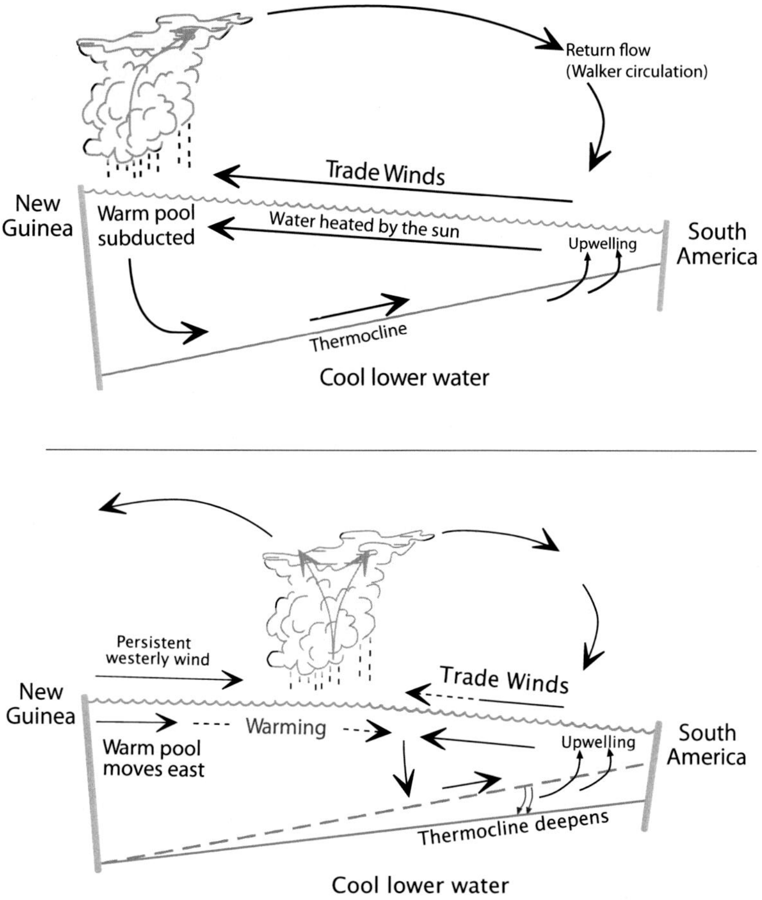

Fig. 16.10: Sketch of the end members of the ENSO cycle. The top panel shows a cross-section along the equator during non-El Niño, years, and the bottom panel at the peak of El Niño. Not to scale, and slopes are exaggerated. (Adapted from drawings by Dr. Billy Kessler.)

western Pacific region becomes more convectively unstable with respect to convection, the eastern Pacific is correspondingly cool and, as sketched in Fig. 16.10, an east-west overturning circulation is set up in the atmosphere called the *Walker circulation*. Evidently, the oceanic and the atmospheric states reinforce each other: the westward trade winds give rise to an east–west SST gradient in the ocean which generates the Walker Cell, so the surface winds are stronger than they would be if there were no ocean or if the ocean extended all the way round the globe.

By their nature positive feedbacks reinforce initial tendencies, whatever those tendencies may be, and such a feedback is at the core of El Niño. If the trade winds weaken then so will the zonal SST gradient, and the trades will weaken further and the ocean will warm more in the East, and soon we have a full-fledged El Niño event. The feedback cannot amplify without bound and, in fact, there are natural damping mechanisms for El Niño. The main one is simply that a warm sea-surface will give up heat to the atmosphere, and another is that the warm pool in the east will 'leak out' along the coast of North and South America. The El Niño event then begins to decay and the whole feedback then occurs in the opposite sense and eventually the system reverts to its normal state. The mechanism that gives rise to the ENSO cycle is known as the *Bjerknes feedback*, and the feedback is generally regarded as being stronger in the eastern ocean because there the thermocline is shallower. Thermocline depth and SST are correlated, with a shallow thermocline corresponding to low SST because cold water can then upwell through the thermocline, with the correlation getting weaker as the thermocline thickens.

The Bjerknes feedback is named for Jacob Bjerknes (1897–1975), son of the equally well-known Vilhelm Bjerknes (1862–1951). In the early twentieth century Vilhelm did important work on vorticity, proposed a method for numerical weather forecasting, and was the senior founder of the 'Bergen School' of meteorology in 1917, which marked the beginning of the transformation of meteorology from a rather descriptive subject to a quantitative one based on mathematics representing physical laws. As for Jacob, as well as his work on El Niño he was one of the discoverers of the North Atlantic Oscillation, one of the main modes of variability in the midlatitude atmosphere.

16.4.3 Why the Tropics?

Why do these dynamics occur in the tropics, and not in midlatitudes? The tropics are different because there is a close and reasonably direct connection between SST and the winds, so enabling a feedback to occur that reinforces initial tendencies, and that arises because of a combination of the following factors:

(i) Equatorial sea surface temperatures are generally high, around 27°C (Fig. 16.7), and convection is readily triggered if the temperature further increases (although there is nothing magical about 27°C). Ascending motion occurs over warm regions with associated low-level convergence, and the surface winds thus directly respond to a changing SST gradient.

(ii) The equatorial thermocline is quite shallow, varying from 200 m in the west to 50 m in the east, and is thus sensitive to changing wind patterns. Furthermore, because the thermocline is shallow upwelling can occur *through* it (Fig. 16.10), leading to a close connection between thermocline depth and SST, especially in the east.

(iii) In equatorial regions the main large-scale waves, namely equatorial Rossby and Kelvin waves are quite fast, certainly compared to advective motion, so allowing cross-basin communication to occur on short, but not immediate, timescales. In midlatitudes the advective terms tend to be more dominant, drowning out the signal from the linear waves.

In midlatitudes none of the above are as effective. Here the atmosphere is internally highly variable and its response to an SST anomaly has small signal to noise ratio and the induced winds may have little immediate local correspondence with the anomaly. Adding to this, the midlatitude thermocline is deep and anomalous winds do not necessarily reinforce an existing anomaly. For all of these reasons midlatitude ocean–atmosphere coupling is weaker and slower than its tropical counterpart. None of this discussion, of course, is to imply that midlatitude SST anomalies do not affect the atmosphere; indeed, such anomalies may be the main cause of seasonal anomalies in the weather.

When described in the terms above the El Niño phenomenon seems readily understandable, but it glosses over a host of issues. If the above feedback is robust why doesn't it occur in the tropical Atlantic? Why doesn't the system find a stable fixed point, or oscillate in a regular manner? What determines the interval between events, and the magnitude of

El Niño and the Southern Oscillation

- *El Niño* is the name given to the aperiodic warming of the ocean surface in the eastern equatorial Pacific. The interval between warm events is typically from two to seven years but is quite irregular (Fig. 16.8).
- El Niño events are associated with a weakening of the trade winds and an eastward shift of the region of convection, conveniently measured by a pressure difference between Tahiti and Darwin and known as the *Southern Oscillation*. The combined phenomenon is known as the El Niño–Southern Oscillation, or ENSO.
- El Niño, and its complement La Niña (an anomalous, weaker, warming in the western equatorial Pacific) are caused by the mutual interaction between the atmosphere and ocean in the equatorial Pacific. The key ingredients are:
  - *(i)* A close correlation between the thermocline thickness and surface temperature, especially in the east, with a shallow thermocline associated with a cool surface. A shallow thermocline allows water to upwell through it, bringing cold abyssal water to the surface (Fig. 16.10).
  - *(ii)* A positive feedback between the winds and the SST. Surface warming in the eastern Pacific leads to convergence of the winds, a deepening of the thermocline, and a further warming of the surface.
  - *(iii)* The back-and-forth 'sloshing' of the thermocline depth anomalies, mediated by Kelvin and Rossby waves propagating quickly eastward and more slowly westward, respectively, with corresponding basin crossing times of about 70 and 200 days.
  - *(iv)* The interaction of these timescales with the annual cycle of the trade winds and thermodynamic forcing, and with the natural ('stochastic') variability of the atmosphere on shorter timescales.
  - *(v)* Damping by way of leakage in the west via coastal Kelvin waves and loss of heat to the atmosphere and deep ocean, and a delayed negative feedback because of the finite crossing time of the waves.

the events? These questions are all areas of active research, and in the rest of the chapter we just illustrate some of the essential dynamics.

16.5 Ocean Dynamics in El Niño

The oceanic dynamics during an ENSO cycle are greatly affected by Kelvin and Rossby waves propagating westward and eastward respectively, across the basin. The equations of motion describing these waves are just the familiar linear shallow water equations, which in our usual notation are

$$\frac{\partial u}{\partial t} - \beta y v = -g'\frac{\partial h}{\partial x} - ru + F^x, \qquad \frac{\partial v}{\partial t} + \beta y u = -g'\frac{\partial h}{\partial y} - ru + F^y, \quad (16.22\text{a,b})$$

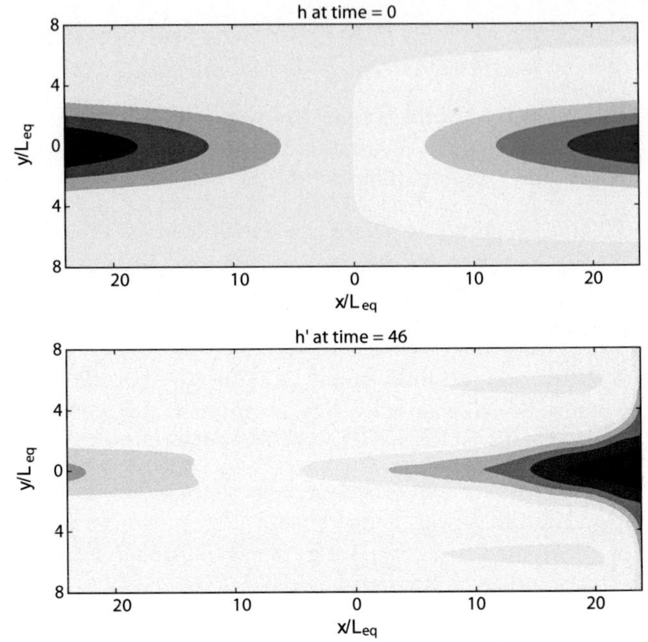

Fig. 16.11: A numerical solution of the shallow water equations showing the initial thermocline depth and the perturbation depth (i.e., final minus initial) at $t = 46/\sqrt{\beta c} \approx 70$ days, which is approximately the Kelvin wave crossing time for the basin. Red shading indicates deeper, warmer, water.

The initial depth field increases linearly toward the west, and the later state has a tongue of deep (and therefore warm) water in the east, a little like that seen in the bottom panel of Fig. 16.7.

$$\frac{\partial h}{\partial t} + H\left(\frac{\partial u}{\partial x} + \frac{\partial v}{\partial y}\right) = 0. \tag{16.22c}$$

In particular, F^x, F^y is the force due to the wind stress and h is the layer thickness, corresponding to the thermocline.

If the trade winds blow steadily westward, with no y-variation, then (as before) there is a stationary, steady state situation with

$$u = 0, \quad v = 0, \qquad g\frac{\partial h}{\partial x} = F^x, \quad \frac{\partial h}{\partial y} = 0. \tag{16.23}$$

For negative τ_x (westward winds) the thermocline thickness deepens going west. If the trade winds slacken (as during an El Niño) then the slope of the thermocline will, in its equilibrium state, diminish (as in Fig. 16.10) but the new equilibrium is not achieved instantly; rather, it is mediated by Kelvin and Rossby waves (as described in Section 4.3.3 and Section 6.3), just as the passage to geostrophic balance is mediated by gravity waves. One simple case is illustrated in Fig. 16.11, which shows a numerical integration of the shallow water equations. This example shows the evolution of the ocean from a state in which water is piled up against the western boundary (as is the normal state of affairs in the equatorial Pacific) and then released, just as might happen if the westward winds were to collapse. Kelvin waves propagate eastward, and about 70 days later (the time it takes for a Kelvin wave to cross the basin) the water is piled up in the west, rather as occurs during an El Niño event.

A more complete example is shown in Fig. 16.12, in which we start with a hump in the middle of the basin and then let it go. Kelvin waves

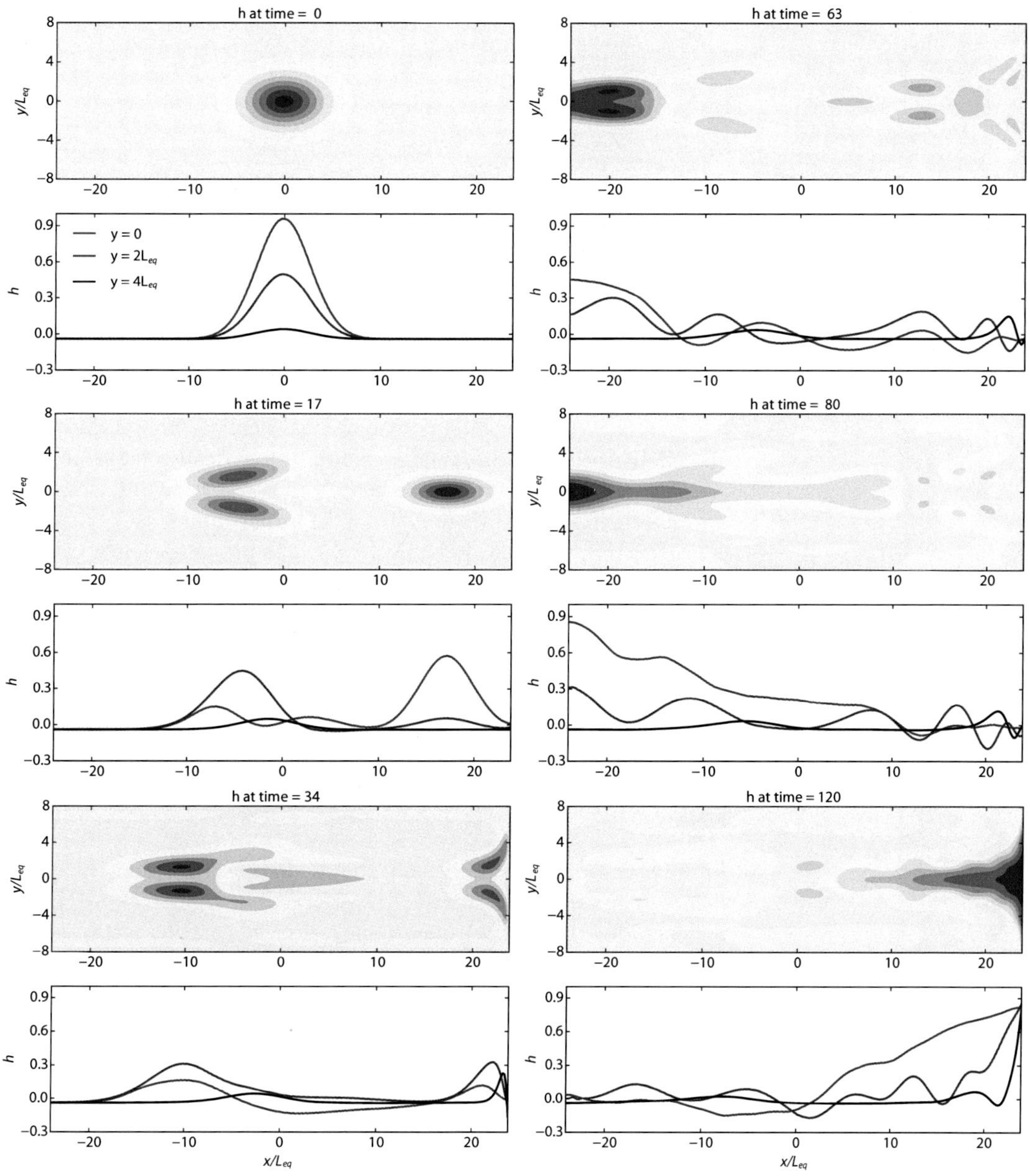

Fig. 16.12: Numerical evolution of the height field on an equatorial beta plane shown at the nondimensional times indicated, starting from a hump centred at the equator. The blue curve shows the height field along the equator and the red and black curves at two and four deformation radii poleward. The Kelvin wave propagates eastward and the off-equatorial Rossby waves move more slowly westward. At time 34 the equatorial Kelvin wave is partially reflected back as Rossby waves and partially propagating polewards as coastal Kelvin waves. At times 63 and 80 the Rossby waves are reflected back as an equatorial Kelvin wave.

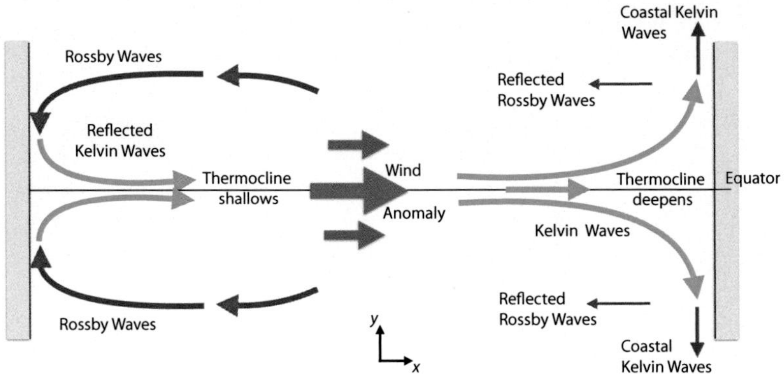

Fig. 16.13: Oceanic waves induced by a wind anomaly on the equator, leading to a local thickening of the thermocline at the equator and a shallowing off the equator. The Kelvin wave carries the deep anomaly westward, where it is eventually partially propagated away as coastal Kelvin waves. Rossby waves carry the shallow anomaly westward, and it is then reflected back as Kelvin waves.

propagate eastward and Rossby waves propagate westward, slightly off the equator. When the Kelvin waves reach the eastern boundary they generate Kelvin coastal waves that move along the coast away from the boundary and are eventually dissipated, as well as rather weak reflected Rossby waves. When the Rossby waves (from the initial hump) reach the western boundary they reflect as Kelvin waves, which propagate eastward all the way to the eastern boundary where they in turn generate coastal Kelvin waves, and eventually all the waves die out. The sequence is illustrated schematically in Fig. 16.13, where we assume that the wind anomaly produces a deepening of the thermocline on the equator and a shallowing off the equator (generated by the *curl* of the wind stress) which are propagated away by Kelvin and Rossby waves respectively.

16.6 ♦ UNSTABLE AIR–SEA INTERACTIONS

In the ocean-only example above the oscillations eventually die out. In reality, they can be sustained by the interaction with the atmosphere, by the Bjerknes feedback we described earlier and in this section we will construct a very simple mathematical model of that feedback. In order to illustrate the feedback in its simplest form we will make very severe approximations, eliminating all terms that are not essential to the mechanism, even if they are as large as some of the terms retained.

16.6.1 Equations of Motion

Ocean

We will suppose that the dynamics of the upper equatorial ocean are given by the linear shallow water equations. If we neglect the Coriolis term the zonal momentum and mass continuity equations become

$$\frac{\partial u}{\partial t} = -\frac{\partial \phi}{\partial x} - r_o u + F^x, \qquad \frac{\partial \phi}{\partial t} + c_o^2 \left(\frac{\partial u}{\partial x} + \frac{\partial v}{\partial y} \right) = 0, \qquad (16.24\text{a,b})$$

where r_o is a drag coefficient, ϕ is the pressure (proportional to the perturbation thickness of the thermocline), and c_o is the speed of gravity waves, which is around 1–2 m s^{-1}. The term F^x is the zonal force due to the wind stress at the surface, which we come to shortly, and we henceforth neglect variations in the y direction; the v-momentum equation is then not needed,

Upwelling is a dominant factor in determining the sea-surface temperature near the equator, since if the thermocline is thin upwelling brings cold water to the surface. We therefore suppose that the sea-surface temperature, T, is linearly related to the thermocline thickness, and so to ϕ, by

$$T = A\phi, \tag{16.25}$$

where A is a constant.

Atmosphere

On the atmospheric side we use a very similar set of equations. We again neglect the Coriolis force and nonlinear terms, and using capital letters for the variables we have

$$\frac{\partial U}{\partial t} = -\frac{\partial \Phi}{\partial x} - r_a U, \qquad \frac{\partial \Phi}{\partial t} + c_a^2\left(\frac{\partial U}{\partial x} + \frac{\partial V}{\partial y}\right) = Q_a, \tag{16.26a,b}$$

where r_a is a drag coefficient and c_a is the gravity wave speed in the atmosphere, which is around 30 m s^{-1}, and Q_a represents a thermodynamic forcing by SST anomalies. We further simplify (16.26) by supposing that the response time of the atmosphere is much shorter than that of the ocean so that, neglecting y-derivatives here also, we have

$$r_a U = -\frac{\partial \Phi}{\partial x}, \qquad c_a^2 \frac{\partial U}{\partial x} = Q_a. \tag{16.27a,b}$$

Coupled equations and unstable Interactions

The coupling between the ocean and atmosphere arises because it is the atmospheric wind that provides the stress on the ocean, and it is the sea-surface temperature (and hence the thermocline thickness) that provides the heating to the atmosphere. We therefore let $F^x = -\gamma U$ and $Q_a = BT$ where γ and B are constants. Using (16.25) we obtain $Q_a = \alpha\phi$ where $\alpha = AB$, and so in summary we have

$$F^x = -\gamma U, \qquad Q_a = \alpha\phi. \tag{16.28}$$

Using (16.28) in (16.24) and (16.27), and neglecting any remaining y-derivatives, gives a complete set of equations for the coupled, equatorial, ocean–atmosphere system:

$$\frac{\partial u}{\partial t} = -\frac{\partial \phi}{\partial x} - r_o u - \gamma U, \qquad \frac{\partial \phi}{\partial t} + c_o^2 \frac{\partial u}{\partial x} = 0, \tag{16.29a,b}$$

$$0 = -\frac{\partial \Phi}{\partial x} - r_a U, \qquad c_a^2 \frac{\partial U}{\partial x} = \alpha\phi. \tag{16.29c,d}$$

The most ad hoc aspects of these equations are the coupling relations in (16.28), which appear on the right-hand sides of (16.29a) and (16.29d).

Taken together these terms represent the Bjerknes feedback and their magnitude determines whether the system is unstable.

After a little algebra the four equations in (16.29) equations combine into a single equation for ϕ (a proxy for ocean surface temperature), namely

$$\frac{\partial^2 \phi}{\partial t^2} - c_o^2 \frac{\partial^2 \phi}{\partial x^2} + r_o \frac{\partial \phi}{\partial t} - \frac{c_o^2}{c_a^2} \gamma \alpha \phi = 0. \qquad (16.30)$$

Seeking a harmonic solution of the form $e^{i(kx-\omega t)}$ yields the dispersion relation

$$\omega^2 = -i r_o \omega + c_0^2 k^2 - \gamma \alpha \frac{c_o^2}{c_a^2}. \qquad (16.31)$$

The terms on the right-hand side give rise to damping (the term involving r_o, the drag on the oceanic flow), waves (the term $c_o^2 k^2$), and, potentially, exponential growth (the term $\gamma\alpha$, which is the feedback between atmosphere and ocean). If this feedback is large enough the flow will be unstable, and this is a simple mathematical representation of the feedback described verbally in Section 16.4.2. The solution of (16.31) is

$$\omega = -\frac{i r_o}{2} \pm \sqrt{c_o^2 k^2 - \frac{r_o^2}{4} - \gamma \alpha \frac{c_o^2}{c_a^2}} \quad \text{or} \quad \sigma = -\frac{r_o}{2} \pm \sqrt{\gamma \alpha \frac{c_o^2}{c_a^2} + \frac{r_o^2}{4} - c_o^2 k^2}, \qquad (16.32a,b)$$

where $\sigma \equiv -i\omega$ is the growth rate, and depending on the size of $\alpha\gamma$ the solution may be decaying or growing, and if decaying it may be oscillatory. Note that the frequency of the oscillations is reduced (from its pure oceanic value, $c_o k$) by the interaction with the atmosphere.

The above analysis is very instructive in revealing that coupling between the atmosphere and ocean can lead to an instability, and it is this instability that leads to the phenomenon that is ENSO. Needless to say (but we shall say it) the above analysis is overly-simple and omits a host of effects known to be important — the propagation of Kelvin and Rossby waves for example! A summary of the key ingredients of ENSO is provided in the box on page 339, and the incorporation of such phenomena into a mathematical and quantitative theory of El Niño and the Southern Oscillation is a topic of research that the reader is invited to pursue.

We have come to the end of this chapter, and indeed the end of the book. I hope that you, the reader, will be motivated to delve more into the fascinating dynamics of the atmosphere and ocean on Earth and, perhaps, other planets, and I thank you for reading.

Notes and References

To read about the remote effects of winds on the undercurrent the reader should consult Pedlosky (1987b) and McCreary & Lu (1994). For historical and paleo records of El Niño see, for example, Tudhope *et al.* (2001) and Wittenberg (2009). Books specializing on El Niño include those of Clarke (2008) and Sarachik & Cane (2010).

Problems

16.1 Carry through the algebra to obtain the solution of the Munk model, (14.38). Repeat the calculation with free-slip boundary conditions.

16.2 The Munk model also requires boundary conditions on the zonal boundaries at the Northern and Southern edges, in order to satisfy no slip or free slip conditions. Obtain estimates of the thickness of these boundary layers by scaling arguments. (For the keen: confirm your estimates by a detailed calculation of the Munk solution with zonal boundary layers.)

16.3 Consider the steady nonlinear Stommel problem,

$$J(\psi, \zeta) + \beta \frac{\partial \psi}{\partial x} = \mathrm{curl}_z \boldsymbol{\tau}_T - r\zeta.$$

If we use linear scaling we estimate that the streamfunction has magnitude $\psi \sim \tau/\beta$. Show that the ratio of the magnitude of the nonlinear term to the linear term is given by $\tau/(\beta^2 L^3)$, where L is a representative length scale. When τ is large the linear scaling must be incorrect — what then is the correct scaling for the streamfunction magnitude? In the real ocean, estimate whether the nonlinear term is in fact likely to be as large as the linear term, assuming the Stommel model is valid and using observed values of wind stress. (Be careful with density.)

16.4 In our considerations of flow in the ACC we did not use Sverdrup balance as an estimate for the vertically integrated meridional flow. Why not? Explain carefully why Sverdrup balance is not a good estimate of the meridional flow in a zonally re-entrant channel. Conversely, explain why estimates of the form (15.21) cannot be used in a basin.

16.5 Suppose that we know from observations that the meridional overturning circulation in the Atlantic Ocean has a magnitude of 20 Sv (that is $20 \times 10^9\,\mathrm{kg\,s^{-1}}$).

(*a*) Suppose this circulation is maintained by mixing. Using the diffusive scaling of Section 15.3.1 and an estimate of the pole–equatorial buoyancy difference Δb associated with a temperature difference of 40 K, estimate what value of κ is required to maintain such an overturning circulation. If we are now told that $\kappa = 10^{-5}\,\mathrm{m^2\,s^{-2}}$, what value of temperature difference is required to maintain 20 Sv.

(*b*) Suppose the circulation is maintained by winds in the Southern Ocean, as in Section 15.4.2. What value of wind stress is required to maintain 20 Sv, and what approximate surface wind values does this correspond to?

(*c*) Discuss what these calculations tell you about how the MOC is maintained. Are either of the above calculations robust enough to draw conclusions from, and if so what are those conclusions?

16.6 The undercurrent in the Pacific is about 4° wide and about $1\,\mathrm{m\,s^{-1}}$ strong. Using the model of Section 16.3 with a linear drag, estimate the values of the drag coefficient and the wind stress that give these values, approximately.

16.7 Calculate the approximate crossing time of Kelvin and Rossby waves in the Atlantic and Pacific, respectively. Be clear about any assumptions you may make about the ocean properties. Given this, and supposing that these waves are an essential part of the El Niño cycle, comment on whether the Atlantic or Pacific is more likely to display phenomena that are locked to the seasonal cycle.

Bibliography

Acheson, D. J., 1990. *Elementary Fluid Dynamics.* Oxford University Press, 406 pp.

Ambaum, M. H. P., 2010. *Thermal Physics of the Atmosphere.* Wiley, 239 pp.

Andrews, D. G. & McIntyre, M. E., 1976. Planetary waves in horizontal and vertical shear: the generalized Eliassen–Palm relation and the mean zonal acceleration. *J. Atmos. Sci.,* **33**, 2031–2048.

Batchelor, G. K., 1969. Computation of the energy spectrum in homogeneous two-dimensional turbulence. *Phys. Fluids Suppl.,* **12**, II–233–239.

Bengtsson, L., Bonnet, R.-M., Grinspoon, D., Koumoutsaris, S. *et al.*, Eds., 2013. *Towards Understanding the Climate of Venus: Applications of Terrestrial Models to Our Sister Planet.* Springer Science & Business Media, 185 pp.

Bohren, C. F. & Albrecht, B. A., 1998. *Atmospheric Thermodynamics.* Oxford University Press, 416 pp.

Boussinesq, J., 1903. Théorie analytique de la chaleur (Analytic theory of heat). *Tome, Paris, Gauthier-Villars,* **II**, 170–172.

Boyd, J. P., 1976. The noninteraction of waves with the zonally averaged flow on a spherical Earth and the interrelationships of eddy fluxes of energy, heat and momentum. *J. Atmos. Sci.,* **33**, 2285–2291.

Brewer, A. W., 1949. Evidence for a world circulation provided by the measurements of helium and water vapour distribution in the stratosphere. *Quart. J. Roy. Meteor. Soc.,* **75**, 251–363.

Brillouin, L., 1926. La mécanique ondulatoire de Schrödinger; une méthode générale de resolution par approximations successives (The wave mechanics of Schrödinger: a general method of solution by successive approximation). *Comptes Rendus,* **183**, 24–26.

Bühler, O., 2009. *Waves and Mean Flows.* Cambridge University Press, 370 pp.

Carillo, C. N., 1892. Desertacion sobre las corrientes y estudios de la corriente Peruana de Humboldt. (Dissertation on currents and studies of the Peruvian Humboldt current). *Bol. Soc. Geogr. Lima,* **11**, 72–110.

Charney, J. G., 1947. The dynamics of long waves in a baroclinic westerly current. *J. Meteor.,* **4**, 135–162.

Charney, J. G., 1948. On the scale of atmospheric motion. *Geofys. Publ. Oslo,* **17** (2), 1–17.

Charney, J. G., 1963. A note on large-scale motions in the tropics. *J. Atmos. Sci.*, **20**, 607–609.

Charney, J. G., 1971. Geostrophic turbulence. *J. Atmos. Sci.*, **28**, 1087–1095.

Charney, J. G. & Drazin, P. G., 1961. Propagation of planetary scale disturbances from the lower into the upper atmosphere. *J. Geophys. Res.*, **66**, 83–109.

Charney, J. G. & Stern, M. E., 1962. On the stability of internal baroclinic jets in a rotating atmosphere. *J. Atmos. Sci.*, **19**, 159–172.

Clarke, A. J., 2008. *An Introduction to the Dynamics of El Niño and the Southern Oscillation*. Elsevier, 308 pp.

Cushman-Roisin, B. & Beckers, J.-M., 2011. *Introduction to Geophysical Fluid Dynamics*. 2nd edn. Academic Press, 828 pp.

Davidson, P., 2015. *Turbulence: An Introduction for Scientists and Engineers*. Oxford University Press, 688 pp.

Dickinson, R. E., 1969. Theory of planetary wave–zonal flow interaction. *J. Atmos. Sci.*, **26**, 73–81.

Dobson, G. M. B., 1956. Origin and distribution of the polyatomic molecules in the atmosphere. *Proc. Roy. Soc. Lond. A*, **236**, 187–193.

Döös, K. & Coward, A., 1997. The Southern Ocean as the major upwelling zone of the North Atlantic. *Int. WOCE Newsletter*, **27**, 3–4.

Drazin, P. G. & Reid, W. H., 1981. *Hydrodynamic Stability*. Cambridge University Press, 527 pp.

Durran, D. R., 1993. Is the Coriolis force really responsible for the inertial oscillations? *Bull. Am. Meteor. Soc.*, **74**, 2179–2184.

Durran, D. R., 2015. Lee waves and mountain waves. *Encycl. Atmos. Sci.*, 2nd edn, **4**, 95–102.

Durst, C. S. & Sutcliffe, R. C., 1938. The effect of vertical motion on the "geostrophic departure" of the wind. *Quart. J. Roy. Meteor. Soc.*, **64**, 240.

Eady, E. T., 1949. Long waves and cyclone waves. *Tellus*, **1**, 33–52.

Eady, E. T., 1950. The cause of the general circulation of the atmosphere. In *Cent. Proc. Roy. Meteor. Soc.* (1950), pp. 156–172.

Egger, J., 1999. Inertial oscillations revisited. *J. Atmos. Sci.*, **56**, 2951–2954.

Ekman, V. W., 1905. On the influence of the Earth's rotation on ocean currents. *Arch. Math. Astron. Phys.*, **2**, 1–52.

Eliassen, A. & Palm, E., 1961. On the transfer of energy in stationary mountain waves. *Geofys. Publ.*, **22**, 1–23.

Emanuel, K. A., 1994. *Atmospheric Convection*. Oxford University Press, 580 pp.

Fofonoff, N. P., 1954. Steady flow in a frictionless homogeneous ocean. *J. Mar. Res.*, **13**, 254–262.

Gill, A. E., 1982. *Atmosphere–Ocean Dynamics*. Academic Press, 662 pp.

Gnanadesikan, A., 1999. A simple predictive model for the structure of the oceanic pycnocline. *Science*, **283**, 2077–2079.

Haynes, P., 2005. Stratospheric dynamics. *Ann. Rev. Fluid Mech.*, **37**, 263–293.

Heimpel, M., Gastine, T. & Wicht, J., 2016. Simulation of deep-seated zonal jets and shallow vortices in gas giant atmospheres. *Nature Geos.*, **9**, 19–23.

Held, I. M., 2000. The general circulation of the atmosphere. In *Woods Hole Program in Geophysical Fluid Dynamics* (2000), pp. 66.

Held, I. M. & Hou, A. Y., 1980. Nonlinear axially symmetric circulations in a nearly inviscid atmosphere. *J. Atmos. Sci.*, **37**, 515–533.

Holmes, M. H., 2013. *Introduction to Perturbation Methods*. 2nd edn. Springer, 436 pp.

Holton, J. R. & Hakim, G., 2012. *An Introduction to Dynamic Meteorology*. 5th edn. Academic Press, 552 pp.

Hoskins, B. J. & Karoly, D. J., 1981. The steady linear response of a spherical atmosphere to thermal and orographic forcing. *J. Atmos. Sci.*, **38**, 1179–1196.

Hough, S. S., 1898. On the application of harmonic analysis to the dynamical theory of the tides. Part II: On the general integration of Laplace's dynamical equations. *Phil. Trans. (A)*, **191 (V)**, 139–186.

Ingersoll, A. P., 2013. *Planetary Climates*. Princeton University Press, 278 pp.

IOC, SCOR & IAPSO, 2010. The international thermodynamic equation of seawater – 2010: Calculation and use of thermodynamic properties. Technical report, Intergovernmental Oceanographic Commission, Manuals and Guides No. 56, UNESCO (English).

Jeffreys, H., 1924. On certain approximate solutions of linear differential equations of the second order. *Proc. London Math. Soc.*, **23**, 428–436.

Johnson, G. C. & Bryden, H. L., 1989. On the size of the Antarctic Circumpolar Current. *Deep-Sea Res.*, **36**, 39–53.

Juckes, M. N., 2001. A generalization of the transformed Eulerian-mean meridional circulation. *Quart. J. Roy. Meteor. Soc.*, **127**, 147–160.

Kaspi, Y. & Showman, A. P., 2015. Atmospheric dynamics of terrestrial exoplanets over a wide range of orbital and atmospheric parameters. *Astrophys. J*, **804**, 18pp.

Khatuntsev, I., Patsaeva, M., Titov, D., Ignatiev, N. *et al.*, 2013. Cloud level winds from the Venus Express monitoring camera imaging. *Icarus*, **226**, 140–158.

Kolmogorov, A. N., 1941. The local structure of turbulence in incompressible viscous fluid for very large Reynolds numbers. *Dokl. Acad. Sci. USSR*, **30**, 299–303.

Kraichnan, R., 1967. Inertial ranges in two-dimensional turbulence. *Phys. Fluids*, **10**, 1417–1423.

Kraichnan, R., 1971. Inertial range transfer in two- and three-dimensional turbulence. *J. Fluid Mech.*, **47**, 525–535.

Kramers, H. A., 1926. Wellenmechanik und halbzahlige Quantisierung (Wave mechanics and semi-integral quantization). *Zeit. fur Physik A*, **39**, 828–840.

Kundu, P., Cohen, I. M. & Dowling, D. R., 2015. *Fluid Mechanics*. Academic Press, 928 pp.

Kuo, H.-I., 1949. Dynamic instability of two-dimensional nondivergent flow in a barotropic atmosphere. *J. Meteorol.*, **6**, 105–122.

Kuo, H.-I., 1951. Vorticity transfer as related to the development of the general circulation. *J. Meteorol.*, **8**, 307–315.

Laplace, P., 1832. *Elementary illustrations of the celestial mechanics of Laplace*. John Murray.

LeBlond, P. H. & Mysak, L. A., 1980. *Waves in the Ocean*. Elsevier, 616 pp.

Leith, C. E., 1968. Diffusion approximation for two-dimensional turbulence. *Phys. Fluids*, **11**, 671–672.

Limaye, S. S., 2007. Venus atmospheric circulation: Known and unknown. *J. Geophys. Res. (Planets)*, **112**, E04S094.

Lorenz, E. N., 1967. *The Nature and the Theory of the General Circulation of the Atmosphere*. Vol. 218, World Meteorological Organization.

Lumpkin, R. & Speer, K., 2007. Global ocean meridional overturning. *J. Phys. Oceanogr.*, **37**, 2550–2562.

Luyten, J. R., Pedlosky, J. & Stommel, H., 1983. The ventilated thermocline. *J. Phys. Oceanogr.*, **13**, 292–309.

Marcus, P. S., 1993. Jupiter's Great Red Spot and other vortices. *Ann. Rev. Astron. Astrophys*, **31**, 523–573.

Marshall, J. C. & Radko, T., 2003. Residual-mean solutions for the Antarctic Circumpolar Current and its associated overturning circulation. *J. Phys. Oceanogr.*, **22**, 2341–2354.

McCreary, J. P. & Lu, P., 1994. Interaction between the subtropical and equatorial ocean circulations: the subtropical cell. *J. Phys. Oceanogr.*, **24**, 466–497.

Munk, W. H., 1950. On the wind-driven ocean circulation. *J. Meteorol*, **7**, 79–93.

Nikurashin, M. & Vallis, G. K., 2011. A theory of deep stratification and overturning circulation in the ocean. *J. Phys. Oceanogr.*, **41**, 485–502.

Nikurashin, M. & Vallis, G. K., 2012. A theory of the interhemispheric meridional overturning circulation and associated stratification. *J. Phys. Oceanogr.*, **42**, 1652–1667.

Obukhov, A. M., 1941. Energy distribution in the spectrum of turbulent flow. *Izv. Akad. Nauk. SSR, Ser. Geogr. Geofiz.*, **5**, 453–466.

Olbers, D., Willebrand, J. & Eden, C., 2012. *Ocean Dynamics*. Springer, 704 pp.

Pedlosky, J., 1964. The stability of currents in the atmosphere and ocean. Part I. *J. Atmos. Sci.*, **21**, 201–219.

Pedlosky, J., 1987a. *Geophysical Fluid Dynamics*. 2nd edn. Springer-Verlag, 710 pp.

Pedlosky, J., 1987b. An inertial theory of the equatorial undercurrent. *J. Phys. Oceanogr.*, **17**, 1978–1985.

Pedlosky, J., 2003. *Waves in the Ocean and Atmosphere: Introduction to Wave Dynamics*. Springer-Verlag, 260 pp.

Phillips, N. A., 1963. Geostrophic motion. *Rev. Geophys.*, **1**, 123–176.

Pierrehumbert, R. T., 2010. *Principles of Planetary Climate*. Cambridge University Press, 652 pp.

Pope, S. B., 2000. *Turbulent Flows*. Cambridge University Press, 754 pp.

Porco, C. C., West, R. A., McEwen, A., Del Genio, A. D. *et al.*, 2003. Cassini imaging of Jupiter's atmosphere, satellites, and rings. *Science*, **299**, 5612, 1541–1547.

Proudman, J., 1916. On the motion of solids in liquids. *Proc. Roy. Soc. Lond. A*, **92**, 408–424.

Rayleigh, Lord, 1880. On the stability, or instability, of certain fluid motions. *Proc. London Math. Soc.*, **11**, 57–70.

Read, P., Lewis, S. & Vallis, G. K., 2018. Atmospheric dynamics of terrestrial planets. In H. Deeg & J. Belmonte, Eds., *Handbook of Exoplanets*, pp. 1–31. Springer.

Rhines, P. B., 1975. Waves and turbulence on a β-plane. *J. Fluid. Mech.*, **69**, 417–443.

Rhines, P. B., 1977. The dynamics of unsteady currents. In E. A. Goldberg, I. N. McCane, J. J. O'Brien, & J. H. Steele, Eds., *The Sea*, Vol. 6, pp. 189–318. J. Wiley and Sons.

Richardson, L. F., 1920. The supply of energy from and to atmospheric eddies. *Proc. Roy. Soc. Lond. A*, **97**, 354–373.

Riehl, H. & Fultz, D., 1957. Jet stream and long waves in a steady rotating-dishpan experiment: structure of the circulation. *Quart. J. Roy. Meteor. Soc.*, **82**, 215–231.

Robinson, A. R. & Stommel, H., 1959. The oceanic thermocline and the associated thermohaline circulation. *Tellus*, **11**, 295–308.

Rossby, C.-G., 1939. Relations between variation in the intensity of the zonal circulation and the displacements of the semi-permanent centers of action. *J. Marine Res.*, **2**, 38–55.

Salmon, R., 1980. Baroclinic instability and geostrophic turbulence. *Geophys. Astrophys. Fluid Dyn.*, **10**, 25–52.

Salmon, R., 1990. The thermocline as an internal boundary layer. *J. Mar. Res.*, **48**, 437–469.

Salmon, R., 1998. *Lectures on Geophysical Fluid Dynamics*. Oxford University Press, 378 pp.

Samelson, R. M., 2004. Simple mechanistic models of middepth meridional overturning. *J. Phys. Oceanogr.*, **34**, 2096–2103.

Samelson, R. M., 2011. *The Theory of Large-Scale Ocean Circulation*. Cambridge University Press, 193 pp.

Samelson, R. M. & Vallis, G. K., 1997. Large-scale circulation with small diapycnal diffusion: the two-thermocline limit. *J. Mar. Res.*, **55**, 223–275.

Sanchez-Lavega, A., 2011. *An Introduction to Planetary Atmospheres*. CRC Press, 587 pp.

Sanchez-Lavega, A. & Heimpel, M., 2018. Atmospheric dynamics of giants and icy planets. In H. Deeg & J. Belmonte, Eds., *Handbook of Exoplanets*, pp. 1–32. Springer.

Sarachik, E. S. & Cane, M. A., 2010. *The El Niño-Southern Oscillation Phenomenon*. Cambridge University Press, 369 pp.

Schneider, E. K., 1977. Axially symmetric steady-state models of the basic state for instability and climate studies. Part II: Nonlinear calculations. *J. Atmos. Sci.*, **34**, 280–297.

Seiff, A., Kirk, D., Blanchard, R. C., Findlay, J. T. & Kelly, G. M., 1979. Thermal contrast in the atmosphere of Venus: Initial appraisal from Pioneer Venus probe data. *Science*, **205**, 46–49.

Seiff, A., Kirk, D. B., Knight, T. C., Young, R. E. *et al.*, 1998. Thermal structure of Jupiter's atmosphere near the edge of a 5-μm hot spot in the north equatorial belt. *J. Geophys. Res. (Planets)*, **103**, E10, 22857–22889.

Showman, A. P., Cho, J. Y. & Menou, K., 2010. Atmospheric circulation of exoplanets. *Exoplanets*, **526**, 471–516.

Simmonds, J. G. & Mann, J. E., 1998. *A First Look at Perturbation Theory*. Dover Publications, 139 pp.

Simon, A. A., Wong, M. H. & Orton, G. S., 2015. First results from the Hubble OPAL program: Jupiter in 2015. *Astrophys. J*, **812**, 1, 55.

Simon-Miller, A. A. & Gierasch, P. J., 2010. On the long-term variability of Jupiter's winds and brightness as observed from Hubble. *Icarus*, **210**, 258–269.

Sobel, A. H., Nilsson, J. & Polvani, L., 2001. The weak temperature gradient approximation and balanced tropical moisture waves. *J. Atmos. Sci.*, **58**, 3650–3665.

Stommel, H., 1948. The westward intensification of wind-driven ocean currents. *Trans. Amer. Geophys. Union*, **29**, 202–206.

Stommel, H. & Arons, A. B., 1960. On the abyssal circulation of the world ocean—I. Stationary planetary flow patterns on a sphere. *Deep-Sea Res.*, **6**, 140–154.

Sutcliffe, R. C., 1947. A contribution to the problem of development. *Quart. J. Roy. Meteor. Soc.*, **73**, 370–383.

Sutherland, B., 2010. *Internal Gravity Waves*. Cambridge University Press, 394 pp.

Swallow, J. C. & Worthington, V., 1961. An observation of a deep countercurrent in the western North Atlantic. *Deep-Sea Res.*, **8**, 1–19.

Talley, L. D., Pickard, G., Emery, W. J. & Swift, J. H., 2011. *Descriptive Physical Oceanography: An Introduction*. Academic Press, 555 pp.

Taylor, G. I., 1921. Experiments with rotating fluids. *Proc. Roy. Soc. Lond. A*, **100**, 114–121.

Thompson, R. O. R. Y., 1971. Why there is an intense eastward current in the North Atlantic but not in the South Atlantic. *J. Phys. Oceanogr.*, **1**, 235–237.

Thompson, R. O. R. Y., 1980. A prograde jet driven by Rossby waves. *J. Atmos. Sci.*, **37**, 1216–1226.

Thomson, W. (Lord Kelvin), 1869. On vortex motion. *Trans. Roy. Soc. Edinburgh*, **25**, 217–260.

Thomson, W. (Lord Kelvin), 1879. On gravitational oscillations of rotating water. *Proc. Roy. Soc. Edinburgh*, **10**, 92–100.

Titov, D. V., Piccioni, G., Drossart, P. & Markiewicz, W. J., 2013. Radiative energy balance in the Venus atmosphere. In L. Bengtsson *et al.*, Eds., *Towards Understanding the Climate of Venus*, pp. 23–53. Springer.

Toggweiler, J. R. & Samuels, B., 1998. On the ocean's large-scale circulation in the limit of no vertical mixing. *J. Phys. Oceanogr.*, **28**, 1832–1852.

Tudhope, A. W., Chilcott, C. P., McCulloch, M. T., Cook, E. R. *et al.*, 2001. Variability in the El Niño Southern Oscillation through a glacial-interglacial cycle. *Science*, **291**, 1511–1517.

Vallis, G. K., 2000. Large-scale circulation and production of stratification: effects of wind, geometry and diffusion. *J. Phys. Oceanogr.*, **30**, 933–954.

Vallis, G. K., 2017. *Atmospheric and Oceanic Fluid Dynamics*. 2nd edn. Cambridge University Press, 946 pp.

Vallis, G. K., Colyer, G., Geen, R., Gerber, E. *et al.*, 2018. Isca, v1.0: A framework for the global modelling of the atmospheres of Earth and other planets at varying levels of complexity. *Geosci. Model Dev*, **11**, 843–859.

Vallis, G. K. & Maltrud, M. E., 1993. Generation of mean flows and jets on a beta plane and over topography. *J. Phys. Oceanogr.*, **23**, 1346–1362.

Veronis, G., 1966. Wind-driven ocean circulation – Part 2: Numerical solutions of the non-linear problem. *Deep-Sea Res.*, **13**, 30–55.

Watson, A., Vallis, G. K. & Nikurashin, M., 2015. Southern Ocean buoyancy forcing of ocean ventilation and glacial atmospheric CO_2. *Nature Geos.*, **8**, 861–864. doi:10.1038/ngeo2538.

Welander, P., 1959. An advective model of the ocean thermocline. *Tellus*, **11**, 309–318.

Wentzel, G., 1926. Eine Verallgemeinerung der Quantenbedingungen für die Zwecke der Wellenmechanik (A generalization of the quantum conditions for the purposes of wave mechanics). *Zeit. fur Physic A*, **38**, 518–529.

White, A. A., 2002. A view of the equations of meteorological dynamics and various approximations. In J. Norbury & I. Roulstone, Eds., *Large-Scale Atmosphere-Ocean Dynamics I*, pp. 1–100. Cambridge University Press.

Wittenberg, A. T., 2009. Are historical records sufficient to constrain ENSO simulations? *Geophys. Res. Lett.*, **36**, L12702. doi:10.1029/2009GL038710.

Wolfe, C. L. & Cessi, P., 2011. The adiabatic pole-to-pole overturning circulation. *J. Phys. Oceanogr.*, **41**, 1795–1810.

Wunsch, C., 2015. *Modern Observational Physical Oceanography*. Princeton University Press, 481 pp.

Index

Printed in the United States
by Baker & Taylor Publisher Services